Lecture Notes in Computer Science 11695

Founding Editors

Gerhard Goos
  Karlsruhe Institute of Technology, Karlsruhe, Germany
Juris Hartmanis
  Cornell University, Ithaca, NY, USA

Editorial Board Members

Elisa Bertino
  Purdue University, West Lafayette, IN, USA
Wen Gao
  Peking University, Beijing, China
Bernhard Steffen
  TU Dortmund University, Dortmund, Germany
Gerhard Woeginger
  RWTH Aachen, Aachen, Germany
Moti Yung
  Columbia University, New York, NY, USA

More information about this series at http://www.springer.com/series/7409

Tatjana Welzer · Johann Eder ·
Vili Podgorelec · Aida Kamišalić Latifić (Eds.)

Advances in Databases and Information Systems

23rd European Conference, ADBIS 2019
Bled, Slovenia, September 8–11, 2019
Proceedings

 Springer

Editors
Tatjana Welzer 🆔
University of Maribor
Maribor, Slovenia

Johann Eder 🆔
Alpen-Adria Universität Klagenfurt
Klagenfurt, Austria

Vili Podgorelec 🆔
University of Maribor
Maribor, Slovenia

Aida Kamišalić Latifić 🆔
University of Maribor
Maribor, Slovenia

ISSN 0302-9743   ISSN 1611-3349   (electronic)
Lecture Notes in Computer Science
ISBN 978-3-030-28729-0   ISBN 978-3-030-28730-6   (eBook)
https://doi.org/10.1007/978-3-030-28730-6

LNCS Sublibrary: SL3 – Information Systems and Applications, incl. Internet/Web, and HCI

© Springer Nature Switzerland AG 2019
This work is subject to copyright. All rights are reserved by the Publisher, whether the whole or part of the material is concerned, specifically the rights of translation, reprinting, reuse of illustrations, recitation, broadcasting, reproduction on microfilms or in any other physical way, and transmission or information storage and retrieval, electronic adaptation, computer software, or by similar or dissimilar methodology now known or hereafter developed.
The use of general descriptive names, registered names, trademarks, service marks, etc. in this publication does not imply, even in the absence of a specific statement, that such names are exempt from the relevant protective laws and regulations and therefore free for general use.
The publisher, the authors and the editors are safe to assume that the advice and information in this book are believed to be true and accurate at the date of publication. Neither the publisher nor the authors or the editors give a warranty, expressed or implied, with respect to the material contained herein or for any errors or omissions that may have been made. The publisher remains neutral with regard to jurisdictional claims in published maps and institutional affiliations.

This Springer imprint is published by the registered company Springer Nature Switzerland AG
The registered company address is: Gewerbestrasse 11, 6330 Cham, Switzerland

Preface

The European Conference on Advances in Databases and Information Systems (ADBIS) celebrated its 23rd anniversary this year. Previous ADBIS conferences were held in St. Petersburg (1997), Poznan (1998), Maribor (1999), Prague (2000), Vilnius (2001), Bratislava (2002), Dresden (2003), Budapest (2004), Tallinn (2005), Thessaloniki (2006), Varna (2007), Pori (2008), Riga (2009), Novi Sad (2010), Vienna (2011), Poznan (2012), Genoa (2013), Ohrid (2014), Poitiers (2015), Prague (2016), Nicosia (2017), and Budapest (2018). After 20 years, the conference returned to Slovenia. It was organized at Bled.

ADBIS established itself as a highly recognized conference in Europe in the broad field of databases and information systems. The conference aims at: (1) providing an international forum for presenting research achievements on database theory and practice, development of advanced DBMS technologies, and their applications; (2) promoting the interaction and collaboration of the database and information systems research communities both within European and with the rest of the world; (3) offering a forum for a less formal exchange of research ideas by means of affiliated workshops; and (4) providing guidance, motivation, and feedback for young researchers from all over the world by means of a doctoral consortium.

This volume contains 27 full research papers from the main conference, which were selected by the international Program Committee in a tough reviewing process out of a total of 103 submissions (acceptance rate 26%). The selected full papers span a wide spectrum of topics related to the ADBIS conference from different areas of research in database and information systems technologies and their advanced applications from theoretical foundations to optimizing index structures. Major focal areas are data mining and machine learning, data warehouses and big data technologies, semantic data processing, and data modeling.

We would like to express our sincere gratitude to everyone who contributed to make ADBIS 2019 successful:

- All the organizers of the previous ADBIS workshops and conferences. They made ADBIS a valuable trademark and we are proud to continue their work.
- The authors, who submitted papers of high quality to the conference.
- The members of the international Program Committee for dedicating their time and expertise for assuring the high quality of the program.
- The members of ADBIS Steering Committee for proven trust and conferred organization of the conference.
- Springer for publishing these proceedings.
- Last but not least, to all the helping hands from the webmaster, programmers to technicians and administration without whom the organization of such a conference would not be possible.

- Finally, we would like to express our special thanks to the local chair, Lili Nemec Zlatolas, for her continuous and coordinating activities that ensured the success of ADBIS 2019.

July 2019

Tatjana Welzer
Johann Eder
Vili Podgorelec
Aida Kamišalić Latifić

Organization

Steering Committee Chair

Yannis Manolopoulos   Open University of Cyprus, Cyprus

Steering Committee

Ladjel Bellatreche	Laboratory of Computer Science and Automatic Control for Systems, France
Andras Benczur	Eötvös Loránd University, Hungary
Maria Bielikova	Slovak University of Technology, Slovakia
Barbara Catania	University of Genoa, Italy
Johann Eder	Alpen-Adria-Universität Klagenfurt, Austria
Theo Haerder	University of Kaiserslautern, Germany
Mirjana Ivanović	University of Novi Sad, Serbia
Hannu Jaakkola	Tampere University, Finland
Marite Kirikova	Riga Technical University, Latvia
Yannis Manolopoulos	Open University of Cyprus, Cyprus
Rainer Manthey	University of Bonn, Germany
Manuk Manukyan	Yerevan State University, Armenia
Tadeusz Morzy	Poznan University of Technology, Poland
Pavol Navrat	Slovak University of Technology, Slovakia
Boris Novikov	Saint Petersburg State University, Russia
George Angelos Papadopoulos	University of Cyprus, Cyprus
Jaroslav Pokorny	Charles University in Prague, Czech Republic
Boris Rachev	Technical University of Varna, Bulgaria
Bernhard Thalheim	Kiel University, Germany
Goce Trajcevski	Iowa State University of Science and Technology, USA
Tatjana Welzer	University of Maribor, Slovenia
Robert Wrembel	Poznan University of Technology, Poland
Ester Zumpano	Università Della Calabria, Italy

Program Committee Chairs

Johann Eder	Alpen-Adria-Universität Klagenfurt, Austria
Vili Podgorelec	University of Maribor, Slovenia

Program Committee

Syed Sibte Raza Abidi	Dalhousie University, Halifax, Canada
Bernd Amann	Sorbonne Université, France
Costin Badica	University of Craiova, Romania
Marko Bajec	University of Ljubljana, Slovenia
Rodrigo Coelho Barros	Pontifícia Universidade Católica do Rio Grande do Sul, Brazil
Andreas Behrend	University of Bonn, Germany
Ladjel Bellatreche	LIAS/ENSMA, France
András Benczúr	Eötvös Loránd University, Hungary
Maria Bielikova	Slovak University of Technology in Bratislava, Slovakia
Nikos Bikakis	University of Ioannina, Greece
Zoran Bosnić	University of Ljubljana, Slovenia
Dražen Brdjanin	University of Banja Luka, Bosnia and Herzegovina
Albertas Caplinskas	Vilnius University, Lithuania
Christos Doulkeridis	University of Piraeus, Greece
Johann Eder	Alpen Adria Universität Klagenfurt, Austria
Markus Endres	University of Passau, Germany
Werner Esswein	TU Dresden, Germany
Flavio Ferrarotti	Software Competence Centre Hagenberg, Austria
Flavius Frasincar	Erasmus University Rotterdam, The Netherlands
Jānis Grabis	Riga Technical University, Latvia
Francesco Guerra	Università di Modena e Reggio Emilia, Italy
Giancarlo Guizzardi	Federal University of Espirito Santo, Brazil
Hele-Mai Haav	Tallinn University of Technology, Estonia
Theo Härder	TU Kaiserslautern, Germany
Tomáš Horváth	Eötvös Loránd University, Hungary
Marko Hölbl	University of Maribor, Slovenia
Andres Iglesias	University of Cantabria, Spain
Mirjana Ivanović	University of Novi Sad, Serbia
Hannu Jaakkola	Tampere University, Finland
Lili Jiang	Umea University, Sweden
Aida Kamišalić Latifić	University of Maribor, Slovenia
Mehmed Kantardzic	University of Louisville, USA
Dimitris Karagiannis	University of Vienna, Austria
Sašo Karakatič	University of Maribor, Slovenia
Zoubida Kedad	University of Versailles, France
Marite Kirikova	Riga Technical University, Latvia
Attila Kiss	Eötvös Loránd University, Hungary
Margita Kon-Popovska	Ss. Cyril and Methodius University in Skopje, North Macedonia
Harald Kosch	Universität Passau, Germany
Michal Kratky	VSB-Technical University of Ostrava, Czech Republic
Ralf-Detlef Kutsche	TU Berlin, Germany

Julius Köpke	Alpen Adria Universität Klagenfurt, Austria
Dejan Lavbič	University of Ljubljana, Slovenia
Sebastian Link	The University of Auckland, New Zealand
Audrone Lupeikiene	Vilnius University, Lithuania
Federica Mandreoli	University of Modena, Italy
Yannis Manolopoulos	Open University of Cyprus, Cyprus
Manuk Manukyan	Yerevan State University, Armenia
Karol Matiasko	University of Žilina, Slovakia
Goran Mauša	University of Rijeka, Croatia
Bálint Molnár	Eötvös University of Budapest, Hungary
Angelo Montanari	University of Udine, Italy
Tadeusz Morzy	Poznan University of Technology, Poland
Boris Novikov	St. Petersburg University, Russia
Kjetil Nørvåg	Norwegian University of Science and Technology, Norway
Andreas Oberweis	Karlsruhe Institute of Technology, Germany
Andreas L Opdahl	University of Bergen, Norway
Eneko Osaba	TECNALIA Research & Innovation, Spain
Odysseas Papapetrou	Eindhoven University of Technology, The Netherlands
András Pataricza	Budapest University of Technology and Economics, Czech Republic
Tomas Pitner	Masaryk University, Czech Republic
Vili Podgorelec	University of Maribor, Slovenia
Jaroslav Pokorný	Charles University in Prague, Czech Republic
Giuseppe Polese	University of Salerno, Italy
Boris Rachev	Technical University of Varna, Bulgaria
Miloš Radovanović	University of Novi Sad, Serbia
Heri Ramampiaro	Norwegian University of Science and Technology, Norway
Stefano Rizzi	University of Bologna, Italy
Peter Ruppel	Technische Universität Berlin, Germany
Gunter Saake	University of Magdeburg, Germany
Petr Saloun	VSB-TU Ostrava, Czech Republic
José Luis Sánchez de la Rosa	University of La Laguna, Spain
Shiori Sasaki	Keio University, Japan
Kai-Uwe Sattler	TU Ilmenau, Germany
Miloš Savić	University of Novi Sad, Serbia
Timos Sellis	Swinburne University of Technology, Australia
Bela Stantic	Griffith University, Australia
Kostas Stefanidis	Tampere University, Finland
Claudia Steinberger	Alpen-Adria-Universität Klagenfurt, Austria
Sergey Stupnikov	Russian Academy of Sciences, Russia
Bernhard Thalheim	Kiel University, Germany
Raquel Trillo-Lado	Universidad de Zaragoza, Spain
Muhamed Turkanović	University of Maribor, Slovenia

Olegas Vasilecas	Vilnius Gediminas Technical University, Lithuania
Goran Velinov	Ss. Cyril and Methodius University, North Macedonia
Peter Vojtas	Charles University Prague, Czech Republic
Isabelle Wattiau	ESSEC and CNAM, France
Tatjana Welzer	University of Maribor, Slovenia
Robert Wrembel	Poznan University of Technology, Poland
Jaroslav Zendulka	Brno University of Technology, Czech Republic

Additional Reviewers

Nabila Berkani	Haridimos Kondylakis
Dominik Bork	Ilya Makarov
Andrea Brunello	Riccardo Martoglia
Loredana Caruccio	Matteo Paganelli
Stefano Cirillo	Marek Rychlý
Victoria Döller	Victor Sepulveda
Peter Gašpar	Paolo Sottovia
Sandi Gec	Nicola Vitacolonna
Yong-Bin Kang	Farhad Zafari
Selma Khouri	

General Chair

Tatjana Welzer	University of Maribor, Slovenia

Honorary Chair

Ivan Rozman	University of Maribor, Slovenia

Proceedings Chair

Aida Kamišalić Latifić	University of Maribor, Slovenia

Workshops Chairs

Robert Wrembel	Poznan University of Technology, Poland
Mirjana Ivanović	University of Novi Sad, Serbia
Johann Gamper	Free University of Bozen-Bolzano, Italy

Doctoral Consortium Chairs

Jerome Darmont	Université Lumière Lyon 2, France
Mikolay Morzy	Poznan University of Technology, Poland
Theodoros Tzouramanis	University of the Aegean, Greece

Local Chair

Lili Nemec Zlatolas   University of Maribor, Slovenia

Organizing Committee

Marko Hölbl   University of Maribor, Slovenia
Luka Hrgarek   University of Maribor, Slovenia
Aida Kamišalić Latifić   University of Maribor, Slovenia
Marko Kompara   University of Maribor, Slovenia
Lili Nemec Zlatolas   University of Maribor, Slovenia
Tatjana Welzer   University of Maribor, Slovenia
Borut Zlatolas   University of Maribor, Slovenia

Abstracts of Invited Talks

Location-in-Time Data: Compression vs. Augmentation

Goce Trajcevski

Department of Electrical and Computer Engineering, Iowa State University,
Ames, IA, USA
gocet25@iastate.edu

1 Introduction and Motivation

Data compression aims at devising efficient methodologies for a compact representation of information [6, 12]. The "raw" information can be a plain text file, numeric descriptors of images/video, social networks, etc. – and one can rely on properties of the structure, semantics, etc., when developing the methodologies for making the underlying representation more compact. While the process is something in-between a science and an art, broadly, data compression is a methodology that takes a dataset D_1 with a size β as an input, and produces a dataset D_1' as a representation of D_1 and having a size β', where $\beta' << \beta$.

The *location-in-time* data – equivalently, *Spatio-temporal Data* – is essential in multiple applications of societal relevance, arises in variety of contexts, and is generated by many heterogeneous sources. Its management gave rise to the fields of Spatio-temporal and Moving Objects Databases (MOD) [3, 9, 10]. The natural quest for compressing such data stems from several facts of life: (1) The GPS-obtained locations of the smart phone users alone generate O(Peta-Bytes) per year – and it is projected that the size of that location data could increase up to 400 times if cell-tower locations are included [8]; (2) daily travel in the US averages 11 billion miles a day (approximately 40 miles per person), and 87% of them take place in personal vehicles – thus, samples from vehicles generated every 10 seconds produce an amount of approx. 275TB daily [7, 11].

2 Contemporary Trends and Challenges

Location-based Social Networks (LBSNs) such as Instagram and Twitter generate large scale geo-spatial datasets capturing human behavior at unprecedented volume and level of detail [2]. This spurred the paradigm of *semantic/activity* trajectories [4, 13]. While providing a semantic enrichment to the location-in-time information, in majority of

Goce Trajcevski—Research supported by NSF grants III-1823279 and CNS-1823267.

applications there is the problem of sparsity of a user-location check-ins. For example, the sparsity of the Gowalla dataset [1] is about 99.98% [15].

One of the popular application domains with high societal relevance is the Point of Interest (PoI) recommendation [5]. In this realm, both the location-in-time and semantic data need to be consideres "in = concert", in order to learn the impact of the transitions across different contexts over time [14].

After the broad introduction, this talk will have three distinct portions: (i) trajectories data compression; (ii) semantic/activity trajectories; (iii) PoI recommendation via fusing the sparse location data with the semantic data.

References

1. Cho, E., Myers, S.A., Leskovec, J.: Friendship and mobility: user movement in location-based social networks. In: ACM SIGKDD (2011)
2. Deville, P., Song, C., Eagle, N., Blondel, V.D., Barabási, A.L., Wang, D.: Scaling identity connects human mobility and social interactions. In: Proceedings of the National Academy of Sciences of the United States of America (PNAS) (2016)
3. Güting, R.H., Schneider, M.: Moving Objects Databases. Morgan Kaufmann (2005)
4. Güting, R.H., Valdés, F., Damiani, M.L.: Symbolic trajectories. ACM Trans. Spat. Algorithms Syst. **1**(2), 7 (2015)
5. He, J., Li, X., Liao, L.: Category-aware next point-of-interest recommendation via listwise bayesian personalized ranking. In: IJCAI (2017)
6. Hirschberg, D., Lelewer, D.A.: Data compression. Comput. Surv. **19**(3) (1987)
7. Jang, J., Kim, H., Cho, H.: Smart roadside server for driver assistance and safety warning: framework and applications. In: CUTE 2010, pp. 1–5, December 2010
8. Mckinsey Global Institute: Big Data: The Next Frontier for Innovation, Competition, and Productivity (2011)
9. Mokbel, M.F., Aref, W.G.: SOLE: scalable on-line execution of continuous queries on spatio-temporal data streams. VLDB J. **17**(5), 971–995 (2008)
10. Di Pasquale, A., et al.: Access methods and query processing techniques. In: Sellis, T.K., et al. (eds.) Spatio-Temporal Databases. LNCS, vol. 2520, pp. 203–261. Springer, Heidelberg (2003). https://doi.org/10.1007/978-3-540-45081-8_6
11. Radaelli, L., Moses, Y., Jensen, C.S.: Using cameras to improve wi-fi based indoor positioning. In: Pfoser, D., Li, K.J. (eds.) W2GIS 2014. LNCS, vol 8470, pp. 166–183. Springer, Heidelberg (2014). https://doi.org/10.1007/978-3-642-55334-9_11
12. Sayood, K.: Introduction to Data Compression. Morgan Kauffman (1996)
13. Zheng, K., Zheng, B., Xu, J., Liu, G., Liu, A., Li, Z.: Popularity-aware spatial keyword search on activity trajectories. World Wide Web **20**(4), 749–773 (2017)
14. Zhou, F., Yin, R., Zhang, K., Trajcevski, G., Zhong, T., Wu, J.: Adversarial point-of-interest recommendation. In: The World Wide Web Conference, WWW (2019)
15. Zhuang, C., Yuan, N.J., Song, R., Xie, X., Ma, Q.: Understanding people lifestyles: construction of urban movement knowledge graph from gps trajectory. In: IJCAI (2017)

Evolution of Data Management Systems: State of the Art and Open Issues

Abdelkader Hameurlain

Informatics Research Institute of Toulouse IRIT, Paul Sabatier University,
118, Route de Narbonne, 31062 Toulouse Cedex, France
abdelkader.hameurlain@irit.fr

Abstract. The purpose of this talk is to provide a comprehensive state of the art concerning the evolution of data management systems from uniprocessor file systems to Big Data Management Systems (BDMS) in cloud environments. In the landscape of database management systems, data analysis systems (OLAP) and transaction processing systems (OLTP) are separately managed. The reasons for this dichotomy are that both systems have very different functionalities, characteristics and requirements. The talk will focus on the first class OLAP systems. In this perspective, firstly, I introduce the main problems of data management systems DMS. Then, for each environment (e.g. uniprocessor, parallel, distributed, cloud computing), I describe synthetically, the underlying concepts and the main characteristics of the proposed DMS. I also explain the relationships between those DMS. In addition, data management based on parallel and cloud systems (i.e., Parallel Relational DBMS versus BDMS) are overviewed and compared by relying on fundamental criterion such as software requirements (Data Independence, Software Reuse), High Performance, Data Availability, Fault-Tolerance, Scalability and Elasticity. I point out their advantages and weaknesses, and the reasons for which the relevant choice of a DMS is very hard. Also, I try to learn some lessons, particularly how can the evolution of these systems help for big data applications? Lastly, I point out some open issues that should be tackled to ensure the viability of the next generation of large-scale data management systems for big data applications.

Keywords: Big Data Management · Data partitioning · Data integration · Parallel database systems · Cloud data management systems · Query processing and optimization · High performance · Scalability · Elasticity · Hadoop MapReduce · Spark · Multistore systems

References

1. Abadi, D. et al.: The Beckman report on database research. Commun. ACM **59**(2), 92–99 (2016)
2. Agrawal, D., El Abbadi, A., Ooi, B.C., Das, S., Elmore, A.J.: The evolving landscape of data management in the cloud. IJCSE **7**(1), 2–16 (2012)
3. Babu, S., Herodotou H.: Massively parallel databases and MapReduce systems. Found. Trends Databases **5**(1), 1–104 (2013)

4. DeWitt, D.J., Gray, J.: Parallel database systems: the future of high performance database systems. Commun. ACM **35**(6), 85–98 (1992)
5. DeWitt, D.J., et al.: Split query processing in polybase. In: ACM SIGMOD Conference, New York, NY, USA, 22–27 June, pp. 1255–1266 (2013)
6. Duggan, J., Elmore, A., Stonebraker, M., et. al.: The BigDAWG polystore system. ACM SIGMOD Rec. **44**(2), 11–16 (2015)
7. Gray, J.: Evolution of data management. IEEE Comput. **29**(10), 38–46 (1996)
8. Hameurlain, A., Morvan, F.: An optimization method of data communication and control for parallel execution of SQL queries. In: Mařík, V., Lažanský, J., Wagner, R.R. (eds.) DEXA 1993. LNCS, vol. 720, pp. 301–312. Springer, Heidelberg (1993). https://doi.org/10.1007/3-540-57234-1_27
9. Hameurlain, A., Morvan, F.: Scheduling and mapping for parallel execution of extended SQL queries. In: CIKM '95, November 28 – December 2 1995, Baltimore, Maryland, USA, pp. 197–204 (1995)
10. Hameurlain, A., Morvan, F.: Big data management in the cloud: evolution or crossroad? In: Kozielski, S., Mrozek, D., Kasprowski, P., Małysiak-Mrozek, B., Kostrzewa, D. (eds.) BDAS 2015, BDAS 2016. CCIS, vol. 613, pp. 23–38. Springer, Cham (2016). https://doi.org/10.1007/978-3-319-34099-9_2
11. Hong, W.: Exploiting inter-operation parallelism in XPRS. In: ACM SIGMOD, San Diego, California, 2–5 June 1992, pp. 19–28 (1992)
12. Indrawan-Santiago, M.: Database research: are we at a crossroad? Reflection on NoSQL. In: NBiS 2012, Melbourne, Australia, 26–28 September 2012, pp. 45–51 (2012)
13. Lee, K., Lee, Y., Choi, H., Chung, Y.D., Moon, B.: Parallel data processing with mapreduce: a survey. SIGMOD Rec. **40**(4), 11–20 (2011)
14. Lu, H., Tan, K.L., O., B.-C.: Query processing in parallel relational database systems. IEEE CS Press (1994)
15. Özsu, M.T., Valduriez, P.: Principles of Distributed Database Systems, Third Edition. Springer, New York (2011). https://doi.org/10.1007/978-1-4419-8834-8. ISBN 978-1-4419-8833-1
16. Pietri, I., Chronis, Y., Ioannidis, Y.: Fairness in dataflow scheduling in the cloud. Inf. Syst. **83**, 118–125 (2019)
17. Schneider, D.A., DeWitt, D.J.: Tradeoffs in processing complex join queries via hashing in multiprocessor database machines. In: VLDB Conference, 13–16 August 1990, Brisbane, Queensland, Australia, pp. 469–480 (1990)
18. Stonebraker, M., et al.: Mapreduce and parallel DBMSs: friends or foes? Commun. ACM **53**(1), 64–71 (2010)
19. Thusoo, A., et al.: Hive - a petabyte scale data warehouse using hadoop. In: IEEE ICDE Conference, 1–6 March 2010, Long Beach, California, USA, pp. 996–1005 (2010)
20. Tos, U., Mokadem, R., Hameurlain, A., Ayav, T., Bora S.: Ensuring performance and provider profit through data replication in cloud systems. Clust. Comput. J. Netw. Softw. Tools Appl. **21**(3), 1479–1492 (2017). Springer, USA
21. Trummer, I., Koch, C.: Multi-objective parametric query optimization. VLDB J. **26**(1), 107–124 (2017)
22. Yin, S., Hameurlain, A., Morvan, F.: Robust query optimization methods with respect to estimation errors: a survey. SIGMOD Rec. **44**(3), 25–36, (2015). ACM Press
23. Yin, S., Hameurlain, A., Morvan, F.: SLA definition for multi-tenant DBMS and its impact on query optimization. IEEE TKDE **30**(11), 2213–2226 (2018)
24. Valduriez, P.: Parallel database systems: open problems and new issues. Distrib. Parallel Databases **1**(2), 137–165 (1993)

Semantic Relational Learning

Nada Lavrač[1,2]

[1] Jožef Stefan Institute, Ljubljana, Slovenia
nada.lavrac@ijs.si
[2] University of Nova Gorica, Vipava, Slovenia

Abstract. Relational Data Mining (RDM) addresses the task of inducing models or patterns from multi-relational data. One of the established approaches to RDM is propositionalization, characterized by transforming a relational database into a single-table representation. The talk provides an overview of propositionalization algorithms, and a particular approach named wordification, all of which have been made publicly available through the web-based ClowdFlows data mining platform. This talk addresses also Semantic Data Mining, characterized by exploiting domain ontologies in the process of model and pattern construction, which are available through the ClowdFlows platform to enable software reuse and experiment replication. The talk concludes by presenting the recent developments, which allow to speed up Semantic Relational Learning by data preprocessing using network analysis approaches.

Keywords: Relational learning · Semantic data mining · Propositionalization

Relational Learning and Semantic Data Mining

Standard machine learning and data mining algorithms induce hypotheses in the form of models or propositional patterns learned from a given data table, where one example corresponds to a single row in the table. Most types of propositional models and patterns have corresponding relational counterparts, such as relational classification rules, relational regression trees, relational association rules. Inductive Logic Programming (ILP) and Relational Data Mining (RDM) algorithms can be used to induce such relational models and patterns from multi-relational data, e.g., data stored in a relational database.

Problems characterized by multiple relations can be tackled in two different ways: (1) by using a relational learner such as Progol [3] or Aleph [5], which can build a model or induce a set of patterns directly, or (2) by constructing and using complex relational features to transform the relational representation into a propositional format and then applying a propositional learner on the transformed single-table representation. This approach is called *propositionalization* [1].

Extensive description of a number of propositionalization algorithms and their experimental evaluation is presented in [4]. In order to make the use of propositionalization algorithms easier for non-experts, as well as to make the experiments shareable and repeatable, a number of freely available propositionalization methods were wrapped as reusable components in the web-based data mining platform ClowdFlows [2], together with the utilities for working with a relational database management system (RDBMS).

A recent relational learning setting, referred to as *semantic relational learning* or *semantic data mining* (SDM), is characterized by exploiting relational background knowledge in the form of domain ontologies in the process of model and pattern construction. The development of SDM techniques is motivated by the availability of large amounts of knowledge and semantically annotated data in all domains of science, and biology in particular, posing requirements for new data mining approaches which need to deal with increased data complexity, the relational character of semantic representations, as well as the reasoning capacities of the underlying ontologies. An example SDM system Hedwig [6] performs semantic subgroup discovery by taking into account background knowledge in the form of RDF triplets and by using a search mechanism tailored to exploit the hierarchical nature of ontologies.

Acknowledgments. This work was supported by ARRS funded research program Knowledge Technologies (grant number P2-0103) and SDM-Open-SLO project Semantic Data Mining for linked open data (grant number N2-0078).

References

1. Kramer, S., Pfahringer, B., Helma, C.: Stochastic propositionalization of non-determinate background knowledge. In: Page, D. (ed.) ILP 1998. LNCS, vol 1446, pp. 80–94. Springer, Heidelberg (1998). https://doi.org/10.1007/BFb0027312
2. Kranjc, J., Podpečan, V., Lavrač, N.: ClowdFlows: a cloud based scientific workflow platform. In: Flach, P.A., De Bie, T., Cristianini, N. (eds.) ECML PKDD 2012. LNCS, vol 7524, pp. 816–819. Springer, Heidelberg (2012). https://doi.org/10.1007/978-3-642-33486-3_54
3. Muggleton, S.: Inverse entailment and Progol. New Gen. Comput. **13**(3–4), 245–286 (1995). Special issue on Inductive Logic Programming
4. Perovšek, M., Vavpetič, A., Kranjc, J., Cestnik, B., Lavrač, N.: Wordification: propositionalization by unfolding relational data into bags of words. Expert Syst. Appl. **42**(17–18), 6442–6456 (2015)
5. Srinivasan, A.: Aleph manual, March 2007. http://www.cs.ox.ac.uk/activities/machinelearning/Aleph/
6. Vavpetič, A., Novak, P.K., Grčar, M., Mozetič, I., Lavrač, N.: Semantic data mining of financial news articles. In: Fürnkranz, J., Hüllermeier, E., Higuchi, T. (eds.) DS 2013. LNCS, vol 8140, pp. 294–307. Springer, Heidelberg (2013). https://doi.org/10.1007/978-3-642-40897-7_20

Contents

Data Mining

Unsupervised Artificial Neural Networks for Outlier Detection
in High-Dimensional Data 3
    Daniel Popovic, Edouard Fouché, and Klemens Böhm

Improving Data Reduction by Merging Prototypes 20
    Pavlos Ponos, Stefanos Ougiaroglou, and Georgios Evangelidis

Keys in Relational Databases with Nulls and Bounded Domains 33
    Munqath Alattar and Attila Sali

Machine Learning

ILIME: Local and Global Interpretable Model-Agnostic Explainer
of Black-Box Decision 53
    Radwa ElShawi, Youssef Sherif, Mouaz Al-Mallah, and Sherif Sakr

Heterogeneous Committee-Based Active Learning for Entity
Resolution (HeALER) 69
    Xiao Chen, Yinlong Xu, David Broneske, Gabriel Campero Durand,
    Roman Zoun, and Gunter Saake

Document and Text Databases

Using Process Mining in Real-Time to Reduce the Number
of Faulty Products 89
    Zsuzsanna Nagy, Agnes Werner-Stark, and Tibor Dulai

Pseudo-Relevance Feedback Based on Locally-Built Co-occurrence Graphs . . . 105
    Billel Aklouche, Ibrahim Bounhas, and Yahya Slimani

Big Data

Workload-Awareness in a NoSQL-Based Triplestore 123
    Luiz Henrique Zambom Santana and Ronaldo dos Santos Mello

nativeNDP: Processing Big Data Analytics on Native Storage Nodes 139
    Tobias Vinçon, Sergey Hardock, Christian Riegger, Andreas Koch,
    and Ilia Petrov

Calculating Fourier Transforms in SQL   151
Dennis Marten, Holger Meyer, and Andreas Heuer

Novel Applications

Finding Synonymous Attributes in Evolving Wikipedia Infoboxes   169
Paolo Sottovia, Matteo Paganelli, Francesco Guerra,
and Yannis Velegrakis

Web-Navigation Skill Assessment Through Eye-Tracking Data   186
Patrik Hlavac, Jakub Simko, and Maria Bielikova

Ontologies and Knowledge Management

Updating Ontology Alignment on the Concept Level Based on Ontology
Evolution   201
Adrianna Kozierkiewicz and Marcin Pietranik

On the Application of Ontological Patterns for Conceptual Modeling
in Multidimensional Models   215
Glenda Amaral and Giancarlo Guizzardi

Process Mining and Stream Processing

Accurate and Transparent Path Prediction Using Process Mining   235
Gaël Bernard and Periklis Andritsos

Contextual and Behavioral Customer Journey Discovery Using
a Genetic Approach   251
Gaël Bernard and Periklis Andritsos

Adaptive Partitioning and Order-Preserved Merging of Data Streams   267
Constantin Pohl and Kai-Uwe Sattler

Data Quality

CrowdED and CREX: Towards Easy Crowdsourcing Quality
Control Evaluation   285
Tarek Awwad, Nadia Bennani, Veronika Rehn-Sonigo, Lionel Brunie,
and Harald Kosch

Query-Oriented Answer Imputation for Aggregate Queries   302
Fatma-Zohra Hannou, Bernd Amann, and Mohamed-Amine Baazizi

Optimization

You Have the Choice: The Borda Voting Rule
for Clustering Recommendations.   321
  Johannes Kastner and Markus Endres

BM-index: Balanced Metric Space Index Based on Weighted
Voronoi Partitioning   337
  Matej Antol and Vlastislav Dohnal

Theoretical Foundation and New Requirements

ProSA—Using the CHASE for Provenance Management.   357
  Tanja Auge and Andreas Heuer

ECHOES: A Fail-Safe, Conflict Handling, and Scalable Data
Management Mechanism for the Internet of Things.   373
  Christoph Stach and Bernhard Mitschang

Transaction Isolation in Mixed-Level and Mixed-Scope Settings.   390
  Stephen J. Hegner

Data Warehouses

Data Reduction in Multifunction OLAP.   409
  Ali Hassan and Patrice Darmon

A Framework for Learning Cell Interestingness from Cube Explorations   425
  Patrick Marcel, Veronika Peralta, and Panos Vassiliadis

Towards a Cost Model to Optimize User-Defined Functions in an ETL
Workflow Based on User-Defined Performance Metrics.   441
  Syed Muhammad Fawad Ali and Robert Wrembel

Author Index   457

Data Mining

Unsupervised Artificial Neural Networks for Outlier Detection in High-Dimensional Data

Daniel Popovic[(✉)] , Edouard Fouché , and Klemens Böhm

Karlsruhe Institute of Technology (KIT), Karlsruhe, Germany
popovic@cognitana.com, {edouard.fouche,klemens.boehm}@kit.edu

Abstract. Outlier detection is an important field in data mining. For high-dimensional data the task is particularly challenging because of the so-called "curse of dimensionality": The notion of neighborhood becomes meaningless, and points typically show their outlying behavior only in subspaces. As a result, traditional approaches are ineffective. Because of the lack of a ground truth in real-world data and of a priori knowledge about the characteristics of potential outliers, outlier detection should be considered an unsupervised learning problem. In this paper, we examine the usefulness of unsupervised artificial neural networks – autoencoders, self-organising maps and restricted Boltzmann machines – to detect outliers in high-dimensional data in a fully unsupervised way. Each of those approaches targets at learning an approximate representation of the data. We show that one can measure the "outlierness" of objects effectively, by measuring their deviation from the learned representation. Our experiments show that neural-based approaches outperform the current state of the art in terms of both runtime and accuracy.

Keywords: Unsupervised learning · Outlier detection · Neural networks

1 Introduction

Outliers are objects that deviate significantly from others as to arouse the suspicion that a different mechanism has generated them [17]. The search for outliers has interested researchers and practitioners for many years, with applications such as the detection of fraud or intrusions, and medical diagnosis.

In real-world use cases, the characteristics of outliers are unknown beforehand. One can only obtain a ground truth with the help of domain experts, who produce explicit labels on the nature of data points. However, generating this ground truth is costly or even impossible. In high-dimensional spaces in particular, objects can be outlying in unexpected ways, which the expert does not notice during inspection. For example, in aircraft fault diagnostics, thousands of sensors collect huge amounts of in-flight data. The sensors not only collect airplane

© Springer Nature Switzerland AG 2019
T. Welzer et al. (Eds.): ADBIS 2019, LNCS 11695, pp. 3–19, 2019.
https://doi.org/10.1007/978-3-030-28730-6_1

data (accelerometer, speed sensor, voltage sensors, etc.) but also environmental or weather data (thermocouple, pressure sensors, etc.) [43]. Since the space of all valid sensor-value combinations is unknown a priori, it is impossible to discern between normal and abnormal instances. Thus, one cannot train a classifier in a supervised way, or even obtain a set of instances labelled as "normal". The absence of training data results in a fully unsupervised learning problem.

When the data is high-dimensional, i.e., has hundreds of dimensions, traditional outlier detectors do not work well at all. This is due to a number of effects summarized as the "curse of dimensionality" [3,4]. There exists a number of outlier detectors, which are robust against high dimensionality: ABOD and FastABOD [30] use the angle between data objects as a deviation measure. HiCS [24] and approaches by Aggarwal *et al.* [1], Kriegel *et al.* [29], Müller *et al.* [36] and Nguyen *et al.* [38,39] propose to assess the outlierness of data points only in low-dimensional projections of the full space. While these approaches do solve the problem in high-dimensional spaces to some extent, they often come with high computational complexity and unintuitive parameters.

Several studies propose to use artificial neural networks (ANNs) for outlier detection [9,10,15,18,23,34,37]. However, these studies have only considered relatively low-dimensional settings, i.e., fewer than 100 dimensions. Thus, the performance of these approaches in high-dimensional spaces is so far unknown. Next, they often treat outlier detection as a supervised or semi-supervised problem, which contradicts our view on it as unsupervised.

According to Bishop [5], discrepancies between the data used for training and testing is one of the main factors leading to inaccurate results with neural networks. ANN-based outlier detection approaches make use of this erroneous behavior on novel data, interpreting the deviations from the expected results – or "errors" – of the neural networks as an indication for "outlierness". The idea is to train the network to learn a good representation of the majority of the data objects. Since outliers are assumed to be "few and different" [33], the hypothesis is that they do not fit the representation learned by the model. Thus, one can detect them by measuring the respective error of the neural network.

To our knowledge, this study is the first to describe and compare the specifics of a range of ANN models for such an unsupervised detection of outliers in high-dimensional spaces, together with an extensive empirical evaluation. We articulate our contributions as follows:

- **We describe the principles of three unsupervised ANN-based approaches for outlier detection**, either based on autoencoders (AEs), self-organizing maps (SOMs) or restricted Boltzmann machines (RBMs).
- **We study the effects of different parameter settings for the ANN-based approaches empirically**. Based on this evaluation, we recommend parameter values to tune each approach for the outlier detection task.
- **We compare our approaches to state-of-the-art outlier detectors, using 26 real-world data sets**. The results show a significantly better detection quality of AE and SOM on most data sets for a reduced runtime.

– **We release our implementation** on GitHub[1], to ensure the reproducibility
of the experiments and further use of the algorithms.

Paper outline: Sect. 2 features the related work. Section 3 presents our adaption of three families of ANNs for outlier detection. Section 4 is our evaluation.
Section 5 summarizes our results and possible further research questions.

2 Related Work

2.1 High-Dimensional Outlier Detection

Numerous approaches exist to detect outliers in high-dimensional spaces. One
can classify them as density-based [1,36], deviation-based [29,35], distance-based
[38] or angle-based [30]. Other approaches, such as Isolation Forest [33], use
decision tree ensembles to identify outliers. Alternatively, HiCS [24] decouples
the search for subspaces from the actual outlier ranking.

Although methods explicitly targeting at high-dimensional data yield better
results than most traditional methods, they come with certain drawbacks. First,
most methods have high computational complexity and therefore do not scale
well to large data sets. Second, each of these detectors requires at least one
parameter, and the detection quality strongly depends on the parameter values
for a given data set, as observed in [8]. There is little or no indication on how
to choose suitable parameter values for these detectors. We in turn provide
recommendations for suitable values for the approaches we investigate.

As an example, the time complexity of the original ABOD algorithm is in
$O(n^3)$, which is not efficient with large data sets [30]. Even though there exists a
faster version, FastABOD [30], with complexity in $O(n^2 + nk^2)$, the complexity
remains quadratic with the number of objects, and determining a "good" value
for parameter k is not straightforward.

2.2 ANNs for Outlier Detection

There also exists a number of approaches based on artificial neural networks.
Japkowicz *et al.* [23] and Hawkins *et al.* [18] propose approaches based on the
Autoencoder, sometimes named *Replicator Neural Network*. The work was followed by [9,12,34]. Muñoz and Muruzábal [37] were the first to propose an
approach based on SOM and Sammon's mapping [45]. Fiore *et al.* [15] use the
RBM to detect outliers in a semi-supervised way, as well as [10]. However, each
of these contributions has at least one of the following issues:

– **Low-dimensional:** The evaluation of the approach is restricted to data with
few dimensions, i.e., typically less than 100.
– **(Semi-)supervised:** Outlier detection operates only in a supervised or semi-supervised way, so that the unsupervised setting remains unaddressed.

---

[1] https://github.com/Cognitana-Research/NNOHD.

– **Specialized:** The approach is tailored to a specific scenario, such as time series, or assumes the availability of prior knowledge, i.e., it is not applicable to the general outlier detection problem.

To our knowledge, the effectiveness of ANNs for general outlier detection is unknown so far in high-dimensional spaces under the unsupervised setting. In this work we correct this and compare ANN-based models to the state-of-the-art.

3   ANN-Based Approaches

3.1   Requirements and General Idea

Any method mentioned in Sect. 2 comes with at least one disadvantage for high-dimensional outlier detection. Given this, we formulate requirements on new methods for the detection of outliers in high-dimensional spaces:

– **R1: Accuracy.** Superior outlier detection results in high-dimensional spaces, compared to existing approaches.
– **R2: Runtime.** Low computational burden, which allows the deployment of the method on high-dimensional data.
– **R3: Parameterization.** Small range of possible parameter values. Ideally, one should be able to derive default parameter values for high outlier-detection quality or recommendations for parameter-value selection a priori.

We show that ANN-based models fulfill these requirements. We only consider unsupervised ANN models and focus on the three main families: autoencoders (AEs), self-organizing maps (SOMs), and restricted Boltzmann machines (RBMs).

Unsupervised approaches have in common that they learn a representation of the data. One can measure the deviation of each object from this representation and use it as a score of the "outlierness" of this object. The implicit assumption here is that outliers are "few and different", so that they do not fit the learned representation and have a greater outlier score OS:

$$OS(x_{inlier}) < OS(x_{outlier}) \qquad (1)$$

This score in turn determines a "confidence" that a given point x is an outlier: The higher the score, the higher this confidence. Note that, as a standard preprocessing step, one may scale the values of each dimension for each data set to $[0, 1]$. This limits the effect of different scales in different dimensions.

3.2   Autoencoder

Model Description. The autoencoder (AE) is a multi-layer neural network that learns a lower-dimensional representation of a data set, from which this data can be reconstructed approximately [6,21]. To achieve this, the AE has an input and an output layer, with a number of neurons n that matches the number of dimensions in the data set, and one or more hidden layers with different numbers of neurons m_i, $m_i \neq n$. It is a combination of an *encoder* part that transforms the data into a representation called the *code* and a *decoder* part that transforms the code back to the original data space. Figure 1(a) graphs the typical AE architecture.

In general, no transformation exists that leads to a perfect reconstruction of all data objects. As a result, the output of the AE is an approximate reconstruction of the input data.

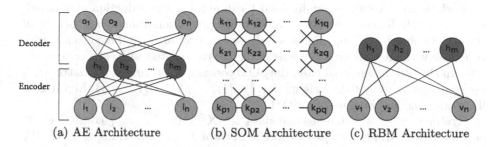

(a) AE Architecture   (b) SOM Architecture   (c) RBM Architecture

Fig. 1. ANN architectures

The encoder and decoder can be represented as functions $\mathbf{y} = e(\mathbf{x})$ and $\hat{\mathbf{x}} = d(\mathbf{y})$ with $\mathbf{x}, \hat{\mathbf{x}} \in \mathbb{R}^n$, $\mathbf{y} \in \mathbb{R}^m$, which are learned in order to minimize the sum of the squared error of the reconstruction

$$argmin_{e,d} \|\mathbf{x} - d\left(e(\mathbf{x})\right)\|^2 \qquad (2)$$

Our Outlier Detection Approach. For the outlier detection we use an AE with one hidden layer using the so-called ReLU activation function [16] for the hidden layer, and the sigmoid activation function for the output layer. Such a design choice is considered standard, as it alleviates the so-called vanishing gradient problem [22] and gives way to fast computation [31].

The number of neurons in the hidden layer is set to a fraction of the number of dimensions in the data set examined. We call this parameter the encoding factor ϵ. Learning the weights of the model is done via the widely-known back-propagation algorithm [25,32], with the AdaDelta gradient optimizer [51], in a number n^e of training epochs. Since the data is scaled to $[0, 1]$, we use the binary cross entropy loss function [44] between the input $\mathbf{i}$ and the output $\mathbf{o}$, defined as

$$l(\mathbf{i}, \mathbf{o}) = -(\mathbf{i} \ln \mathbf{o} + (1 - \mathbf{i}) \ln (1 - \mathbf{o})) \qquad (3)$$

As stated above, the AE learns an approximated reconstruction of the input data. The expectation is that the reconstruction of "abnormal" objects will be less accurate than for "normal" objects. We use the outlier score OS_{AE} of an object x as the Euclidean distance between its actual values x_j and its reconstruction $\hat{x}_j$, similarly as in [18]:

$$OS_{AE}(x) = \sum_{j=1}^{n} \sqrt{(x_j - \hat{x}_j)^2} \tag{4}$$

3.3 Self-Organising Maps

Model Description. A Self-Organising Map (SOM), also known as Kohonen network, is an ANN traditionally used for dimensionality reduction and visualization of high-dimensional data [26,27]. It is a projection from an n-dimensional set of data objects to a low-dimensional, usually two- or three-dimensional, grid. A neuron with an n-dimensional weight vector is associated with each node in the grid. In the two-dimensional case, the SOM consists of a $p \times q$ neuron matrix with neurons described by their weights $w_{ij}, i \in \{1, \ldots, p\}, j \in \{1, \ldots, q\}$. Figure 1(b) illustrates the architecture of a SOM.

Training using a n-dimensional data set $\mathbf{X} = \{x_1, \ldots, x_m\}$ consists of n^e epochs with m training steps per epoch, in total $T = n^e * m$ training steps. We let $\mathbf{w_{ij}}(t)$ denote the weight vectors after the tth training step, $\mathbf{w_{ij}}(0)$ the initial weight vectors, $\alpha(t)$ a learning rate decreasing with t and $h_{ij,cd}(t)$ a neighborhood function for neurons w_{ij} and w_{cd}, instantiated as a smoothing kernel whose width decreases with t. We use a Gaussian neighborhood kernel.

Training the SOM is done by finding for each data object x_k the neuron that has the closest distance, usually the Euclidean distance, to this data object, also called the *best matching unit* (BMU), and updating the weight vector of each neuron as follows:

$$w_{ij}(t+1) = w_{ij}(t) + \alpha(t)h_{ij,cd}(t)(x_k - w_{ij}(t)) \tag{5}$$

So the weight vectors of the BMU and its neighboring neurons in the grid are moved closer to the data object, and we repeat this process iteratively.

Our Outlier Detection Approach. Our approach is based on the idea that the trained SOM forms a map that is adjusted to the majority of data objects. Outliers are assumed to be located farther away from their BMUs than inliers.

As is common in the literature, we use a 2-dimensional SOM with n rows and columns, where n is called the *topology size*. We initialize the weights by selecting the first two subspaces spanned by the first two eigenvectors of the correlation matrix, as in [2,11].

The number of training epochs n^e is a parameter that has an effect on outlier detection quality. The outlier score for the SOM, OS_{SOM}, is the Euclidean distance of a data object to its BMU. We refer to the BMU for object $\mathbf{x}$ as $bmu_{\mathbf{x}}$.

$$OS_{SOM}(x) = \sum_{j=1}^{n} \sqrt{(x_j - bmu_{x_j})^2} \tag{6}$$

3.4   Restricted Boltzmann Machine

Model Description. The restricted Boltzmann machine (RBM) is a stochastic ANN that learns a probability distribution over a training data set. It is a special case of the Markov random field [20, 49]. The RBM consists of a layer of n "visible" neurons v_i and a layer of m "hidden" neurons h_j that form a bipartite graph with a connection weight matrix $\mathbf{W}_{n \times m} = (w_{ij})_{n \times m}$, a bias vector $\mathbf{a} = < a_i >_{i \in \mathbb{N}[1,n]}$ for the visible neurons and a bias vector $\mathbf{b} = < b_j >_{j \in \mathbb{N}[1,m]}$ for the hidden neurons. The probability distribution is defined using the energy function

$$E(\mathbf{v}, \mathbf{h}) = -\mathbf{a}^T \mathbf{v} - \mathbf{b}^T \mathbf{h} - \mathbf{v}^T \mathbf{W} \mathbf{h} \tag{7}$$

which assigns a scalar energy to each configuration, i.e., to each pair of visible and hidden neuron values. A high energy for a configuration corresponds to a low probability of that configuration to appear in the model. The objective of the training is to find a configuration of weights and biases that lead to a high probability for the training data objects and a low probability for other data. So the energy for data objects from the training data set is sought to be minimized in the training process. Using the gradient descent algorithm to minimize this objective function would involve the computation of the expectation over all possible configurations of the input data object, which is not feasible in practice. Hence, the training usually is performed with the contrastive divergence algorithm [19] using Gibbs sampling, which approximates the gradient descent. This algorithm simplifies and speeds up training compared to gradient descent. It performs three learning steps on each training data object. First, all hidden units are updated in parallel from the training data object at the visible neurons. Then, the visible neurons are updated in parallel to get a reconstruction of the training data object. Finally, the hidden neurons are updated in parallel again. Figure 1(c) illustrates the architecture of a RBM.

Our Outlier Detection Approach. We use a RBM with Gaussian visible neurons. As it assumes that the data is normally distributed, we standardize each dimension by subtracting the mean and dividing by the standard deviation as a preprocessing step. After the training, the energy is expected to be low for normal data objects and high for rare or unknown data objects. The outlier score for a data object, OS_{RBM}, is its so-called free energy:

$$OS_{RBM}(x) = -\sum_i a_i x_i - \sum_i \frac{x_i^2}{2} + \sum_i \ln \sum_{h_i} e^{b_i + \mathbf{w_i} x} \tag{8}$$

The proportion δ of hidden units w.r.t. visible units, and the number of training epochs n^e are free parameters. The share of the data set used for training is referred to as γ.

4   Evaluation

In this section, we pursue two separate evaluations. First, we evaluate the parameter ranges of each of our approaches on high-dimensional data. This leads to the recommendation of "good" parameters. Second, we compare the approaches against the state of the art and evaluate them using high-dimensional data.

We implement the models in Python 3 using Keras, Tensorflow and the SOM implementation Somoclu [50]. All experiments run on a quad-core processor at 3.20 GHz with 8 GB RAM. As mentioned earlier, we publish the source code for our experiments on GitHub[2], to ensure reproducibility.

4.1   Parameter Selection

Campos *et al.* [8] compare various parametrized outlier detection approaches, with a large range of parameter values. Their results indicate that the entire parameter-value range is needed to achieve the highest outlier detection quality over different data sets, and that there exists no obvious way to find good parameter values a priori for a given data set.

To verify whether this applies to NN-based approaches as well, we investigate the range of parameter combinations for each approach, based on 26 data sets. Table 1 lists the characteristics of each data set in the corpus. It contains the same data sets as in [9] except for KddCup99, for which there are no exact results in [9], plus an assortment of high-dimensional data sets: Arrhythmia [8], InternetAds [8], ISOLET [14], MNIST and Musk [42]. Arrhythmia, InternetAds and ISOLET have several variants with different proportions of outliers. In this work we present the results for the variants with approximately 2% outliers. Because of the restricted number of pages, we present the evaluation results for the other variants in our GitHub repository, evaluating in total 26 data sets.

Table 1. List of evaluated data sets.

Data set	Dimensions	Data objects	Outliers	Outlier ratio
Arrhythmia-2	259	248	4	1.61%
Cardio	21	1,831	176	9.61%
Ecoli	7	336	9	2.68%
InternetAds-2	1,555	1,630	32	1.96%
ISOLET-2	617	2,449	50	2.00%
Lympho	18	148	6	4.05%
MNIST	100	7,603	700	9.21%
Musk	166	3,062	97	3.12%
Optdigits	64	5,216	150	2.88%
P53	5,408	16,592	143	0.86%
Pendigits	16	6,870	156	2.27%
Seismic	11	2,584	170	6.58%
Thyroid	6	3,772	93	2.4%
Waveform	21	3,509	166	4.73%
Yeast	8	1,364	65	4.77%

[2] https://github.com/Cognitana-Research/NNOHD.

We evaluate the goodness of each parameter combination on the whole data-set assortment. The aim of the evaluation is to derive parameters that lead to results which are *best on average*. To this end, we define a notion of deviation $\mathcal{D}_p$ that should ideally be minimized:

$$\mathcal{D}_p(\mathcal{A}) = \frac{\sum_{d \in D} (S_d^{\max} - S_d^p)}{|D|} \tag{9}$$

Intuitively, the deviation $\mathcal{D}_p$ of an algorithm $\mathcal{A}$ is the average difference between the best achievable score $S_d^{\max}$ over the parameters $p \in \mathcal{P}$ and the actual score S_d^p obtained with parameter combination p for all data sets $d \in D$. In the end, choosing the parameter combination minimizing $\mathcal{D}_p$ means maximizing the normalized average score for each data set in the assortment. Our hope is that those parameters will lead to good outlier detection on data sets that are not part of this assortment as well, so they can be useful to others.

We instantiate the score S as the commonly used ROC AUC. For the AE, we investigate encoding factors from 0.5 to 0.9 in steps of 0.1. For the SOM, we use quadratic maps with columns and rows from the range $\{1, \ldots, 20\}$. We test the RBM with values for δ, the number of hidden neurons as share of the visible neurons, of 0.1 to 0.9 in steps of 0.1, and for γ, the share of data used for training, of 0.1 to 0.9 in steps of 0.1. For all three approaches we use 10, 20, 50, 100 and 1,000 training epochs. The final result for a parameter combination is computed as the average of 20 runs. As there is no obvious way to select these parameter values a priori, we test all $p \in \mathcal{P}$ for each of our approaches in a brute-force fashion. This results in a total number of 10,500 experiments for the AE, 42,000 for the SOM, and 170,000 for the RBM.

Table 2 is an excerpt of the parameter evaluation of the AE, SOM and RBM approaches. The best parameter value for each approach is in boldface, the parameter values that are within 0.01 of the best result are in gray boldface. We publish the complete list in our GitHub repository. The evaluation leads to the following recommendations:

- For the Autoencoder:
    - Encoding factor $\epsilon = 0.8$
    - Number of training epochs $n^e = 20$
- For the Self-organizing map:
    - Topology size $n = 2$
    - Number of training epochs $n^e = 10$
- For the Restricted Boltzmann machine:
    - Proportion of hidden neurons $\delta = 0.8$
    - Number of training epochs $n^e = 100$
    - Proportion of training data $\gamma = 0.9$

Interestingly, the number of training epochs n^e minimizing $\mathcal{D}_p$ for the ANN-based approaches is relatively low, between 10 and 20 epochs for AE and SOM and 100 for RBM, and this observation is consistent even if we vary the encoding factor ϵ. This means that, in contrast to other application domains of neural networks such as image recognition, good outlier detection is feasible with low computational effort, as few training epochs are required.

We can also see that SOM achieves the best results for very low-dimensional maps, i.e., only 2 or 3 columns/rows, which also stands for a low computational burden. For the RBM we notice that the best results are achieved for a larger number of epochs, namely 100 and 1,000, while the robustness is rather stable over the number of hidden neurons. Finally, we see that the average deviation for the large majority of the parameter values is not greater than 5%. Thus, the performance of neural-based methods seems to be relatively independent from the chosen parameters, i.e., they fulfill Requirement **R3**.

Table 2. Parameter evaluation of AE and SOM (excerpt).

Parameters			AE	SOM	Parameters			RBM
ϵ	n	n^e	$\mathcal{D}_{\epsilon,n^e}$	$\mathcal{D}_{n,n^e}$	δ	n^e	γ	$\mathcal{D}_{\delta,n^e,\gamma}$
0.6	2	10	0.0478	**0.0213**	0.7	100	0.5	0.1148
0.6	2	20	0.0411	0.0320	0.7	100	0.9	0.0750
0.6	2	100	0.0476	0.0214	0.7	1000	0.5	0.0940
0.6	2	1,000	0.0844	0.0214	0.7	1000	0.9	0.0724
0.7	3	10	0.0501	0.0306	0.8	100	0.5	0.1173
0.7	3	20	0.0412	0.0457	0.8	100	0.9	**0.0597**
0.7	3	100	0.0527	0.0327	0.8	1000	0.5	0.0905
0.7	3	1,000	0.0788	0.0323	0.8	1000	0.9	0.0701
0.8	4	10	0.0409	0.0390	0.9	100	0.5	0.1027
0.8	4	20	**0.0403**	0.0500	0.9	100	0.9	0.0734
0.8	4	100	0.0512	0.0535	0.9	1000	0.5	0.0868
0.8	4	1,000	0.0827	0.0545	0.9	1000	0.9	0.0769

4.2   Outlier Detection Quality Evaluation

We now compare our approaches to the state of the art. We consider RandNet [9], which is – to our knowledge – the most recently published ANN-based contribution. The authors did not publish their implementation nor enough information for reproducibility, so we simply compare to the same data sets except for the 41-dimensional KddCup99, and the baseline methods LOF [7], Hawkins [18], HiCS [24] and LODES [46] as in [9]. The authors also have set $k = 5$ for LOF and HiCS, which yields suboptimal results for these approaches. Investigating all parameter values k in the range $k \in 1, \ldots, 100$ for LOF and HiCS, we find that $k = 100$ leads to the best results on average. Thus, we repeat the evaluation on these data sets with $k = 100$ for a fair comparison. We further set HiCS parameters to $M = 50$, $\alpha = 0.1$ and $candidate\_cutoff = 100$, which is in line with the recommendations by the authors [24]. For any competing algorithm, we use the implementation from ELKI [48].

Table 3 lists the ROC AUC values for the data sets in [9]. The best values in the table are highlighted in boldface, values within 1% of the best value are highlighted in italics. We see that AE and SOM are at least competitive using the recommended parameter values. SOM even outperforms all approaches in three data sets. The RBM stands behind for most data sets, only having competitive results for the Pendigits and Thyroid data sets. Surprisingly, in a few cases (e.g., [Optdigits, LODES]), the score falls way below 0.5, i.e., it is worse than random guessing. We notice that this never occurs with neural-based approaches.

In Table 4, we observe similar results for the high-dimensional data sets Arrhythmia, ISOLET, MNIST, Musk and P53. We compare our approaches against LOF [7], HiCS [24], FastABOD [30], LoOP [28]. In addition, we use one-class SVM [47] and KNN Outlier [41] with $k = 1$ as a baseline. AE and SOM yield the highest ROC AUC for 4 of 6 data sets. For the ISOLET data set where LOF has the highest ROC AUC, AE and SOM are within reach of the best results. Only for InternetAds, which consists only of binary attributes, the ANN-based approaches fall behind the best results. Figure 2 graphs the ROC AUC comparisons for the high-dimensional data sets.

Table 3. ROC AUC comparison for data sets used in [9] (in %).

Data set	AE	SOM	RBM	RandNet	LODES	HiCS	LOF	Hawkins
Cardio	*92.10*	**93.01**	53.75	*92.87*	78.90	85.59	91.41	*92.36*
Ecoli	87.19	86.65	76.62	85.42	**91.81**	88.25	90.35	82.87
Lympho	90.33	*99.77*	58.57	*99.06*	78.16	92.94	**99.88**	98.70
Optdigits	71.00	71.90	48.90	*87.11*	2.00	97.94	38.94	**87.03**
Pendigits	66.53	**95.83**	91.25	93.44	87.88	72.78	51.51	89.81
Seismic	*71.88*	**71.99**	33.03	*71.28*	66.71	68.19	65.58	68.25
Thyroid	89.46	92.99	94.35	90.42	72.94	91.74	**96.31**	87.47
Waveform	59.22	69.39	60.35	70.05	62.88	71.82	**76.56**	61.57
Yeast	**83.81**	81.81	49.37	*82.95*	77.70	78.21	78.19	82.12

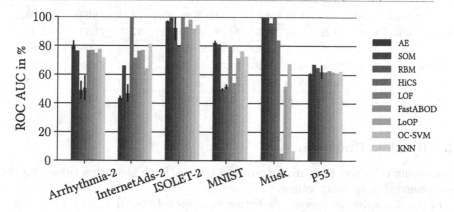

Fig. 2. ROC AUC for high-dimensional data sets

Provost and Foster [40] argue that ROC AUC is not an appropriate performance measure for the classification of highly skewed data sets, which certainly is the case with outliers. Thus, we use the area under the precision-recall curve (PR AUC) as a complementary measure for the high-dimensional data-set evaluation [13]. Table 5 lists the corresponding PR AUC values.

As we can see, AE and SOM have the best results in terms of PR AUC for 2 out of 6 data sets and are close to the best results for the other data sets except for InternetAds. This also indicates that AE and SOM have fewer false positives in most data sets. A significant insight is that SOM, and to a certain degree also AE, deliver competitive results over all data sets, while all reference algorithms fall behind the top group by much, at least for some data sets. This stability of results is a great advantage for the SOM and the AE. The RBM again has a mixed performance: It is competitive on the ISOLET and Musk data sets, but close to guessing for the Arrhythmia and MNIST data sets. It is noticeable that the RBM yields results similar to random guesses for all sparse data sets that are evaluated, namely the Arrhythmia, InternetAds, Lympho, MNIST and Optdigits data sets.

Table 4. ROC AUC comparison for high dimensional data sets (in %).

Data set	AE	SOM	RBM	HiCS	LOF	FastABOD	LoOP	OC-SVM	KNN
Arrhythmia-2	**80.22**	76.33	49.04	50.56	76.74	76.84	75.00	77.66	71.88
InternetAds-2	43.36	66.12	46.93	**99.84**	71.64	76.49	77.14	64.18	81.23
ISOLET-2	96.85	*99.28*	92.28	79.71	**99.58**	93.09	98.23	92.05	94.66
MNIST	**82.06**	*81.07*	49.87	51.74	80.34	54.35	71.66	76.46	72.74
Musk	**100**	**100**	95.60	*99.60*	84.00	5.11	51.86	67.60	7.11
P53	60.63	**67.17**	64.76	62.09	61.99	62.92	61.99	61.27	62.56

Table 5. PR AUC comparison for high dimensional data sets (in %).

Data set	AE	SOM	RBM	HiCS	LOF	FastABOD	LoOP	OC-SVM	KNN
Arrhythmia-2	**29.37**	27.94	1.66	1.64	3.83	3.98	3.51	4.90	3.04
InternetAds-2	1.58	32.78	1.91	**94.52**	34.86	32.79	37.09	24.83	32.24
ISOLET-2	44.45	66.93	29.51	4.47	**70.95**	29.54	51.78	25.65	42.91
MNIST	30.13	27.40	9.26	9.99	**33.95**	14.05	24.74	25.49	27.96
Musk	**100**	**100**	85.58	97.46	14.68	1.65	3.71	4.54	1.91
P53	1.04	*1.29*	1.20	1.25	1.06	1.13	1.06	1.06	**1.30**

4.3   Runtime Evaluation

We measure the execution time of our approaches. Each algorithm runs with the recommended parameter values.

Figure 3 graphs the average execution time for selected data sets with a logarithmic scale. We see that SOM is very fast for all data sets. This is particularly obvious for the P53 data set, which consists of 5,408 dimensions and 16,592 data objects. While the fastest of the compared algorithms needs more than 18 min, the SOM needs less than 2 min. The RBM with less than 9 min and the AE with less than 20 min still are very fast. Note that the runtimes were measured without GPU support. For the AE in particular, the runtime using a GPU would be much smaller.

We could not compare the runtime of our approaches with RandNet, because of the missing implementation. However, since it consists of an ensemble of up to 200 AEs with 3 hidden layers, it should be clear that it requires much more computational effort than any of our neural-based approaches.

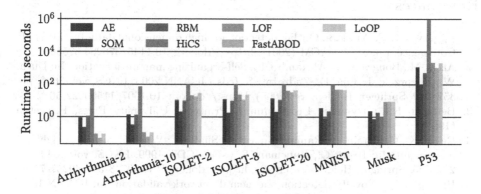

Fig. 3. Runtime comparison

5   Conclusions

This paper studies the application of ANN-based models, namely autoencoder (AE), self-organizing map (SOM) and restricted Boltzmann machine (RBM), to high-dimensional outlier detection. For each of these approaches, we propose to use a model-specific outlier score. Nonetheless, the scores have in common that they quantify in a fully unsupervised way the deviation from the expected output for each data point w.r.t. the learned model.

We evaluate the models on an assortment of high-dimensional data sets and compare the results to state-of-the-art outlier detection algorithms. The SOM and AE approaches show superior performance in terms of detection quality (Requirement **R1**) and runtime (Requirement **R2**) compared to the state of the art. SOM clearly outperforms them all and yields very high result quality in large high-dimensional data sets. At the same time, the range of relevant parameter values (Requirement **R3**) for AE and SOM is significantly smaller than for the state-of-the-art algorithms.

All in all, this study also shows that "simple" is often better in the case of outlier detection. When used properly, well-known ANN-based approaches such as AE and SOM outperform recently proposed approaches for unsupervised outlier-detection tasks in high-dimensional data, both in terms of accuracy and runtime, while being less sensitive to parameter tuning.

In the future, it will be interesting to investigate whether the extension of the RBM to deep belief networks (DBNs) [20] leads to better results for this class of algorithms, since RBM has shown a relatively low detection quality. In this study, we have determined good parameter values for each approach in the general case, but finding the optimal values for each data set for the AE and SOM might improve performance even more. Thus, our goal would be to come up with a method to set parameters automatically in a data-driven way.

References

1. Aggarwal, C.C., Yu, P.S.: Outlier detection for high dimensional data. In: SIGMOD Conference, pp. 37–46. ACM (2001). https://doi.org/10.1145/376284.375668
2. Attik, M., Bougrain, L., Alexandre, F.: Self-organizing map initialization. In: Duch, W., Kacprzyk, J., Oja, E., Zadrożny, S. (eds.) ICANN 2005. LNCS, vol. 3696, pp. 357–362. Springer, Heidelberg (2005). https://doi.org/10.1007/11550822_56
3. Bellman, R.E.: Dynamic Programming. Princeton University Press, Princeton (1957)
4. Beyer, K., Goldstein, J., Ramakrishnan, R., Shaft, U.: When is "nearest neighbor" meaningful? In: Beeri, C., Buneman, P. (eds.) ICDT 1999. LNCS, vol. 1540, pp. 217–235. Springer, Heidelberg (1999). https://doi.org/10.1007/3-540-49257-7_15
5. Bishop, C.M.: Novelty detection and neural network validation. In: ICANN 1993, pp. 789–794 (1993). https://doi.org/10.1007/978-1-4471-2063-6_225
6. Bourland, H., Kamp, Y.: Auto-association by multilayer perceptrons and singular value decomposition. Biol. Cybern. **59**(4), 291–294 (1988). https://doi.org/10.1007/BF00332918
7. Breunig, M.M., Kriegel, H., Ng, R.T., Sander, J.: LOF: identifying density-based local outliers. In: SIGMOD Conference, pp. 93–104. ACM (2000). https://doi.org/10.1145/335191.335388
8. Campos, G.O., et al.: On the evaluation of unsupervised outlier detection: measures, datasets, and an empirical study. Data Min. Knowl. Discov. **30**(4), 891–927 (2016). https://doi.org/10.1007/s10618-015-0444-8
9. Chen, J., Sathe, S., Aggarwal, C.C., Turaga, D.S.: Outlier detection with autoencoder ensembles. In: SDM, pp. 90–98. SIAM (2017). https://doi.org/10.1137/1.9781611974973.11
10. Chen, Y., Lu, L., Li, X.: Application of continuous restricted boltzmann machine to identify multivariate geochemical anomaly. J. Geochem. Explor. **140**, 56–63 (2014). https://doi.org/10.1016/j.gexplo.2014.02.013
11. Ciampi, A., Lechevallier, Y.: Clustering large, multi-level data sets: an approach based on Kohonen Self Organizing Maps. In: Zighed, D.A., Komorowski, J., Żytkow, J. (eds.) PKDD 2000. LNCS (LNAI), vol. 1910, pp. 353–358. Springer, Heidelberg (2000). https://doi.org/10.1007/3-540-45372-5_36
12. Dau, H.A., Ciesielski, V., Song, A.: Anomaly detection using replicator neural networks trained on examples of one class. In: Dick, G., et al. (eds.) SEAL 2014. LNCS, vol. 8886, pp. 311–322. Springer, Cham (2014). https://doi.org/10.1007/978-3-319-13563-2_27

13. Davis, J., Goadrich, M.: The relationship between precision-recall and ROC curves. In: ICML, ACM International Conference Proceeding Series, vol. 148, pp. 233–240. ACM (2006). https://doi.org/10.1145/1143844.1143874
14. Dua, D., Graff, C.: UCI machine learning repository (2019). http://archive.ics.uci.edu/ml
15. Fiore, U., Palmieri, F., Castiglione, A., Santis, A.D.: Network anomaly detection with the restricted boltzmann machine. Neurocomputing **122**, 13–23 (2013). https://doi.org/10.1016/j.neucom.2012.11.050
16. Hahnloser, R.R., Sarpeshkar, R., Mahowald, M.A., Douglas, R.J., Seung, S.H.: Digital selection and analogue amplification coexist in a cortex-inspired silicon circuit. Nature **405**(6789), 947–951 (2000). https://doi.org/10.1038/35016072
17. Hawkins, D.M.: Identification of Outliers, Monographs on Applied Probability and Statistics, vol. 11. Springer, Dordrecht (1980). https://doi.org/10.1007/978-94-015-3994-4
18. Hawkins, S., He, H., Williams, G., Baxter, R.: Outlier detection using replicator neural networks. In: Kambayashi, Y., Winiwarter, W., Arikawa, M. (eds.) DaWaK 2002. LNCS, vol. 2454, pp. 170–180. Springer, Heidelberg (2002). https://doi.org/10.1007/3-540-46145-0_17
19. Hinton, G.E.: Training products of experts by minimizing contrastive divergence. Neural Comput. **14**(8), 1771–1800 (2002). https://doi.org/10.1162/089976602760128018
20. Hinton, G.E., Osindero, S., Teh, Y.W.: A fast learning algorithm for deep belief nets. Neural Comput. **18**(7), 1527–1554 (2006). https://doi.org/10.1162/neco.2006.18.7.1527
21. Hinton, G.E., Zemel, R.S.: Autoencoders, minimum description length and helmholtz free energy. In: NIPS, pp. 3–10. Morgan Kaufmann (1993). http://papers.nips.cc/paper/798-autoencoders-minimum-description-length-and-helmholtz-free-energy
22. Hochreiter, S.: The vanishing gradient problem during learning recurrent neural nets and problem solutions. Int. J. Uncertainty Fuzziness Knowl. Based Syst. **6**(2), 107–116 (1998). https://doi.org/10.1142/S0218488598000094
23. Japkowicz, N., Myers, C., Gluck, M.A.: A novelty detection approach to classification. In: IJCAI, pp. 518–523. Morgan Kaufmann (1995). http://ijcai.org/Proceedings/95-1/Papers/068.pdf
24. Keller, F., Müller, E., Böhm, K.: HiCS: high contrast subspaces for density-based outlier ranking. In: ICDE, pp. 1037–1048. IEEE Computer Society (2012). https://doi.org/10.1109/icde.2012.88
25. Kelley, H.J.: Gradient theory of optimal flight paths. ARS J. **30**(10), 947–954 (1960). https://doi.org/10.2514/8.5282
26. Kohonen, T.: Self-organized formation of topologically correct feature maps. Biol. Cybern. **43**(1), 59–69 (1982). https://doi.org/10.1007/bf00337288
27. Kohonen, T.: Self-Organizing Maps. Springer Series in Information Sciences. Springer, Heidelberg (1995). https://doi.org/10.1007/978-3-642-97610-0
28. Kriegel, H., Kröger, P., Schubert, E., Zimek, A.: Loop: local outlier probabilities. In: CIKM, pp. 1649–1652. ACM (2009). https://doi.org/10.1145/1645953.1646195
29. Kriegel, H.-P., Kröger, P., Schubert, E., Zimek, A.: Outlier detection in axis-parallel subspaces of high dimensional data. In: Theeramunkong, T., Kijsirikul, B., Cercone, N., Ho, T.-B. (eds.) PAKDD 2009. LNCS (LNAI), vol. 5476, pp. 831–838. Springer, Heidelberg (2009). https://doi.org/10.1007/978-3-642-01307-2_86

30. Kriegel, H.P., Schubert, M., Zimek, A.: Angle-based outlier detection in high-dimensional data. In: Proceeding of the 14th ACM SIGKDD International Conference on Knowledge Discovery and Data Mining - KDD 2008, ACM Press, New York, NY, USA, pp. 444–452 (2008). https://doi.org/10.1145/1401890.1401946
31. Krizhevsky, A., Sutskever, I., Hinton, G.E.: Imagenet classification with deep convolutional neural networks. In: NIPS, pp. 1106–1114 (2012). https://doi.org/10.1145/3065386
32. Linnainmaa, S.: Taylor expansion of the accumulated rounding error. BIT Numer. Math. **16**(2), 146–160 (1976). https://doi.org/10.1007/BF01931367
33. Liu, F.T., Ting, K.M., Zhou, Z.: Isolation forest. In: ICDM, pp. 413–422. IEEE Computer Society (2008). https://doi.org/10.1109/ICDM.2008.17
34. Marchi, E., Vesperini, F., Eyben, F., Squartini, S., Schuller, B.W.: A novel approach for automatic acoustic novelty detection using a denoising autoencoder with bidirectional LSTM neural networks. In: ICASSP, pp. 1996–2000. IEEE (2015). https://doi.org/10.1109/ICASSP.2015.7178320
35. Müller, E., Schiffer, M., Seidl, T.: Adaptive outlierness for subspace outlier ranking. In: CIKM, pp. 1629–1632. ACM (2010). https://doi.org/10.1145/1871437.1871690
36. Müller, E., Schiffer, M., Seidl, T.: Statistical selection of relevant subspace projections for outlier ranking. In: ICDE, pp. 434–445. IEEE Computer Society (2011). https://doi.org/10.1109/ICDE.2011.5767916
37. Muñoz, A., Muruzábal, J.: Self-organising maps for outlier detection. Neurocomputing **18**(1), 33–60 (1998). https://doi.org/10.1016/S0925-2312(97)00068-4
38. Nguyen, H.V., Gopalkrishnan, V., Assent, I.: An unbiased distance-based outlier detection approach for high-dimensional data. In: Yu, J.X., Kim, M.H., Unland, R. (eds.) DASFAA 2011. LNCS, vol. 6587, pp. 138–152. Springer, Heidelberg (2011). https://doi.org/10.1007/978-3-642-20149-3_12
39. Nguyen, H.V., Müller, E., Vreeken, J., Keller, F., Böhm, K.: CMI: an information-theoretic contrast measure for enhancing subspace cluster and outlier detection. In: SDM, pp. 198–206 (2013). https://doi.org/10.1137/1.9781611972832.22
40. Provost, F.J., Fawcett, T.: Analysis and visualization of classifier performance: comparison under imprecise class and cost distributions. In: KDD, pp. 43–48. AAAI Press (1997), http://www.aaai.org/Library/KDD/1997/kdd97-007.php
41. Ramaswamy, S., Rastogi, R., Shim, K.: Efficient algorithms for mining outliers from large data sets. In: SIGMOD Conference, pp. 427–438. ACM (2000). https://doi.org/10.1145/342009.335437
42. Rayana, S.: ODDS library (2016). http://odds.cs.stonybrook.edu
43. Reddy, K.K., Sarkar, S., Venugopalan, V., Giering, M.: Anomaly detection and fault disambiguation in large flight data: a multi-modal deep autoencoder approach. In: Proceedings of the Annual Conference of the Prognostics and Health Management Society, Denver, Colorado. PHMC 2016, PHM Society, Rochester, NY, USA, vol. 7, pp. 192–199 (2016). http://www.phmsociety.org/node/2088/
44. Rubinstein, R.: The cross-entropy method for combinatorial and continuous optimization. Methodol. Comput. Appl. Probab. **1**(2), 127–190 (1999). https://doi.org/10.1023/A:1010091220143
45. Sammon, J.W.: A nonlinear mapping for data structure analysis. IEEE Trans. Comput. **18**(5), 401–409 (1969). https://doi.org/10.1109/T-C.1969.222678
46. Sathe, S., Aggarwal, C.C.: LODES: local density meets spectral outlier detection. In: SDM, pp. 171–179. SIAM (2016). https://doi.org/10.1137/1.9781611974348.20
47. Schölkopf, B., Platt, J.C., Shawe-Taylor, J., Smola, A.J., Williamson, R.C.: Estimating the support of a high-dimensional distribution. Neural Comput. **13**(7), 1443–1471 (2001). https://doi.org/10.1162/089976601750264965

48. Schubert, E., Koos, A., Emrich, T., Züfle, A., Schmid, K.A., Zimek, A.: A framework for clustering uncertain data. PVLDB **8**(12), 1976–1979 (2015). http://www.vldb.org/pvldb/vol8/p1976-schubert.pdf
49. Smolensky, P.: Information processing in dynamical systems: Foundations of harmony theory. In: Rumelhart, D.E., McClelland, J.L., PDP Research Group, C. (eds.) Parallel Distributed Processing: Explorations in the Microstructure of Cognition, Vol. 1, pp. 194–281. MIT Press, Cambridge (1986). http://dl.acm.org/citation.cfm?id=104279.104290
50. Wittek, P.: Somoclu: an efficient distributed library for self-organizing maps. CoRR abs/1305.1422 (2013). http://arxiv.org/abs/1305.1422
51. Zeiler, M.D.: ADADELTA: an adaptive learning rate method. CoRR abs/1212.5701 (2012). http://arxiv.org/abs/1212.5701

Improving Data Reduction by Merging Prototypes

Pavlos Ponos[1]([✉]), Stefanos Ougiaroglou[1,2], and Georgios Evangelidis[1]

[1] Department of Applied Informatics, School of Information Sciences,
University of Macedonia, 54636 Thessaloniki, Greece
{pponos,stoug}@uom.edu.gr, gevan@uom.gr
[2] Department of Information Technology, Alexander TEI of Thessaloniki,
57400 Sindos, Greece

Abstract. A well-known and adaptable classifier is the k-Nearest Neighbor (kNN) that requires a training set of relatively small size in order to perform adequately. Training sets can be reduced in size by using conventional data reduction techniques. Unfortunately, these techniques are inappropriate in streaming environments or when executed in devices with limited resources. dRHC is a prototype generation algorithm that works in streaming environments by maintaining a condensed training set that can be updated by continuously arriving training data segments. Prototypes in dRHC carry an appropriate weight to indicate the number of instances of the same class that they represent. dRHC2 is an improvement over dRHC since it can handle fixed size condensing sets by removing the least important prototypes whenever the condensing set exceeds a predefined size. In this paper, we exploit the idea that dRHC or dRHC2 prototypes could be merged whenever they are close enough and represent the same class. Hence, we propose two new prototype merging algorithms. The first algorithm performs a single pass over a newly updated condensing set and merges all prototype pairs of the same class under the condition that each prototype is the nearest neighbor of the other. The second algorithm performs repetitive merging passes until there are no prototypes to be merged. The proposed algorithms are tested against several datasets and the experimental results reveal that the single pass variation performs better for both dRHC and dRHC2 taking into account the trade-off between preprocessing cost, reduction rate and accuracy. In addition, the merging appears to be more appropriate for the static version of the algorithm (dRHC) since it offers higher data reduction without sacrificing accuracy.

Keywords: k-NN classification · Data reduction ·
Prototype merging · Data streams · Clustering

1 Introduction

The attention of the Data Mining and Machine Learning communities has been attracted by the problem of dealing with fast data streams [1] and large datasets

© Springer Nature Switzerland AG 2019
T. Welzer et al. (Eds.): ADBIS 2019, LNCS 11695, pp. 20–32, 2019.
https://doi.org/10.1007/978-3-030-28730-6_2

that cannot fit in main memory. What is more, researchers focus on how to perform data mining tasks on devices with limited memory, instead of transferring
data to powerful processing servers. Classification, being a typical data mining
task, has many applications on all above-mentioned environments.

Classification algorithms (or classifiers) can be categorized to eager (model
based) or lazy (instance based). Both eager and lazy classifiers assign unclassified
items to a predefined set of class values, with their difference being on how they
work. Eager classifiers use a training set to build a model that is used to classify
new items. On the other hand, lazy classifiers do not build any models and
classify a new item by examining the whole training set. What is of utmost
importance for both types of classifiers is the size and the quality of the training
set, as both dictate the classifier's effectiveness and efficiency.

A well known and widely used lazy classifier is the k-Nearest Neighbors
(k-NN) [2]. Once a new unclassified item arrives, its k nearest neighbors are
retrieved from the training data. This is achieved by using a distance metric, i.e.
Euclidean distance. Then, the unclassified item is assigned to the most common
class among the classes of the k nearest neighbors.

k-NN is an effective classifier, especially when it is used on small training
sets. In the event of large training sets, all distances between a new item and the
training data have to be computed, and as a result its performance degrades.
Opposite to an eager classifier that discards the training data after the construction of its classification model, k-NN classifier has higher storage requirements
as it must have the training set always available. Another drawback of k-NN is
that noise can negatively affect its accuracy. A preprocessing step that builds a
small condensing set through a Data Reduction Technique (DRT) can cope with
these weaknesses.

The Dynamic RHC (dRHC) algorithm proposed in [3] is a DRT that is based
on RHC [8] (Reduction through Homogeneous Clusters) and can gradually build
its condensing set. Whenever a new training data segment becomes available,
the existing condensing set gets updated incrementally without needing to keep
the complete training set and regenerate the prototypes. dRHC is applicable
to dynamic environments where training data progressively becomes available
and for the cases where data cannot fit in main memory. Despite the fact that
dRHC is a fast DRT that achieves high reduction rates with no significant loss
in accuracy when applying k-NN, the condensing set that it builds may outpace
the available physical memory. dRHC2 [4] is an improvement over dRHC in that
it keeps the size of the condensing set fixed. The experimental study in [4] shows
that dRHC2 is faster than dRHC while keeping accuracy at high levels.

The motivation of the current work is the scenario shown in Fig. 1. The figure
depicts a condensing set produced by dRHC2. One can notice that there exist
prototypes that could be merged, for example the two prototypes that belong
to class "circle". The paper examines the conditions under which prototypes
could be merged without inhibiting the performance of the k-NN classifier. Two
variations are introduced for both dRHC and dRHC2. The main idea is to merge
pairs of prototypes that belong to the same class, only when certain criteria are

met. The difference between the two variations lies on whether after the arrival of a data segment that updates the condensing set, the merging phase is executed only once or repetitively, until there are no more prototypes found to be merged.

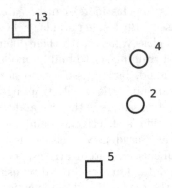

Fig. 1. Items of the same class that could be merged. The two prototypes of class "circle" could be merged. What about the two prototypes of class "square"?

The rest of this paper is structured as follows: Sect. 2 discusses the background knowledge on DRTs and their limitations. Section 3 reviews the dRHC and dRHC2 algorithms. Section 4 considers in detail the proposed algorithms. In Sect. 5, the four new algorithms are experimentally compared to dRHC and dRHC2 on fourteen datasets. Section 6 concludes the paper and proposes directions for future work.

2   Background Knowledge

In the literature, Data Reduction Techniques (or DRTs in short) can be classified into two main categories: (i) Prototype Generation (PG) algorithms that generate prototypes to summarize similar items [5] and (ii) Prototype Selection (PS) algorithms that collect prototypes from the initial training set [6]. Prototype selection algorithms can be further categorized into condensing or editing algorithms. PS-condensing and PG algorithms are used for data condensation, i.e., construction of a condensing set from the initial training data. On the other hand, PS-editing algorithms are used for noise and outlier removal from the training data.

The basic idea behind both PG and PS algorithms is that without loss in accuracy we can remove items that do not delineate decision boundaries between classes. Therefore, PG algorithms generate a few prototypes for the internal areas and many more for the close-class borders, whereas PS algorithms try to collect items the are close to decision boundaries. A point worth mentioning is that both are sensitive to noise, hence an editing algorithm must be applied beforehand.

PS and PG algorithms have been reviewed, and compared to each other in [5–7]. A prevalent feature of both is that the whole training set must reside in main memory, which in general, makes DRTs improper for very large datasets, especially for the cases where algorithms are executed in devices with limited resources or when the training set cannot fit in memory.

In addition, as soon as the condensing sets are constructed, these DRTs cannot contemplate new items. In other words, they cannot update their condensing set in a dynamic manner. What makes DRTs inappropriate for streaming environments is that training items must always be available once the condensing set is built. For each new training item (D) that becomes available, the algorithm must run from scratch in order to calculate the new condensing set. In order to tackle this issue the Dynamic RHC and Dynamic RHC2 algorithms [3,4] can be used in dynamic and/or streaming environments.

3   The dRHC and dRHC2 Algorithms

The dRHC algorithm maintains all properties of RHC (Reduction through Homogeneous Clusters) algorithm [3,8], and in addition it can also manage large or streaming datasets.

The idea behind RHC is to apply k-Means clustering on the training set in order to form as many clusters as the distinct values of the class variable, using as initial seeds the corresponding class representatives. Homogeneous clusters, i.e., clusters with all items belonging to the same class, are replaced by their centroid, whereas, the clustering procedure is applied recursively to all non-homogeneous clusters. RHC is shown to be a fast and effective DRT that outperforms other well-known DTRs in terms of data reduction and accuracy [3,8].

In an analogous fashion, dRHC engages two stages: (i) *initial condensing set construction* and (ii) *condensing set update*. As soon as the first data segment arrives, the *initial condensing set construction* phase is executed. The only difference with RHC's condensing set is that a weight attribute that denotes the number of training items that are represented is stored for each prototype. All the subsequent data segments that arrive, are processed by the *condensing set update* phase. In this phase, the prototypes of the current condensing set and the items of the incoming data segment are used, so that a new set of initial clusters is built. Then, dRHC algorithm proceeds alike to RHC.

An example of the execution of the *condensing set update* phase is depicted in Fig. 2. More specifically, in Fig. 2a we can see a condensing set with three prototypes and their corresponding weights. When a new data segment with seven items arrives (Fig. 2b), each item is assigned to its nearest prototype (Fig. 2c). Cluster A is homogeneous, therefore the prototype's attributes are updated so that it slightly "moves" towards the new items (Fig. 2d), and its weight is updated in order to be the sum of the weights of all items that it now represents (all items in the arriving data segment have weight equal to 1). For cluster B, there is no new item assigned to it, hence the corresponding prototype remains unchanged. On the other hand, cluster C becomes non-homogeneous and RHC is applied on it. k-means creates two homogeneous clusters (Fig. 2e). Finally, a new cluster centroid is computed for each cluster and the final updated condensing set is depicted in Fig. 2f.

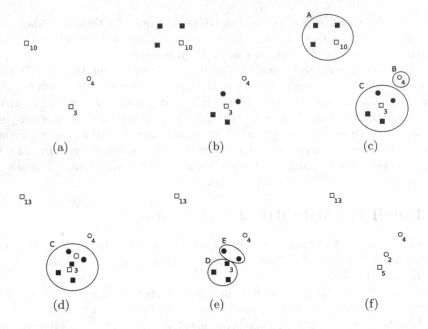

(a)   (b)   (c)

(d)   (e)   (f)

Fig. 2. Example of execution of the *condensing set update* phase of dRHC

Notwithstanding the fact that dRHC seems to be a good fit for data streams or large training datasets, after repetitive condensing set update phases, the condensing set may become too large. An algorithm that confronts with this drawback is dRHC2 [4], in which the size of the condensing set is maintained to a fixed pre-specified threshold. Practically, the only difference between dRHC and dRHC2 is the post-processing step that is presented in [4]. In dRHC2, the condensing set never exceeds a pre-specified size that is determined as a trade-off between computational cost, accuracy, system limitations as well as the level of noise in the data.

dRHC2 encapsulates a mechanism where prototypes are ranked according to their importance. The highest the importance of a prototype, the more likely to survive a condensing set update phase. In order to judiciously rank prototypes, dRHC2 takes into account not only the prototype's weight, but also its age.

Lastly, in case of data with noise, dRHC2 performs better than dRHC due to the fact that the noisy prototypes have lower weight and *AnA* values, and are eventually removed.

4   The Proposed Algorithms

As discussed in Sect. 3, both dRHC and dRHC2 algorithms perform well in streaming environments, with the latter algorithm having a clear advantage after repetitive condensing set update phases. As demonstrated in Fig. 1, we may encounter cases where prototypes that belong to the same class are close enough to be considered for merging. In effect, the size of the condensing set can be further reduced.

In this paper, we propose two new prototype merging algorithms that produce new variations of dRHC and dRHC2. More specifically:

- in case of variations based on dRHC, the merging step is applied after each condensing set update phase, and
- in case of variations based on dRHC2, the merging step is applied between the condensing set update phase and the prototype removal via ranking phase.

After examining various merging options, we propose a strategy that is not very aggressive and manages to improve data reduction while maintaining accuracy at acceptable levels. The first prototype merging algorithm performs a single pass over the condensing set and *merges prototype pairs belonging to the same class where each prototype is the nearest neighbor of the other*. The notation "sm", that stands for single pass merging, is used to denote the new variations of dRHC and dRHC2, namely, dRHCsm and dRHC2sm.

The second prototype merging algorithm performs multiple passes over the condensing set until no more prototype pairs can be merged. The notation "mm", that stands for multiple pass merging, is used to denote the new variations of dRHC and dRHC2, namely, dRHCmm and dRHC2mm.

dRHCsm and dRHC2sm being descendants of dRHC and dRHC2 respectively, retain all their properties, with the difference being that once a new data segment arrives and updates the condensing set, the merging phase described in Algorithm 1 is performed.

The merging algorithm accepts as input a condensing set CS. Initially, the *newCS* is empty. Then, the algorithm for each prototype x checks whether the nearest neighbor of the nearest neighbor y of x is x itself and whether both x and y belong to the same class (line 4). If this is the case, x and y are merged to prototype m, m is added to the *newCS* and x, y are removed from the CS (lines 5–7). Otherwise, x moves to the *newCS* (line 9).

A visual representation of the execution of the single pass merging phase is shown in Fig. 3. More particularly, the initial condensing set is depicted in Fig. 3a. Prototype pairs (E, F) and (C, G) satisfy the merging requirements (Fig. 3b), and are merged to form prototypes EF and CG respectively (Fig. 3c).

In the case of dRHCmm and dRHC2mm, the merging phase is performed again after the condensing set update phase. Algorithm 2 is essentially an extension of Algorithm 1, where merging is applied repetitively until there are no more pairs of prototypes to be merged. Taking into account the condensing set as depicted in Fig. 3c, with two additional merging passes, first CG is merged with D and then CGD with H. As a result, the final condensing set can be seen in Fig. 4.

Algorithm 1 Single Pass Merging Phase

Input: *CS*
Output: *newCS*

 1: newCS ← ∅
 2: **for each** x in CS **do**
 3:     y = NN(x)
 4:     **if** NN(y) == x and class(x) == class(y) **then**
 5:         m = merge (x,y)
 6:         add m to newCS
 7:         remove x,y from CS
 8:     **else**
 9:         move x to newCS
10:     **end if**
11: return newCS
12: **end for**

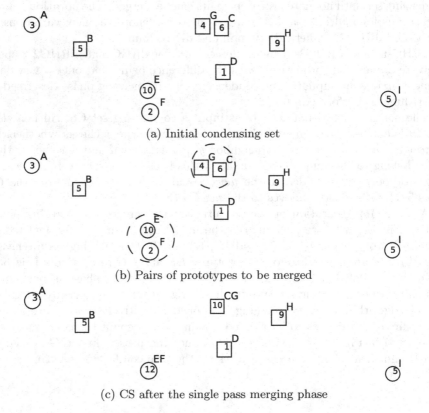

(a) Initial condensing set

(b) Pairs of prototypes to be merged

(c) CS after the single pass merging phase

Fig. 3. Single Pass Merging Algorithm

Algorithm 2 Multiple Pass Merging Phase

Input: CS
Output: $newCS$

1: newCS ← ∅
2: mergeflag ← True
3: **while** mergeflag == True **do**
4:     mergeflag ← False
5:     **for each** x in CS **do**
6:         y = NN(x)
7:         **if** NN(y) == x and class(x) == class(y) **then**
8:             m = merge (x,y)
9:             mergeflag ← True
10:             add m to newCS
11:             remove x,y from CS
12:         **else**
13:             move x to newCS
14:         **end if**
15:     **end for**
16: **end while**
17: **return** newCS

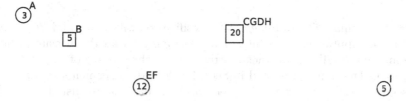

Fig. 4. Multiple Pass Merging Algorithm: condensing set after the multiple pass merging phase

5 Performance Evaluation

5.1 Experimental Setup

The performance of dRHCsm, dRHCmm, dRHC2sm and dRHC2mm was tested against dRHC and dRHC2 by using fourteen datasets distributed by the KEEL dataset repository[1] [9]. Table 1 summarizes the datasets used.

The chosen distance metric was the Euclidean distance. All algorithms were implemented in C. All datasets except KddCup were not normalized. We randomized the datasets that were distributed sorted on the class label (last value in each row of the datasets). For each algorithm and dataset, we measured three average values via five-fold cross-validation. These values are *Accuracy (Acc)*, *Reduction Rate (RR)* and *Preprocessing cost (PC)*.

---

[1] http://sci2s.ugr.es/keel/datasets.php.

Table 1. Dataset description

Dataset	Size	Attributes	Classes	Data segment
Letter Image Recognition (LIR)	20000	16	26	2000
Magic G. Telescope (MGT)	19020	10	2	1902
Pen-Digits (PD)	10992	16	10	1000
Landsat Satellite (LS)	6435	36	6	572
Shuttle (SH)	58000	9	7	1856
Texture (TXR)	5500	40	11	440
Phoneme (PH)	5404	5	2	500
Balance (BL)	625	4	3	100
Pima (PM)	768	8	2	100
Ecoli (ECL)	336	7	8	200
Yeast (YS)	1484	8	10	396
Twonorm (TN)	7400	20	2	592
MONK 2 (MN2)	432	6	2	115
KddCup (KDD)	141481	36	23	4000

Acc was estimated by running k-NN classification with $k = 1$ and *PC* in terms of distance computations. A point worth mentioning is that PC measurements do not include the small cost overhead introduced by the ranking of the prototypes.

All four algorithms presented in Sect. 4 accept data segments as input. The data segment sizes that were adopted for each dataset are listed in the last column in Table 1, therefore the initial training sets were split into specific data segments. In the scenario of limited main memory, data segment size can be related with the size of the available memory, or to the buffer size accepting data from a streamer. Experiments with data segments of different size were not conducted in this study, due to the fact that in [3] dRHC's performance was not found to be influenced at all by the chosen segment size.

In dRHC2, dRHC2sm and dRHC2mm the maximum condensing set size that is allowed is provided as an input in form of the T parameter. In order to be comparable with [4], the T parameter was adjusted to the 85%, 70%, 55% and 40% of the size of the condensing sets constructed by dRHC.

5.2   Results and Discussion

In Table 2, the performance of the dRHC was compared against dRHCsm and dRHCmm. The prevalent values are highlighted in bold. The preprocessing cost measurements are in million distance computations while values of accuracy and reduction rate are reported as percentages.

Due to the extra cost that is introduced by Algorithm 1, preprocessing cost in dRHC was lower compared to dRHCsm and dRHCmm. In addition to that, since dRHCmm may perform many passes over the condensing set in order to merge all eligible prototype pairs, the preprocessing cost of this algorithm was the highest in all datasets. On the other hand, reduction rate with dRHCsm and, especially, dRHCmm was improved at the cost of a small loss in accuracy in some datasets. Interestingly, in some other datasets (BL, MN2, TN, ECL, YS) the accuracy is increased, signifying an improvement in the quality of the condensing set after the merging phase.

Table 2. Comparison of dRHC, dRHCsm and dRHCmm in terms of Accuracy (ACC (%)), Reduction Rate (RR (%)) and Preprocessing Cost (PC (millions of distance computations))

Dataset	ACC (%)			RR (%)			PC (M)		
	dRHC	dRHCsm	dRHCmm	dRHC	dRHCsm	dRHCmm	dRHC	dRHCsm	dRHCmm
BL	70.56	**70.88**	**70.88**	78.12	80.88	**81.24**	**0.029**	0.051	0.084
KDD	**99.42**	99.30	99.28	99.22	99.32	**99.33**	**54.70**	56.54	67.94
LS	**88.50**	88.17	88.14	88.35	89.29	**89.42**	**1.53**	2.63	4.74
LIR	**93.92**	92.59	92.53	88.18	90.68	**90.78**	**19.57**	28.14	53.56
MGT	**72.97**	72.23	72.26	74.62	76.46	**76.48**	**26.03**	67.23	162.04
MN2	97.68	**97.91**	**97.91**	96.88	**97.17**	**97.17**	**0.004**	**0.004**	**0.004**
PD	**98.49**	97.19	96.63	97.23	98.07	**98.28**	**1.44**	1.57	1.68
PH	**85.38**	84.49	84.55	82.34	83.61	**83.70**	**1.64**	3.68	7.05
SH	**99.70**	99.36	99.32	99.50	99.60	**99.62**	**7.98**	8.09	8.92
TXR	**97.60**	95.91	95.96	94.95	96.40	**96.63**	**0.68**	0.74	1.01
TN	93.08	**93.34**	91.22	95.37	95.68	**97.29**	**0.695**	0.893	1.691
ECL	71.46	71.73	**71.74**	68.92	69.67	**69.74**	**0.015**	0.029	0.035
PM	**63.93**	63.40	63.41	65.11	66.33	**66.63**	**0.064**	0.210	0.325
YS	48.38	**48.51**	**48.51**	51.23	52.58	**52.63**	**0.306**	0.779	1.394

Similarly, in Table 3 one can compare the performance of dRHC2 against dRHC2sm and dRHC2mm for the different values of T that is provided as an input to the algorithms. We omit the measurements of the reduction rate (RR) since the different values of T set a threshold (in form of a percentage) to the size of the condensing set that is generated by the dRHC. To understand the concept behind the T value, take for example the LIR dataset. In Table 2, one can observe that for the LIR dataset, dRHC achieves Acc = 93.92 with RR = 88.18 (it practically generates only 2364 prototypes out of the 20000 instances of the dataset). In Table 3, we observe that for $T = 40$ (or by fixing the max size of the condensing set to be 40% of the condensing set produced by dRHC, i.e., the top ranked 946 prototypes) dRHC2, dRHCsm and dRHCmm achieve accuracies 90.08, 90.41 and 90.00 respectively.

As depicted in Table 3, for both dRHC2sm and dRHC2mm higher preprocessing cost values were measured, which is justified by the extra costs introduced with the merging step. Other than this, the accuracy was in most cases slightly affected (negatively or positively), similarly to the results presented in Table 2.

Table 3. Comparison of dRHC2, dRHC2sm and dRHC2mm in terms of Accuracy (ACC (%)) and Preprocessing Cost (PC (millions of distance computations)) taking into account four different values of T

Data	T %	ACC (%)			PC (M)		
		dRHC2	dRHC2sm	dRHC2mm	dRHC2	dRHC2sm	dRHC2mm
LIR	85	**93.40**	92.59	92.53	**19.18**	28.14	53.56
	70	**92.84**	92.44	92.33	**17.72**	27.65	53.47
	55	**91.85**	91.82	91.79	**15.36**	24.06	45.31
	40	90.08	**90.41**	90.00	**12.29**	18.58	32.64
MGT	85	**74.19**	72.89	72.91	**25.85**	67.23	162.04
	70	**74.64**	74.05	74.00	**24.41**	63.21	156.08
	55	**75.11**	74.55	74.13	**21.71**	53.07	124.89
	40	**75.97**	75.48	75.43	**17.73**	39.75	89.54
PD	85	**98.60**	97.19	96.63	**1.41**	1.56	1.68
	70	**98.63**	97.22	96.63	**1.32**	1.37	1.68
	55	**98.34**	97.13	96.53	**1.16**	1.27	1.64
	40	**97.73**	96.95	96.76	**0.93**	1.04	1.33
LS	85	**88.61**	88.07	88.13	**1.51**	2.62	4.72
	70	**88.53**	88.33	88.35	**1.42**	2.46	4.49
	55	**88.58**	87.94	87.96	**1.27**	2.11	3.74
	40	**87.89**	87.62	87.69	**1.03**	1.63	2.72
SH	85	**99.69**	99.38	99.31	**7.61**	7.65	7.71
	70	**99.61**	99.37	99.36	**6.91**	6.92	7.77
	55	**99.56**	99.26	99.36	**5.95**	6.14	6.75
	40	**99.37**	99.28	99.23	**4.73**	4.93	5.23
TXR	85	**97.38**	95.91	95.96	**0.67**	0.74	1.01
	70	**97.00**	96.00	95.98	**0.62**	0.74	1.01
	55	**96.46**	95.87	95.66	**0.53**	0.67	0.94
	40	**95.76**	95.44	95.33	**0.43**	0.53	0.72
PH	85	**86.14**	84.96	85.03	**1.62**	3.65	7.00
	70	85.62	85.59	**85.70**	**1.52**	3.34	6.73
	55	**85.21**	84.97	85.18	**1.33**	2.77	6.04
	40	84.88	**85.36**	84.60	**1.08**	2.06	3.92
BL	85	**71.84**	70.40	70.88	**0.029**	0.050	0.083
	70	73.12	**73.92**	72.32	**0.027**	0.050	0.081
	55	**77.28**	74.56	73.76	**0.025**	0.043	0.070
	40	**81.60**	78.56	78.56	**0.021**	0.035	0.052
PM	85	65.23	**65.62**	65.49	**0.063**	0.200	0.324
	70	**67.96**	65.88	65.88	**0.060**	0.180	0.320
	55	**68.09**	67.44	67.31	**0.055**	0.150	0.260
	40	**68.23**	67.58	68.23	**0.046**	0.110	0.194
ECL	85	74.73	**72.95**	72.95	**0.015**	0.029	0.035
	70	**76.22**	75.62	75.03	**0.015**	0.027	0.040
	55	**78.28**	70.81	70.51	**0.014**	0.024	0.034
	40	**79.75**	77.70	78.30	**0.013**	0.022	0.026
YS	85	48.31	48.72	**48.78**	**0.306**	0.779	1.394
	70	48.65	**48.92**	48.92	**0.306**	0.779	1.394
	55	48.99	**49.46**	49.26	**0.278**	0.682	1.146
	40	**52.83**	51.62	51.62	**0.244**	0.567	1.004
TN	85	**94.03**	94.00	91.22	**0.688**	0.892	1.691
	70	94.54	**94.69**	91.22	**0.654**	0.850	1.691
	55	**95.45**	95.24	91.54	**0.590**	0.746	1.752
	40	**95.93**	95.34	91.54	**0.495**	0.604	1.471
MN2	85	**96.28**	96.28	96.28	**0.0039**	0.0041	0.0041
	70	96.29	**96.99**	96.99	**0.0038**	0.0040	0.0040
	55	**94.45**	90.50	91.66	**0.0038**	0.0040	0.0040
	40	93.52	93.52	**93.52**	**0.0040**	0.0042	0.0042
KDD	85	**99.47**	99.31	99.27	**53.56**	56.45	68.00
	70	**99.51**	99.38	99.36	**49.81**	53.80	66.53
	55	**99.50**	99.38	99.39	**43.48**	47.32	57.97
	40	**99.48**	99.38	99.40	**34.60**	37.34	43.85

6   Conclusions and Future Work

This paper introduces four new algorithms (dRHCsm, dRHCmm, dRHC2sm and dRHC2mm) that are variations of dRHC and dRHC2. The newly proposed algorithms inherit the characteristics of the dRHC and dRHC2 algorithms while trying to further condense the training set by merging pairs of prototypes. In theory, we would expect higher reduction rates while maintaining the accuracy at the same level and a slight increase in preprocessing costs, especially for the "mm" variations.

The experimental study demonstrates that preprocessing costs are significantly higher in the case of the "mm" variations of the merging algorithms compared to the "sm" variations, an increase that is not justified by the small improvement in reduction rate, in the case of dRHC. Overall, the merging of prototypes appears to make more sense in the case of dRHC where it offers increased data reduction without affecting accuracy. In the case of dRHC2, where condensing sets are eventually truncated via ranking, merging of prototypes could be skipped.

In the near future, we plan to further investigate alternative prototype merging algorithms, for example based not only to the proximity of the prototypes but on their weights or age as well. We will examine the effect of the application of prototype merging on additional data reduction algorithms.

Acknowledgments. This research is funded by the University of Macedonia Research Committee as part of the "Principal Research 2019" funding program.

We thank Prof. Yannis Manolopoulos for his excellent remarks during ADBIS 2017 that led to the ideas presented in this paper.

References

1. Aggarwal, C.: Data Streams: Models and Algorithms. Advances in Database Systems Series, Springer, Boston (2007). https://doi.org/10.1007/978-0-387-47534-9
2. Cover, T., Hart, P.: Nearest neighbor pattern classification. IEEE Trans. Inf. Theory **13**(1), 21–27 (2006). http://dx.doi.org/10.1109/TIT.1967.1053964
3. Ougiaroglou, S., Evangelidis, G.: RHC: a non-parametric cluster-based data reduction for efficient k-NN classification. Pattern Anal. Appl. **19**(1), 93–109 (2014). http://dx.doi.org/10.1007/s10044-014-0393-7
4. Ougiaroglou, S., Arampatzis, G., Dervos, D.A., Evangelidis, G.: Generating fixed-size training sets for large and streaming datasets. In: Kirikova, M., Nørvåg, K., Papadopoulos, G.A. (eds.) ADBIS 2017. LNCS, vol. 10509, pp. 88–102. Springer, Cham (2017). https://doi.org/10.1007/978-3-319-66917-5_7
5. Triguero, I., Derrac, J., Garcia, S., Herrera, F.: A taxonomy and experimental study on prototype generation for nearest neighbor classification. Trans. Sys. Man Cyber Part C **42**(1), 86–100 (2012). http://dx.doi.org/10.1109/TSMCC.2010.2103939
6. Garcia, S., Derrac, J., Cano, J., Herrera, F.: Prototype selection for nearest neighbor classification: taxonomy and empirical study. IEEE Trans. Pattern Anal. Mach. Intell. **34**(3), 417–435 (2012). http://dx.doi.org/10.1109/TPAMI.2011.142

7. Lozano, M.: Data Reduction Techniques in Classification Processes. Ph.D. Thesis, Universitat Jaume I (2007)
8. Ougiaroglou, S., Evangelidis, G.: Efficient dataset size reduction by finding homogeneous clusters. In: Proceedings of the Fifth Balkan Conference in Informatics, BCI 2012, ACM, New York, NY, USA, pp. 168–173 (2012). http://doi.acm.org/10.1145/2371316.2371349
9. Alcalá-Fdez, J., Fernández, A., Luengo, J., Derrac, J., García, S.: KEEL data-mining software tool: data set repository, integration of algorithms and experimental analysis framework. Multiple Valued Logic and Soft Comput. **17**(2–3), 255–287 (2011)

Keys in Relational Databases with Nulls and Bounded Domains

Munqath Alattar[1] and Attila Sali[1,2]($\boxtimes$)

[1] Department of Computer Science and Information Theory,
Budapest University of Technology and Economics, Budapest, Hungary
m.attar@cs.bme.hu
[2] Alfréd Rényi Institute of Mathematics,
Hungarian Academy of Sciences, Budapest, Hungary
sali.attila@renyi.mta.hu

Abstract. Missing data value is an extensive problem in both research and industrial developers. Two general approaches are there to deal with the problem of missing values in databases, they either could be ignored (removed) or imputed (filled in) with new values [10]. For some SQL tables it is possible that some candidate key of the table is not null-free and this needs to be handled. Possible keys and certain keys to deal with this situation were introduced in [17]. In the present paper we introduce an intermediate concept called strongly possible keys that is based on a data mining approach using only information already contained in the SQL table. A strongly possible key is a key that holds for some possible world which is obtained by replacing any occurrences of nulls with some values already appearing in the corresponding attributes. Implication among strongly possible keys is characterized and Armstrong tables are constructed. An algorithm to verify a strongly possible key is given applying bipartite matching. Connection between matroid intersection problem and system of strongly possible keys is established.

Keywords: Strongly possible keys · Null values · Armstrong tables · Data imputation · Matroid intersection · Matchings in bipartite graphs

1 Introduction

Keys have always been fundamental for database management, in particular for understanding the structure and semantics of data. For a given collection of entities a key is a set of attributes whose values enable us to uniquely identify each entity. A standard example is a relational table, where a key is a set of

Research of the second author was partially supported by the National Research, Development and Innovation Office (NKFIH) grant K–116769. This work is also connected to the scientific program of the "Development of quality-oriented and harmonized R+D+I strategy and functional model at BME" project, supported by the New Hungary Development Plan (Project ID: TÁMOP 4.2.1/B-09/1/KMR-2010-0002).

© Springer Nature Switzerland AG 2019
T. Welzer et al. (Eds.): ADBIS 2019, LNCS 11695, pp. 33–50, 2019.
https://doi.org/10.1007/978-3-030-28730-6_3

columns such that there are no two distinct rows that are the same restricted to the given columns. Of course, keys are significant to other data models, e.g., XML, RDF, object models and higher order data models, as well.

Many systems today allow entering incomplete tuples into a database. For example, in case of data warehousing if different sources of raw data are merged, some attributes may exist in some of the sources while not available in some of the others. This makes it necessary to treat keys over incomplete tables. It is common to encounter databases having up to half of the entries missing, making it very difficult to mine them using data analysis methods that can work only with complete data [11].

There are different reasons why incompleteness occurs in database tables. Date [7] determined more than one kind of missing data and identified seven distinct types of null as follows: value not applicable, value unknown, value does not exist, value undefined, value not valid, value not supplied, and value is the empty set. The present paper deals with the data consumption with missing values in a database table, we take the second, third, and seventh types, that is we essentiallly follow the no information NULL approach. For the other types of missing data we assume that symbol N/A belongs to each domain, and we treat it as regular domain element in comparisons.

Missing values issue complicates data analysis for the analysts. Other problems are usually associated with missing values such as loss of data efficiency and effectiveness [10]. Although some methods of data analysis may overcome missing values problem, many others require complete databases. Two general approaches are there to deal with the problem of missing values in databases, they either could be ignored (removed) or imputed (filled in) with new values [10]. Ignoring rows is more useful for relations with a few incomplete rows compared to the total number of rows that would not have a vast effect on the data analysis inferences.

In relational databases, a key over a table is satisfied, if no two distinct tuples have the same values on all the attributes of the key. Codd formulated the principle of a key uniqueness and totality that for a key K of any relational schema R, any table with nulls over R must be K null-free [6,15]. For some tables, it is possible that there is no null-free key attribute set possible and this violates Codd's condition of keys. So the occurrences of nulls in the key set of attributes need to be handled. For example, in Table 1a the candidate key $(CourseName\ Year)$ has a null in the last tuple. This occurrence of the null made it difficult to uniquely identify the first and last tuples.

The first approach of handling the missing values in the key attributes involves ignoring any tuple that has a null in any of its values in the key part. This may lead to loss of a large amount of data and may change the original data pattern and integrity if a large number of tuples need to be ignored compared with the total number of tuples. Another approach is an imputation operation for each occurrence of a null in the key part of the data with a value from the attribute domain as explained by [17]. We investigate the situation when the attributes' domains are not known. For that we only consider what we have in the given data and extract the values to be imputed from the data itself for each attribute so that the resulting complete dataset after the imputation would not contain two tuples having the same value in their key. Köhler et al. [17] used possible worlds by replacing each occurrence of a null with a value from the corresponding attribute's (possibly infinite) domain. Each possible world is considered as a table of total tuples that may contain duplicated tuples. They defined a possible key as a key that is satisfied by some possible world of a non total database table and a certain key as a key that is satisfied by every possible world of the table. For example, Table 1a has some possible world that satisfies the possible key {*Course Name*} while there is no possible world of the table that satisfies key {*Lecturer*} and, furthermore, every possible world of the table satisfies the certain key {*Course Name, Year, Semester*}.

In many cases we have no proper reason to assume existence of any other attribute value than the ones already existing in the table. Such examples could be types of cars, diagnoses of patients, applied medications, dates of exams, course descriptions, etc. We define a strongly possible key as a key that satisfied by some possible world that is obtained by replacing each occurrence of null value from the corresponding attribute's existing values. We call this kind of a possible world a strongly possible world. This is a data mining type approach, our idea is that a we are given a raw table with nulls and we would like to identify possible key sets based on the data only.

We treat the implication problem for strongly possible keys and find that it behaves similarly to keys in complete datasets, that makes a difference from possible and certain keys. Also, it is proven that any system of strongly possible keys enjoys Armstrong instances.

We also point out a connection of matroid intersection problem and satisfaction of a system of strongly possible keys.

We introduce an algorithm to verify a single strongly possible key using bipartite graph matching.

The organization of this paper is as follows. In Sect. 2 related work is reviewed, Sect. 3 contains preliminaries and definitions. Strongly possible keys over relational data with null occurrences in the key attributes studied in Sect. 4. The related implication problem is completely characterized. Also, an Armstrong instance for a system of strongly possible keys that satisfies necessary condition is constructed. In Sect. 5, the existence of a system of strongly possible keys and the use of matchings to discover strongly possible keys are studied. An algorithmic aspects of strongly possible keys are discussed in Sect. 6. Results and the future research directions are concluded in Sect. 7.

2   Related Work

Keys are important constraints that enforce the semantics of relational database systems. A key K satisfied by a total table over a relation schema R if there are no two tuples in the table that agree on all the attributes of K. Database tables of real database systems usually contain occurrences of null values and for some cases this includes candidate key columns. Various studies have been done for the purpose of handling missing values.

Sree Dhevi [3] shows that it is necessary to impute the missing values based on other information in the dataset to overcome the biased results that affect the accuracy of classification generated by missing values. Similarly, we use the attribute's existing values for each null in that attribute. Cheng et al. [5] utilize clustering algorithms to cluster data, and calculate coefficient values between different attributes by generating minimum average error.

Aliriza et al. introduced a framework of imputation methods in [10] and evaluates how the choice of different imputation methods affects the performance in [11]. Experimental analyses of several algorithms for imputation of missing values were performed by [1,4,9,16]. Our imputation method adopts the concept of graph matching by assigning for each incomplete record a complete one from the complete set of records constructed by combination of all attribute values of visible domains. An approach introduced by Zhang et al. [24] discusses and compares several strategies that utilize only known values.

Köhler et al. [17] introduced possible and certain keys. A set K of attributes is possible key if there is a possible world where K is a key. On the other hand, K is a certain key if it is a key in every possible world. The main concept of the present paper is between these two, since a strongly possible world is a possible world, as well. Possible worlds may use any value from an attribute domain to replace a null. This effectively allows an infinite pool of values. Strongly possible worlds are created from finite attribute domains. Some of the results in [17] essentially use that some attribute domains are infinite. In the present paper we investigate what can be stated without that assumption.

3   Preliminaries

We start with summarizing some basic definitions and terminologies. Let $R = \{A_1, A_2, \ldots A_n\}$ be a relation schema. The set of all the possible values for each attribute $A_i \in R$ is called the domain of A_i and denoted as $D_i = dom(A_i)$ for $i = 1,2,\ldots n$. For $X \subseteq R$ then $D_X = \prod_{\forall A_i \in K} D_i$.

An instance $T = (t_1, t_2, \ldots t_s)$ over R is a list of tuples that each tuple is a function $t : R \rightarrow \bigcup_{A_i \in R} dom(A_i)$ and $t[A_i] \in dom(A_i)$ for all A_i in R. By taking list of tuples we use the bag semantics that allows several occurrences of the same tuple. For a tuple $t_r \in \overline{T \text{ and } X \subset R}$, let $t_r[X]$ be the restriction of t_r to X.

In practice, database tables may have missing information about the value of some entry $t_j[A_i]$. This is denoted by Codd's null marker $\perp$. Note that null is not a value but it's an absence of value. Codd's null marker is included in each domain to represent the missing information [12]. t_r is called V-total for a set V of attributes if $t_r[A] \neq \perp, \forall A \in V$. Also a tuple t_r is a total tuple if it is a R-total. t_1 and t_2 are weakly similar on $X \subseteq R$ denoted as $t_1[X] \sim_w t_2[X]$ defined by Köhler et al. [17] if:

$$\forall A \in X \quad (t_1[A] = t_2[A] \text{ or } t_1[A] = \perp \text{ or } t_2[A] = \perp).$$

Furthermore, t_1 and t_2 are <u>strongly similar</u> on $X \subseteq T$ denoted by $t_1[X] \sim_s t_2[X]$ if:

$$\forall A \in X \quad (t_1[A] = t_2[A] \neq \perp).$$

For the sake of convenience we write $t_1 \sim_w t_2$ if t_1 and t_2 are weakly similar on R and use the same convenience for strong similarity. For a null-free table (a table with R-total tuples), a set of attributes $K \subset R$ is a <u>key</u> if there are no two distinct tuples in the table that share the same values in all the attributes of K:

$$t_a[K] \neq t_b[K] \,\forall\, 0 \leq a,b \leq s \text{ such that } a \neq b$$

Here we introduce the concept of possible and certain keys defined by Köhler et al. [17]. Let $T' = (t'_1, t'_2, \dots t'_s)$ be a table that represents a total version of T which obtained by replacing the occurrences of $\perp$ in all attributes $t[A_i]$ with a value from the domain D_i different from $\perp$ for each i. T' is called a <u>possible world</u> of T. In a possible world T', t'_i is weakly similar to t_i and T' is completely null-free table. A <u>possible key</u> K denoted as $p\langle K \rangle$, is a key for some possible world T' of T. Similarly, a <u>certain key</u> K denoted as $c\langle K \rangle$, is a key for every possible world T' of T.

4 Strongly Possible Keys

A database attribute domain is a predefined set of values that are allowed to be used for all the tuples under that attribute part of the table. For a given set of tuples T, let A be an attribute that has a domain of some values range. It is possible that all the tuples in T use only a specific group of values from all the possible domain values. For example, in Table 1a, the attribute Course Name has a predefined domain of all computer science course names but it only uses two values Mathematics and Datamining along with $\perp$ in the last tuple.

Table 1. Complete and incomplete datasets

(a) Incomplete Dataset

Course Name	Year	Lecturer	Credits	Semester
Mathematics	2019	$\perp$	5	1
Datamining	2018	Sarah	7	$\perp$
$\perp$	2019	Sarah	$\perp$	2

(b) Complete Dataset

Course Name	Year	Lecturer	Credits	Semester
Mathematics	2019	Sarah	5	1
Datamining	2018	Sarah	7	2
Datamining	2019	Sarah	7	2

Definition 4.1. *The* <u>*visible domain*</u> *of an attribute* A_i *(VD_i) is the set of all distinct values except* $\perp$ *that are already used by tuples in* T:

$$VD_i = \{t[A_i] : t \in T\} \setminus \{\perp\} \text{ for } A_i \in R$$

Then the VD_1 in Table 1a is {Mathematics, Datamining}. The term visible domain refers to the data that already exist in a given dataset. For example, if we have a dataset with no information about the attributes' domains definitions, then we use the data itself to define their own structure and domains. This may provide more realistic results when extracting the relationship between data so it is more reliable to consider only what information we have in a given dataset.

While a possible world is obtained by using the domain values instead of the occurrence of null as defined in Sect. 3, a strongly possible world is obtained by using the visible domain values.

Definition 4.2. *A possible world* T' *is called* <u>*strongly possible world*</u> *if* $t'[A_i] \in VD_i$ *for all* $t' \in T'$ *and* $A_i \in R$.

That allows us to get a possible world of a set of data with some missed values by using only the available information. We define a <u>strongly possible key</u> as a key for some strongly possible world of T.

Definition 4.3. *A subset* $K \subseteq R$ *is a* <u>*strongly possible key*</u> *(in notation* $sp\langle K \rangle$*) of* T *if* $\exists\, T' \subseteq VD_1 \times VD_2 \times \ldots \times VD_n$ *a strongly possible world such that* K *is a key in* T'.

Note that although we use the bag semantics, multiple copies of the same tuple rule out the existence of keys. Recall the same instance in Table 1a, implies $sp\langle CourseName\,Year\rangle$ as a strongly possible key because there is a strongly possible world in Table 1b where $\{CourseName\,Year\}$ is a key. On the other hand, the table implies neither $sp\langle CourseName\,Lecturer\rangle$ nor $sp\langle Year\,Leturer\rangle$, because there are no strongly possible worlds T' that have $\{CourseName\,Leturer\}$ or $\{Year\,Leturer\}$ as keys.

Proposition 4.1. *Let* $T = \{t_1, t_2, \ldots t_p\}$ *be a table instance over* R. $K \subseteq R$ *is a* $sp\langle K \rangle \iff \exists T' \subseteq VD_1 \times VD_2 \times \ldots \times VD_n$ *s.t.* $T' = \{t'_1, t'_2, \ldots t'_p\}$ *where* $t_i[K] \sim_w t'_i[K]$ *and* $t'_i \neq t'_j$ *if* $i \neq j$ *and* K *is a key in* T'.

Note that if $t_i[K] \sim_s t_j[K]$ for $i \neq j$ then K is not a strongly possible key, but the reverse is not necessarily true. For example take the instance T with the two tuples $t_1 = (1,1)$ and $t_2 = (\perp, 2)$. Then the only strongly possible world is $T' = \{(1,1), (1,2)\}$ and A_1 is not a key in T', nevertheless $t_1[A_1] \not\sim_s t_2[A_1]$. This shows that in contrast to possible and certain keys, the validity of $sp\langle K \rangle$ cannot be checked or charcterized by pairwise comparisons of tuples.

In the relational model, any subset of attributes that are not keys called as antikeys [22].

Definition 4.4. *We say that* K *is* <u>*strongly possible anti-key*</u> $\neg sp\langle K \rangle$ *if* $\nexists T'$ *strongly possible world such that* K *is a key in* T'.

Definition 4.5. *An attribute* $A \in K$ *is called* <u>*redundant*</u> *if* $K \setminus A$ *is a strongly possible key in* T. *And the key* $sp\langle K \rangle$ *is called* <u>*minimal*</u> *if* $\neg sp\langle Y \rangle$ *holds* $\forall Y \subset X$.

4.1   Implication Problem

Integrity constraints determine the way the elements are associated to each other in a database. The implication problem asks if a given set of constraints entails further constraints. In other words, given an arbitrary set of constraints, the implication problem is to determine whether a single constraint is satisfied by all instances satisfying that set of constraints. In our context, to define the implication, let Σ be a set of strongly possible keys and θ be a single strongly possible key over a relation schema R. Σ logically implies θ, denoted as $\Sigma \models \theta$ if for every table instance T over R satisfying every strongly possible key in Σ we have that T satisfies θ. The next theorem characterizes the implication problem for strongly possible keys.

Theorem 4.1. $\Sigma \models sp \langle K \rangle \iff \exists Y \subseteq K \ s.t. \ sp \langle Y \rangle \in \Sigma.$

Proof. $\Leftarrow$: $\exists T'$ s.t. $t_i'[Y] \neq t_j'[Y], \forall i \neq j$, so $t_i'[K] \neq t_j'[K], \forall i \neq j$ holds, as well.
   $\Rightarrow$: Suppose indirectly that $sp \langle Y \rangle \notin \Sigma \ \forall Y \subseteq K$. Consider the following instance consisting of two tuples $t_1 = (0, 0, \ldots, 0)$, $t_2[K] = (\bot, \bot, \ldots, \bot)$, and $t_2[R \setminus K] = (1, 1, \ldots 1)$ as in Table 2. Then the only possible t_2' in T' is $t_2'(0, 0, \ldots, 0, 1, 1, \ldots, 1)$. Furthermore, $\forall Z$ where $sp \langle Z \rangle \in \Sigma$, there must be $z \in Z \setminus K$, thus $t_1'[Z] \neq t_2'[Z]$ but $t_1'[K] = t_2'[K]$ showing that (t_1, t_2) satisfies every strongly possible key constraints from Σ, but does not satisfy $sp \langle K \rangle$.

Note 4.1. If $\Sigma \models \neg sp \langle K \rangle$ and $Y \subseteq K$ then $\Sigma \models \neg sp \langle Y \rangle$.

Note 4.2. If $\Sigma \models sp \langle K \rangle$, then $\Sigma \models p \langle K \rangle$ but the reverse is not necessarily true, since $D_K \supseteq V D_K$ could be proper containment so K could be made a key by imputing values from $D_K \setminus V D_K$. For example, in Table 2, it is shown that $\neg sp \langle K \rangle$ holds, but $p \langle K \rangle$ may hold in some T' if there is at least one other value in the domain of K rather than the zeros to be placed instead of the nulls in the second tuple so that $t_1'[K] \neq t_2'[K]$ results.

Table 2. Incomplete data

	K	$R \setminus K$
t_1	0 0 0 0	00000000
t_2	$\bot\bot\bot\bot$	11111111

Note 4.3. If $\Sigma \models c \langle K \rangle$, then $\Sigma \models sp \langle K \rangle$. As certain keys hold in any possible world, they hold also if this possible world is created using visible domains.

Note 4.4. For a single attribute A, $sp \langle A \rangle \iff t[A] \sim_s t'[A] \ \forall t, t'$ s.t. $t \neq t'$, i.e. if nulls do not occur in A.

In other words, single attribute with a null value cannot be a strongly possible key. That is because replacing an occurrence of null with a visible domain value results in duplicated values for that attribute.

4.2   Armstrong Tables

Armstrong tables are useful tools to represent constraint sets in a user friendly way [2,8,14,21]. Following [17], we introduce concept of null-free subschema (NFS). Let R be a schema, an NFS R_S over R is a set such that $R_S \subseteq R$. An instance T satisfies NFS R_S if it is R_S-total, that is each tuple $t \in T$ is R_S-total. This corresponds to NOT NULL constraints of SQL.

Definition 4.6. *An instance* T *over* (R, R_S) *is an Armstrong table for* (R, R_S, Σ) *if for every strongly possible key* θ *over* R θ *holds in* T *iff* $\Sigma \models \theta$, *and for every attribute* $A \in R \setminus R_S$ *there exist a tuple* $t \in T$ *with* $t[A] = \bot$.

Let us suppose that $\Sigma = \{sp\langle K \rangle : K \in \mathcal{K}\}$ is given. By Note 4.4 if $|K| = 1$, then $K \subseteq R_S$ must hold. If this restriction is staisfied, then Σ enjoys an Armstrong table.

Theorem 4.2. *Suppose that* $\Sigma = \{sp\langle K \rangle : K \in \mathcal{K}\}$ *is a collection of strongly possible key constraints such that if* $|K| = 1$, *then* $K \subseteq R_S$. *Then there exists an Armstrong table for* (R, R_S, Σ).

Proof. Let $\mathcal{A}$ be the collection of strongly possible antikeys, that is $\mathcal{A} = \{A \subset R \colon \Sigma \not\models sp\langle A \rangle\}$. According to Theorem 4.1 and Note 4.1 $\mathcal{A}$ is a downset and $\Sigma \models sp\langle K \rangle \iff K \setminus A \neq \emptyset$ for all $A \in \mathcal{A}$. Let $H = \{K_1, K_2, \ldots, K_u\}$ be the set of singleton attribute keys, note that $K_i \not\subseteq A$ for all $i = 1, 2, \ldots, u$ and $A \in \mathcal{A}$. Let the maximal (under containment) elements of $\mathcal{A}$ be $\{A_1, A_2, \ldots, A_p\}$ and assume that $R = \{K_1, K_2, \ldots, K_u, X_1, X_2, \ldots, X_n\}$ with $R \setminus R_S = \{X_1, X_2, \ldots, X_m\}$. Construct table $T = \{t_0, t_1, \ldots, t_{p(m+1)}\}$ as follows. $t_0[X] = 0 \ \forall X \in R$. For $i = 1, 2, \ldots, p$ let $t_i[X] = \bot \ \forall X \in R \setminus R_S$, $t_i[X] = 0$ for $X \in R_S \cap A_i$ and $t_i[X] = i$ for $X \in R_S \setminus A_i$. Note that $H \subseteq R_S \setminus A_i$ for all i. Let $t_{ip+j}[X_j] = i$ and $t_{ip+j}[X_\ell] = ip+j$ for $i = 1, 2, \ldots, p$, $j = 1, 2, \ldots, m$ and $\ell \neq j$. Furthermore, let $t_{ip+j}[K_g] = ip + j$ for $g + 1, 2, \ldots u$. Observe that t_z is R-total for $z > p$ and that $t_z[X] \neq t_u[X]$ for $u, z > p$ and for all $X \in R$.

Create a strongly possible world T' from T by replacing the null of $t_i[X]$ by 0 if $X \in A_i$ and by i otherwise. We claim that no two tuples of T' agree on all attributes of K if $\Sigma \models sp\langle K \rangle$. Indeed, this latter property happens iff $K \setminus A_i \neq \emptyset$ for all $i = 1, 2, \ldots, p$, hence for all $0 \leq j < i \leq p$ there exists an attribute $X \in K \setminus A_i$ such that $t_j[X] \neq i$ but $t_i[X] = i$. Furthermore, if $0 \leq i \leq p$ and $j > p$, then t_i and t_j can agree in at most one attribute, but that attribute is not a singleton attribute key. On the other hand, if $\Sigma \not\models sp\langle L \rangle$, then there exists i such that $L \subseteq A_i$, which implies that $t_0[L] = t_i[L]$, that is L is not a key in table T'.

4.3   Weak Similarity Graph

There is a natural way to assign a graph to a subset $X \subseteq R$ of attributes that gives information on key possibilities of X.

Definition 4.7. *Let $T = \{t_1, t_2, \ldots t_p\}$ be a table (instance) over schema R. The weak similarity graph $G_w[X]$ with respect to X is defined as $G_w[X] = (T, E)$, where $\{t_i, t_j\} \in E \iff t_i[X] \sim_w t_j[X]$.*

Note 4.4 says that for $|K| = 1$ $sp\langle K \rangle$ holds iff $G_w[K]$ is the empty graph. For example, Fig. 1 shows the weak similarity graph of Tables 3a and b.

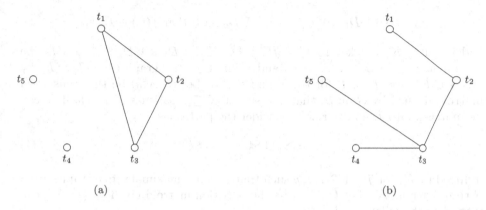

(a)   (b)

Fig. 1. Weak similarity graph

The following is a sufficient condition for $sp\langle X \rangle$ to hold in case of $|X| > 1$.

Proposition 4.2. *Assume that no two tuples are strongly similar in X. If each connected component C in the weak similarity graph $G_w[X]$ of $X \subseteq R$, is a chordless cycle of length ≥ 4, then $sp\langle X \rangle$ holds.*

Proof. $sp\langle X \rangle$ holds if there exists a strongly possible world T' of T such that no two tuples agree on X in T'. Two tuples that are not connected by an edge in $G_w[X]$ are distinct on X in any (strongly) possible world of T, hence it is enough to concentrate on tuples in the same component. If this single component is a circle of k elements for $k \geq 4$ ($t_1[X] \sim_w t_2[X] \sim_w t_3[X] \sim_w \ldots \sim_w t_k[X] \sim_w t_1[X]$), then t_i is distinct on X from $t_{i+2} \ldots t_k, t_1, \ldots, t_{i-2}$ and there exist $A_{j_i} \in X$ such that $t_{i-1}[A_{j_i}] \neq t_{i+1}[A_{j_i}]$ and $t_{i-1}[A_{j_i}], t_{i+1}[A_{j_i}] \neq \bot$. Since $t_{i-1} \sim_w t_i \sim_w t_{i+1}$, we obtain that $t_i[A_{j_i}] = \bot$ (indices of tuples are understood modulo k). We need to make t_i different from t_{i-1} and t_{i+1}, so that we can make $t'_i[A_{j_i}] = t_{i+1}[A_{j_i}]$ and this distinguishes it from t_{i-1}. In this way we made any two neighbouring tuples along the cycle distinct on X, non-neighbouring ones are distinct in any strongly possible world.

5   Matroids, Matchings and Strongly Possible Keys

In this section we show how the existence of a system of strongly possible keys is equivalent with the existence of a given sized common independent set of several matroids (for basic definitions and properties of matroids the reader is referred to Welsh's book [23]). Let us be given schema $R = \{A_1, A_2, \ldots, A_n\}$ and let $\mathcal{K} = \{K_1, K_2, \ldots K_p\}$ be a collection of attribute sets and $T = \{t_1, t_2, \ldots, t_s\}$ be an instance with possible null occurrences. Our main question here is whether $\Sigma = \{sp\langle K_1\rangle, sp\langle K_2\rangle, \ldots, sp\langle K_p\rangle\}$ holds in T? Let $E_i = \{t' \in VD_1 \times VD_2 \times \ldots \times VD_n \colon t' \sim_w t_i\}$. A strongly possible world that satisfies Σ is given by an injective mapping

$$f \colon T \to VD_1 \times VD_2 \times \ldots \times VD_n \text{ such that } f(t_i) \in E_i \ \forall i$$

and for all j, K_j is a key in $T' = f(T)$. Let $S \subseteq VD_1 \times VD_2 \times \ldots \times VD_n$ be the union $S = E_1 \cup E_2 \cup \ldots \cup E_s$ and define bipartite graph $G = (T, S; E)$ by $\{t, t'\} \in E \iff t \sim_w t'$ for $t \in T$ and $t' \in S$. Let $(S, \mathcal{M}_0)$ be the transversal matroid defined by G on S, that is a subset $X \subseteq S$ satisfies $X \in \mathcal{M}_0$ if X can be matched into T. Furthermore consider the partitions

$$S = S_1^j \cup S_2^j \cup \ldots \cup S_{p_j}^j \tag{1}$$

induced by K_j for $j = 1, 2, \ldots, p$ such that S_i^j's are maximal sets of tuples from S that agree on K_j. Let $(S, \mathcal{M}_j)$ be the partition matroid given by (1). We can formulate the following theorem.

Theorem 5.1. *Let T be an instance over schema $R = \{A_1, A_2, \ldots, A_n\}$ and let $\mathcal{K} = \{K_1, K_2, \ldots K_p\}$ be a collection of attribute sets. $\Sigma = \{sp\langle K_1\rangle, sp\langle K_2\rangle, \ldots, sp\langle K_p\rangle\}$ holds in T if and only if the matroids $(S, \mathcal{M}_j)$ have a common independent set of size $|T|$ for $j = 0, 1, \ldots p$.*

Proof. An independent set T' of size $|T|$ in matroid $(S, \mathcal{M}_0)$ means that tuples in T' form a strongly possible world for T. That they are independent in $(S, \mathcal{M}_j)$ means that K_j is a key in T', that is $sp\langle K_j\rangle$ holds.

Conversely, if $\Sigma = \{sp\langle K_1\rangle, sp\langle K_2\rangle, \ldots, sp\langle K_p\rangle\}$ holds in T, then there exists a strongly possible world $T' = \{t'_1, t'_2, \ldots, t'_s\} \subseteq VD_1 \times VD_2 \times \ldots \times VD_n$ such that $t_i \sim_w t'_i$. This means that $T' \subseteq S$ and that T' is independent in transversal matroid $(S, \mathcal{M}_0)$. $sp\langle K_j\rangle$ holds implies that tuples t'_i are pairwise distinct on K_j, that is T' is independent in partition matroid $(S, \mathcal{M}_j)$.

Unfortunately, Theorem 5.1 does not give good algorithm to decide the satisfaction of a system Σ of strongly possible keys, because as soon as Σ contains at least two constraints, then we would have to calculate the size of largest common independent set of at least three matroids, known to be a NP-complete problem in general [13].

In case of a single strongly possible key $sp\langle K\rangle$ constraint Theorem 5.1 requires computing the largest common independent set of two matroids, which can be solved in polynomial time [19]. However, we can reduce the problem to the somewhat simpler problem of matchings in bipartite graphs.

If we want to decide whether $sp\langle K \rangle$ holds or not, we can forget about the attributes that are not in K since we need distinct values on K as a matching from $VD_{A_1} \times VD_{A_2} \times \ldots \times VD_{A_b}$ to $T = \{t_1, t_2 \ldots t_r\}|_K$ where $K = \{A_1, A_2 \ldots A_b\}$. Thus, we may construct a table T' that formed by finding all the possible combinations of the visible domains of $T|_K$ that are weakly similar to some tuple in $T|_K$.

$$T' = \{t' : \exists t \in T : t'[K] \sim_w t[K]\} \subseteq VD_1 \times VD_2 \times \ldots \times VD_b$$

Finding the matching between T and T' that covers all the tuples in T (if exist) will result in the set of tuples in T' that needs to be replaced in T so that K is a strongly possible key.

Table 3. Data samples

	(a)			(b)			(c)	
	Lecturer	Course		Lecturer	Course		Lecturer	Course
t_1	Sarah	Mathematics	t_1	Sarah	Mathematics	t'_1	Sarah	Mathematics
t_2	Sarah	$\perp$	t_2	Sarah	$\perp$	t'_2	Sarah	Data Mining
t_3	$\perp$	Data Mining	t_3	$\perp$	Mathematics	t'_3	James	Mathematics
t_4	James	Data Mining	t_4	James	Data Mining	t'_4	James	Data Mining
t_5	David	Data Mining	t_5	David	Data Mining	t'_5	David	Mathematics
						t'_6	David	Data Mining

Example 5.1. Table 3b shows an incomplete set of tuples Here $K = \{Lecturer, Course\}$. Visible domain can be identified for each attribute to construct tuples of T' by finding the combinations of all the visible domain values as shown in Table 3c (here we included all tuples for tables a and b together). Bipartite graphs between tuples with null(s) in T and tuples in T' excluding those tuples that agree on K to any total tuple in T are constructed. Figure 2a illustrates the graph for Table 3b which contain a complete matching to assign a total tuple to each non-total tuple in T and K is a key. While for Table 3a, Fig. 2b shows there is no matching that covers all the tuples in T.

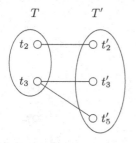

(a) The Graph of Table 3b

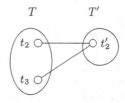

(b) The Graph of Table 3a

Fig. 2. Data graphs

5.1   Necessary Conditions

Let $c_v(A)$ denote the number of tuples that have value v in attribute A, that is $c_v(A) = |\{t \in T \colon t[A] = v\}|$. Next are some necessary conditions to have a strongly possible key.

Proposition 5.1. *Let $K \subseteq R$ be a set of attributes. If $sp\langle K \rangle$ holds, then*

1. *No two tuples t_i, t_j are strongly similar in K.*
2. $|T| \leq \prod_{\forall A \in K} |VD_A|.$
3. $\forall B \in K$, *number of nulls in $B \leq \sum_{\forall v \in VD_B} \left(\frac{\prod_{\forall A \in K} |VD_A|}{|VD_B|} - c_v(B) \right).$*
4. *For all $v \in VD_B$ we have $c_v(B) \leq \frac{\prod_{\forall A \in K} |VD_A|}{|VD_B|}$*

Proof. First condition is obviously required. In addition to that, for any set of attributes, the maximum number of distinct combination of their values is the size of the multiplication of their visible domain, and this proves (2). Moreover, to prove conditions (3) and (4), when K is $sp\langle K \rangle$ in T then there should exist a T' with no two tuples having the same values in all attributes of K after filling all their nulls. So for each set of tuples S that has the same value v in the attribute B, the number of distinct combinations of the other attributes is the multiplication of their VD's, means the number of tuples in S should not be more than $\prod_{\forall A \in (K \setminus B)} VD_A$. Thus the number of times value v can be used to replace a null in attribute B is at most $\frac{\prod_{\forall A \in K} |VD_A|}{VD_B} - c_v(B)$.

Note that $sp\langle K \rangle$ holds if a matching covering T exists in the bipartite graph $G = (T, T'; E)$ defined as above, $\{t, t'\} \in E \iff t[K] \sim_w t'[K]$. We can apply Hall's Theorem to obtain

$$\forall X \subseteq T, \text{ we have } |N(X)| \geq |X| \text{ for } N(X) = \{t' : \exists t \in X \text{ such that } t'[K] \sim_w t[K]\}$$

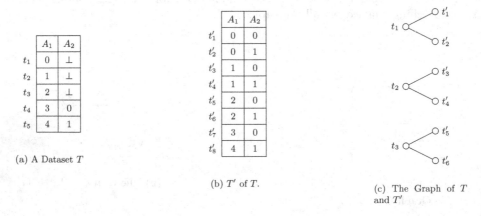

	A_1	A_2
t_1	0	$\bot$
t_2	1	$\bot$
t_3	2	$\bot$
t_4	3	0
t_5	4	1

(a) A Dataset T

	A_1	A_2
t'_1	0	0
t'_2	0	1
t'_3	1	0
t'_4	1	1
t'_5	2	0
t'_6	2	1
t'_7	3	0
t'_8	4	1

(b) T' of T.

(c) The Graph of T and T'

Fig. 3. A certain key graph

The conditions in Proposition 5.1 are implied by Hall's condition, as well. Let assume that Hall's condition is true for a set of tuples T and K has b n attributes. If t_i, t_j are strongly similar, then the set $X = \{t_i, t_j\}$ has $1 = |N(X)| < |X| = 2$ that proves (1). For condition (2), the graph here is $G = (T, T'; E)$ such that $T' \leq \prod_{\forall A \in K} |VD_A|$ and it is the neighboring tuples to the tuples in T so that it is always true that $T \leq T'$. Condition (3) implied as follows. Let X be the set of tuples that have null value in B. Then the number tuples $t' \in N(X)$ such that $t'[B] = v$ is at most $\frac{\prod_{\forall A \in K} |VD_A|}{|VD_B|}$ Finally, for an attribute $B \in K$, let X_v be the set of all tuples that have value v in the attribute B. Then $|N(X_v)| \leq \frac{|VD_1 \times VD_2 \times ... \times VD_b|}{|VD_B|}$ proving (4).

As defined in Sect. 3, certain key is a key for any possible world, i.e. all the tuples are distinct after filling the nulls regardless what values been used.

Theorem 5.2. *Let T be a table instance over schema R such that a strongly possible world of T exists. $K \subseteq R$ is a certain key iff K is a key in any strongly possible world of T.*

Proof. $\Rightarrow$: If $c\langle K \rangle$ holds, then K is a key in any possible world by definition, so in particular in any strongly possible world, as well.

$\Leftarrow$: Let us assume that K is a key in any strongly possible world, but there exists a possible world T', and two distinct tuples $t'_1 \neq t'_2$ of T' such that $t'_1[K] = t'_2[K]$. Let $A \in K$ be an attribute. There are three possibilities. If neither t_1 nor t_1 has $\perp$ in A, then let $t_1[A] = t_2[A] = t''_1[A] = t''_2[A]$. If one of them has $\perp$ in A, say $t_1[A] = \perp$, then we can set $t''_1[A] = t''_2[A] = t_2[A] \neq \perp$ such that $t_2[A] \in VD_A$. Finally, if $t_1[A] = t_2[A] = \perp$, then pick any $x \in VD_A$ and set $t''_1[A] = t''_2[A] = x$. Such an x exists, since T has a strongly possible world. For attributes not in K extend t''_1 and t''_2 arbitrarily from the visible domains of the attributes. Also, fill up the nulls of other tuples of T from the visible domains to obtain a strongly possible world, where distinct tuples t''_1 and t''_2 agree on K, contradicting to the assumption that K is a key in any strongly possible world.

Thus certain keys can also be recognized from the bipartite graph $= (T, T'; E)$ defined above, since all the tuples are distinct after filling the nulls regardless what visible domain values are used. Then, every incomplete tuple in T has a distinct set of weakly similar tuples in T' and any one of these tuples can be assigned to that incomplete tuple. The bipartite graph of a certain key would contain a connected component for each tuple in T with a set of tuples in T' as illustrated in Fig. 3.

Concept of strongly possible keys lies in between the concepts of possible and certain keys. Every certain key is a strongly possible key and every strongly possible key is a possible key. Figure 4 shows that $\{K : c \langle K \rangle \text{ holds}\} \subseteq \{K : sp \langle K \rangle \text{ holds}\} \subseteq \{K : p \langle K \rangle \text{ holds}\}$.

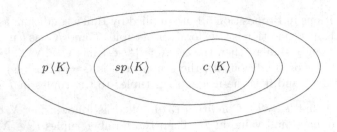

Fig. 4. Possible, strongly possible, and certain keys scopes

6   Strongly Possible Key Discovery

$sp\langle K\rangle$ holds if and only if there exists a matching between T and T' that covers all tuples of T as explained earlier. We introduce Algorithm 1 to find a strongly possible world if exist for a given T which verifies that $sp\langle K\rangle$ holds. The pseudo code is in the Appendix. We start by generating T' from T. T' contains all the total weakly similar tuples to each non-total tuple in T using the usable visible domain values for each null in the attributes of K to reduce complexity. For that we define $UVD_B = \left\{ v \in VD_A : \frac{\Pi_{\forall A \in K}|VD_A|}{VD_B} - c_v(B) > 0 \right\}$. Generating T' is done by taking the non-total tuples of T one by one and finding the total weakly similar tuples using the UVD's for each attribute instead of the nulls. This process may result in some duplicates in T' because it is possible that more than one non-total tuple in T can be weakly similar to same total tuple. For example, in Table 1a, the first and the last tuples are both non-total and are both weakly similar to the total tuple (Mathematics, 2019, Sarah, 5) generated by using UVD for each null. We need to remove these duplicated tuples in T' after generating the weakly similar tuples for each $t \in T$.

After calculating T', we find the maximum matching between $T \setminus T_{total}$ (T_{total} is the set of all the total tuples in T) and $T' \setminus T_{total}$. Function BipartiteMatching finds such a matching by recursively finding the shortest augmenting path of the bipartite graph and increase the matching by one each time. If this matching covers all the tuples in $T \setminus T_{total}$ then T has a strongly possible world verifying $sp\langle K\rangle$, which is $T' \cup T_{total}$ otherwise $\neg sp\langle K\rangle$ holds.

The running time depends on the size of T', which could be exponential in the size of the input. Sorting using Radix Sort takes $O(|R|(|T| + |T'|))$ time, while finding the largest matching in $G = (T \setminus T_{total}, T' \setminus T_{total}; E)$ takes $O((|T \setminus T_{total}| + |T' \setminus T_{total}|)|E|)$ time by the augmenting path method.

7   Conclusion and Future Directions

The main contributions of this paper are as follows:

- We introduced and defined strongly possible keys over database tables that contain some occurrences of nulls.
- We characterized the implication problem for strongly possible key constraints and showed that they enjoy Armstrong tables.
- We gave some necessary conditions for strongly possible keys. We showed that deciding whether a given set of attributes is a strongly possible key can be done by application of matchings in bipartite graph, so Hall's condition is naturally applied.
- We provided an algorithm to validate a strongly possible key by finding a proper strongly possible world for that key if there is any.
- We showed that deciding whether a given system of sets of attributes is a system of possible keys for a given table can be done using matroid intersection. However, we need at least three matroids, and matroid intersection of three or more matroids is NP-complete, which suggests that our problem is also NP-complete.

Strongly possible keys are special case of possible keys of relational schemata with each attribute having finite domain. So future research is needed to decide what properties of implication, axiomatization of inference remain valid in this setting. Note that the main results in [17] use that at least one attribute has infinite domain.

We plan to extend research from keys to functional dependencies. Weak and strong functional dependencies were introduced in [20]. A wFD $X \rightarrow_w Y$ holds if there is a possible world T' that satisfies FD $X \rightarrow Y$, while sFD $X \rightarrow_s Y$ holds if every possible world satisfies FD $X \rightarrow Y$. Our strongly possible world concept naturally induces an intermediate concept of functional dependency. Köhler and Link defined c-FD's, that is certain functional dependencies and showed their usefulness in [18]. Strongly possible worlds and the functional dependencies obtained from them could be used in similar ways. Future research on possible keys of finite domains might extend our results on strongly possible keys.

Finally, Theorem 5.1 defines a matroid intersection problem. It would be interesting to know whether this particular question is NP-complete, which we strongly believe it is.

A Appendix

Algorithm 1. Discovering a Strongly Possible Keys

Input: Dataset T on relation schema R with a candidate key K of b attributes
Output: A strongly possible world F in which K is a key if exist

1: **procedure spKeyDiscovery**(item T, item K)
2:      $T_{total} = \{t : t[i] \neq \bot \; \forall i = (1, 2, \ldots b)\}$
3:      $T' = \emptyset$
4:      **for all** $t \in T \setminus T_{total}$ **do**
5:          $T'_{temp} = \{t'_{temp} : t'_{temp} \sim_w t \quad \underset{\forall A \in K}{\text{Using}} \begin{cases} UVD_A, & \text{if } t[A] = \bot \\ t[A], & \text{if } t[A] \neq \bot \end{cases}\}$
6:          $T' = T' \cup T'_{temp}$
7:          $T'_{temp} = \emptyset$
8:      **end for**
9:      Sort T' using RadixSort
10:     Remove the duplicated tuples in T'
11:     Graph $G = \{V = (T \setminus T_{total}, T' \setminus T_{total}); E\}$
12:     BipartiteMatching(G)
13:     **if** size($Match$) $= |T \setminus T_{total}|$ **then**
14:         $F = Match \cap T$
15:         $F = F \cup T_{total}$
16:         **return** "A strongly possible world that verifies $sp\langle K \rangle$ holds in T is:" $+ F$
17:     **else**
18:         **return** "$sp\langle K \rangle$ does not hold in T"
19:     **end if**
20: **end procedure**
21: **procedure BipartiteMatching**(item G)
22:     $Match = \emptyset$
23:     **repeat**
24:         $P =$(AugmentingPath(G))
25:         Match $=$ Match $\triangle P$
26:     **until** $P = \emptyset$
27:     **return** $Match$
28: **end procedure**
29: **procedure AugmentingPath**(item G)
30:     Direct unmatched edges $T \rightarrow T'$, matched $T' \rightarrow T$
31:     Add s, t and connect them to non-matched vertices in T and T', respectively
32:     Run BFS on G from source s.
33:     **if** t is reached **then**
34:         **return** $P \setminus \{s, t\}$ for a shortest path P from s to t
35:     **else**
36:         **return** $\emptyset$
37:     **end if**
38: **end procedure**

References

1. Acuña, E., Rodriguez, C.: The treatment of missing values and its effect on classifier accuracy. In: Banks, D., McMorris, F.R., Arabie, P., Gaul, W. (eds.) Classification, Clustering, and Data Mining Applications. Studies in Classification, Data Analysis, and Knowledge Organisation, pp. 639–647. Springer, Heidelberg (2004). https://doi.org/10.1007/978-3-642-17103-1_60
2. Beeri, C., Dowd, M., Fagin, R., Statman, R.: On the structure of Armstrong relations for functional dependencies. J. ACM **31**(1), 30–46 (1984)
3. Sree Dhevi, A.T.: Imputing missing values using Inverse Distance Weighted Interpolation for time series data. In: Sixth International Conference on Advanced Computing (ICoAC), Chennai, pp. 255–259 (2014). https://doi.org/10.1109/ICoAC.2014.7229721
4. Chang, G., Ge, T.: Comparison of missing data imputation methods for traffic flow. In: Proceedings 2011 International Conference on Transportation, Mechanical, and Electrical Engineering (TMEE), Changchun, pp. 639–642 (2011). https://doi.org/10.1109/TMEE.2011.6199284
5. Cheng, C., Wei, L., Lin, T.: Improving relational database quality based on adaptive learning method for estimating null value. In: Second International Conference on Innovative Computing, Informatio and Control (ICICIC 2007), Kumamoto, p. 81 (2007). https://doi.org/10.1109/ICICIC.2007.350
6. Codd, E.F.: The Relational Model for Database Management, Version 2. Addison-Wesley Publishing Company, Boston (1990)
7. Date, C.J.: NOT Is Not "Not"! (Notes on Three-Valued Logic and Related Matters) in Relational Database Writings 1985–1989. Addison-Wesley Reading, Boston (1990)
8. Fagin, R.: Horn clauses and database dependencies. J. ACM **29**(4), 952–985 (1982)
9. Farhangfar, A., Kurgan, L.A., Pedrycz, W.: Experimental analysis of methods for imputation of missing values in databases. In: Proceedings of SPIE 5421, Intelligent Computing: Theory and Applications II, 12 April 2004. https://doi.org/10.1117/12.542509
10. Farhangfar, A., Kurgan, L.A., Pedrycz, W.: A novel framework for imputation of missing values in databases. IEEE Trans. Syst. Man Cybern. Part A Syst. Hum. **37**(5), 692–709 (2007)
11. Farhangfar, A., Kurgan, L.A., Dy, J.: Impact of imputation of missing values on classification error for discrete data. Pattern Recogn. **41**(12), 3692–3705 (2008)
12. Ferrarotti, F., Hartmann, S., Le, V.B.T., Link, S.: Codd table representations under weak possible world semantics. In: Hameurlain, A., Liddle, S.W., Schewe, K.-D., Zhou, X. (eds.) DEXA 2011. LNCS, vol. 6860, pp. 125–139. Springer, Heidelberg (2011). https://doi.org/10.1007/978-3-642-23088-2_9
13. Garey, M.R., Johnson, D.S.: Computers and Intractability. A guide to the Theory of NP-Completeness. Freeman, New York (1979)
14. Hartmann, S., Kirchberg, M., Link, S.: Design by example for SQL table definitions with functional dependencies. VLDB J. **21**(1), 121–144 (2012)
15. Hartmann, S., Leck, U., Link, S.: On Codd families of keys over incomplete relations. Comput. J. **54**(7), 1166–1180 (2010)
16. Grzymala-Busse, J.W., Hu, M.: A comparison of several approaches to missing attribute values in data mining. In: Ziarko, W., Yao, Y. (eds.) RSCTC 2000. LNCS (LNAI), vol. 2005, pp. 378–385. Springer, Heidelberg (2001). https://doi.org/10.1007/3-540-45554-X_46

17. Köhler, H., Leck, U., Link, S., Zhou, X.: Possible and certain keys for SQL. VLDB J. **25**(4), 571–596 (2016)
18. Köhler, H., Link, S.: SQL schema design: foundations, normal forms, and normalization. Inf. Syst. **76**, 88–113 (2018)
19. Lawler, E.L.: Matroid intersection algorithms. Math. Program. **9**, 31–56 (1975)
20. Levene, M., Loizou, G.: Axiomatisation of functional dependencies in incomplete relations. J. Theor. Comput. Sci. **206**(1–2), 283–300 (1998)
21. Mannila, H., Rähä, K.-J.: Design of Relational Databases. Addison-Wesley, Boston (1992)
22. Sali, A., Schewe, K.-D.: Keys and Armstrong databases in trees with restructuring. Acta Cybernetica **18**(3), 529–556 (2008)
23. Welsh, D.J.A.: Matroid Theory. Academic Press, New York (1976)
24. Zhang, S., Qin, Z., Ling, C.X., Sheng, S.: "Missing is Useful": missing values in cost-sensitive decision trees. IEEE Trans. Knowl. Data Eng. **17**(12), 1689–1693 (2005)

Machine Learning

ILIME: Local and Global Interpretable Model-Agnostic Explainer of Black-Box Decision

Radwa ElShawi[1], Youssef Sherif[2], Mouaz Al-Mallah[3], and Sherif Sakr[1(✉)]

[1] Tartu University, Tartu, Estonia
{radwa.elshawi,Sherif.Sakr}@ut.ee
[2] Zewail City, Cairo, Egypt
[3] Houston Methodist Heart and Vascular Center, Houston, TA, USA

Abstract. Despite outperforming humans in different supervised learning tasks, complex machine learning models are criticised for their opacity which make them hard to trust especially when used in critical domains (e.g., healthcare, self-driving car). Understanding the reasons behind the decision of a machine learning model provides insights into the model and transforms the model from a non-interpretable model (black-box) to an interpretable one that can be understood by humans. In addition, such insights are important for identifying any bias or unfairness in the decision made by the model and ensure that the model works as expected. In this paper, we present ILIME, a novel technique that explains the prediction of any supervised learning-based prediction model by relying on an interpretation mechanism that is based on the most influencing instances for the prediction of the instance to be explained. We demonstrate the effectiveness of our approach by explaining different models on different datasets. Our experiments show that ILIME outperforms a state-of-the-art baseline technique, LIME, in terms of the quality of the explanation and the accuracy in mimicking the behaviour of the black-box model. In addition, we present a global attribution technique that aggregates the local explanations generated from ILIME into few global explanations that can mimic the behaviour of the black-box model globally in a simple way.

Keywords: Machine learning · Interpretability · Model-agnostic

1 Introduction

Complex predictive machine learning models have been used intensively in different areas including banking, health care sector, personality profiles, and marketing. Despite the fact that such complex (black-box) models are able to achieve high accuracy, they lack an explanation of the prediction outcome. Thus, stakeholders find it hard to trust such complex models specially when used in critical domains such as self-driving cars, and disease diagnosis systems. Predictions of

© Springer Nature Switzerland AG 2019
T. Welzer et al. (Eds.): ADBIS 2019, LNCS 11695, pp. 53–68, 2019.
https://doi.org/10.1007/978-3-030-28730-6_4

black-box models suffer not only from transparency but also from possible biases that may lead to unfair decisions or disasters [7,14]. For example, over the period between 1970 and 1980, St. George's Hospital Medical School in London used a program to invite candidates to interview, it has been discovered that the program selection was based on the place of birth and surnames of candidates which make the selection procedure unfair [18]. Ribeiro et al. [20] gave an example for a highly accurate classifier that turned to be un-trusted one. The classifier used a deep learning model trained to classify wolves and huskies and after interpreting the model outcome, it has been discovered that the classifier classifies an image as wolf based on the snow in the image background. Caruana et al. [4] proposed a machine learning model for predicting the risk of readmission for patients with pneumonia. Counter-intuitively, the trained machine learning model learned that patients with asthma are at lower risk of readmission. If this model was used in production, it would have resulted in severe safety risk.

Since May 2018, the interpretability of machine learning models has been receiving huge attention as the General Data Protection Regulation (GDPR) requested industries to explain any automated decision in a meaningful way. In particular, it stated that "*a right of explanation for all individuals to obtain meaningful explanations of the logic involved*"[1]. In general, one way to define machine learning interpretability is the degree to which machine learning stakeholders can interpret and understand the decisions made by machine learning models [17]. In principle, interpretability aims to provide insights into the black-box model to be explained and answers questions such as (1) How a specific automated decision is taken? (2) What are the most critical features in the input data that leads to the decision? In general, making black-box machine learning models interpretable contributes positively toward establishing trust and confidence in their predictions.

In practice, decision trees have been used intensively to mimic the behaviour of black-box models such as neural networks models and tree ensembles [12]. Decision rules have been used widely to explain complex models such as support vector machines and neural networks models [3]. LIME is a local interpretable model-agnostic explanation [20] that explains the decision of an instance from any black-box model by fitting an interpretable model around the instance to be explained. Anchors [21] is an extension of LIME that uses the bandit algorithm to generate decision rules with high precision and coverage. Decision rules have used intensively to mimic the behaviour of a black-box model globally or locally given that the training data is available when providing local explanations [11,16,19]. Decision rules have been used intensively to mimic the behaviour of a black-box model globally or locally given that the training data is available when providing local explanations [11]. Koh and Liang [15] used influence functions to find the most influential training examples that lead to a particular decision. This method requires access to the training dataset used in training the black-box model.

In this paper, we focus on the open challenge of how to find meaningful explanation of the decision of a complex black-box model. The LIME technique

---

[1] https://ec.europa.eu/commission.

provides locality faithful explanation by fitting an interpretable model on per-turbed instances weighted by their distances from the instance to be explained. However, the quality of the provided explanation by LIME depends heavily on the weights assigned to the perturbed instances. Our proposed techniques, ILIME, tackles this challenge by weighting the perturbed instances using their distance from the instance to be explained and using their influence on the instance to be explained. In addition, we focus on providing the least number of local represen-tative instances with their corresponding explanations to characterize the whole model globally. Unlike LIME global attribution, SP-LIME [20], our proposed global attribution technique identify the best number of local instances that can give global understanding for the black-box model. In particular, the proposed local interpretability technique introduced in this work satisfies the following cri-teria (1) *Model-agnostic*, means that it can be used to explain the decision made by any black-box model, (2) *Understandable*, means that the explanation of the decision should be comprehensible to humans, (3) *Local*, means it provides local explanation for the decision of a particular instance. (4) *High fidelity*, means that it provides reliable and good approximation of the black-box model.

The main contribution of this paper can be summarized as follows:

- We present a model-agnostic local explanation technique, ILIME, that has been evaluated on different black-box models and has been compared to one of the state of art model agnostic instance explainer (LIME).
- We present a global attribution technique based on the aggregation of the local explanations provided by ILIME to characterize the whole model.
- The results of our experimental evaluation show that ILIME and our proposed global attribution are more faithful to the black-box models being explained and more trusted.

The remainder of this paper is organized as follows. Section 2 provides a back-ground on the concepts and techniques on which we rely on our work. In Sect. 3, we present our local explainer, ILIME, while our global attribution technique is presented in Sect. 4. We present a detailed experimental evaluation for our proposed techniques in Sect. 5 before we finally conclude the paper in Sect. 6.

2   Background

2.1   Influence Functions

Perturbing data while retraining the black-box model is computationally exhaus-tive and hence using influence functions [5] can overcome this problem. Instead of retraining the model to get the influence of a particular instance, influence functions work by upweighting an instance in the loss function by a very small amount ϵ in the empirical risk, the new model parameters θ is described as follows:

$$\hat{\theta}_{\epsilon,z} = argmin_{\epsilon \in \Theta} \frac{1}{n} \sum_{i=1}^{n} L(z_i, \theta) + \epsilon L(z, \theta) \tag{1}$$

Where θ is the model parameters vector and $\hat{\theta}_{\theta,z}$ is the new model parameters vector after upweighting the training instance z by a small amount ϵ. L is the loss function in which the black-box model is trained. Cook et al. [6] show that the influence of upweighting instance z on the model parameters $\hat{\theta}$ is as follows:

$$I_{up,params}(z) = \frac{d\hat{\theta}_{\epsilon,z}}{d\epsilon}|_{\epsilon=0} = -H_{\hat{\theta}}^{-1} \nabla_\theta L(z,\hat{\theta}) \tag{2}$$

Where $H_{\hat{\theta}} = \frac{1}{n} \sum_{i=1}^{n} \nabla_\theta^2 L(z,\hat{\theta})$ is the Hessian and $\nabla_\theta L(z,\hat{\theta})$ is the gradient of the loss function with respect to the model parameters given upweighting training instance z.

Approximating the loss using gradient and Hessian matrix capture the effect of a particular instance on the model [15]. For instance $z = (x,y)$, where x is the input features and y is the class label, the approximate influence of z on the model loss function at the instance to be explained $z_{explained}$ is as follows [15]:

$$I_{pert,loss}(z, z_{explained}) = - \nabla_\theta L(z_{explained},\hat{\theta})^T H_{\hat{\theta}}^{-1} \nabla_x \nabla_\theta L(z,\hat{\theta}) \tag{3}$$

One way to interpret $I_{pert,loss}(z, z_{explained})$ is that the influence of a training instance z on the instance $z_{explained}$ is proportionate to the gradient of the model loss function; the higher the gradient of the loss, the higher the influence of z on $z_{explained}$.

2.2   LIME

LIME [20] has been introduced as a local interpretability technique that relies on the assumption that the decision boundary of a black-box model is almost linear around the instance being explained. LIME explains an instance by fitting an interpretable model on perturbed sample around the input instance of interest. The methodology to build a locally interpretable model to explain the decision of instance x is as follows. First, generate perturbed sample around x. Next, get the prediction of the black-box model for each instance in the perturbed sample. Assign a weight to each instance in the perturbed sample according to its proximity to x. Train an interpretable model such as a linear regression model on the perturbed samples with their predictions. Finally, select the most important k features from the interpretable model to explain x.

SP-LIME [20] is a method that selects a set of representative instances with their corresponding explanations from the local explanations obtained from LIME as a way to characterize the entire model and gives a global understanding for the whole model. The SP-LIME works by selecting a set of non-redundant sample explanations based on a budget B identified by the user, where B is the number of explanations that a user can examine. More precisely, given a set X of instances, the SP-LIME pick a maximum of B instances that effectively capture the model behaviour in a non-redundant way. SP-LIME starts by constructing a matrix W of size $n \times d$ that contains the explanations of all instances in X, where $n = |X|$ and d' is the number of features used to explain instances in

X. Each row in W represents the local importance of features of a particular instance. Further, for each column j in W, let I_j denote the global importance of this feature in explanation space. Intuitively, I denotes the global importance of all features in the explanation space. The main goal of the picking step is to pick the least number of samples V that are different and have high coverage. The coverage is defined as the set function c that, given the explanation matrix W and global importance I, computes the total importance of the features that appear at least once in V as follows:

$$c(V, W, I) = \Sigma_{j=1}^{d'} 1_{[\exists i \in V : W_{ij} > 0]} I_j \qquad (4)$$

3   Local Explainer ILIME

In this section, we present a local interpretable model-agnostic explanation technique without making any assumption about the black-box model to be explained. Our proposed local explainer focuses on extending LIME and the influence functions-based explainer [15] into a new technique called ILIME that is more faithful than LIME and with more focus on tabular data.

Let the model being explained be denoted $b : X^{(m)} - > y$ where $X^{(m)}$ is the feature space of m features, and y is the output space. If b is a probabilistic classifier, we use $b(x)$ to denote the probability that an instance x belongs to certain class and to denote the most likely class otherwise. We define an explanation e of the decision of a black-box model to be a model g, which is a linear model. Let $\Omega(g)$ be the complexity of linear model g and is measured by the number of non-zero weights of the linear model. Let L be a measure of how unfaithful model g approximate a black-box b locally, where locality is captured by using a kernel function that is used to assign high weights to the most influential instances that are close to the instance to be explained. ILIME aims to minimize L which ensure local fidelity by keeping $\Omega(g)$ low enough to ensure interpretability (having g with hundreds of non-zero coefficients will not be comprehensible to humans).

ILIME is similar to LIME in the way it generates the perturbed sample around the instance to be explained x. ILIME explains x by fitting a linear model on the perturbed samples. To better understand x, we identify the most influential instances from the perturbed sample in a linear model that affects the prediction of x. ILIME provides an explanation that is faithful locally through using the kernel function that captures the influence of each perturbed instance on x and the distance of each perturbed instance from x.

The algorithm steps of ILIME for explaining an instance is summarized in Algorithm 1. Let x' be an interpretable representation of the instance to be explained x. First, we generate synthetic dataset $Z' = z_1', ..z_N'$, where N is the number of instances in Z' and z' is a binary representation obtained from drawing uniformly fraction of nonzero elements in x'. Next, we get the prediction of the black-box model b on each instance in $Z = z_1, ..z_N$, where Z contains instances in Z' in the feature domain. Given the dataset Z' of perturbed instances along

---

Algorithm 1: Influence-based Local Interpretable Model-Agnostic Explanation (ILIME)

Input   : instance x to be explained, black-box model b, Length of explanation k
Output: explanation e of instance x

1 $x' \longleftarrow$ interpretable representation of x

2 $Z'_c \longleftarrow \{\}$

3 Z' $\longleftarrow$ Generate synthetic dataset of size N obtained from x'

4 Z $\longleftarrow$ Recoverer samples in Z' in the feature space

5 **foreach** $i \in N$ **do**

6 $\quad\Big|\quad Z'_c \longleftarrow Z'_c \cup \langle z'_i, b(z_i) \rangle$

7 **end**

8 Train a linear model D on Z'_c

9 **foreach** $i \in N$ **do**

10 $\quad\Big|\quad$ Calculate the $I_{pert,loss}(z'_i, x')$ from model D

11 $\quad\Big|\quad \pi_x(z_i) \longleftarrow exp(-d(x, z_i)) * I_{pert,loss}(z'_i, x')$

12 **end**

13 $e \longleftarrow$ K-Lasso(Z'_c,k,$\pi_x(z)$)

14 **return** e

---

with the class labels obtained from the black-box b (from the previous step) stored in Z'_c, the main intuition of ILIME is to get the most influential instances in Z' that contribute to the prediction of instance x in a linear model trained on instances in Z'. The influence of each instance in z'_i is obtained by computing $I_{pert,loss}(z'_i, x')$, for each $i \in [1, N]$ in a linear model. Next, we weight each instance in Z by a kernel function $\pi_x(z)$ that captures the distance between z_i and x (the closer z to x, the greater the weight assigned to z) multiplied by $I_{pert,loss}(z'_i, x')$ obtained from the previous step. The kernel function $\pi_x(z) = exp(-d(x, z)) * I_{pert,loss}(z', x')$, where d is the Euclidean distance between the instance to be explained x and z. Finally, we train a linear model g on Z'_c that minimize the following loss function. $L(b, g, \pi_x(z)) = \sum_{z \in Z, z' \in Z'_c} (\pi_x(z)(f(z) - g(z'))^2)$. Choosing k features is achieved using Lasso regression [9].

4   Global Attribution Using ILIME

Although explaining the prediction of a single instance gives insights to the most important features that contribute to that prediction, it is not sufficient to access the trust in the whole model. One of the limitations of the SP-LIME is that the user must identify the number of explanations that can be inspected to have a global understanding of the model. In this section, we propose a global model agnostic attribution that characterizes the whole model and gives a better understanding to the model behaviour. More precisely, given the set of local explanations of a set of instances X using ILIME, the goal is to find the best set of instances in X with their corresponding explanations to explain the black-box model globally.

In general, clustering algorithms aim to detect interesting global patterns in the input instances by grouping similar instances into the same cluster. We apply the same idea on the set of local explanations provided by ILIME to transform them into a few number of global attributions. Our approach is a bottom-up approach starting from all local explanations obtained from ILIME. The main idea is to construct a *dendrogram* from all local explanations to find an optimal number of explanations to proxy the overall behaviour of the model being explained. Given a set X of instances of size n, we construct an $n \times d$ matrix E of local explanations, where d is the number of features used to explain instances in X. When using the linear model as a local explanation for instance x_i, we set $E_{i,j} = |w_{i,j}|$, where $w_{i,j}$ is the weight of feature j in instance x_i obtained from the linear model. Our proposed global attribution algorithm steps for characterizing the whole model is summarized in Algorithm 2.

First, we get the explanation of each instance in X using local explainer ILIME. Second, we normalize each local explanation in $E = e_1, ..., e_n$ to ensure that distances reflect the similarity between explanations appropriately. Third, we construct the pair-wise distance matrix for all explanations in E. The distance function between two explanations e_i and e_j is defined to be weighted Spearman's Rho squared rank distance [25] $d(e_i, e_j)$. From the computational perspective, Spearman's Rho squared is a well-suited distance metric especially when the number of features or instances is large, due to its ability to make differences between ranking more noticeable. Fourth, we merge the most similar two explanations e and e' in E and replace e, e' with $e \cup e'$ and then update the distance matrix. Repeat the third and fourth steps until the set of explanations contains only one explanation. In order to identify the optimal number of clusters k that represents the representative explanations, we use their Bayesian Information Criterion (BIC). More precisely, we compute the BIC for all E results by cutting the dendrogram at all possible cutting points and select the cut that maximizes the BIC. From each of the k clusters, we get their corresponding medoids $V = v_1, ..v_k$ which are used to summarize the patterns detected in each of the clusters formed in the previous step.

5   Experimental Results

In our experiments, we focus on evaluating how faithful the local explanations of ILIME to the model being explained. In addition, we evaluate the quality of the explanation provided by ILIME. To evaluate our technique, we compare it to LIME on different datasets from the UCI repository including iris, wine, breast cancer, and cervical cancer [8]. In addition, we use other medical datasets including diabetes, hypertension and mortality. These medical dataset were collected from patients who underwent treadmill stress testing by

---

Algorithm 2: Global attributation algorithm

Input   : Set of instances $X = x_1, ...x_n$
Output: Set of explanations V

1 $E \longleftarrow$

2 **foreach** $x_i \in X$ **do**
3 $\quad$ $e_i \longleftarrow$ explain using ILIME(x_i) $\triangleright$ using Algorithm 1
4 $\quad$ $E \longleftarrow E \cup e_i$
5 **end**

6 $E \leftarrow$ normalize the set of explanations E

7 **foreach** $e_i \in E$ **do**
8 $\quad$ **foreach** $e_j \in E$ **do**
9 $\quad \quad$ | $d(e_i, e_j) \longleftarrow$ distance between e_i and e_j
10 $\quad$ **end**
11 **end**
12 **while** *only one cluster remains* **do**
13 $\quad$ select e and e' from E such that $d(e, e')$ is minimal
14 $\quad$ set $E = E \backslash \{e, e'\} \cup \{e \cup e'\}$
15 $\quad$ update the distant matrix d
16 **end**
17 generate the clusters by cutting the dendrogram at an appropriate level based on BIC

18 $V \leftarrow$ medoid of clusters generated in previous step

19 **return** V

---

physician referrals at Henry Ford Affiliated Hospitals in metropolitan Detroit, MI in the U.S. The data has been obtained from the electronic medical records, administrative databases, and the linked claim files and death registry of the hospital [1]. For more details about the datasets and the process of developing the prediction model for predicting the risk of hypertension, diabetes and mortality, we refer the readers to [2, 23, 24].

5.1   How Faithful ILIME to the Model Being Explained?

In this experiment, we focus on measuring how faithful ILIME is to the model being explained. More precisely, we want to answer the following question: *are the relevant features returned by the ILIME truly relevant?* One way to answer this question is to assume that the model being explained is a white-box model (interpretable) such as logistic regression. For each of the hypertension, wine and breast cancer dataset, we follow the same pipeline:

1. Set the size of perturbed sample $N = 5000$, and $k = 7$ for both of LIME and ILIME.
2. Partition the dataset into 70% for training and 30% for testing the model.

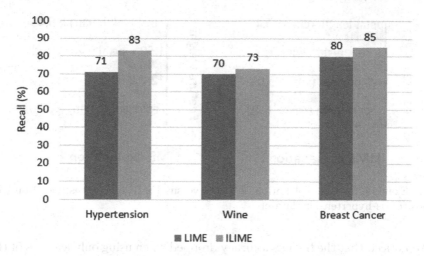

Fig. 1. Recall on truly important features for logistic regression classifier on different datasets

3. Train a logistic regression model with the training dataset such that the maximum number of features used for any instance in the dataset is not more than 7 features. Now we know the most important features for the model globally and we call them the set of important features.
4. For each instance in the test dataset, we get the explanation from both LIME and ILIME based on the top k features.
5. For each explanation retrieved by LIME and **ILIME**, compute the fraction of features contained in the set of important features.

We report the recall averaged over the instances in the test dataset as shown in Fig. 1. The results show that ILIME consistently achieves higher recall than LIME across all the datasets. In particular, these results demonstrate that ILIME provides explanations that are more faithful to the model than LIME.

We conducted another experiment to measure how faithful the explanations obtained from ILIME compared LIME. In particular, for each pair of the following datasets Iris-hypertension, Iris-breast cancer, wine-diabetes, and diabetes-mortality, we used the following pipeline:

1. Construct a composite dataset by concatenating all the instances in class 1 and 2 from the first dataset (Iris in the first pair of datasets) with the same number of instances in class 1 and 2 from the second dataset (hypertension in the first pair of datasets)
2. Partition the second dataset and the composite data into 2/3 for training and 1/3 for testing
3. Train a random forest classifier on the second dataset and the composite dataset

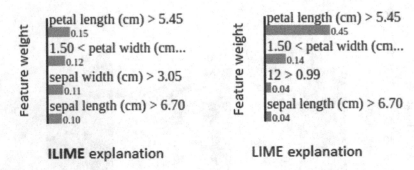

ILIME explanation   **LIME explanation**

Fig. 2. Example for the explanation of ILIME versus LIME on one instance from the composite Iris-hypertension dataset

We noticed that the testing accuracy dropped when using only features of the second dataset which implies that the model learns to depend on the features of the first dataset. From the accuracy drop when using only features of the second dataset in prediction, we know that features of the first dataset are important to the model's prediction. In order to verify that ILIME explanations reveal this relation, we do the following for each pair of datasets. For each instance in the composite dataset, we get the explanations from both ILIME and LIME using k features, where k equals to the total number of features in the composite dataset. We compute the fraction of features of the first dataset recovered by the explanations of LIME and ILIME as shown in Fig. 3. The results show that ILIME successfully capture more important features than LIME across all datasets. Figure 2 shows an example for the explanation of ILIME versus LIME on one instance from the composite Iris-hypertension dataset. Figure 2 shows that ILIME identifies four features from the iris dataset while LIME identifies three.

5.2 Can We Trust the Explanations of ILIME?

In this experiment, we focus on measuring the quality of the explanations of ILIME and measure how trusted the explanations obtained from ILIME when compared to LIME. We follow the following pipeline on each of the diabetes, mortality and hypertension datsets:

- Partition the dataset into 2/3 for training and 1/3 for testing
- Train a black-box classifier such as random forest (RF) and support vector machine (SVM)
- For each test instance w_i in the test dataset, we get the prediction of the black-box model
- Randomly select 25% of features of instance w_i and create an instance w_i' with the same feature values of w_i but without the randomly selected features
- Get the prediction of instance w_i' from the black-box model

We assume that we have a trustworthiness oracle that label a test instance as trusted if the prediction of w_i equals to the prediction of w_i' and untrusted

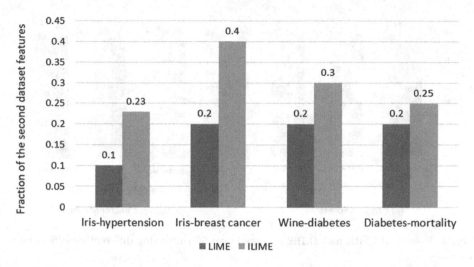

Fig. 3. Fraction of the second dataset features in all pairs of datasets recovered by the explanations of LIME and ILIME

otherwise. For each instance w_i in the test dataset, we do the following for both LIME and ILIME. Get the explanation of instance w_i and then get the prediction from the linear approximation model for instance w_i' and check that if the prediction of w_i is not equal to the prediction of w_i', then this instance is untrusted for the explanation method used and trusted otherwise. Now, we compare the trusted and untrusted instances for both LIME and ILIME against the trustworthiness oracle. Using this set up, we report the overall F-score of each explainer averaged over 50 runs using different classifiers and datasets as shown in Fig. 4. The results show that ILIME outperforms LIME across all datasets using random forest and support vector machine classifiers.

We conducted another experiment to assess the quality of the explanation obtained from ILIME when compared LIME. We used a toy dataset [22], which is $5 \times 5 \times 3$ RGB images with four possible collection of colours. Images are categorized into two distinct groups. Images in group 1 satisfy the following two rules: (1) All the pixels in the four corners are the same. (2) The middle three pixels in the top row have different colors. The images belong to group 2 satisfy neither rules. We partitioned the dataset into $2/3$ for training and $1/3$ for testing. We train multilayer perceptron such that the number of hidden layers is 2 of size 50 and 30, ReLU nonlinearities with softmax output. The network has been trained using Adam optimizer [13]. For each test image w_i in the test dataset that satisfy the two rules mentioned above, we get prediction of the black-box model. An accurate model should capture the corner pixels and the middle three pixels in the top row, we call these pixels important pixels. If w_i is correctly predicted by the model as group 1, then we get its explanation using LIME and ILIME based on the top 7 features. We return the fraction of important pixels averaged over all instances in the test dataset, averaged over 10

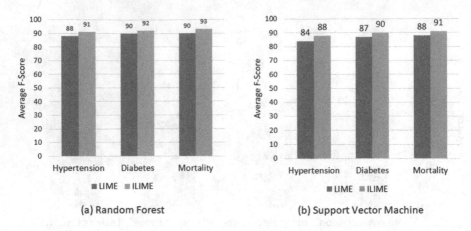

(a) Random Forest   (b) Support Vector Machine

Fig. 4. F-score of LIME and ILIME averaged over 50 runs using different classifiers and datasets

runs. ILIME achieves an overall recall of 80% while LIME achieves 76%. Figure 5 shows an example for the explanations obtained from both LIME and ILIME based on the top 6 features. Figure 5 shows that ILIME can successfully identify more important pixels compared to LIME.

5.3 Can We Trust the Whole Model?

In order to evaluate the proposed global attribution technique that characterizes the whole model, we compare it with global interpretable classifiers such as logistic regression (LR) and decision trees (DT). We follow the following pipeline for the cervical cancer, diabetes and breast cancer datasets for each of logistic regression and decision trees:

1. Partition the dataset into 2/3 for training and 1/3 for testing
2. Train a classifier using the training dataset such that the maximum number of features used for any instance is 10 and hence, we know the set of important features identified by the model
3. Get the explanations of all test instances in the testing dataset using LIME and ILIME local explainers
4. Get the set of explanations that best characterize the model using SP-LIME and our proposed global attribution technique, we set the budget B in LIME to be equals to the number of explanations retrieved by our global attribution framework
5. Get the fraction of the important features that are recovered by SP-LIME and our proposed framework

We report the recall averaged over all the explanations retrieved by SP-LIME and our proposed framework as shown in Fig. 6. The overall recall reported by our proposed framework outperforms SP-LIME across all datasets. In particular, our proposed framework consistency achieves recall over 85% across all datasets.

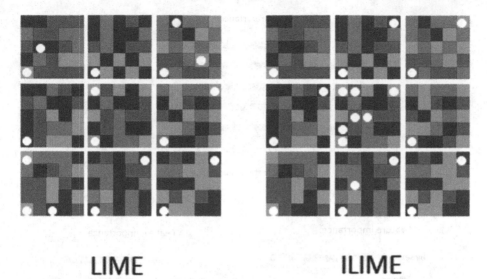

LIME   ILIME

Fig. 5. Example for the explanation of ILIME versus LIME on toy dataset based on the top 6 features

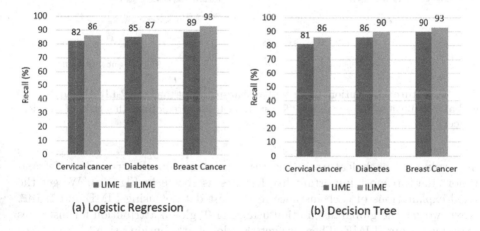

(a) Logistic Regression   (b) Decision Tree

Fig. 6. The recall of trustworthiness for SP-LIME and ILIME based on different classifiers using different datasets

We conducted another experiment to validate our proposed global attribution by comparing it against the permutation feature importance algorithm [10] in which the importance of a feature is measured by calculating the increase in the model's prediction error when the feature is permuted. A feature is considered important if shuffling its values resulted in increasing the model error. A feature is considered unimportant if shuffling its value does not have any impact on the model error. We train a random forest model on cervical cancer dataset such that 70% used for training and 30% used for testing. We get the feature

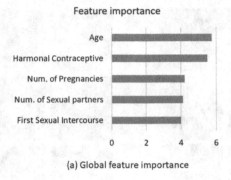

(a) Global feature importance

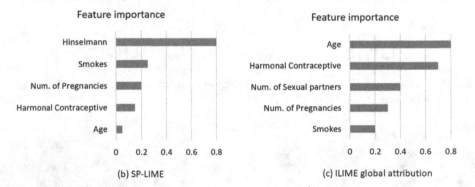

(b) SP-LIME   (c) ILIME global attribution

Fig. 7. Feature attribution using global feature importance, SP-LIME and ILIME global attribution in a random forest model for predicting the risk of cervical cancer

importance of each feature in the testing dataset on the trained model and report the top most important five features as shown in Fig. 7(a). We get the local explanations of each instance in the test dataset using LIME and ILIME. Next, we get the global attribution using ILIME global attribution for instances explained using ILIME. Then, we get the global attribution using SP-LIME for instances explained using LIME such that the budget is identified to be the same number of clusters identified by the ILIME global attribution technique. For each of SP-LIME and ILIME global attribution, we report the top five most important features as shown in Fig. 7(b) and (c). The results show that four out of the five most important features obtained by the global feature importance technique appear in the global attribution of ILIME while only three out of five appear in the SP-LIME. Furthermore, the top two features in the ILIME global attribution appear in the same order of the global feature importance technique. For the global attribution of SP-LIME, neither of the features appear in the same order of the global feature importance technique.

6   Conclusion and Future Work

In this paper, we presented, ILIME, a local model agnostic explainer for explaining the prediction of any machine learning black-box by fitting interpretable model on synthetic dataset around the instance to be explained. Each instance in the synthetic dataset is weighted by its influence in a linear model on the instance to be explained and distance to the instance to be explained. The proposed explainer ILIME has been evaluated and demonstrated its effectiveness by outperforming the state-of art local explainer, LIME, in terms of mimicking the behaviour of the model to be explained locally and globally. In addition, we proposed a global attribution technique based on local explanations obtained from ILIME to characterize the model being explained and to provide a global understanding for the whole model. In this work, we have been mainly focusing on tabular data. As a future work, we are planning to extend our approaches to images and text data.

Acknowledgment. The work of Radwa Elshawi is funded by the European Regional Development Funds via the Mobilitas Plus programme (MOBJD341). The work of Sherif Sakr is funded by the European Regional Development Funds via the Mobilitas Plus programme (grant MOBTT75).

References

1. Al-Mallah, M.H., et al.: Rationale and design of the Henry Ford Exercise Testing project (the FIT project). Clin. Cardiol. **37**(8), 456–461 (2014)
2. Alghamdi, M., Al-Mallah, M., Keteyian, S., Brawner, C., Ehrman, J., Sakr, S.: Predicting diabetes mellitus using SMOTE and ensemble machine learning approach: the Henry Ford ExercIse Testing (FIT) project. PLoS One **12**(7), e0179805 (2017)
3. Augasta, M.G., Kathirvalavakumar, T.: Reverse engineering the neural networks for rule extraction in classification problems. Neural Process. Lett. **35**(2), 131–150 (2012)
4. Caruana, R., et al.: Intelligible models for healthcare: Predicting pneumonia risk and hospital 30-day readmission. In: KDD (2015)
5. Cook, R.D., Weisberg, S.: Characterizations of an empirical influence function for detecting influential cases in regression. Technometrics **22**(4), 495–508 (1980)
6. Cook, R.D., Weisberg, S.: Residuals and Influence in Regression. Chapman and Hall, New York (1982)
7. Danks, D., London, A.J.: Regulating autonomous systems: beyond standards. IEEE Intell. Syst. **32**(1), 88–91 (2017)
8. Dua, D., Karra Taniskidou, E.: UCI machine learning repository (2017). http://archive.ics.uci.edu/ml
9. Efron, B., Hastie, T., Johnstone, I., Tibshirani, R., et al.: Least angle regression. Ann. Stat. **32**(2), 407–499 (2004)
10. Fisher, A., Rudin, C., Dominici, F.: Model class reliance: variable importance measures for any machine learning model class, from the rashomon perspective. arXiv preprint arXiv:1801.01489 (2018)

11. Guidotti, R., Monreale, A., Ruggieri, S., Turini, F., Giannotti, F., Pedreschi, D.: A survey of methods for explaining black box models. ACM Comput. Surv. **51**(5), 93 (2018)

12. Hara, S., Hayashi, K.: Making tree ensembles interpretable. arXiv preprint arXiv:1606.05390 (2016)

13. Kingma, D.P., Ba, J.: Adam: a method for stochastic optimization. arXiv preprint arXiv:1412.6980 (2014)

14. Kingston, J.K.C.: Artificial intelligence and legal liability. In: Bramer, M., Petridis, M. (eds.) Research and Development in Intelligent Systems XXXIII, pp. 269–279. Springer, Cham (2016). https://doi.org/10.1007/978-3-319-47175-4_20

15. Koh, P.W., Liang, P.: Understanding black-box predictions via influence functions. arXiv preprint arXiv:1703.04730 (2017)

16. Lakkaraju, H., Bach, S.H., Leskovec, J.: Interpretable decision sets: a joint framework for description and prediction. In: Proceedings of the 22nd ACM SIGKDD International Conference on Knowledge Discovery and Data Mining, pp. 1675–1684. ACM (2016)

17. Lim, B.Y., Dey, A.K., Avrahami, D.: Why and why not explanations improve the intelligibility of context-aware intelligent systems. In: SIGCHI (2009)

18. Lowry, S., Macpherson, G.: A blot on the profession. Br. Med. J. (Clin. Res. Ed.) **296**(6623), 657 (1988)

19. Malioutov, D.M., Varshney, K.R., Emad, A., Dash, S.: Learning interpretable classification rules with boolean compressed sensing. In: Cerquitelli, T., Quercia, D., Pasquale, F. (eds.) Transparent Data Mining for Big and Small Data. SBD, vol. 11, pp. 95–121. Springer, Cham (2017). https://doi.org/10.1007/978-3-319-54024-5_5

20. Ribeiro, M.T., Singh, S., Guestrin, C.: Why should i trust you?: Explaining the predictions of any classifier. In: KDD (2016)

21. Ribeiro, M.T., Singh, S., Guestrin, C.: Anchors: high-precision model-agnostic explanations. In: AAAI Conference on Artificial Intelligence (2018)

22. Ross, A.S., Hughes, M.C., Doshi-Velez, F.: Right for the right reasons: training differentiable models by constraining their explanations. arXiv preprint arXiv:1703.03717 (2017)

23. Sakr, S., et al.: Using machine learning on cardiorespiratory fitness data for predicting hypertension: the henry Ford Exercise Testing (FIT) project. PLoS ONE **13**(4), e0195344 (2018)

24. Sakr, S., et al.: Comparison of machine learning techniques to predict all-cause mortality using fitness data: the henry Ford Exercise Testing (FIT) project. BMC Med. Inform. Decis. Mak. **17**(1), 174 (2017)

25. Shieh, G.S., Bai, Z., Tsai, W.Y.: Rank tests for independence–with a weighted contamination alternative. Statistica Sinica **10**, 577–593 (2000)

Heterogeneous Committee-Based Active Learning for Entity Resolution (HeALER)

Xiao Chen[✉], Yinlong Xu, David Broneske, Gabriel Campero Durand,
Roman Zoun, and Gunter Saake

Otto-von-Guericke-University of Magdeburg, Magdeburg, Germany
{xiao.chen,yinlong.xu,david.broneske,campero,roman.zoun,saake}@ovgu.de

Abstract. Entity resolution identifies records that refer to the same real-world entity. For its classification step, supervised learning can be adopted, but this faces limitations in the availability of labeled training data. Under this situation, active learning has been proposed to gather labels while reducing the human labeling effort, by selecting the most informative data as candidates for labeling. Committee-based active learning is one of the most commonly used approaches, which chooses data with the most disagreement of voting results of the committee, considering this as the most informative data. However, the current state-of-the-art committee-based active learning approaches for entity resolution have two main drawbacks: First, the selected initial training data is usually not balanced and informative enough. Second, the committee is formed with homogeneous classifiers by comprising their accuracy to achieve diversity of the committee, i.e., the classifiers are not trained with all available training data or the best parameter setting. In this paper, we propose our committee-based active learning approach HeALER, which overcomes both drawbacks by using more effective initial training data selection approaches and a more effective heterogenous committee. We implemented HeALER and compared it with passive learning and other state-of-the-art approaches. The experiment results prove that our approach outperforms other state-of-the-art committee-based active learning approaches.

Keywords: Entity resolution ·
Query-by-committee-based active learning ·
Learning-based entity resolution · Record linkage

1 Introduction

Entity resolution (ER) is the task of identifying digital records that refer to the same real-world entity [6]. The classification step in an ER process can be considered as a binary classification problem [9]. Supervised learning can be

This work was partially funded by the DFG [grant no.: SA 465/50-1], China Scholarship Council [No. 201408080093] and Graduiertenförderung des Landes Sachsen-Anhalt.

© Springer Nature Switzerland AG 2019
T. Welzer et al. (Eds.): ADBIS 2019, LNCS 11695, pp. 69–85, 2019.
https://doi.org/10.1007/978-3-030-28730-6_5

adopted to solve this problem. However, in order to reach a satisfactory accuracy, a high amount of training data has to be provided, which is usually not available, and has to be labeled by domain experts. The training dataset labeled by domain experts for ER tasks is even more difficult to get than a normal classification problem, since for each labeling, experts have to work on both records of pairs and all their attribute pairs to make the final decision. Therefore, reducing the number of required training data is of great importance for ER.

As a means to reduce human effort, active learning (AL), which is a specific branch of machine learning (ML), is proposed to deal with this problem. Compared to a normal one-off ML process, an AL process is interactive and iterative. It reduces the number of required training data to achieve a desired accuracy by querying experts to label only the most informative data for each iteration and adding these into the training data (those pairs that are intrinsically difficult to classify based on available features are informative data). Then classifiers are retrained on updated training data and after each iteration, the stopping criteria are checked to see whether more iterations are required. So far, there have been different AL approaches proposed, which differ on the strategies to choose the most informative data. *Query by committee (QBC)* is an effective approach that has been successfully applied to many applications [16]. It gets the most informative data by selecting those data that result in the most disagreement in the votes of a committee of multiple classifiers [25]. However, applying QBC approaches for AL-based entity resolution (AL-ER) problems faces two challenges:

Diversified Committee: The key challenge to make QBC work in common is to generate a diversified committee, which can insightful voting disagreements so that the informativeness of data can be represented and distinguished [16] [14]. In order to achieve the diversity, for the vast majority of proposed AL approaches, they consider how to get multiple models with only one single type of classification algorithm. So far, several methods have been proposed for the diversity purpose, such as query by bagging, query by boosting [15]. However, for all those ensemble-based approaches, the accuracy of each model is compromised to get this diversity. For instance, in the bagging approach, the initial training dataset is divided into several smaller subsets, then different models are built based on different subsets [15]. Those trained models cannot be expected to achieve such accuracy as the model trained on the whole training dataset. Besides, nowadays, data is also quite variable in their types and there is no universally best model for all types of data. If a system completely relies on a single type of model, accuracy could not be acceptable for the worst cases.

Imbalanced ER Classification: The second challenge specialized for an AL-ER solution is the generation of the initial training dataset. The binary classification task for ER is a special task because of the imbalance of its two groups. In our real world, there are much fewer match pairs than non-match pairs, e.g., for the well-known Scholar-DBLP dataset, the imbalance ratio is 1 match but 3273 non-matches [28]. If the initial training dataset is randomly selected from all candidate pairs, the possibility to contain match pairs would be quite low, which may lead to a very low starting accuracy of trained models or even fail in training a model. Facing imbalanced data, oversampling and undersampling are commonly-used. However, except for their intrinsic shortcomings (overfitting for oversampling and discarding potentially useful data for undersampling [12]), they also contradict the goal of AL: saving labeling effort as much as possible. Facing both challenges, we propose in this paper a novel **H**eterogeneous **A**ctive **L**earning **E**ntity **R**esolution (HeALER) solution. We specifically detail our contributions as follow:

- We design a specialized technique to generate the initial training dataset, which is suitable for the inherent class imbalance in ER;
- We propose to construct the AL committee with different types of classification algorithms, through which we can achieve diversity, accuracy and robustness requirements of a committee;
- We prototype our solution and evaluate it with two well-known ER benchmarking datasets, and comparing with passive ML and two state-of-the-art AL-ER approaches (ATLAS [27] and ALIAS [24]). The evaluation results show that HeALER is faster to converge and can reach a higher final F-measure, which also indicates that with fewer labels a satisfactory F-measure can be achieved.

The remainder of this paper is organized as follows: In Sect. 2, we introduce our HeALER approach. Subsequently, we evaluate our approach and discuss the experiment results in Sect. 3. Before we conclude and list future work in Sect. 5, we also compare our method to other related work in Sect. 4.

2   Heterogeneous Committee-Based Active Learning for Entity Resolution

In this section, we introduce our designed QBC AL method for ER, which is characterized by its initial training data selection approach and its heterogeneous committee. We start with a global picture of our approach in Sect. 2.1, then we represent our initial training data selection method, heterogeneous committee in the following sections.

2.1   The Global Workflow

Figure 1 represents the global workflow of our method HeALER. It is separated into two parts, the left green area describing the preparation steps, and to the side the light red area corresponds to the AL stage.

Preparation for Active Learning. As we can see from the left green area of Fig. 1, several preparation steps are required to start the ER process. At first, input data is preprocessed if necessary, which may include data cleaning, formatting, standardization. Afterwards, blocking is performed to omit unnecessary comparisons, which are obvious non-matches based on predefined blocking keys [5]. Then candidate pairs are generated based on the blocking result. Subsequently, for each attribute pair, one or more similarity functions are chosen

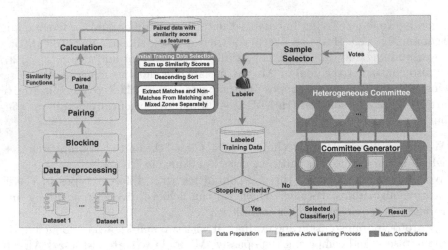

Fig. 1. The global workflow of HeALER

to best calculate similarities between each attribute pair in order to get similarity scores as features for the following learning-based classification step [4]. For the above-introduced steps, proper techniques should be employed based on ER task requirements, our contributions are reflected on the AL part, which will be briefly introduced next.

Iterative Active Learning Process. The first step of HeALER is to select pairs from the candidate pairs to be labeled by domain experts for an initial training dataset. As mentioned in Sect. 1, the classification step of ER is an imbalanced binary classification problem, i.e., there are much fewer match pairs than non-match pairs [9]. In order to reach a relatively high starting point with the initial training dataset, the training data is required to be balanced and informative. Balanced means the initial training dataset should contain sufficient percentages of match and non-match pairs, which is hard to achieve when one randomly picks pairs from the entire input data, since too high percentage of non-matches would be selected. Informative means the initial training data could involve useful information, which can benefit classifiers. The details how we achieve both goals will be introduced in Sect. 2.2.

Based on the initial training dataset, different classifiers are trained on them and then all classifiers together form the required committee. Notably, our classifiers are trained by different classification algorithms, which means our committee is heterogeneous. Compared to the majority of state-of-the-art QBC-AL approaches, our heterogeneous committee has the following advantages: First, the fundamental requirement - diversity of the committee - is achieved in a natural way without any other efforts. Second, each member of the committee is trained with the best or full ability without any compromise, which is more promising to provide a more accurate vote. Last, the committee analyzes training data and provides the result from multiple perspectives, no matter which kind of data the committee is facing, it can provide relatively stable and acceptable results. The methods to form our committee, including how to define the number of required classifiers and how to select classification algorithms as committee members, will be explained in Sect. 2.3.

After the committee is formed, they are employed to vote each pair from the unlabeled pool into match or non-match. The calculation of the disagreement of voting results for pairs will be firstly represented in Sect. 2.4. Then this process is iterated until the stopping criteria are reached.

2.2 Initial Training Dataset Generation

As explained in the last section, a good initial training dataset should be balanced and informative. In order to achieve both criteria, we analyzed a learning-based ER process. The resources that we have for the classification step are the candidate pairs and already calculated similarity scores for each attribute pair as features. Figure 2 is a histogram formed for the benchmarking bibliography dataset ACM-DBLP [13], which describes how the percentages of matching and non-matching pairs varies along with different similarity score levels. There are four attributes in this dataset, in total there are 16 similarity scores calculated as features (five similarity scores for the first three attributes: title, author, venue with different string similarity calculation functions; and one similarity score calculated for the attribute: publication year), each separate similarity score is normalized between zero and one, then the total similarity scores of all pairs should be between zero and sixteen by summing up all similarity scores. Based on this, we divided all candidate pairs into 15 groups and each group is an interval between n and $n+1$ (n is from 0 to 15). As we can see from it, globally the whole pairs are located in three zones. For areas with the lowest similarity scores the vast majority of pairs are non-matching pairs (the non-matching zone). Then the percentage of matching pairs increases in relatively middle levels (the mixed zone), and for the last levels with highest similarity scores, the vast majority of pairs become matching pairs (the matching zone). Dealing with variable datasets, the concrete ranges of the three zones may vary, however, globally speaking, those three zones and their trends should be valid for almost all datasets.

From the perspective of balance, the difficulty for the imbalanced classification step of ER is to find a sufficient number of matching pairs, while non-match pairs are quite easy to get, because there are much more non-matching pairs than matching pairs in our real world. The percentages shown in the figure can indicate the difficulty to get matching and non-matching pairs. In order to get sufficient matching pairs, the matching zone has to be focused. In order to get sufficient non-matching pairs, both the non-matching zone and the mixed zone can be the candidates.

From the other perspective of being informative, those pairs that are intrinsically difficult to classify based on available features, can be considered as informative data, since the classifier would be significantly improved if informative pairs are labeled and added to help the classifier training. Hence, those error-prone pairs should be true match pairs with relatively *low* similarity scores and true non-match pairs with relatively *high* similarity scores. True match pairs with relatively *low* similarity scores should be located in the mixed zone, but it is not possible to get them, since the matching pairs account for very small percentages in the mixed zone. Therefore, for achieving both balance and informativeness, we have to pick matching pairs from the matching zone. On the other hand, for non-match pairs, true non-match pairs with relatively *high* similarity scores locate in the mixed zone, by combining the conclusion from above (the non-matching zone and the mixed zone for getting non-matching pairs from the perspective of balance), the mixed zone is the aiming zone for high quality non-matching pairs.

Based on the above considerations, we conclude our method to generate the initial training dataset for learning-based ER in the following way:

1. First, there can be many similarity scores calculated and for different attributes, values of similarity scores may vary much. Hence, it is difficult to look into each separate similarity score and judge the possibility based on them separately. Therefore, we calculate a total score of each pair by summing up all similarity scores of attributes.

2. Next, we sort all candidate pairs based on their total scores in descending order.

3. Last, we divide all sorted pairs into k groups, then we can get the initial training dataset by randomly picking $n/2$ number of pairs from the top k_1 groups (the matching zone) for getting sufficient matching pairs and $n/2$ number of pairs from the next k_2 groups (the mixed zone) for getting sufficient and informative non-matching pairs (n is the preset number of initial training data). There is no accurate method to determine which k, k_1, k_2 are the best. The following hypotheses can be used. If the ER problem is between two data sources and the linkage is one-to-one linkage, the highest number of matches is the number of records in the smaller dataset. This number can be used as the size of the matching and mixed zones. If the linkage is one to many, even many to many linkages, an information that can be used is the approximate percentage of matching pairs, then this can be the basis to locate the matching zone and the same percentage of pairs can counted for the mixed zone. If even the percentage of matching pairs is unknown, as a rule of thumb, 10 groups should be a good number to averagely divide all pairs with a proper blocking step, then the matching zone is the top group with highest similarity scores and the mixed zone corresponds to the second group for getting non-match pairs.

With the above-introduced strategy the interesting areas analyzed above are established. With the first top groups, we are able to get sufficient matching pairs, and with the next groups, sufficient and informative non-matching pairs can be obtained.

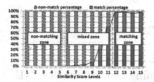

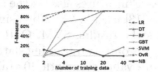

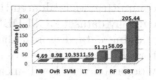

Fig. 2. Distribution of similarity scores

Fig. 3. F-measures of 7 classification algorithms

Fig. 4. Efficiency of 7 classification algorithms

2.3 Heterogeneous Committee

As introduced in Sect. 1, our committee is heterogeneous, which means that classifiers of the committee are trained with different classification algorithms. The method designation focuses on two aspects: how many classifiers and which classifiers to choose.

Generally speaking, our heterogeneous committee is allowed to contain any number of classifiers. Based on the result in [24], the performance of the classifier is not too sensitive to how many members a committee has and with four classifiers the aggregated accuracy is already satisfactory enough. On the one hand, each additional member in the committee means one more training process per iteration, which can heavily increase time needed for generating one round committee and has negative impact on efficiency. On the other hand, having more than four members for the committee achieves even lower accuracy [24]. Therefore, in our evaluation, four classifiers are generated to form the committee. Next, we present which candidate algorithms are suitable to be committee members. In general, we considered the following factors:

Accuracy with Little Training Data: The selected classifiers should have relatively high accuracy. Particularly, because the purpose of using AL is to reduce required human labeling efforts, we assume that for the training dataset, not so much training data is required to achieve high accuracy, which means that the selected classifiers should still work when only little training data is available. This is the main factor we use to choose classification algorithms.

Efficiency: Efficiency also requires consideration, since a learning-based classification is much more time-consuming than a simple threshold-based classification and such factor can be expected to have a large impact on the performance, as data grows.

Interpretability: Interpretability is also of great importance for choosing the learning algorithms, because we can use machine learning responsibly to ensure that our values are aligned and our knowledge is reflected [8].

We considered the following seven common binary classification algorithms: logistic regression (LR), decision tree (DT), random forest (RF), gradient-boosted tree (GBT), support vector machine (SVM), one-vs-rest logistic regression (OvR) and naive bayes (NB). In order to select classifiers for our committee, we evaluated their F-measures also on the same benchmarking bibliography dataset ACM-DBLP [13] used in Fig. 2 by using different sizes of initial training data. Figure 3 shows the results. SVM, OvR and LR have a satisfactory F-measure value even with only two training data pairs. NB and RF provide still a very low F-measure value even with 40 training data instances. NB classifiers are generative models that need prior probabilities. The probabilities are inaccurate for our case, because our initial training data is chosen by our proposed method in the last section, which normally generates relatively balanced training data. This state of training data does not conform to the test data [21]. Besides, it assumes that all features are independent [23]. However, our features are actually not independent, which may lead to the low F-measure for NB classifiers. The RF classifier cannot perform well, because it trains an ensemble of decision tree classifiers by splitting the training dataset into multiple subsets, then chooses subsets of features for each decision tree classifier [26]. This leads to a low F-measure especially when there is not enough training data. DT overall performs well except for the case with two training data pairs, in which DT classifier is not possible to be trained. GBT is in a similar situation as DT. However, its F-measure values are always lower than DT. We also evaluate the efficiency of all seven classification algorithms. All candidate pairs are divided into roughly two equal groups. Training data is generated by randomly picking four matching and non-matching pairs from the first group and then test data is the other entire group. The results are shown in Fig. 4. As we can see from it, results show that DT, GBT and RF need obviously more time than the other algorithms. NB runs the fastest, OvR, SVM and LR follows. However, all three tree-based classification algorithms DT, RF and GBT are quite slow. By combining the perspectives of interpretability and efficiency with the accuracy result, SVM, OvR, LR, and DT are selected to form our heterogeneous committee.

Above we provided guidelines on how to choose classification algorithms to form the heterogeneous committee. Facing different implementations of algorithms with different adopted libraries, the best choices of classification algorithms may change case by case.

2.4   Training Data Candidate Selection

After our heterogeneous committee is formed based on the above introduced approach, it is used to vote unlabeled pairs as matches or non-matches. Then those pairs with the most disagreement are those interesting pairs that we may select to be labeled by domain experts and added to the training dataset. The disagreement value of the voting results for pairs is calculated with the following equation:

$$Disagreement(pair) = \sum_{(a_m, a_n) \in committee} Difference(result(a_m), result(a_n)) \quad (1)$$

where (a_m, a_n) are the combinations of results from any two classification algorithms from the committee and the $Difference(x, y)$ function returns zero or one,

Table 1. Datasets used in experiments

Datasets	#Input	#Records in DBLP	#Records in ACM/Scholar	#Pairs (#match pairs)	#For training data selection	#For testing
ACM-DBLP2	2	2616	2294	21095 (2189)	10547	10548
Scholar-DBLP1	2	2616	64263	44999 (4351)	22500	22499

depending on whether x equals to y or not. With this equation, we sum up all the differences between any two combination of classification algorithms as the final disagreement value of the votes. However, the pair with a high disagreement value has also a high possibility that it is an outlier. If an outlier is selected and added to the training dataset, it will negatively impact the performance of classifiers. In order to reduce the possibility that outliers are selected, the random sampling proposed in [24] randomly picks the pair from the *top-n* pairs to alleviate the probability that an outlier is selected to be labeled, n can be set manually, such as 10, 20, 30.

Then the training data is updated in the above-introduced way iteratively and after the iteration process is completed according to preset termination conditions, the committee or a specific classifier can be used to identify duplicates for any unlabeled data.

3   Evaluation

In this section, we evaluate HeALER from three aspects: first, we solely conducted experiments to evaluate the balance and accuracy of our initial training data selection method (Sect. 3.2). Second, we evaluate our heterogeneous committee and compare it to the passive learning, committees formed by ALIAS and ATLAS (Sect. 3.3). Last, we evaluate our entire HeALER approach against a ML process and two state-of-the-art QBC-AL approaches: ALIAS and ATLAS (Sect. 3.4). For all results, the accuracy is measured using F-measure.

3.1   Experimental Setting

Datasets: We evaluate HeALER on two commonly-used real-world datasets: ACM-DBLP and Scholar-DBLP citation datasets [13]. Both datasets include two parts, one part is from the DBLP citation database and the other one is from ACM or google scholar citation databases, respectively. All of them have four attributes, including title, authors, venue and publication year. In order to prepare data for HeALER, we have done the following steps based on the two original citation databases: We first preprocess both databases by removing stop words and null values. Then we generate blocking keys (the first five letters of the title) for each record. Subsequently, we join two database tables with the blocking key as the join attribute, so that we get all candidate pairs. Afterwards, similarity functions are performed on each attribute to get corresponding features.

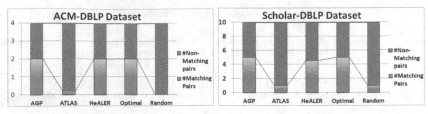

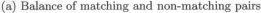

(a) Balance of matching and non-matching pairs

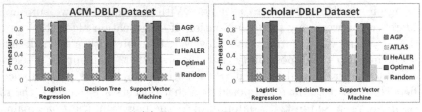

(b) F-measure Comparison

Fig. 5. Initial training dataset selection approaches evaluation

For attributes "title", "author", we apply cosine, Jaccard, Jaro-Winkler, metric longest common subsequence, N-Gram, normalized Levenshtein and Sorensen-Dice similarity functions[1]. For attribute "venue", the Jaccard similarity function is used. For the last attribute "year", the similarity between two values is one or zero based on whether they equal or not. In this way, we obtained 16 features. For the preparation of our initial training dataset selection method, total similarity scores are calculated and appended to data as well. With the above introduced steps, for the ACM-DBLP dataset, we got 21095 pairs after blocking (including 2189 true match pairs). We randomly divide all pairs into two parts: the first half 10547 pairs as the first part form the dataset to select training data and the remaining pairs for testing. For the Scholar-DBLP dataset, we got 44999 pairs after blocking (including 4351 true match pairs). We also randomly separate it into two parts in the same way as the ACM-DBLP dataset. The details of datasets are summarized in Table 1.

Implementation Related: Since learning-based classification is much more time-consuming than threshold-based classification, we implemented HeALER with Apache Spark (version 2.4), which is a general framework supporting distributed computation, as a preparation for big data processing. However, this paper focuses only on the quality side of ER results. The classification algorithms used are implemented with Spark MLlib. The programming language is Scala with the version 2.11.12.

---

[1] Implemented by the Debatty library (version 1.1.0).

3.2   Initial Training Dataset Evaluation

Experimental Design. This experiment is to evaluate different strategies to select the initial training dataset by getting the average results over five runs. We use both datasets in Table 1. The tested initial dataset sizes are four and ten, which are proved to be the least to function selected classifiers (Fig. 3). The following strategies are evaluated:

Random Selection: It means we randomly select the required number of pairs.

Optimal Selection: The optimal selection means that training data is optimally balanced, i.e., because we have the ground truth for our datasets, we pick half matches and half non-matches from the unlabeled data. However, this is not practical, since before labeling, we have no idea which pairs are matches or non-matches. In [24], they selected initial training data in this unpractical way.

Initial Training Data Selection of ATLAS [27]: ATLAS ranks all pairs on their total similarity scores, then divides the whole pool to n groups (4 or 10 groups for two tested dataset sizes respectively), at last the initial training dataset is obtained by randomly selecting one data pair from each group.

Initial Training Data Selection of AGP [7]: In order to get both matching pairs and non-matching pairs, the initial training dataset of AGP is obtained by selecting half number of pairs with highest total similarity scores (2 or 5 pairs for two tested dataset sizes respectively) and the other half number of pairs with lowest total similarity scores (2 or 5 pairs for two tested dataset sizes respectively).

Initial Training Data Selection of HeALER: Our own method HeALER selects the initial training dataset in the way of the hypotheses described in Sect. 2.2. Since the linkage for ACM-DBLP dataset is one-to-one linkage, the highest number of matches is the number of records in the smaller dataset, i.e., 2294 records from ACM library. As the whole dataset is almost equally split to two datasets. Then the matches contained in the first dataset to select training data should be 1147. This number can be used to get the matching and mixed zones, i.e., two pairs randomly picked from the first 1147 pairs with the highest similarity scores, and two pairs randomly picked from the next 1147 pairs. Regarding the other dataset Scholar-ACM, it is not one-to-one linkage, but we know that the approximate percentage of its matching pairs is 10, therefore, we divide all pairs into 10 groups, and the first top group with the highest total similarity scores is the matching zone, where we randomly get 5 pairs, and the second group is the mixed zone, where we randomly get the rest 5 pairs.

We evaluate those above-introduced selection methods with balance and F-measure metrics. For the balance metric, how many matching and non-matching pairs in the training dataset is shown. For the F-measure metric, F-measures values are calculated by testing the classifiers trained on different training datasets with LG, DT, SVM classification algorithms respectively on the test dataset.

Results and Discussion. As can be seen from Fig. 5a, with random and ATLAS approaches, the training data selected is quite skewed, no sufficient matching pairs are picked, especially the random selection for the ACM-DBLP dataset selects no matching pairs, which may make the training data unusable, since some classifier algorithms cannot work with only one class of data for a binary classification problem. HeALER can achieve relatively balanced training data, but not as completely balanced as AGP and Optimal selection. The F-measures using LR, DT, and SVM calculated on the training data selected with different approaches are shown in Fig. 5b. Therein, the training data selected using ATLAS and the random approach works only for DT and SVM on the Scholar-DBLP dataset. For all other cases, no classifiers are successfully trained and used for the later test classification because of exceedingly skewed training data. The other three approaches work apparently better. With the training data they selected, it is always possible to complete the classification tasks using the trained classifiers. Particularly, HeALER outperforms AGP and the optimal case with DT due to the more informative training data, which makes the splitting closer to the truth. However, it achieves a bit lower F-measure for LR and SVM. By concluding the results, we can say that the quality of HeALER training data is high when the number of divided groups can be correctly defined. Otherwise, the AGP strategy can be applied to achieve acceptable F-measure.

3.3   Heterogeneous-Committee Evaluation

Experimental Design. This experiment is designed to specially evaluate our heterogeneous committee and compare it to other approaches (committees formed in [27] and [24] and passive learning to randomly pick pairs without basing on committees' decisions). Both datasets in Table 1 are used. We fix the initial training data selected by our own strategy for all approaches, which provides them fair and good starting points. And the strategy used to reduce the possibility to get outliers is fixed with the Sampling20 approach (It is evaluated as the best strategy by comparing Sampling10, Sampling20 and Sampling30 using the random sampling method (n is set as 10, 20 or 30) introduced in Subsect. 2.4. The evaluation results are omitted due to limited space). After each iteration of the AL process, the F-measure is calculated on the classification results of the test data in Table 1 obtained by using the DT classifier trained on the updated training datasets by each approach. The AL process terminates after 199 rounds. Each experiment is repeated three times to get the final average result. The details how different approaches perform are introduced as follows:

Passive Learning: This approach randomly picks pairs to be labeled by humans and added to the training dataset without relying on any committee votings.

ALIAS Committee [24]: ALIAS forms its committee by randomizing parameters while training classifiers with the selected algorithm. In our experiments for both datasets, SVM algorithm is used. We vary the value of its parameter for maximum number of iterations with 4, 6, 8 and 10. Then four classifiers are trained respectively and form its committee. In the ALIAS paper, for their experiments, they applied DT algorithm and varied the parameter where to split. However,

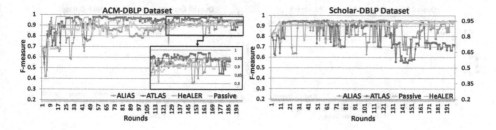

Fig. 6. Different committee comparison

as our implementation depends on the Spark MLlib, it is not possible to adjust this parameter. Therefore, we apply SVM algorithm for our experiments.

ATLAS Committee [27]: ATLAS partitions the training dataset to four subsets, then each subset of the training data is used to train its classifier to form its committee. The classification algorithm used here is the same as ALIAS: SVM for the purpose of comparison. For both datasets, each time 80 percent of pairs are randomly chosen to constitute the training dataset. Four subsets are required to get four classifiers of the committee.

HeALER Committee: As explained in Sect. 2.3, our heterogenous committee includes four classifiers, which are trained with SVM, OvR, LR, and DT algorithms, using the complete training dataset.

Results and Discussion. Figure 6 shows the comparison results of different committees and the passive learning. As we can see from the results of ACM-DBLP dataset, the F-measures of all approaches fluctuate much in the first 50 rounds, then becomes more stable later on. After about 140 rounds, our heterogeneous committee keeps F-measures higher than 0.92 and reaches its rough convergence. In contrast, ALIAS and ATLAS committees still cannot achieve their convergences till 199 runs. They show even less stable and lower results than passive learning. However, the highest F-measures they are able to reach during the experiments are much higher than passive learning, which proves the effectiveness of the committee to explore which are more informative pairs. Since the passive learning randomly chooses more pairs to be labeled, the informative pairs are hard to be selected to really cover the shortages of the classifiers. From the result of the other Scholar-DBLP dataset, we get similar conclusions. Our heterogeneous committee converges already after about 110 rounds and keeps the F-measure 0.95 afterwards. ALIAS and ATLAS committees are far from their convergence even with 199 rounds. The passive learning works quite good for this dataset due to the high initial F-measure. However, it requires much labeling effort to improves its F-measure. To summarize the results, our heterogeneous committee shows its advantage in picking informative data to improve the F-measure of the classifier and reach the convergence with much less labeling efforts than passive learning, ALIAS and ATLAS committees.

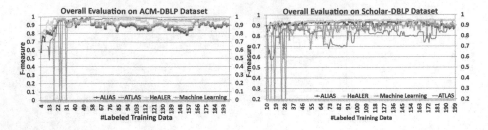

Fig. 7. Overall evaluation

3.4   Overall Evaluation and Comparison

Experimental Design. After we evaluate our initial training data selection approach and our heterogenous committee separately, in this section, we evaluate our entire HeALER approach by comparing the F-measures using a one-off ML approach, ALIAS and ATLAS approaches based on a same number of training data. For this overall evaluation, the same datasets are used as in the last two sections. The F-measures are all calculated on the classification results of the test data in Table 1 obtained by using the DT classifier. Except the normal machine learning approach, other approaches follow the iteration process of AL and terminate after 199 rounds. For the normal one-off ML approach, we randomly picked the corresponding number of training data of each iteration and calculate the F-measure of the test data using the DT classifier. ATLAS has no strategy to reduce the possibility to get outliers but it chooses the pair with the highest similarity value among all pairs with the highest disagreement value. Therefore, in the overall evaluation, for ATLAS, this approach choosing the pair with the highest similarity value is used. For HeALER and ALIAS, the sampling20 strategy is used as in the committee comparison experiment. The final result is averaged by three times' repetition (Fig. 7).

Results and Discussion. Figure 7 shows the comparison results of different AL approaches and a normal ML process. As we can see from the results of ACM-DBLP dataset, HeALER has the highest initial F-measure and keeps a F-measure around 0.9 with 20 or more training data. ALIAS and ML perform the worst and fluctuate their F-measures from the beginning to the end. ML starts to function stably with at least 33 labeled data and cannot significantly improve its F-measure when labeling more data. ATLAS starts to work with 10 labeled data and hardly varies its F-measure. The reason can be because its strategy always selects data with the highest total similarity score, weakens the effects of the disagreement values of data, and often chooses same data for different iterative rounds, which leads to changeless F-measure for several or even dozens of iterative rounds. Although it seems that ATLAS performs quite good, the results of the Scholar-DBLP dataset, in which ATLAS performs the worst, shows that ATLAS is not reliable, more research on the strategy of selecting the highest similarity score from the data with highest disagreement values is required. For the results of ALIAS, HeALER and ML on Scholar-DBLP, similar conclusions can be made. The results show that HeALER works better than the other compared approaches.

4   Related Work

AL related approaches for ER include the common AL with the goal of selecting the most informative data for classifiers to be labeled by humans (single-model-based [18], committee-based [7,11,19,20,24,27]), and special AL approaches for the purpose of getting the best rules (like classifiers) that are able to provide high precision without considering the quality of training data [1–3,10,22]. Therein, Ngomo et al. [18] identify the most informative data to be labeled and added into the training dataset with the maximized convergence of the used classifier. The proposed committee-based AL approaches differ from each other globally with different committee forming approaches. The approaches [7,11,19,20] use genetic programming algorithms to learn multi-attribute functions. However, the quality of those functions cannot be guaranteed. The research approaches [24,27] are the most similar to ours. They form their committees with several classifiers, which are trained on a single type of classification algorithm. However, in order to achieve diversity of classifiers in the committee to make AL work with the most disagreement strategy, their classifier qualities are compromised, which restricts the ability of the committee to identify the most informative data. Moreover, the initial training dataset selection problem is not correctly handled. Sarawagi and Bhamidipaty [24] directly assume that the AL process starts with an initial training dataset including five matching and non-matching pairs, which is not realistic, since it cannot be known whether a pair is matching or non-matching before labeling. Although in the other paper [27], this reality is considered, however, the initial training dataset they selected is quite biased with the number of matching and non-matching pairs, which leads to a very low quality of classifiers for the beginning iterations. In contrast to them, our proposed HeALER can provide a high-qualified initial training dataset and the heterogenous committee can select more informative data to improve the classifiers faster.

5   Conclusions and Future Work

To conclude this paper, we propose our AL approach HeALER for ER, which could select relatively balanced and informative initial training dataset and use its heterogeneous committee to select informative pairs to be labeled by human in order to improve the classifier. We evaluated and compared it with the passive (machine) learning and two state-of-the-art AL-ER approaches ATLAS and ALIAS. The evaluation results show that HeALER is faster to converge and can reach a higher final F-measure than other approaches. In addition, the results also indicate that it requires less training data to reach a satisfactory F-measure, which conforms to the purpose of using AL approach: reducing human labeling effort. However, we also observed the fluctuations during the early rounds, which are caused by choosing outliers to the training dataset. For future work, techniques to exploit the local density to handle imbalanced data and recognize outliers [17] should be studied in order to improve HeALER and make it reach the convergence faster.

References

1. Arasu, A., Götz, M., Kaushik, R.: On active learning of record matching packages. In: SIGMOD, pp. 783–794 (2010)
2. Bellare, K., Iyengar, S., Parameswaran, A.G., Rastogi, V.: Active sampling for entity matching. In: SIGKDD, pp. 1131–1139 (2012)
3. Bellare, K., Iyengar, S., Parameswaran, A.G., Rastogi, V.: Active sampling for entity matching with guarantees. In: TKDD, pp. 12:1–12:24 (2013)
4. Chen, X., Durand, G.C., Zoun, R., Broneske, D., Li, Y., Saake, G.: The best of both worlds: combining hand-tuned and word-embedding-based similarity measures for entity resolution. In: BTW (2019)
5. Chen, X., Schallehn, E., Saake, G.: Cloud-scale entity resolution: current state and open challenges. In: OJBD, pp. 30–51 (2018)
6. Christen, P.: Data Matching: Concepts and Techniques for Record Linkage, Entity Resolution, and Duplicate Detection. Springer Science & Business Media, Heidelberg (2012)
7. de Freitas, J., Pappa, G.L., da Silva, A.S., et al.: Active learning genetic programming for record deduplication. In: CEC, pp. 1–8 (2010)
8. Doshi-Velez, F., Kim, B.: Towards a rigorous science of interpretable machine learning. arXiv preprint arXiv:1702.08608 (2017)
9. Elmagarmid, A.K., Ipeirotis, P.G., Verykios, V.S.: Duplicate record detection: a survey. In: IEEE TKDE, pp. 1–16 (2007)
10. Fisher, J., Christen, P., Wang, Q.: Active learning based entity resolution using Markov logic. In: Bailey, J., Khan, L., Washio, T., Dobbie, G., Huang, J.Z., Wang, R. (eds.) PAKDD 2016. LNCS (LNAI), vol. 9652, pp. 338–349. Springer, Cham (2016). https://doi.org/10.1007/978-3-319-31750-2_27
11. Isele, R., Bizer, C.: Active learning of expressive linkage rules using genetic programming. J. Web Semant. 23, 2–15 (2013)
12. Kotsiantis, S., Kanellopoulos, D., Pintelas, P., et al.: Handling imbalanced datasets: a review. GESTS Int'l. Trans. Comp. Sci. Eng. 30(1), 25–36 (2006)
13. Leipzig, D.G.: Benchmark datasets for entity resolution (2017). Accessed 27 Nov 2017
14. Lu, Z., Wu, X., Bongard, J.: Active learning with adaptive heterogeneous ensembles. In: ICDM, pp. 327–336 (2009)
15. Mamitsuka, N.A.H., et al.: Query learning strategies using boosting and bagging. In: ICML (1998)
16. Melville, P., Mooney, R.J.: Diverse ensembles for active learning. In: ICML (2004)
17. Nanopoulos, A., Manolopoulos, Y., Theodoridis, Y.: An efficient and effective algorithm for density biased sampling. In: CIKM, pp. 398–404. ACM (2002)
18. Ngomo, A.N., Lehmann, J., Auer, S., Höffner, K.: RAVEN - active learning of link specifications. In: Proceedings of the International, Workshop on Ontology Matching (2011)

19. Ngonga Ngomo, A.-C., Lyko, K.: EAGLE: efficient active learning of link specifications using genetic programming. In: Simperl, E., Cimiano, P., Polleres, A., Corcho, O., Presutti, V. (eds.) ESWC 2012. LNCS, vol. 7295, pp. 149–163. Springer, Heidelberg (2012). https://doi.org/10.1007/978-3-642-30284-8_17

20. Ngomo, A.-C.N., Lyko, K., Christen, V.: COALA – correlation-aware active learning of link specifications. In: Cimiano, P., Corcho, O., Presutti, V., Hollink, L., Rudolph, S. (eds.) ESWC 2013. LNCS, vol. 7882, pp. 442–456. Springer, Heidelberg (2013). https://doi.org/10.1007/978-3-642-38288-8_30

21. Nguyen, H.T., Smeulders, A.: Active learning using pre-clustering. In: ICML, p. 79 (2004)

22. Qian, K., Popa, L., Sen, P.: Active learning for large-scale entity resolution. In: CIKM, pp. 1379–1388 (2017)

23. Rennie, J.D., Shih, L., Teevan, J., Karger, D.R.: Tackling the poor assumptions of Naive Bayes ext classifiers. In: ICML, pp. 616–623 (2003)

24. Sarawagi, S., Bhamidipaty, A.: Interactive deduplication using active learning. In: SIGKDD, pp. 269–278 (2002)

25. Seung, M.O., Sebastian, H., Sompolinsky, H.: Query by committee. In: Proceedings of the Workshop on Computational Learning Theory (1992)

26. Spark. Spark.mllib documentation. https://spark.apache.org/docs/latest/mllib-ensembles.html. Accessed 29 Nov 2018

27. Tejada, S., Knoblock, C.A., Minton, S.: Learning object identification rules for information integration. Inf. Syst. **26**, 607–633 (2001)

28. Wang, Q., Vatsalan, D., Christen, P.: Efficient interactive training selection for large-scale entity resolution. In: Cao, T., Lim, E.-P., Zhou, Z.-H., Ho, T.-B., Cheung, D., Motoda, H. (eds.) PAKDD 2015. LNCS (LNAI), vol. 9078, pp. 562–573. Springer, Cham (2015). https://doi.org/10.1007/978-3-319-18032-8_44

Document and Text Databases

Using Process Mining in Real-Time to Reduce the Number of Faulty Products

Zsuzsanna Nagy[(⊠)], Agnes Werner-Stark, and Tibor Dulai

Department of Electrical Engineering and Information Systems,
University of Pannonia, Egyetem Str. 10, Veszprém 8200, Hungary
nagyzsuzsi25@gmail.com,
{werner,dulai.tibor}@virt.uni-pannon.hu

Abstract. Process mining is a field of research whose tools can be used to extract useful hidden information about a process, from its execution log files. The current problem is that there is no solution available to track the formation of faulty products in real-time, both in time and space, to make it possible to reduce their number. The aim of this study is to find an effective solution for real-time analysis of manufacturing processes. The solution is considered to be effective if it helps to detect the error source points as soon as possible, and thus helping to eliminate them, it contributes in reducing the number of faulty products. Our previous solution, the "Time and Space Distribution Analysis" (TSDA), can analyze production data in time and space, but not in real-time. As a further development, we created the "Real-Time and Space Distribution Analysis" (RTSDA), which is capable of observing manufacturing process log data in real-time. It was implemented in software and tested with real process data. Real-time process mining can increase the productivity by quickening the detection process of the potential error source points, thus reducing the number of faulty products.

Keywords: Process mining · Real-time data processing ·
Production log data analysis · Fault source detection

1 Introduction

1.1 Real-Time Data Processing

Real-time data processing means that the processing of the input data is done in such a short time period, that the output can be obtained nearly instantaneously. The time that passes between the birth of the input data and the generation of the output depends on many factors (e.g.: the size and variety of the input data, the qualities of the data transmission network, the used data storing and processing techniques, etc.). To obtain continuous output, continuous stream of input data is needed.

Real-time data processing is widely used by monitoring systems to detect notable changes in the subject of the observation as soon as possible. The data processing system collects data about the subject of the observation and processes these data to detect anomalies in them. There are various real case studies available which are discussing the usage of real-time data processing at specific cases. For instance, CERN

© Springer Nature Switzerland AG 2019
T. Welzer et al. (Eds.): ADBIS 2019, LNCS 11695, pp. 89–104, 2019.
https://doi.org/10.1007/978-3-030-28730-6_6

operates a Beam Loss Monitoring (BLM) system to observe beam energy of an accelerator. The system uses real-time signal processing to detect negative anomalies in the measurement values [1]. An another instance, the Palomar Transient Factory (PTF), a synoptic sky survey in operation utilized an enormous camera on a telescope to survey the sky primarily at a single wavelength (R-band) at a rate of 1000–3000 square degrees a night. The data were processed to detect and study transient and moving objects (e.g.: gamma ray bursts, supernovae, etc.) [2].

There is diverse literature available about designing real-time data processing systems and choosing the most suitable data processing technique. For instance, Chen et al. identified the most important design decisions that must be made while designing a real-time stream processing system [3] and Nasir gave an overview of the existing data processing technologies regarding their abilities for real-time data processing [4]. They were dealing with the problem of processing big and varied data (i.e. Big Data) in real-time. For smaller and invariant data, a simpler solution is sufficient, too. Bertin et al. developed a method for archiving monitoring data by enabling real-time analysis within a live database. Their method uses two database schemas: a current schema and a historical schema. The current schema is storing the currently recorded data in its original form and the historical schema is storing the aggregated version of the old data [5].

As for manufacturing processes with automatized data collection, the output data usually are in text form (i.e. the data is not varied) and (assuming the output is about the manufactured products) the amount of newly recorded data is always limited by the number of newly manufactured products (i.e. the data is not big). If we know the minimum possible throughput time of manufacturing one product, and the maximum possible size of the data recorded about one product, then we can estimate the size of the newly generated data per minute or even second. With today's technology it is possible to store the whole data of one work shift within memory. It significantly facilitates the data managing process, but for data safety and archiving purposes, it is highly recommended to save the processed data into a remote server. Consequently, in our case, the most suitable solution is an in-memory database (for recent data) supplemented with a remote database (for archive data).

1.2   Process Mining

Process mining is a young field of research that provides mining of useful hidden information. It is well recognized as a valuable tool for observing and diagnosing inefficiencies in processes based on event data. Van der Aalst et al. presented techniques to identify anomalies in executions of business processes, by using available process mining tools in ProM. The problem is that these are not really applicable on automated manufacturing processes and none of these are able to process the data in real-time [6].

In our previous study, we introduced our developed set of methods that can be used to analyze a manufacturing process from multiple perspectives. These methods are based on time and space distribution of faulty products, so they were named "Time and Space Distribution Analysis" (TSDA). TSDA can give a comprehensive view of the production process, but not in real-time [7, 8].

1.3 Real-Time Process Mining

Real-time process mining means executing process mining methods in real-time. If the process model is known, then with the help of real-time process mining we can detect deviations in the events faster. Because of the requirements of real-time data processing, it is suited only for processes with short throughput time.

In the case of manufacturing processes, real-time process mining can increase the productivity by speeding up the process of detecting potential sources of errors, thereby reducing the number of faulty products. It allows the machine operators to notice the possible error source points sooner, so they can resolve them earlier.

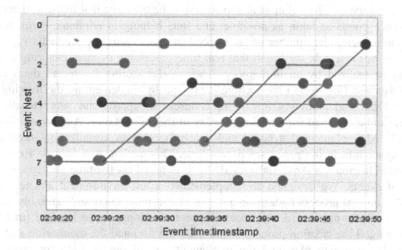

Fig. 1. A screenshot of an output chart of the DCA. The different events are marked with differently colored dots, and the consecutive events from the same case are connected with straight lines. In this example, the resource (Nest) is the user-selected attribute.

Among the tools in ProM, the "Dotted Chart Analysis" (DCA) and the "Performance Sequence Diagram Analysis" (PSDA) modules provide visualization methods that also could be used in real-time processing of manufacturing process data [9]. PSDA is capable of displaying the process executions (cases) by events as a function of time and a user-selected attribute, but only if the event name is the selected attribute. Besides that, PSDA has many other disadvantages. For instance, the time is always displayed as time duration on the y-axis, which is not advantageous for displaying process data in real-time. Compared to PSDA, DCA gives a better solution. DCA is capable of displaying the cases by events as a function of time and any user-selected attribute, and it displays the time on the x-axis as timestamp (Fig. 1). However, DCA is not a perfect solution either. The greatest disadvantage of it is that there is no possibility to scroll in the enlarged chart or to change the logic behind how it the events are colored. If a longer time period is observed with many relatively short events, then it can make the investigation of the output difficult.

By implementing the missing features of DCA, an appropriate data visualization solution can be obtained for real-time displaying of process data. Also, it would be useful to have the option of splitting the chart according to a user-selected attribute (e.g.: physical location of the resource). The separation of the chart has real importance, if some of the resources have the same name but they are at different location. In addition to avoiding possible name conflicts, it can also make the output more comprehensible, if the events that occurred at different locations are shown on separate charts.

1.4   The Content of This Study

The aim of this study is to find an effective solution for real-time analysis of manufacturing processes. The solution is considered to be effective if it helps to detect the error source points as soon as possible, and thus helping to eliminate them, it contributes to reducing the number of faulty products, increasing the productivity.

During our research, we found that real-time processing of manufacturing process data can be realized by an in-memory database (for storing the recent data) and a remote database (for storing the archive data), and by implementing the missing features of DCA and adding a few more features, an appropriate data visualization solution can be obtained for real-time display of the process data. As our solution, we designed and implemented a novel real-time process data visualization method for observing manufacturing processes in real-time. This method was named "Real-Time and Space Distribution Analysis" (RTSDA). It is based on the time and space distribution of events. We designed and implemented a test environment and tested the RTSDA with real process data of an automated coil-production and assembly line.

In Sect. 2, the manufacturing process is presented alongside the log files (including the content and the creation process of the main log file) and the monitoring software. In Sect. 3, the design plan and the implementation of our solution for processing manufacturing process data in real-time are described. In Sect. 4, the developed method, RTSDA, is presented, and finally, in Sect. 5, the new method is tested on real production data.

2   Background

2.1   The Manufacturing Process

The studied manufacturing process is executed by two production lines: one coil producer and one assembly line. The assembly line has a total of 8 stations and is connected to the coil production line at Station 4 [7].

On the assembly line the products are placed on nests of a round Table (8 in total), to make it easier for the machine to hold them and move them from station to station. The product is assembled from Station 1 to 6. At the last two stations only quality checks are done. The finished product is subjected to electrical inspection at Station 7 and to optical inspection at Station 8. The rejected unfinished products are discarded at Station 7 and finished products are placed on trays (based on their quality) at Station 8 [7]. A simplified version of the manufacturing process model can be seen on Fig. 2.

There are cameras at almost every station. They perform quality control tasks in the manufacturing process. The camera pictures get processed for searching deviations on the products. If the result is bad, the product will get declared as faulty with the corresponding error code. Once a product is declared as faulty, it will be handled as a faulty product until the end of the process.

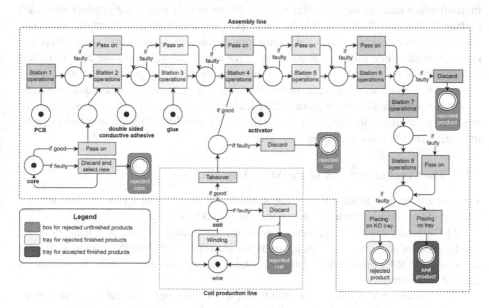

Fig. 2. A simplified version of the manufacturing process model. Different stations are marked with different colors. The red boxes indicate discarding points. The rejected commodities and the rejected unfinished products are dropped into red boxes (which are colored medium gray on the figure). The finished products are placed on trays. The light gray one is for rejected (faulty) products, and the dark gray one is for accepted (good) products.

2.2 The Log Files

The data is stored in CSV files (data separated by semicolons) and one file contains information about one day production of one type of product. Each row of the file contains process and measurement data of one product. So one row is one trace (a sequence of events) [8].

In our case the data of the manufactured main products, the data of the manufactured coils, and the output data of the electrical inspections at Station 7 are stored in separate files and on separate computers. In this study we are taking into account only the log files of the manufactured main products (i.e. main log file).

The main log file has 166 columns in total. The first few columns contain data for **product identification** such as product type identifier number, order identifier number, product identifier number and identification numbers of the commodities built into the product (e.g.: coil number). The next columns contain **process data** for each station such as quality indicator bit, start timestamp and work time duration of the station operations. The quality indicator bit indicates the quality of the product (good or faulty) when the product left the station. The next columns contain **measurement data** for each station such as the values obtained from processed camera images, and the tool identifier values (e.g.: the identifier number of the gripper which placed the coil on the product) from sensors. The type and amount of measured values vary between the stations. The last three columns are very important. Two of them contain data for **place identification** such as the identifier number of the nest in which the product was until Station 7 and after Station 7. The last but most important column contains the **error code**. The error code identifies the station and the quality control operation where the product failed. Good products also get an "error code". In such cases the code indicates that there was no error during the manufacturing process, the product is good.

The data rows with the same order identifier are collected into the same file. The maximum possible file size is around 2 MB. If the current main log file exceeds the size limit, then the new data rows will be written into a new file. It means that there can be multiple files for the same order identifier.

2.3   The Creation Process of the Main Log File

The control system maintains a string variable for each product (i.e. product string). During the production process, it expands these variables.

At the beginning of the process, it adds the data values for product identification. As the process goes on, it always appends the data collected from the current station to the string variable. From every station it obtains the process data and the measurement data (if there are any).

Assume an error has occurred at Station 4. For example, the adhesive dispenser puts too much adhesive on the core of a product. In this case the error handling and data generating works the way described below.

Detecting the Error. The control system detects the error, when it is processing the camera image. The results of the camera image processing show that there is too much adhesive on both sides of the core. The system calculates the area of the adhesive surface and compares it with the minimum and maximum allowed value. In this example, the calculated result exceeds the maximum allowed value. The system determines that an error occurred during the manufacturing process of the product.

Alarming the Machine Operator. The system alarms the machine operator with a sound and displays an error message on the connected monitor with the error code and the short description of the error.

Recording the Error. The system sets the error indicator bit to 0 (faulty) for the station data and appends the corresponding error code to the end of the product string. In this example, this code will be 401.

Managing the Faulty Product. The system carries the product until the next dropping point, then discards it there.

Carrying the Product. The machine will no longer carry out more manufacturing operations on the product, but will continue carrying it through the stations until the next discarding point. The system continues to add the process information of the stations, but the error indicator bit will stay constant 0 (because the product is unalterably faulty).

Removing the Product. In the example, the next dropping point is at the beginning of Station 7. The system will no longer record any data; it throws the product into a collector box. The system sets null values for the data values about the following stations, and then appends these values to the product string in the correct order.

Documenting the Product String. The expansion of the product string ends with this. The system writes the finished string into the output log file.

2.4 Monitoring Software

The monitoring software is an important part of the control system. Its main purpose is to provide an interface for the machine operator, where they can monitor the manufacturing process or change the light setting of the cameras (if it is needed). It provides several types of information during the production period and it alarms the machine operator in case of error occurrences.

During the production period the program shows the camera images of the currently examined products and also shows the examined part of the image and the result of the image processing. If it detects an error during the processing, it highlights the problematic area on the image in question.

The main problem with this software is that although it provides information about manufacturing process in real-time, it does not provide enough information to help identifying the possible sources of errors.

3 Design and Development of the Test Environment

As it was mentioned in Sect. 1, in case of a manufacturing process, the most suitable solution for real-time data processing is an in-memory data storage supplemented with a remote database. The in-memory database stores the recently processed input data, and the remote database stores the archive data. The input is read from the newest log file at specified time intervals (if there are new rows) into the memory, then it gets processed and finally the results are saved into the in-memory data storage. The new content of the in-memory data storage is saved into the remote database at specified time intervals (if there are new records). The output is created from the content of the in-memory data storage and is presented in visual form for the user.

We developed a simulator and an analyzer software. The purpose of the simulator software is to make it possible to test the analyzing software outside industrial environment. The purpose of the analyzer software is to realize the previously described tasks: reading and processing the input data, managing the in-memory data storages, and generating and visualizing the output for the user. The remote database is not implemented yet, but when it will be, then it will have connection with the analyzer software.

Both software were created in Microsoft Visual Studio Community 2017, in C# programming language. WPF (Windows Presentation Foundation) was used for the user interface, and OxyPlot graphical library was used for displaying the charts.

A detailed description can be read below about the simulator software (Sect. 3.1) and the analyzer software (Sect. 3.2).

3.1 Simulator Software

The simulator software was developed to make it possible to test the analyzing software outside industrial environment. This software simulates the process of generating log files of the manufacturing process.

The software generates the output file from a log file of a previous production. It reads the whole content of the chosen log file into the memory, then it updates the datetime values of the data, and then starts writing the data rows one by one, at specified time points, into the generated output file.

The algorithm implemented within the software is described below. It is a case-specific solution, but it can easily be applied to other cases with minimal modification of the code.

Reading the Input. It reads the entire content of the chosen log file of a previous production.

Checking the Data. It checks the number of columns, the type and value of the data, and keeps only complete and completely correct rows (i.e.: rows that have formally good values in all columns).

Updating the Timestamps. It updates all datetime values (timestamps) in the data. First it calculates the difference between the earliest datetime value of the data and the current datetime value, then it adds the difference to all timestamps.

Calculating the Time Points of Writing into the Output. It calculates for each data row when to write them into the output file. The method of calculation also depends on whether the log file stores an event or a complete process execution (trace) in one row, and stores complete timestamp or time duration for each start timestamp. In case of "1 row = 1 trace" file structure, the complete timestamp of the last event of the trace will determine the time point of writing the data row into the output file. In case of "1 row = 1 event" file structure, the complete timestamp of the event will determine the time point of writing the data row into the output file. If the file contains time duration instead of complete timestamp of the event, the complete timestamp can be obtained from the sum of the start timestamp and the time duration.

Simulating. It starts the simulation right after it is finished with recalculating the timestamp values. At the beginning of the simulation process it creates an output file with the log file header as first row. During the simulation process it writes the data rows into the output file one by one, at the specified time points. The output rows are exactly in the same form as the input rows (in the original file), this way the only difference between the input and the output rows are the date and time values.

3.2 Analyzer Software

The analyzer software was developed to analyze the manufacturing process in real-time. It can function as a complement to the production process monitoring system (it can provide plus information) in an industrial environment.

The software stores the processed data in memory. For testing, the output file of the simulation software was used as the input file, but it can read the input in real industrial environment as well, without any modification of the source code.

The algorithm implemented within the software is described below. Similarly to the simulator software, it is a case-specific solution, but it can easily be applied to other cases with minimal modification of the code.

Reading the Input. At the beginning of the analyzing process the software starts searching for the input file. The found input file will be observed until its size exceeds the log file size limit, then the software starts searching for a new file again. The software is either in file searching or in file checking state. In both cases the task is repeated at specified time intervals: If it is in the file searching state, then it is searching for the next input file at specified time intervals, or if it is in the file checking state, then it is checking the current input file at specified time intervals. Sometimes the size of the log file exceeds the set limit by few bytes. In that case, the software will keep switching between file searching state and file checking state until a new input file appears (or the software gets paused).

File Searching State. The software in this state keeps searching for the newest file in the given folder. If it finds a file with the right format (a CSV file with the expected log file header), then it will choose this file as the new input file. The software will record the name of this file, reads the entire content of it, records the datetime of the last read (which is the current datetime), and finally steps into file checking state. If the software does not find a file with the right format, then it will stay in the file searching state.

File Checking State. The software in this state keeps checking the datetime of the last modification of the current input file. If it is greater than the datetime of the last read (i.e.: the file has been modified since the last read, there are new lines), it will read the new lines and records the current datetime as the datetime of the last read. If the datetime of the last modification of the current input file is less than the datetime of the last read (i.e.: the file has not been modified since the last read, there are not any new lines) and the size of the file is greater than or equal to the log file size limit, then the software will step into file searching state.

Checking the Data. It checks the number of columns, and the type and value of the data, and keeps only complete and completely correct rows.

Preprocessing the Data. It filters the data and keeps only the important data values (that can be used for analysis). The following process data is retained in all cases: case identifier, event name, event timestamps (start timestamp and complete timestamp or time duration), and resource.

Processing the Data. It processes the filtered data and then, depending on the type of the output, saves the data into separate data stores (e.g.: data tables, lists, or single value variables). For new inputs, the data tables and the lists get extended and the values of the single value variables get updated. Also, the affected aggregated values are recalculated. The exact way of processing the data is described in Sect. 4.4.

Visualizing the Output. It visualizes the output in form of tables, diagrams and text (depending on the form of the output). If the number of elements of a data table or a list grows too high, the visualization of these data can result in lagging of the software. For this reason, the number of elements in that kind of data storages is restricted to a specified number. If the number of elements exceeds the limit, then the oldest element or elements will get removed to make the number of elements equal to the limit.

4   The New Method, the RTSDA

We developed a process data visualization method and named it "Real-Time and Space Distribution Analysis" (RTSDA). RTSDA displays the events as a function of time (x-axis) and resource (y-axis). It is designed to process log data in real-time, so it always processes only the new lines, then adds the newly processed data to previously processed ones, and finally it updates the affected parts of the output charts.

4.1   Criteria for Applying the Method

This method can be applied to other manufacturing processes as well, if they meet the following criteria:

- The process throughput time is small, or at least new event information can be obtained in short time periods (every second or minute). Otherwise there is no point in observing the process in real-time.
- A resource (person or machine/tool) can perform only one activity at a time. It means that the resource can start the next action or event only if they have finished the previous one.
- Each event has time information recorded, or more precisely, each event has at least one timestamp. If an event has a start timestamp, then it either has a complete timestamp or an event time duration value, too. If an event has only one timestamp, then it is both the start and the complete timestamp of the event.
- In case of "1 line = 1 event" log file content structure, there is a case identifier for each event. This is necessary, because the case identifier identifies the events which belong to the same case (trace).
- For "1 line = 1 event" log file content structure, event names are recorded for each event. This is necessary, because the corresponding events can be identified by the event name. There is no formal restriction on the name.
- For "1 line = 1 trace" log file content structure, all possible events are known. In this case, event data are stored in columns or column groups, so it is important to know which column contains data for which event.

- The name of the utilized resource (person or machine/tool) is recorded for each event. If the resource can be linked to a location (e.g.: a tool to a machine, or a person to a class), it is worthwhile to record the name of the place in a separate attribute. There is no formal restriction on the name of the resource or place; it can be an identification number, too.

4.2 Data Visualization Considerations

DCA attaches importance to the process executions. It connects the events which belong to the same case and colors them based on the values of a user-selected attribute. RTSDA rather attaches importance to the resource utilization, so it does not connect the events. It separates the events into multiple diagrams based on the values of a user-selected attribute (**separating attribute**), and colors them based on the values of another user-selected attribute (**coloring attribute**).

In the case of RTSDA, the y-axis is fixed regarding that the source of its values is expected to be the resource attribute, but the user is free to decide which attribute is considered to be the resource. The separating attribute is expected to be an attribute which has values that can group the values of the resource attribute into distinct sets. For example, if the resources are tools of different machines, the separating attribute can be the machine. That way each machine will have its own chart with their tools as the values of the y-axis.

In addition, the user can select a list of attributes as the source of **plus information**. The values of these attributes are displayed within one textbox, when the user selects an event on a chart of the output.

RTSDA also distinguishes between short and long events. **Long events** have a start datetime point (start timestamp) and a complete datetime point (complete timestamp or time duration from which it can be calculated). These are usually complex events that could be broken down into additional events, but it is either not possible (because of lack of information) or just not important (because it has no significance). These events can be visualized as thick lines or bars. For example, in our case station processes are long events. For example, in case of a manufacturing process, station processes are long events. **Short events** have only one datetime point value (occurrence timestamp). These events can be important parts of a long event, so they need to be highlighted. They can be visualized as markers (e.g.: dot, rectangle, etc.). For example, in case of a manufacturing process, declaring the product as faulty, throwing away or placing the product on tray are short events.

4.3 Data Storing

Dynamic solutions have been used for data storage.

A data table (DataTable) type variable was used to store the **preprocessed data**. The original names of the column headers (the attribute names) are kept in a separate string array to avoid the possible name conflicts between them. The data role labels for the attributes (e.g.: case identifier, event name, etc.) are stored as column identifier numbers in separate single variables.

For **visualization**, the data of the **events** are recorded in individual objects. These objects are from a **custom class**, the Event class. This class stores the case identifier (*case_id*), alongside with the name of the event (*event_name*), the timestamp of the event (*timestamp*), the name of the resource who executed the event (*resource*) and some plus information about the event (*plus_info*) as well. The timestamp for long events is either the start or complete timestamp, and for short events it is always the occurrence timestamp. The plus information is generated from the selected attribute values and stored in form of a string value.

The **events** are collected into **lists** according to the value of the two selected attributes, the **separating attribute** and the **coloring attribute**. Each separator attribute value has an event list for each coloring attribute value. It means that each value of the separating attribute has its own diagram, and within the diagrams, each value of the coloring attribute has its own dot or line series. It also can be imagined as a dynamic two-dimensional array, where the separating attribute and the coloring attribute are the two dimensions, and the event lists are the elements of the array. For the short and long events, separate lists are reserved, because the chart drawing library expects the data in different form in case of different visualizations. The short events are visualized with dot series, and the long events are visualized as line series.

4.4 Data Processing

The way the data is processed depends heavily on the structure and composition of the content of the log file. The data rows have to be processed differently in the case of "1 line = 1 event" and "1 line = 1 trace" file structure. For "1 line = 1 event" content structure, the steps described below have to be executed after reading the new data rows. In the case of "1 line = 1 trace" file structure, every step must be performed for each column group (event) in each new data row. The only difference is that in this case the empty values (which mean the event did not get executed) should be deleted.

Sorting the Data. The new events are sorted in ascending order by the start timestamp or the occurrence timestamp.

Adding New Events. Add new events one by one to the corresponding event list. In the case of a short event, one event is added to the corresponding event list which stores data for a dot series. In the case of a long event, three events are added to the corresponding event list which stores data for a line series: one event with the start timestamp, one event with the complete timestamp, and one blank event. (Adding a blank event is required to break the line. If it was not added, the end of this event would visually get connected to the start of the next event.)

Maintaining the Number of Events. As it was mentioned before, if the number of elements in the list exceeds the limit, then the oldest element or elements will get removed to make the number of elements equal to the limit.

5 Application of RTSDA to Real Data

We applied the method to the main log files of the automated coil-production and assembling line.

In this example, the **nests** are the **resources**, so the events on the output diagrams are plotted against the nest (y-axis) and time (x-axis). As the manufacturing runs in parallel between the nests, the diagrams make the entire manufacturing process visible, but also provides picture of the nest utilization and the station efficiency.

The **rotary tables** were selected as **separator attributes**. There are two round tables: one bigger and one smaller. The bigger one has 8 nests and the smaller one has 4 nests. The bigger one is used by the entire machine line to transport products between stations, while the smaller one is used only by Station 7 to move finished products to execute electrical quality control tests. Therefore, Station 7 received its own diagram.

The names of the **events** were selected as the **coloring attribute**. In this example, station operations are long events and error occurrences are short events. The station operations are represented by horizontal bars, and the error occurrences are indicated by black diamonds.

The results of the application of RTSDA to real data are presented through three types of output cases (Figs. 3, 4 and 5).

On Fig. 3 the beginning of a production period can be seen, where the same error occurred repetitively, one after another. This error is identified by the "402" error code, which means that there is not enough glue on the product. For the machine operator this output is informative. They can see that the error occurred more than once, so they can assume that it was not just a random one-time event, it is an actual problem that needs to be solved. In this case the problem is probably with the glue (there is no more in the dispenser), but it is also possible that the problem is with the camera (it could not detect the glued area on the product well, because of bad light circumstances). In such cases, the machine operator can recheck the camera pictures themselves and make a decision according to it.

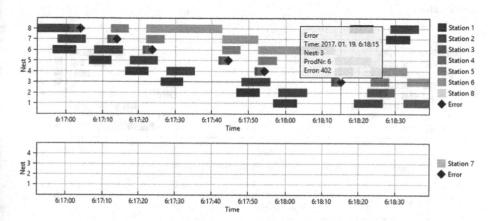

Fig. 3. The same error occurs repetitively, one after another.

On Fig. 4 a continuous production can be seen. It could be the optimal case, but it is not, because it has error occurrences at Station 8, at Nest 7. Station 8 is the place of optical inspection of the finished product, so if the nest is dirty, it can cause an actually good product to get classified as faulty. If the same error occurs more than once at the same nest, then it is highly possible that the nest is the actual source of errors. In such cases, the machine operator can clean the nest.

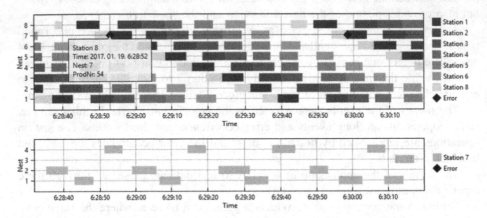

Fig. 4. Continuous production with a few error occurrences at the same station and nest.

On Fig. 5 a short outage time can be seen. The outage time is a time period where the production line did not produce any new products, the production stopped. Longer stoppages are not beneficial for the company so it is important to detect and eliminate them as soon as possible [7]. In this case the outage time is really short (one minute long), so it does not cause serious loss in production. It could be because of maintenance. If it was more minutes long, intervention would be needed. If there is no information about the possible causes, the machine operator has to carry out a complete maintenance.

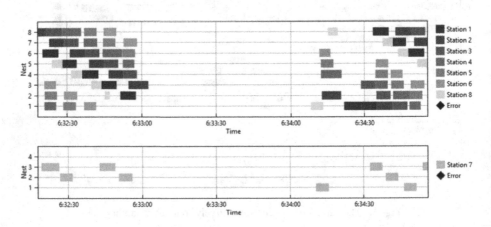

Fig. 5. A short outage time.

5.1 Evaluation of the Method

RTSDA is capable of visualizing the log data of a manufacturing process in real-time. It provides a comprehensive view of the production, with a tool-level image of the utilization of the resources and the error cases that have occurred, thus helping the machine operator to observe the evolution of the production.

The current version is a rudimentary solution for detecting error sources, because it only implies the possible sources of the problems, it does not report them. Notwithstanding, it still can provide sufficient support for an experienced machine operator.

It must be noted, that if the available data does not have enough information to make it possible to detect the exact source of errors, the source of the errors can only be presumed. To validate the results and to determine the exact locations, physical examination of the presumed locations is necessary.

Presently, the log data of only one manufacturing process was available for the research, so RTSDA was applied only to log data of one kind of process and only within a test environment. For better evaluation, the method should be applied to various processes and within the real industrial environment of the processes. Hopefully, there will be more possibilities in the near future to test and improve RTSDA.

The current version of RTSDA can only process log files. The method would be more efficient, if it could get the data directly from the source, but having a process log file with "1 line = 1 event" file structure as the input can be just as good. In such cases, the achieved delay from actual event occurrence (in the manufacturing line) until the event is displayed in the analysis software depends on the length of the time interval (waiting time between time points of checking the file) specified by the user. If the time interval is around 1 s or less, then the delay is not noticeable to the human eye. In case of "1 line = 1 trace" file structure, all the events of the case can be displayed only after the last event was finished. In such cases, the delay for the first event is nearly equal to the throughput time of the case. The log files of the examined manufacturing process originally have "1 line = 1 trace" file structure, so in a real industrial environment it would be more beneficial if the RTSDA could have direct access to each recorded data value, instead of reading in the data row by row from the output files.

The company that we are in cooperation with gave positive feedback about the applicability of this method. Investigations after further analysis have started. After that, the method will be tested in real situations.

6 Conclusion and Future Work

In this paper we introduced a new method to the issue of real-time process mining. The new method allows real-time tracking of causes for faulty products in a manufacturing process based on the available process log data. This method was named RTSDA.

RTSDA can give a comprehensive view of the production, and from the generated diagrams conclusions can be drawn about the state of the production tools and the possible sources of the errors. It can help process engineers and machine operators to design more efficient maintenance, reduce outage time, and increase production time, thus reducing the number of faulty products.

RTSDA was tested only on log data of one manufacturing process and only in test environment, so from the future works, testing RTSDA in real industrial environment and testing it with log data of other manufacturing processes is the most important task to accomplish. It is not easy, but it would bring the deficiencies of the method to the surface, thus helping to progress with the development of the method.

Acknowledgment. We acknowledge the financial support of Széchenyi 2020 under the **EFOP-3.6.1-16-2016-00015**. Supported by the **ÚNKP-18-2** New National Excellence Program of the Ministry of Human Capacities.

References

1. Zamantzas, C.: The real-time data analysis and decision system for particle flux detection in the LHC accelerator at CERN, No. CERN-THESIS-2006-037 (2006)
2. Surace, J., Laher, R., Masci, F., Grillmair, C., Helou, G.: The Palomar Transient Factory: High Quality Realtime Data Processing in a Cost-Constrained Environment, arXiv preprint arXiv:1501.06007 (2015)
3. Chen, G.J., et al.: Realtime data processing at Facebook. In: Proceedings of the 2016 International Conference on Management of Data, pp. 1087–1098 (2016)
4. Nasir, M.A.U.: Mining big and fast data: algorithms and optimizations for real-time data processing. Doctoral dissertation, KTH Royal Institute of Technology (2018)
5. Bertin, L., Borba, R.G., Krishnapillai, A., Tulchinsky, A.: U.S. Patent No. 9,971,777. U.S. Patent and Trademark Office, Washington, DC (2018)
6. van der Aalst, V.M.P., van Dongen, B.F., Günther, C.W., Rozinat, A., Verbeek, E., Weijters, T.: ProM: the process mining toolkit. BPM (Demos) **489**(31), 2 (2009)
7. Nagy, Z., Werner-Stark, Á., Dulai, T.: An industrial application using process mining to reduce the number of faulty products. In: Benczúr, A., et al. (eds.) ADBIS 2018. CCIS, vol. 909, pp. 352–363. Springer, Cham (2018). https://doi.org/10.1007/978-3-030-00063-9_33
8. Nagy, Z., Werner-Stark, A., Dulai, T.: Analysis of industrial logs to reduce the number of faulty products of manufacturing. In: Proceedings of National Conference on Economy-Informatics, pp. 53–57 (2018). ISBN 978-615-81098-1-9
9. Kannan, V., van der Aalst, V.M.P., Voorhoeve, M.: Formal modeling and analysis by simulation of data paths in digital document printers. In: Proceedings of the Nineth Workshop on the Practical Use of Coloured Petri Nets and CPN Tools (CPN 2008), vol. 588 (2008)

Pseudo-Relevance Feedback Based on Locally-Built Co-occurrence Graphs

Billel Aklouche[1,2,4]([envelope]) [iD], Ibrahim Bounhas[1,4] [iD], and Yahya Slimani[1,3,4] [iD]

[1] LISI Laboratory of Computer Science for Industrial System, INSAT,
Carthage University, Tunis, Tunisia
`billel.aklouche@ensi-uma.tn`, `bounhas.ibrahim@gmail.com`,
`yahya.slimani@gmail.com`
[2] National School of Computer Science (ENSI), La Manouba University,
Manouba, Tunisia
[3] Higher Institute of Multimedia Arts of Manouba (ISAMM),
La Manouba University, Manouba, Tunisia
[4] JARIR: Joint group for Artificial Reasoning and Information Retrieval,
Manouba, Tunisia
`http://www.jarir.tn/`

Abstract. In Information Retrieval (IR), user queries are often too short, making the selection of relevant documents hard. Pseudo-relevance feedback (PRF) is an effective method to automatically expand the query with new terms using a set of pseudo-relevant documents. However, a main issue in PRF is the selection of good expansion terms that allow improving retrieval effectiveness. In this paper, we present a new PRF method based on locally-built term co-occurrence graphs. We use a context window-based approach to construct our term co-occurrence graphs over top pseudo-relevant documents. For expansion terms selection, we propose an adapted version of the BM25 model, which allows to measure term-term similarity in co-occurrence graphs. This measure has the advantage of selecting discriminant expansion terms that are semantically related to the query as a whole. We evaluate our PRF method using four TREC collections, including the standard TREC Robust04 collection and the newest TREC Washington Post collection. Experimental results show that our proposal outperforms competitive state-of-the-art baselines and achieves significant improvements.

Keywords: Query expansion · Pseudo-relevance feedback ·
Term co-occurrence graph · BM25 · Context window ·
Term's discriminative power

1 Introduction

The massive growth in data size has made retrieving relevant information a challenging task. When a user makes a query, it is essential to know exactly what he wants and then provide a valid response. However, the user's query is

© Springer Nature Switzerland AG 2019
T. Welzer et al. (Eds.): ADBIS 2019, LNCS 11695, pp. 105–119, 2019.
https://doi.org/10.1007/978-3-030-28730-6_7

often too short and usually omits important terms that allow the selection of relevant documents which satisfy his needs. To overcome this problem, query expansion (QE) is one of the main tasks in information retrieval (IR), which refers to techniques that reformulate the initial query by adding new terms that better express the user's information needs, in order to improve retrieval effectiveness [6].

Among query expansion techniques, pseudo-relevance feedback (PRF) is one of the most popular techniques that has been widely applied. Indeed, PRF is an effective automatic query expansion technique that relies on local analysis of top-ranked documents, which are assumed to be relevant, in order to extract expansion terms. A main challenging problem in PRF is the selection of good expansion terms, which allow the retrieval of relevant documents and avoid query drift. This latter results from the quality of the added terms, which may be unrelated to the original query, and thus do not improve, but harm, retrieval effectiveness.

In general, two main approaches are applied to select expansion terms from top-ranked documents [5]. The first approach relies on the analysis of the distributional differences of terms in the pseudo-relevant documents and in the entire collection. The second approach is based on the analysis of term association and co-occurrence relationships in pseudo-relevant documents. Indeed, the use of term co-occurrence statistics for QE has been extensively applied. In this approach, the co-occurrence of terms is used to express the semantic relationships between terms [6]. However, a fundamental problem with the co-occurrence-based approach is the selection of terms that are very frequent, and thus are unlikely to be discriminative. This limitation is primarily related to how the similarity between terms is measured [18]. Besides, terms' position and proximity, which help capture the context of terms, are usually not considered [6].

Several measures have been used to calculate the similarity between terms from co-occurrence statistics, such as Cosine similarity, Dice coefficient and Mutual Information [6]. The BM25 [22] probabilistic model is one of the most effective and robust IR models, which has been widely used to measure query-document similarity, especially in TREC experiments. The BM25 model is an extension of the Binary Independence Model (BIM) [21]; it reconsiders certain deficiencies of the model which is based solely on the presence or absence of the query terms in documents. It includes statistics about both terms and documents, integrating local and global frequency of terms and document length. Since its inception, a lot of research has been proposed presenting improvements and extensions to the model [3,11,13,14,20,23,25]. In this paper, we propose a PRF method that relies on local, query-specific, analysis of the top-ranked documents using an adapted version of BM25 to measure term similarity in co-occurrence graphs. Specifically, we explore the use of context window-based approach to construct the term co-occurrence graphs over pseudo-relevant documents. For selecting expansion terms, we propose an adapted version of BM25 that allows measuring term-term similarity in the constructed graphs. This measure has the advantage of selecting discriminative terms that are semantically related to the query as a whole, where we define a good expansion candidate as a term that frequently co-occurs with the query terms and has a relatively rare co-occurrence with other terms. That is, terms that tend to co-occur with many other terms are penalized.

We evaluate our method on four TREC collections, namely: the TREC-7 collection, the TREC-8 collection, the standard TREC Robust04 collection and the newest TREC Washington Post collection. Experimental results show that our proposal outperforms the baselines by significant margins.

The remainder of the paper is organized as follows. We provide an overview of some related work in Sect. 2. Section 3 outlines the proposed PRF method. Experimental design and results are presented in Sect. 4. Section 5 concludes the paper and provides insights for future work.

2   Related Work

Providing a relevant response to the user's query has always been challenging, as this latter is usually insufficient to allow the selection of documents that meet the user's needs. To face this problem, several methods of query expansion have been proposed [6]. Two major categories can encompass the wide range of query expansion methods, namely global and local methods.

Global methods allow expanding the query independently from the initial retrieval results. Expansion terms are selected by analyzing the entire document collection (corpus) or from external resources. Corpus-based approaches rely on statistical analysis in order to discover terms associations and co-occurrence relationships [29]. On the other hand, external resources like dictionaries and thesauri (e.g. WordNet) are used to find expansion terms that are the most similar to the initial query [6].

Unlike global methods, local QE methods use the top-ranked documents resulting from an initial retrieval round [29]. Two techniques can be distinguished to select the documents from which expansion terms are extracted: relevance feedback and pseudo-relevance feedback. In the former, the user is involved in the selection process of relevant documents, where he examines the query results and manually specifies relevant documents. Most of relevance feedback approaches are inspired and based on the Rocchio algorithm [24]. A detailed discussion on relevance feedback can be found in [10]. On the other hand, pseudo-relevance feedback is an effective QE technique derived from relevance feedback, where it is assumed that the top n ranked documents in the returned results of the initial query are relevant [4]. The process is done automatically and transparently to the user. This technique was first introduced in [8], which made use of it in a probabilistic model. Indeed, most of local methods make use of the pseudo-relevance feedback technique [15]. Xu and Croft [28] have shown generally better results by using a local QE approach compared to a global approach. However, a main challenge in PRF is the selection of good expansion terms given a set of pseudo-relevant documents.

Xu and Croft [29] proposed an automatic query expansion technique called LCA (Local Context Analysis), which generates expansion terms based on their co-occurrence with the query terms in pseudo-relevant documents. They demonstrated the effectiveness of this technique in different languages. Xu et al. [30] presented a method using Wikipedia as a large document collection for generating expansion terms. They categorized the queries based on their relationship to Wikipedia subjects using the article's titles. Zamani et al. [32] considered PRF as a recommendation task. They proposed a matrix factorization technique for recommending useful expansion terms. In a similar work, Valcarce et al. [26] explored the use of linear methods and employed an inter-term similarity matrix to obtain expanded queries. Clinchant and Gaussier [7] conducted a theoretical analysis of various widely used PRF models. They showed that several models tend to select non-discriminative terms, which are very frequent and therefore are unlikely to improve retrieval effectiveness. They argue that the PRF model should attribute higher scores to terms with lower frequency in the collection. Similarly, Cao et al. [5] observed that many of the most frequent terms in pseudo-relevant documents do not improve, but hurt retrieval effectiveness. In this study, authors proposed to select good expansion terms by integrating a term classification process. In our work, we attempt to select the most discriminative terms by penalizing terms that co-occur with many other terms.

Recently, automatic query expansion methods based on word embedding [1,9, 31] have shown an interesting improvement on retrieval effectiveness by exploring word relationships from word vectors. Indeed, term co-occurrence statistics are used to learn word vector representations based on word embedding algorithms such as Word2vec [17] and Glove [19]. Diaz et al. [9] used Word2vec to learn word vector representations over top 1000 ranked documents. Similarly, Zamani and Croft [31] used top 1000 ranked documents and proposed to train word vectors in an offline setting. In these methods, the co-occurrence of terms in the same context window is used to produce word vectors [31]. Afterward, the obtained vectors are used to select terms that are semantically related to the query. We use the same approach, i.e. a context window-based approach over pseudo-relevant documents in order to construct our term co-occurrence graphs.

3   Proposed Method

In this section, we describe our PRF method. Figure 1 illustrates the general architecture of the proposed method. We follow two steps to obtain semantically related terms to the query. First, term co-occurrence graphs are constructed over the pseudo-relevant documents for each query using a context window-based approach. Then, using an adapted version of BM25 that allows measuring the similarity between terms from the constructed graphs, the candidate expansion terms are sorted according to their similarity score with the entire query.

3.1   Term Co-occurrence Graphs

Exploiting co-occurrence data has been widely used in IR. The context window-based co-occurrence approach, a simple yet effective approach [27], has proven

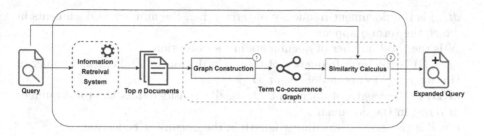

Fig. 1. General architecture of the proposed method.

useful in multiple Natural Language Processing (NLP) and IR applications such as building co-occurrence matrices in Glove [19] and learning word vector representations in Word2vec [16]. Given a target term, the terms on its left and right within a specified window are the co-occurring contextual terms. The co-occurrence of terms is used to capture and express the semantic relationships between them.

In our method, we consider co-occurrences at the sentence level. That is, two terms are considered as co-occurring if they are in the same sentence and within the specified context window. For instance, given the sentence "The ADBIS conference will take place in Slovenia." and taking "conference" as the target term with a window-size equal to 2, its context terms will be "The", "ADBIS", "will" and "take". A sliding-window is applied over the sentences to measure the co-occurrence of terms and construct the corresponding graphs. In our experiments we explored different settings of the window-size parameter to study its impact on retrieval effectiveness.

For graph building, terms are represented as nodes in the graph and an edge between two nodes is created once the two corresponding terms co-occur together in a sentence within the specified window-size. The edges of the graph are undirected and the weight of the edge corresponds to the total number of times the linked terms co-occur in the set of documents on which the graph was built.

3.2   Local Co-occurrence Graphs-Based PRF

The BM25 model calculates the similarity score of document D to query Q as follows [22]:

$$BM25(Q, D) = \sum_{q \in Q} IDF(q) \times \frac{(k_1 + 1)tf(q, D)}{k_1(1 - b + b\frac{dl(D)}{avgdl}) + tf(q, D)} \quad (1)$$

where:

– $IDF(q)$ is the Inverse Term Frequency of term q and it is calculated as follows:

$$IDF(q) = log\frac{N - df(q) + 0.5}{df(q) + 0.5} \quad (2)$$

- $df(q)$ is the document frequency of term q, i.e., the number of documents in which the term q appears.
- N is the total number of documents in the collection.
- $tf(q, D)$ is the term frequency of term q in document D, i.e., the number of times term q occurs in document D.
- $dl(D)$ is the length of document D, usually considered as the total number of terms in the document.
- $avgdl$ is the average document length in the document collection.
- k_1 and b are free tuning parameters.

BM25 integrates term distribution statistics in both individual documents and across the entire collection. In this model, it is assumed that good document descriptors and discriminators are terms that occur frequently in this document and are relatively rare in the rest of the corpus [12]. Following these assumptions, we define good expansion candidates as terms that frequently co-occur with the query terms and have a relatively rare co-occurrence with other terms.

To get an expended query, we first need to project the query onto the corresponding constructed graph to calculate the similarity score of each candidate term. A candidate term is a term that has at least one co-occurrence relationship with one of the query terms. Therefore, the top n terms with the highest scores are added to the initial query.

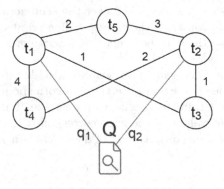

Fig. 2. Example of query projection on co-occurrence sub-graph. t_1, t_2 are the query terms and t_3, t_4, t_5 are expansion candidates.

Let G be a co-occurrence graph built on the set of top k ranked documents from the initial retrieval results. Given a query Q with terms $q1, ..., qm$, the projection of query Q on graph G is the set of nodes C, where $C = \{t_1, ..., t_m\}$. An example of query projection on a graph is depicted in Fig. 2. The proposed adapted version of BM25, which we refer to as $BM25_{cog}$ (BM25 for co-occurrence graphs), allows measuring the similarity between a candidate term and the whole query. The similarity function is the sum of the similarity scores of the candidate term to each of the query terms. $BM25_{cog}$ calculates the similarity score of a candidate node t_c to the set of nodes C as follows:

$$BM25_{cog}(C, t_c) = \sum_{t_i \in C} INF(t_i) \times \frac{(k_1 + 1)e(t_i, t_c)}{k_1(1 - b + b\frac{sum\_e(t_c)}{avgsum\_e}) + e(t_i, t_c)} \quad (3)$$

where:

- $INF(t_i)$ indicates the Inverse Node Frequency. This is analogous to the IDF factor in BM25 and it is computed as follows:

$$INF(t_i) = log\frac{N - co\_degree(t_i) + 0.5}{co\_degree(t_i) + 0.5} \quad (4)$$

- $co\_degree(t_i)$ denotes the number of nodes co-occurring with t_i. In document-term matrix, this is similar to the number of documents which contain a given term.
- N is the total number of nodes in the graph, which is analogous to the total number of documents in the collection.
- $e(t_i, t_c)$ counts the number of co-occurrence of t_i with t_c. That is, the weight of the edge linking the two corresponding nodes. This is analogous to the number of times a term occurs in a document.
- $sum\_e(t_c)$ is the sum of the weights of the edges of the node t_c in the graph. In document-query matching, this is similar to the length of a document.
- $avgsum\_e$ is the average of the $sum\_e(t_c)$ parameter computed for all the nodes in the graph. This is analogous to the $avgdl$ parameter in the original BM25 formula.
- k_1 and b are the usual BM25 free tuning parameters.

This proposed measure has at least the following advantages. First, it can be used for both one-to-one and one-to-many associations. In addition, the INF factor enables to measure the terms' discriminative power: terms that tend to co-occur with many other terms are penalized. Besides, $BM25_{cog}$ has two free tuning parameters, which can be adjusted to improve effectiveness. Furthermore, the proposed PRF method can be implemented on top of any retrieval model.

4 Experiments

In this section, we evaluate the performance of our proposed PRF method against state-of-the-art baselines. We first present our test collections and describe our experimental settings as well as the evaluation metrics. Then we report and discuss the obtained results.

4.1 Experimental Setup

We conducted our experiments on four TREC (Text REtrieval Conference) collections. The first one is the TREC-7 collection, with 50 queries. This test collection was used in the TREC 1998 ad-hoc retrieval task. The second collection is the TREC-8 collection, with 50 queries, which was used in the TREC 1999 ad-hoc retrieval task. The third one is the standard TREC Robust04 collection, with 249 queries. It was used in the TREC 2004 Robust Track, which focused on poor performing queries. The three collections are composed of more than 500k documents, which consist of high-quality newswire articles collected from various newspapers (Financial Times, Federal Register 94, FBIS and LA Times). Finally, we also used the newest TREC Washington Post collection[1], with 50 queries, which was provided by the TREC 2018 Common Core Track. It consists of more than 600k news articles and blog posts published between 2012 and 2017 by Washington Post. The statistics of these four TREC collections are summarized in Table 1.

Table 1. TREC collections statistics.

Collection	Document set	#docs	Size	#query	#qrels
TREC-7	TREC Disks 4 & 5	528k	1.9 GB	50	4,674
TREC-8	(minus Congressional Record)			50	4,728
Robust04				249	17,412
WAPOST	TREC Washington Post Corpus	608k	6.9 GB	50	3,948

All experiments were carried out using the Terrier platform[2]. We used only the title field of the TREC topics as queries in our experiments. In terms of preprocessing, stopwords were removed using the Terrier's standard stopword list. Documents and queries were stemmed using the Porter stemmer. We varied the number of feedback documents (for PRF methods) between {100, 200, 1000} and we empirically set the number of expansion terms to 10.

For the construction of term co-occurrence graphs, we need to specify the size of the context window. We swept the value of the window-size parameter from {2, 3, ..., 10}. Besides, we tested a dynamic window-size equal to sentence length. Best results were obtained using a window-size in {2, 3, 4}.

We employed the BM25 model for performing retrieval. Top 1000 ranked documents are retrieved for each query. To evaluate the effectiveness of our PRF method, we use three standard evaluation metrics: mean average precision (MAP), normalized discounted cumulative gain (nDCG) and precision at top 10 ranked documents (P@10). Statistical significance tests in terms of MAP are performed using the two-tailed paired t-test at a 95% confidence level (i.e., $p\_value < 0.05$).

4.2   Results and Discussion

The baseline of our experiments is the state-of-the-art BM25 model, i.e., using the original queries without feedback. Besides, we also consider three standard

---

[1] https://trec.nist.gov/data/wapost/.
[2] http://terrier.org/.

PRF models, namely: Bo1, Bo2 and KL [2]. The experimental results on the four test collections regarding MAP, nDCG and P@10 are summarized in Tables 2, 3, 4 and 5. According to these tables, the proposed PRF method outperforms the BM25 baseline (without query expansion) in terms of MAP and nDCG in all cases. On all the four collections, the MAP improvements are always statistically significant with up to 12.9% improvement. In terms of precision, we can see that our PRF method also outperforms the unexpanded baseline in all the collections except in one case (TREC-7 with 1000 feedback documents). These results show that our PRF method, which selects semantically related terms to the whole query, leads to improvement in retrieval effectiveness of the state-of-the-art BM25 model w.r.t. using various collections and different number of feedback documents.

By comparing our method and the PRF baselines, we can see that it performs better in terms of MAP and nDCG in the majority of cases. According to the results, the baselines are outperformed by our method on three out of four collections. For instance, our PRF method outperforms all the other methods in terms of MAP on the standard Robust04 collection in all cases. As for the precision, we can see that our method outperforms all the baselines in all the collections except in one case (TREC Washington Post with 200 feedback documents). Furthermore, by comparing the PRF baselines and the unexpanded baseline, it can be observed that the precision is hurt in the majority of cases. This shows that our proposal is better at generating discriminant and good expansion terms, thus improving the precision at top-ranked documents.

Regarding the results when varying the number of feedback documents, we can observe that the performance of the baselines declines monotonically as the number of feedback documents increases. In contrast, the results obtained by our PRF method are quite stable. This shows the ability of our proposal to penalize and filter out bad expansion terms, especially when the data get noisier.

To further investigate the effectiveness of our PRF method, We plot the interpolated precision-recall curves in Figs. 3, 4, 5 and 6. As it can be seen, the proposed method outperforms the baselines in the majority of cases.

Parameter Sensitivity. As discussed in Sect. 3, $BM25_{cog}$ has two free tuning parameters, k_1 and b, which can be adjusted to improve effectiveness. In the next set of experiments, we study the sensitivity of $BM25_{cog}$ to k_1 and b. We tuned the parameter k_1 from 0.1 to 3.0 and the parameter b from 0.1 to 0.9, in increments of 0.1. We report the optimal settings of k_1 and b on the four test collections in Table 6. According to this table, the optimal parameter settings vary across collections, and thus are collection-dependent, which is similar to the behavior of the BM25 model [13].

Table 2. Retrieval results on TREC Robust04 collection. The superscript * indicates that the MAP improvements over the BM25 model are statistically significant (t-test with $p\_value < 0.05$). The highest value in each row is marked in bold.

#docs	Metric	BM25	PRF_BM25_{cog}	PRF_Bo1	PRF_Bo2	PRF_KL
1000	MAP	0.2363	**0.2536***	0.2044	0.1522	0.2072
	nDCG	0.5081	**0.5259**	0.4731	0.3995	0.4748
	P@10	0.4100	**0.4189**	0.3655	0.2920	0.3715
200	MAP	0.2363	**0.2560***	0.2332	0.2097	0.2332
	nDCG	0.5081	**0.5294**	0.5115	0.4770	0.5116
	P@10	0.4100	**0.4301**	0.3807	0.3602	0.3924
100	MAP	0.2363	**0.2567***	0.2532	0.2387	0.2509
	nDCG	0.5081	0.5279	**0.5328**	0.5099	0.5292
	P@10	0.4100	**0.4245**	0.3984	0.3924	0.4068

Table 3. Retrieval results on TREC Washington Post collection. The superscript * indicates that the MAP improvements over the BM25 model are statistically significant (t-test with $p\_value < 0.05$). The highest value in each row is marked in bold.

#docs	Metric	BM25	PRF_BM25_{cog}	PRF_Bo1	PRF_Bo2	PRF_KL
1000	MAP	0.2374	**0.2596***	0.2147	0.1657	0.2190
	nDCG	0.5062	**0.5360**	0.4765	0.4062	0.4834
	P@10	0.4240	**0.4360**	0.3700	0.3080	0.3900
200	MAP	0.2374	0.2537*	0.2566*	0.2407	**0.2615***
	nDCG	0.5062	0.5233	0.5228	0.5048	**0.5253**
	P@10	0.4240	0.4340	0.4300	0.3880	**0.4400**
100	MAP	0.2374	0.2546*	0.2648*	0.2660*	**0.2663***
	nDCG	0.5062	0.5289	**0.5339**	0.5203	0.5281
	P@10	0.4240	**0.4320**	0.4140	0.4100	0.4200

Table 4. Retrieval results on TREC-7 collection. The superscript * indicates that the MAP improvements over the BM25 model are statistically significant (t-test with $p\_value < 0.05$). The highest value in each row is marked in bold.

#docs	Metric	BM25	PRF_BM25_{cog}	PRF_Bo1	PRF_Bo2	PRF_KL
1000	MAP	0.1728	**0.1844***	0.1531	0.0943	0.1543
	nDCG	0.4312	**0.4520**	0.4075	0.3256	0.4060
	P@10	**0.3920**	0.3860	0.3400	0.2400	0.3620
200	MAP	0.1728	**0.1830***	0.1828*	0.1544	0.1793
	nDCG	0.4312	**0.4530**	0.4484	0.4116	0.4455
	P@10	0.3920	**0.4040**	0.3640	0.3540	0.3720
100	MAP	0.1728	0.1951*	**0.1959***	0.1789	0.1905*
	nDCG	0.4312	0.4639	**0.4686**	0.4373	0.4596
	P@10	**0.3920**	**0.3920**	0.3700	0.3760	0.3760

Table 5. Retrieval results on TREC-8 collection. The superscript * indicates that the MAP improvements over the BM25 model are statistically significant (t-test with $p\_value < 0.05$). The highest value in each row is marked in bold.

#docs	Metric	BM25	PRF_BM25$_{cog}$	PRF_Bo1	PRF_Bo2	PRF_KL
1000	MAP	0.2293	**0.2445***	0.1963	0.1459	0.1988
	nDCG	0.5006	**0.5214**	0.4761	0.4055	0.4735
	P@10	0.4360	**0.4680**	0.4100	0.3380	0.4020
200	MAP	0.2293	**0.2471***	0.2375	0.2123	0.2391
	nDCG	0.5006	**0.5290**	0.5263	0.4868	0.5246
	P@10	0.4360	**0.4660**	0.4300	0.3920	0.4420
100	MAP	0.2293	0.2501*	0.2521*	0.2337	**0.2533***
	nDCG	0.5006	0.5288	**0.5440**	0.5200	0.5427
	P@10	0.4360	**0.4600**	0.4440	0.4100	0.4600

Table 6. Optimal settings of k_1 and b in BM25$_{cog}$.

Parameter	Robust04	WAPOST	TREC-7	TREC-8
k_1	1.5	0.8	0.5	1.1
b	0.3	0.1	0.1	0.1

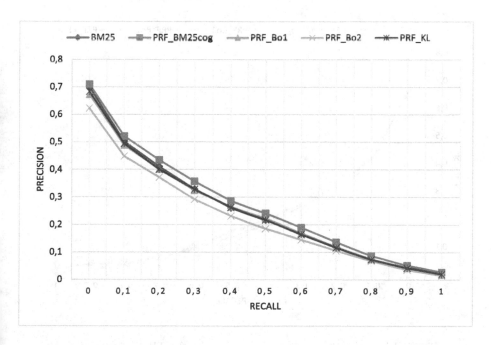

Fig. 3. Interpolated precision-recall curves for the Robust04 collection.

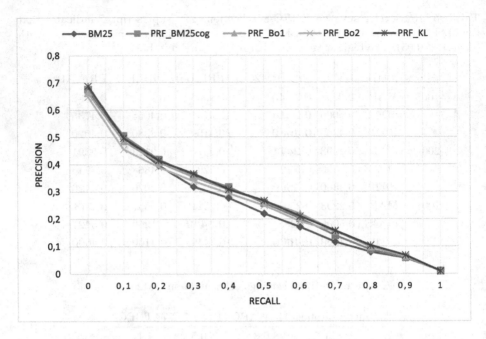

Fig. 4. Interpolated precision-recall curves for the Washington Post collection.

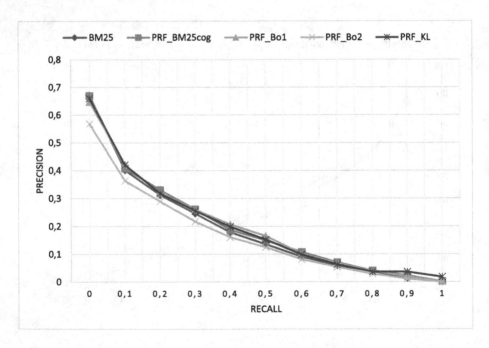

Fig. 5. Interpolated precision-recall curves for the TREC-7 collection.

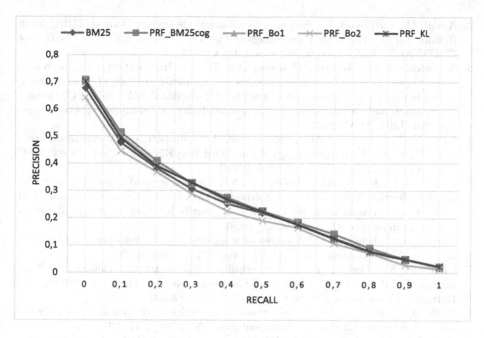

Fig. 6. Interpolated precision-recall curves for the TREC-8 collection.

5   Conclusion and Future Work

In this paper, we proposed a new pseudo-relevance feedback method based on term co-occurrence graphs, which are locally-built over top pseudo-relevant documents. We proposed an adapted version of the BM25 model, which allows to measure the similarity between terms in co-occurrence graphs and to evaluate the discriminative power of terms.

Experimental results on four TREC collections show that our proposal significantly outperforms state-of-the-art baselines in nearly all cases. We showed that the proposed PRF method is able to select good expansion terms and can filter out non-discriminative ones.

As future work, we plan to evaluate our PRF method on other state-of-the-art retrieval models (e.g., the language model and the divergence from randomness model) and compare it to other PRF methods, especially word embedding-based PRF methods. Moreover, we intend to investigate the use of external resources such as Wikipedia for the construction of co-occurrence graphs. Another promising future research direction is to study the use of the proposed similarity measure in other IR tasks, such as Word Sense Disambiguation (WSD) and Query Reweighing.

References

1. Aklouche, B., Bounhas, I., Slimani, Y.: Query expansion based on NLP and word embeddings. In: Proceedings of the Twenty-Seventh Text Retrieval Conference (TREC 2018), Gaithersburg, Maryland, USA (2018)

2. Amati, G., Carpineto, C., Romano, G.: Fondazione ugo bordoni at TREC 2003: Robust and web track. In: Proceedings of The Twelfth Text REtrieval Conference (TREC 2003), Gaithersburg, Maryland, USA (2003)
3. Ariannezhad, M., Montazeralghaem, A., Zamani, H., Shakery, A.: Improving retrieval performance for verbose queries via axiomatic analysis of term discrimination heuristic. In: Proceedings of the 40th International ACM SIGIR Conference on Research and Development in Information Retrieval, Shinjuku, Tokyo, Japan, pp. 1201–1204. ACM (2017)
4. Buckley, C., Salton, G., Allan, J., Singhal, A.: Automatic query expansion using SMART: TREC 3. In: Proceedings of The Third Text REtrieval Conference (TREC 1994), Gaithersburg, Maryland, USA (1994)
5. Cao, G., Nie, J.Y., Gao, J., Robertson, S.E.: Selecting good expansion terms for pseudo-relevance feedback. In: Proceedings of the 31st Annual International ACM SIGIR Conference on Research and Development in Information Retrieval, Singapore, Singapore, pp. 243–250. ACM (2008)
6. Carpineto, C., Romano, G.: A survey of automatic query expansion in information retrieval. ACM Comput. Surv. (CSUR) 44(1), 1:1–1:50 (2012)
7. Clinchant, S., Gaussier, E.: A theoretical analysis of pseudo-relevance feedback models. In: Proceedings of the 2013 Conference on the Theory of Information Retrieval, Copenhagen, Denmark, pp. 6–13. ACM (2013)
8. Croft, W.B., Harper, D.J.: Using probabilistic models of document retrieval without relevance information. J. Documentation 35(4), 285–295 (1979)
9. Diaz, F., Mitra, B., Craswell, N.: Query expansion with locally-trained word embeddings. In: Proceedings of the 54th Annual Meeting of the Association for Computational Linguistics. ACL, Berlin, Germany (2016)
10. Fang, H., Zhai, C.: Web search relevance feedback. In: Liu, L., Özsu, M.T. (eds.) Encyclopedia of Database Systems, pp. 3493–3497. Springer, Boston (2009). https://doi.org/10.1007/978-0-387-39940-9
11. He, B., Huang, J.X., Zhou, X.: Modeling term proximity for probabilistic information retrieval models. Inf. Sci. 181(14), 3017–3031 (2011)
12. Jones, K.S., Walker, S., Robertson, S.E.: A probabilistic model of information retrieval: development and comparative experiments: Part 2. Inf. Process. Manag. 36(6), 809–840 (2000)
13. Lv, Y., Zhai, C.: Lower-bounding term frequency normalization. In: Proceedings of the 20th ACM International Conference on Information and Knowledge Management, Glasgow, Scotland, UK, pp. 7–16. ACM (2011)
14. Lv, Y., Zhai, C.: When documents are very long, BM25 fails! In: Proceedings of the 34th International ACM SIGIR Conference on Research and Development in Information Retrieval, Beijing, China, pp. 1103–1104. ACM (2011)
15. Manning, C.D., Raghavan, P., Schütze, H.: Introduction to Information Retrieval. Cambridge University Press, UK (2008)
16. Mikolov, T., Chen, K., Corrado, G., Dean, J.: Efficient estimation of word representations in vector space. In: International Conference on Learning Representations Workshop Papers (2013)
17. Mikolov, T., Sutskever, I., Chen, K., Corrado, G.S., Dean, J.: Distributed representations of words and phrases and their compositionality. In: Advances in Neural Information Processing Systems, Proceedings of the 26th International Conference on Neural Information Processing Systems, Lake Tahoe, Nevada, United States, pp. 3111–3119 (2013)
18. Peat, H.J., Willett, P.: The limitations of term co-occurrence data for query expansion in document retrieval systems. J. Am. Soc. Inf. Sci. 42(5), 378–383 (1991)

19. Pennington, J., Socher, R., Manning, C.D.: Glove: global vectors for word representation. In: Proceedings of the 2014 Conference on Empirical Methods in Natural Language Processing (EMNLP), Doha, Qatar, pp. 1532–1543. ACL (2014)
20. Rasolofo, Y., Savoy, J.: Term proximity scoring for keyword-based retrieval systems. In: Sebastiani, F. (ed.) ECIR 2003. LNCS, vol. 2633, pp. 207–218. Springer, Heidelberg (2003). https://doi.org/10.1007/3-540-36618-0_15
21. Robertson, S.E., Jones, K.S.: Relevance weighting of search terms. J. Am. Soc. Inf. Sci. **27**(3), 129–146 (1976)
22. Robertson, S.E., Walker, S.: Some simple effective approximations to the 2-poisson model for probabilistic weighted retrieval. In: Proceedings of the 17th Annual International ACM SIGIR Conference on Research and Development in Information Retrieval, Dublin, Ireland, pp. 232–241. Springer-Verlag, New York, Inc. (1994)
23. Robertson, S.E., Zaragoza, H.: The probabilistic relevance framework: BM25 and beyond. Found. Trends Inf. Retrieval **3**(4), 333–389 (2009)
24. Rocchio, J.J.: Relevance feedback in information retrieval. In: Salton, G. (ed.) The SMART retrieval System: Experiments in Automatic Document Processing, pp. 313–323. Prentice-Hall, Englewood Cliffs (1971)
25. Song, R., Taylor, M.J., Wen, J.-R., Hon, H.-W., Yu, Y.: Viewing term proximity from a different perspective. In: Macdonald, C., Ounis, I., Plachouras, V., Ruthven, I., White, R.W. (eds.) ECIR 2008. LNCS, vol. 4956, pp. 346–357. Springer, Heidelberg (2008). https://doi.org/10.1007/978-3-540-78646-7_32
26. Valcarce, D., Parapar, J., Barreiro, A.: Lime: linear methods for pseudo-relevance feedback. In: Proceedings of the 33rd Annual ACM Symposium on Applied Computing, Pau, France, pp. 678–687. ACM (2018)
27. Wei, X., Croft, W.B.: Modeling term associations for ad-hoc retrieval performance within language modeling framework. In: Amati, G., Carpineto, C., Romano, G. (eds.) ECIR 2007. LNCS, vol. 4425, pp. 52–63. Springer, Heidelberg (2007). https://doi.org/10.1007/978-3-540-71496-5_8
28. Xu, J., Croft, W.B.: Query expansion using local and global document analysis. In: Proceedings of the 19th Annual International ACM SIGIR Conference on Research and Development in Information Retrieval, Zurich, Switzerland, pp. 4–11. ACM (1996)
29. Xu, J., Croft, W.B.: Improving the effectiveness of information retrieval with local context analysis. ACM Trans. Inf. Syst. (TOIS) **18**(1), 79–112 (2000)
30. Xu, Y., Jones, G.J., Wang, B.: Query dependent pseudo-relevance feedback based on Wikipedia. In: Proceedings of the 32nd International ACM SIGIR Conference on Research and Development in Information Retrieval, Boston, MA, USA, pp. 59–66. ACM (2009)
31. Zamani, H., Croft, W.B.: Relevance-based word embedding. In: Proceedings of the 40th International ACM SIGIR Conference on Research and Development in Information Retrieval, Shinjuku, Tokyo, Japan, pp. 505–514. ACM (2017)
32. Zamani, H., Dadashkarimi, J., Shakery, A., Croft, W.B.: Pseudo-relevance feedback based on matrix factorization. In: Proceedings of the 25th ACM International on Conference on Information and Knowledge Management, Indianapolis, Indiana, USA, pp. 1483–1492. ACM (2016)

Big Data

Workload-Awareness in a NoSQL-Based Triplestore

Luiz Henrique Zambom Santana[✉] and Ronaldo dos Santos Mello

Universidade Federal de Santa Catarina, Florianópolis, Brazil
luiz.santana@posgrad.ufsc.br, r.mello@ufsc.br

Abstract. RDF and SPARQL are increasingly used in a broad range of information management scenarios. Scalable processing of SPARQL queries has been the main goal for virtually all the recently proposed RDF triplestores. Workload-awareness is considered an important feature for the current generation of triplestores. This paper presents *WA-RDF*, a middleware that addresses workload-adaptive management of large RDF graphs. These graphs are stored into NoSQL databases, which provide high availability and scalability. The focus of this paper is on the Workload-Aware component (WAc) of WA-RDF. WAc was developed to avoid data fragmentation, improve data placement and reduce the intermediate results. Our experimental evaluation shows that the solution is promising, outperforming a recent baseline.

Keywords: RDF · SPARQL · NoSQL · Triplestore · Workload

1 Introduction

In the last decades, RDF, the standardized data model that - along with other technologies like OWL, RDFS, and SPARQL - grounds the vision of the *Semantic Web*, was affected by a wide range of data management problems, like data integration, search optimization, data representation, and information extraction. The main reason for that is the current scale of the applications (*e.g.*, smart cities, sensor networks, and healthcare), which generates huge datasets and need to efficiently store massive RDF graphs that go beyond the processing capacities of existing RDF storage systems. This scenario includes innovations in the frontier of Semantic Web research fields. For example, original works have been proposed semantic trajectories of moving objects [7] and semantic interoperability for healthcare data [12]. The scale of these domains raises the need for triplestores that take advantage of NoSQL databases [4] to store large volumes of RDF data, giving their scaling capabilities.

Workload-awareness is recognized as an important feature for the development of scalable triplestores, as described by Aluç, Ozsu, and Daudjee [2]. The lack of awareness can lead to important performance barriers: *(i)* naive data fragmentation; *(ii)* poor data localization; and *(iii)* unnecessary large intermediate results. Thus, although many triplestores have been proposed in the last

© Springer Nature Switzerland AG 2019
T. Welzer et al. (Eds.): ADBIS 2019, LNCS 11695, pp. 123–138, 2019.
https://doi.org/10.1007/978-3-030-28730-6_8

years, the current solutions fall short because of the RDF structural diversity of SPARQL workloads.

This paper presents *WA-RDF*, a triplestore composed of a middleware and multiple NoSQL databases. The middleware maps pieces of an RDF graph into NoSQL databases with different data models. The main contributions of this paper is a workload-aware component that supports fragmentation, mapping, partitioning, querying and caching processes. Our strong point is the ability to process queries over large RDF graphs stored on multiple NoSQL database servers with zero or a subtle amount of data joining cost. An experimental evaluation shows that WA-RDF scales well, being able to process huge RDF datasets.

The rest of the paper is organized as follows. Section 2 presents the background and related work. Section 3 details the WA-RDF approach. Section 4 reports the experimental evaluation and Sect. 5 concludes the paper.

2 Background and Related Work

The most important pillar of this work is the *Semantic Web*. The effort of developing the Semantic Web was harvested mainly in terms of well-established standards for expressing shared meaning, defined by the WWW Consortium (W3C), like the *Resource Description Framework (RDF)* and the *Simple Protocol and RDF Query Language (SPARQL)*.

RDF is a format to express triples that define a relationship between resources. SPARQL is a query language for RDF data. A query statement in SPARQL consists of triple patterns, conjunctions, disjunctions, and optional patterns, and a *triple pattern*, in particular, defines what have to be searched, *i.e.*, a query predicate. SPARQL queries can be categorized by its shape into *star* and *chain* queries, which heavily influence query performance [11]. Star shape is characterized by subject-subject (S-S), or object-object (O-O) joins. Chain shape is formed by object-subject (O-S) joins, *i.e.*, the join variable is in the subject location in one pattern, and on object location in the other one. Other complex query structures are usually a composition of these two shapes.

The most important limitation of the triplestores proposed in the recent years is caused by the RDF structural diversity and the dynamism of SPARQL workloads. As discussed by Aluç, Güneş and Özsu [2], the lack of SPARQL workload awareness can lead to naive data fragmentation, poor data localization and unnecessarily large intermediate results.

Since the advent of the RDF format, the RDF storage has been evolving to meet the requirements of the Computer Science trends. This subject has been the focus of a great variety of works including relational databases, peer-to-peer systems and, more recently, NoSQL databases. Table 1 shows some related works in terms of type of awareness, storage, fragmentation (Frag.), localization (Loc.) and intermediate results (Inter.). The idea is not to present an exhaustive catalog of the newly proposed triplestores. Instead, we aim to compare solutions and understand how they interact and complement each other. Two main concepts guided the choice for these solutions: *workload-awareness* and *storage* technology.

The *AdaptRDF* approach [9] consists firstly of a vertical partitioning phase that uses the workload information to generate an efficient relational schema that reduces the number of joins. Secondly, in the adjustment phase, any change in the workload is considered to create a sequence of pivoting and unpivoting operations to adapt the underlying schema and maintain the efficiency of the query processing. *WARP* [6] presents a fragmentation and replication method on top of graph-based partitioning that takes the workload into account to create a cost-aware query optimization and provide efficient execution. *Cerise* [8] is a distributed RDF data store that adapts the underlying storage and query execution according to the history of queries. It co-locates (on the same data segment) data that are accessed frequently together to reduce overall disk and network latency. *Partout* [5] uses a fragmentation procedure followed by an allocation process that aims to allocate the fragments that are used together as close as possible and to distribute the fragments evenly. The allocation algorithm sorts fragments in descending order by the load, assigning a fragment to the most beneficial host. The benefit of allocating a new fragment to a host is inversely proportional to the host's current loaded and directly proportional to the number of already allocated fragments that join with the new fragment. The fragmentation problem is discussed by *WARP* and *Partout*, which define expanded fragments to avoid unnecessary joins. On the other hand, the localization is the main contribution of *AdaptRDF* and *Cerise*. Their monitor query changes to better rearrange the localization of the underlying schema.

Table 1. Related Workload-aware triplestores

Work	Awareness	Storage	Frag.	Loc.	Inter.
AdaptRDF (2012)	Query stream	RDBMSs	No	Yes	No
WARP (2013)	Logs	Independent	Yes	No	No
Cerise (2014)	History	Native	No	Yes	No
Partout (2014)	History	Native	Yes	Yes	No
S2RDF (2016)	Star query statistics	Document and Graph	No	No	Yes

Another critical issue in this paper is NoSQL databases, specifically, NoSQL-based triplestores. Among the recent works that propose triplestores and use NoSQL as data storage, we highlight *S2RDF* [11], a scalable query processor. *S2RDF* proposes a *Spark*-based SPARQL query processor that offers speedy response time by extending the vertical partitioning. Based on the selectivity of triple patterns, S2RDF tries to reduce the intermediate results of the queries by optimizing the join order.

Different from related work, our approach, called *WA-RDF*, considers all of these features (fragmentation, localization and intermediate results) to provide an efficient solution for RDF data management. It is detailed in the following.

3   WA-RDF

WA-RDF is a middleware for storing RDF data into multiple NoSQL databases. A workload-aware approach is the cornerstone of WA-RDF. Based on it, WA-RDF decides where and how to place data, which influences fragmentation, mapping, partitioning, and querying strategies. WA-RDF is an evolution of *Rendezvous* [10]. This new version includes several new features, like query relevance-based workload monitoring, active data allocation, merging of fragments, support to update and deletion, replacement of a columnar to a graph NoSQL database for improving chain queries performance, map-reduce-based fragmentation, and workload-aware caching. The focus of this paper is on the WA-RDF *workload-aware component (WAc)*. Thus, first of all, it is essential to explain how this component monitors the workload.

Figure 1 presents a partial architecture of WA-RDF. WAc is in its core, and it is related to *Dictionary, Cache, Partitioner*, and *Fragmenter*. Also, WA-RDF is connected to NoSQL databases.

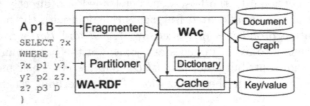

Fig. 1. The WAc component of WA-RDF

WA-RDF registers information about the triple patterns of each incoming SPARQL query. For instance, if we have a SPARQL query SELECT ?x WHERE {?x p1 y?. y? p2 z?. z? p3 D} (q_1) at time t_1, and we have a query SELECT ?x WHERE {?x p1 y?. x? p4 a?. x? p5 b?} (q_2) at time t_2 ($t_1 < t_2$), WA-RDF stores, with the aid of WAc, information about the triple patterns, the shape of the file, the time that the query was most recently received, and how many times WA-RDF received this query, as illustrated in Fig. 2. WAc is composed of the following in-memory structures: *QMap* (registers the queries); *TPMap* (registers the triple patterns); *TP2Q* (maps triple patterns to queries and identifies the workload version for each triple pattern); and *Q2TP* (maps queries to triple patterns).

Next sections detail how WAc can reduce data fragmentation to improve data localization and avoid unnecessary intermediate results.

3.1   Data Fragmentation

To reduce data fragmentation in a distributed database is fundamental to speed up query processing. The goal of the WA-RDF fragmentation process is to minimize the number of joins to be executed by preprocessing these joins during a triple insertion.

```
SELECT ?x WHERE {
?x p1 y?.
y? p2 z?.
z? p3 D
}
SELECT ?x WHERE {
?x p1 y?.
x? p4 a?.
x? p5 b?
}
```

QMap					TPMap		TP2Q		
q1	?x p1 y?. y? p2 z?. z? p3 D	Chain	t1	1	?x p1 y?	tp1	tp1	q1, q2	wa1
					y? p2 z?	tp2	tp2	q1	wa2
q2	?x p1 y?. x? p4 a? x? p5 b?	Star	t2	10	z? p3 D	tp3	tp3	q1	wa3
					x? p4 a?	tp4	tp4	q2	wa4
					x? p5 b?	tp5	tp5	q2	wa5

Q2TP	
q1	tp1, tp2, tp3
q2	tp1, tp4, tp5

WAc

Fig. 2. WAc: workload monitoring

When WA-RDF receives a new RDF triple $t_{new} = \{s, p, o\}$, it persists t_{new} into a document NoSQL database with a temporary JSON format $\{f = s, p-i = o\}$. This mapping format is discussed later in this section, but it allows t_{new} to be queried while WA-RDF creates a fragment. In a background process, WA-RDF checks if t_{new} matches any triple pattern present in *TPMap*. For instance, A p3 D matches z? p3 D (tp3) in Fig. 2, but A p3 C does not match any triple pattern. If a match occurs, based on *TP2Q* information, WA-RDF verifies in *QMap* if the typical workload is star or chain (or both) according to the match. In any case, WA-RDF expands t_{new} to a fragment representing the union of all the queries that hold the match. If t_{new} does not match any workload information, it is solely registered in WAc and the temporary JSON format still remains.

In the example of Fig. 2, a new triple A p1 B matches the triple pattern ?x p1 y?, which, in turn, is related to queries with both shapes in the typical workload. Hence, if Fig. 3 (2) represents the current graph managed by WA-RDF, the fragments formed with the new triple A p1 B will be the star {A p1 B, F p6 A, A p4 D, A p5 E} and the chain {A p1 B, B p2 C, C p3 D} (Fig. 3 (3)). As presented in Fig. 3 (4), after the fragments creation, they are mapped and stored into a document or graph (or both) NoSQL databases.

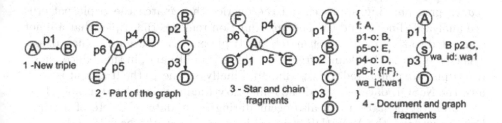

Fig. 3. Fragment creation

RDF-to-NoSQL mapping is crucial for WA-RDF, so we briefly explain how the RDF fragments are mapped to document and graph NoSQL databases. WA-RDF transforms each document fragment into a JSON document. The most connected node, *i.e.*, the node that has more incoming or outcoming predicates, is the value for the document key f. Moreover, for each predicate, it is created a key with the predicate value concatenated with -i for incoming and -o for outcoming. The value is the resource connected to this predicate keys, or a subdocument with the same structure. For instance, a fragment A p1 B, A p2 C, D p3 A, E p4 D is mapped to the JSON document {f:A, p1-o:B, p2-o:C, p3-i:{f:D, p4-i:E}}. The benefit of this document mapping is to solve the typical queries with only one filter access.

For the graph NoSQL database, we could do a straightahead mapping from the fragment but, as presented in Sect. 4, the following map leads to better results. Firstly, WA-RDF finds the longest chain, creates graph nodes for the first subject S_1, the last object O_n of this chain, and a summary node S_i. Further, it maps the predicates linked to S_1 and O_n to edges, and S_i contains, as properties, all the triples between S_1 and O_n. Finally, WA-RDF connects S_i to all the nodes that are not in the longest chain. This procedure is repeated to all the sizes of chains until S_i would have only one triple. As an example, the fragment A p1 B, B p2 C, C p3 D, B p4 E is mapped to the nodes A, D, S_i and E. S_i contains B p2 C and B p4 E as properties. Additionally, the edges p1 (from A to S_i), p3 (from S_i to D) and p4 from (S_i to E) are also created. This mapping strategy is efficient when there are more than one chain query connected. For instance, if a query with the triple patterns ?x p1 y?. ?y p2 z?. ?z p3 D. ?z p4 ?k. ?k p5 ?l. ?l p6 ?m is issued, WA-RDF can solve it with only one access.

As illustrated in Fig. 3 (4), every new fragment receives an identifier from WAc. For document fragments, it is added to the key *wa_id* in the body of the document. For graph fragments, it is a property called *wa_id* of the node S_i. This identifier is used by WA-RDF to register where each fragment is located, as further detailed in this section.

Algorithm 1 describes the fragmentation process accomplished by WA-RDF. The algorithm adapts the *Apache Spark* syntax[1]. Its input is a new triple t_{new}, and its return is the list of triples $\{t_1, t_2, ..., t_n\}$ for a created fragment f. Firstly, the algorithm creates a list of triple patterns (line 1). If the list is empty then the temporary fragment is returned (lines 2 and 3). Otherwise, by using a *map-reduce* programming approach, all the queries that match the triple patterns are analyzed (lines 5 and 6). The *map* function removes the triples that are not connected to t_{new} or does not match any triple pattern. During the map phase, f is created from the cleaned fragments (lines 7 and 8) and, in the reduce phase, the triples are connected to t_{new} (line 9). Finally, in line 10 the fragment is stred into the NoSQL databases and the temporary fragment is deleted in line 11.

Data fragmentation can also occur during an update or delete of a triple. For the update, the WA-RDF strategy is to create, in the background, a new fragment and remove the old one. To avoid inconsistency during this process, WA-RDF manages an updated/deleted list that assures that the fragment triples are updated or deleted before they are sent to the client, as illustrated in Fig. 4 (1). In this case, the triple A p1 B was updated to A p1 Z. The update process

---

[1] https://spark.apache.org/docs/2.2.0/rdd-programming-guide.htmlrdd-operations.

Algorithm 1: Map-reduce-based fragmentation

Input: t_{new}
Output: f = $\{t_1, t_2, ..., t_n\}$
1 triplePatterns=TPMap.matches(t_{new});
2 **if** *triplePatterns.size()= 0* **then**
3   f=getTemporaryFragment(t_{new});
4 **else**
5   queries=QMAP.getQueries(triplePatterns);
6   fragment=analyze(queries);
7   f = fragment;
8   .map(cleaned: (fragment, triplePatterns));
9   .reduce((cleaned, t_{new}), f: connect(cleaned, t_{new}));
10   persistToNoSQL(f);
11   deleteTemporaryFragment(t_{new});
12 **return** f;

---

has two phases, as also shown in Fig. 4. Firstly, WA-RDF matches this triple to the WAc triple partners. In the following, it removes the fragments that have this triple (Fig. 4 (2)). Secondly, it runs an insertion of the new value of the triple (Fig. 4 (3)). Section 4 shows that the update time is not prohibitive. The deletion process is similar to the update. In this case, WA-RDF creates a new fragment that holds all the existing triples in the old fragment except the deleted triple.

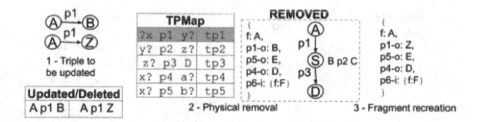

Fig. 4. Triple update

WA-RDF also manages multiple nodes of each NoSQL database, where each node maintains a partition of the data. As presented in Fig. 5, a *Dictionary* component maintains the workload identifiers of the persisted fragments and the number of triples for each workload identifier (wa_{id}) maintained in data partitions into the NoSQL databases. The goal of the Dictionary is to help on the querying process, avoiding unnecessary searches and reducing the calls and round trips to the NoSQL databases. For instance, as illustrated in Fig. 5, when a WA-RDF installation with three partitions receives a query, it matches the triple patterns in *TPMap*, get the wa_{id} from *TP2Q*, and finds out which partitions to query from the Dictionary.

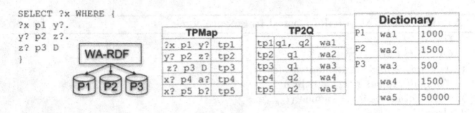

Fig. 5. Fragment distribution

The choice for using wa_{id} to distribute the fragments into partitions can lead to skewed data placement. We try to balance it with a *round-robin* process on the partitions when new fragments are inserted or updated. Moreover, periodically WA-RDF checks the *Dictionary* to find skewed partitions. In the example of Fig. 6 (left), partition *P3* has too many triples for the workload *wa5*. In this case, WA-RDF checks *TP2Q* to see the queries causing the skewness, get the fragments, and replicate or move data. WA-RDF moves data (replication and deletion) to evenly distribute the storage size and the query load among the servers. This is not the case of the example, where deleting the fragments from *P3* would only move the skewness to the other partitions. In this case, WA-RDF only replicates data and balances the query load (Fig. 6 (right)). As presented in Sect. 4, the additional storage need generated by this replication is usually irrelevant in comparison to the whole dataset.

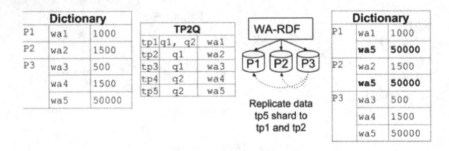

Fig. 6. Fragments redistribution

In a dynamic architecture, the workload of SPARQL queries changes over time. To deal with it, WA-RDF maintains only the most relevant queries in WAc. The relevance R for each query is calculated by the sum of how frequent (f) this query is, the number of placeholders (p) in the query (the more placeholders a query has, the more triples it potentially matches), the size (s) of the query (the number of triple patterns), divided by the novelty (n) of the query (how long since it was lastely received). During WA-RDF development, other measures of relevance were considered. However, the experiments of Sect. 4 revealed that the following formula obtained the best results.

$$R = (f + p) * s/n \tag{1}$$

Instead of adding immediately the new queries in the WAc, WA-RDF keeps them in a queue and waits until the size of this queue passes a soft dynamic threshold T. This threshold is calculated by multiplying the number of querying threads (q) to the number of queries in the last $10\,s$ (t), divided by the average of queries in the last minute (m). The relevance R is calculated if T is bigger than the number of querying threads. This threshold is used to avoid the process to be fired for every new query.

$$T = q * t/m \tag{2}$$

Also, when the threshold is reached, WAc defines a new wa_{id} as well as deletes and recreates all the fragments stamped with the old workload identifier. This change leads to modifications on how the data is placed in the architecture by changing the fragments, the partitions and the data storage in the NoSQL databases. During fragment recreation (it is better explored in Sect. 4), WA-RDF performs in a sub-optimum placement. However, this task does not affect the consistency of the architecture. It is important to notice that the workload version only changes if queries with new structures gain relevance, which is usually seldom. In our experimental evaluation only four versions were created for around 10,000 queries derived from 20 templates.

3.2 Data Querying

To reduce the intermediate results during a query execution it is important to be assertive in the querying process by returning only fragments that contain the desired query answer. Moreover, ideally, the fragments retrieved from the NoSQL databases should not have repeated triples. Thus, WAc heavily influences querying and caching.

A query is processed by the WA-RDF component *QProc*, as shown in Fig. 7. Its design goal is to avoid joins between partitions, reduce the unnecessary intermediate results, and dynamically choose the best NoSQL node to query. *QProc* is formed by a *Decomposer*, which uses WAc information to decompose the query into star and chain queries. The *Decomposer* forwards these queries to *Planner*, which generates different plans to execute the queries, sending these plans to the *Optimizer/Executor*. Each query is tested in parallel against the *Cache* by the *Optimizer/Executor*. For the queries that are not in the *Cache*, the *Optimizer/Executor* estimates the *Cost* and, if necessary, rewrite the query (*Rewritter*). Finally, the rewritten query is executed in the NoSQL databases. For each answered query, the *Optimizer/Executor* updates the cost of this action (*Phys.Cost* in Fig. 7).

For each SPARQL query, WA-RDF decomposes it into star and chain queries. For each decomposed query, it checks if this query matches any triple pattern registered on WAc. If so, WA-RDF knows what data partitions potentially store the fragments by getting the wa_{id} from the *Dictionary*. Otherwise, if the query

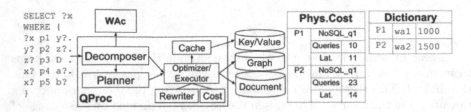

Fig. 7. Overview of QProc

matches no triple patterns, WA-RDF returns an empty result. If multiple partitions are found, *QProc* has to join the results.

The processing of joins occurs when WA-RDF cannot execute a query on a single partition. In this case, the query must be decomposed into a set of subqueries, being each subquery evaluated separately and further joined. For example, supposing that query *q1* (the same query in Fig. 7) in the following is not able to be completed accessing only one partition, WA-RDF divides it into subqueries *s1q1* and *s2q1*, and issues them to the partitions that hold the data (*P1* and *P2*, for example - see Fig. 7). Next, it joins the result sets by matching the subqueries by the triple x? p1 y? (connection between the partitions).

q1: SELECT ?x WHERE {x? p1 y?. y? p2 z?. z? p3 D. x? p4 a?. x? p5 b?}
s1q1: SELECT ?x WHERE {x? p1 y?. y? p2 z?. z? p3 D.}
s2q1: SELECT ?x WHERE {x? p1 y?. x? p4 a?. x? p5 b?}

Query optimization is made along the *QProc* execution. Firstly, the *Decomposer* tries to find all the star shapes present in a query q_i. If it finds only one star or chain shape, it forwards q_i directly to the *Optimizer/Executor*. However, if it detects star and chains, or more than one star or chain in q_i, it forwards q_i to the *Planner* component. The *Planner* tries to define options to solve the query and forwards it to the *Optimizer/Executor*. For each query, the *Optimizer/Executor* tries to run it on the *Cache*. WA-RDF also consider WAc information to manage caching. When it is time to evict data from the *Cache*, it searches *QMap* to find out and remove the least common fragments. If the query is not found, the *Optimizer/Executor* asks for the *Cost* component to find out the less costly plan, and if it works, it asks for the *Rewriter* component to modify it in order to improve the query processing.

The first strategy of the *Optimizer/Executor* is to foster the early execution of triples with low selectivity to reduce the number of intermediate results. The selectivity of a triple pattern is an estimation of the percentage of accessed data. This information can be obtained with the aid of the *Dictionary*. As shown in Fig. 7, the *Dictionary* maintains, for each typical workload, the number of triples for each partition. For instance, the selectivity of *wa1* is 1000/54500, and 1500/54500 for *wa2* (54500 is total number of triples present in the Dictionary). When WA-RDF receives *q1*, it processes *wa1* first. Secondly, WA-RDF considers the historical latency and the number of queries running for each NoSQL query in the table *PhysicalCost* maintained by the *Cost* component. This component also maintains a *matrix of joins* between fragments. This matrix is based on the work of Chawla, Singh, and Pilli [3], but, in our case, it contains joins of fragments instead of joins of triples. The cost matrix is updated after each query. Finally, WA-RDF performs the join based on the cost of each partial query.

Star queries are converted to queries over NoSQL document databases. For instance, the *star* queries *q1* (O-O) and *q2* (S-S) in the following are converted to the access methods *m1* and *m2*, respectively (MongoDB[2] NoSQL database syntax). The `$exists` function of MongoDB filters the JSON documents that have all the predicates of each query. WA-RDF also filters by the object A in *m1* and by the subject B in *m2*.

```
q1: SELECT ?x WHERE {x? p1 y?. x? p2 z? . x? p3 A}
q2: SELECT ?x WHERE {x? p1 y? . z? p2 y? . B p3 y? .}
m1: db.p1.find({p1-o:{$exists:true}, p2-o:{$exists:true}, p3-o:A}})
m2: db.p1.find({p1-i:{$exists:true}, p2-i:{$exists:true}, p3-i:B}})
```

Chain queries, in turn, are converted to queries over NoSQL graph databases. For example, the query *q3* in the following, with O-S joins, is translated to the query *g1* according to the *Cypher*[3] query language of the Neo4J NoSQL database.

```
q3: SELECT ?x WHERE {x? p1 y?. y? p2 z?. z? p3 w?.}
g1: MATCH (f:Fragment)
WHERE ANY(item IN f.p WHERE item = p1 OR
item = p2 OR item = p3)
RETURN p
```

Algorithm 2: Map-reduce-based fragment cleaning

Input: $f = \{f_1, f_2, ..., f_n\}$, q
Output: $R = \{t_1, t_2, ..., t_n\}$
1 R = f
2 .map(match: (f, q))
3 .reduce((t_1, t_2), if => !t_1.equals(t_2));
4 **return** R;

With the fragmentation approach, WA-RDF has to assure that the client will not receive additional triples. This checking is made after all resulting fragments come from the partitions. Algorithm 2 is executed to remove unnecessary triples of the result set. Its input is the list of fragments $f = \{f_1, ..., f_n\}$ and the user query, and the output is the final result set of triples R. The algorithm also follows the *map-reduce* paradigm. During the *map* phase, all triples of each fragment that are not desired are removed by matching to the query. During the *reduce* phase, the triples are deduplicated. Finally, the result set is returned.

---

[2] https://docs.mongodb.com/manual/tutorial/query-documents/.
[3] https://neo4j.com/developer/cypher-query-language/.

To reduce intermediate results, every time a query is responded, WA-RDF also tries to merge fragments. This task is necessary due to the storage problems of *Rendezvous*, where the storage would grow exponentially. The merging process drastically reduces storage size. It occurs when a fragment has only a small difference in their triples than another existing one. In our current version, this difference between the fragments must be less than 30%, but WA-RDF is flexible to set other thresholds, and this is a subject for future research.

For instance, suppose that the query presented in Fig. 8 (1) returns the fragments illustrated in Fig. 8 (2) from a document NoSQL database. In this case, a new document is generated by the union of the triples from both fragments, as shown in Fig. 8 (3), and the two previous fragments are deleted. A queue manages this process, which permits that, instead of recreating the whole database, we refragment only the "warm" portions of the database without impacting the rest of the WA-RDF processes.

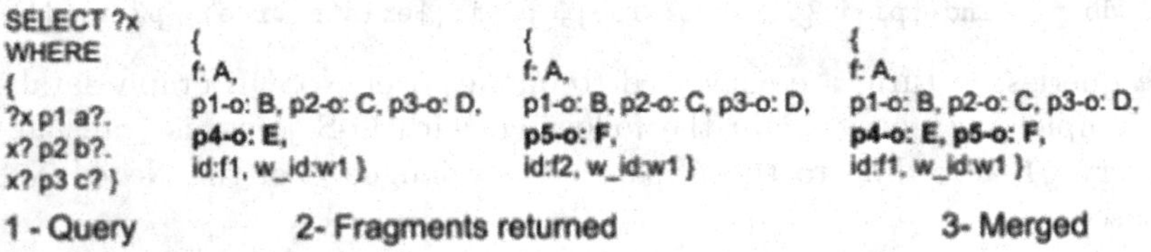

```
SELECT ?x
WHERE          {                    {                     {
{              f: A,                f: A,                 f: A,
?x p1 a?.      p1-o: B, p2-o: C, p3-o: D,   p1-o: B, p2-o: C, p3-o: D,   p1-o: B, p2-o: C, p3-o: D,
x? p2 b?.      p4-o: E,             p5-o: F,              p4-o: E, p5-o: F,
x? p3 c? }     id:f1, w_id:w1 }     id:f2, w_id:w1 }      id:f1, w_id:w1 }

1 - Query      2- Fragments returned                      3- Merged
```

Fig. 8. Merging of fragments

4 Experimental Evaluation

We execute a set of experiments to evaluate the efficacy of WA-RDF. All of them consider around a billion RDF triples generated on the e-commerce domain using the *WatDiv* benchmark [1]. The goal of the experiments is to verify if the decisions made during WA-RDF development reach the work purposes (to avoid data fragmentation, to improve data placement, to check the scalability of WA-RDF, and to reduce the intermediate results). Additionally, we compare WA-RDF with Rendezvous and another baseline (S2RDF).

The current version of WA-RDF was developed using Apache Jena version 3.2.0 with Java 1.8, and we use MongoDB 3.4.3, Neo4J 3.2.5, and Redis 5.0.3[4] as the document, graph, and key/value NoSQL databases, respectively. The choice for these solutions was based on their current high popularity in *DB-engines ranking*[5]. We also used Apache Spark 2.4.0[6] as the processing framework for map/reduce algorithms and Apache Kafka 2.1.0[7] as the queue manager. The tests were executed in the Azure cloud[8] using nodes with 7.5 GB of memory and

---

[4] https://redis.io/.
[5] https://db-engines.com/en/ranking.
[6] https://spark.apache.org/.
[7] https://kafka.apache.org/.
[8] https://azure.microsoft.com/en-us/.

1×32 SSD capacity, with Ubuntu 18.04 server. WatDiv randomly generated a workload consisting of 20 query templates, which involves almost 10,000 queries at each experiment. These queries are approximately 40% simple, 40% star-shaped and 20% chain or complex.

Figure 9(a) compares our previous middleware version (*Rendezvous*), which uses the columnar NoSQL database *Cassandra*, with our current version (WA-RDF), which uses Neo4j as the graph NoSQL database. It shows that WA-RDF is more than 300 ms faster on average, mainly because Rendezvous had to perform multiple calls to *Cassandra*. This change speed up the overall architecture in around 180 ms. We also evaluate the performance with only key/value, document, and graph NoSQL databases. The key/value main problem is the need for executing multiple calls during query response, and the unnecessary returned triples, mainly for the limitations on filtering only by the key. The document is bad on solving queries with multiple chain queries given its hierarchical data access. The graph database is slow for the simple and star queries, also due the multiple calls. This result shows that using document and graph databases achieves the best mix in terms of performance.

Figure 9(b) shows the average times for update and delete operations. They were calculated from the time between the request and the final update or deletion. The workload for the update and delete was created by modifying the triples to be inserted. During the deletion test, we just inserted all the triples and then deleted them one-by-one. In this experiment, we could not compare with *Rendezvous* and *S2RDF* because they do not support these operations. For the 1 GB (*i*), 10 GB (*ii*) and 100 GB (*iii*) dataset sizes, the graph shows that the deletion time is uniform (around 1200 ms on average), but the update time increases as the dataset grows (from 1300 ms to 1600 ms on average). This time increase is due to bigger fragments that can be created with a bigger dataset. It makes the fragment creation process takes longer as the dataset grows.

Figure 9(c) lists how much fragment triples were not necessary during the processing of four different query shapes. We consider here scenarios with and without the merging process. In case (*i*), the triples are all complex, and the percentage of unnecessary triples was 20%. In case (*ii*), we have chain queries, and 16% was discarded. In case (*iii*), we have star queries, and 9% was not necessary, and in case (*iv*), where all queries were solved in the cache, only 2% was unnecessary. Without the merging process, around 35% of the triples were unnecessary for all the tests. It highlights the importance of having a proper access plan, mainly for the chain queries, and the benefit of merging the fragments.

Figure 9(d) shows the spent time to merge fragments. The time calculation begins after a query response until the old fragment is deleted. The experiments were run for sizes of 500 thousand and 1 million triples. The only phase of this process that is influenced by the size of the dataset is the merging analysis (when WA-RDF compares the queries), mainly due to the memory usage. Figure 9(e) presents the time spent by each fragmentation creation phase (temporary fragment creation, queue time, querying related fragments and translation to the target NoSQL database) for the dataset sizes of 500 thousand, 1 million and

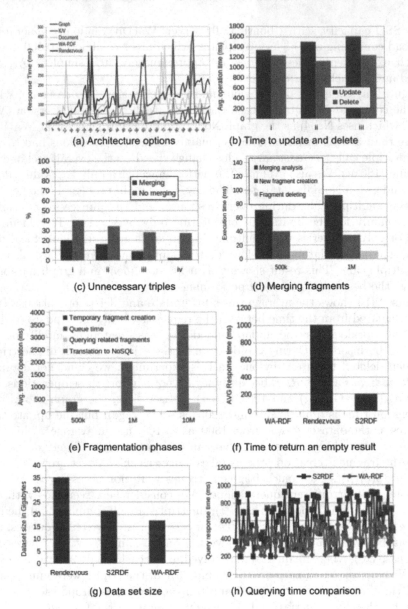

Fig. 9. Experiments with WA-RDF

10 million triples. The queue time is heavily influenced by the dataset size, as well as the time spent on querying the fragments to expand the query. As expected, there is a directly proportional relation between dataset size and processing time for creating the fragment.

The next experiments compare WA-RDF with our previous version and a similar approach (*S2RDF*). Figure 9(f) shows the average time to return an empty value. WA-RDF is much faster because it does not need to access the NoSQL databases if the workload information is not present in the *Dictionary*. Figure 9(g) presents the dataset size for WA-RDF, Rendezvous, and S2RDF. We consider here a dataset with a raw size of around 13 Gigabytes. As illustrated, Rendezvous uses around 35 Gigabytes, S2RDF about 20 Gigabytes and WA-RDF slightly more than 18 Gigabytes. This result is exclusively due to the merging process that avoids unnecessary replication of triples. Finally, Fig. 9(h) compares the average query processing time for WA-RDF against the S2RDF baseline for 100 queries. It shows the superior performance of WA-RDF for the great majority of the queries. It is possible to see that WA-RDF average execution was around 400 ms while S2RDF average was around 600 ms. However, S2RDF presented pretty larger standard deviations (200 ms to 1000 ms) when compared to our solution (200 ms to 600 ms).

5 Conclusion

This paper shows how the workload-awareness is pervasive in the WA-RDF architecture, a NoSQL-based triplestore. WA-RDF presents a novel RDF data distribution approach for persistence purposes. According to the typical shape of main SPARQL queries, it defines RDF fragments and stores them into document and graph NoSQL databases.

WA-RDF is an evolution of *Rendezvous* [10]. The new features of WA-RDF include merging of fragments, support to update and delete operations, usage of map/reduce paradigm on the fragmentation and results cleaning, dynamic data allocation, and relevance-based workload monitoring. The experiments show that WA-RDF solves the Rendezvous storage space problem by merging similar fragments and avoiding unnecessary replication. Also, WA-RDF offers a map/reduce fragmentation process, which allows process distribution using Apache Spark and Kafka. This change permitted to execute the fragmentation in the background without impacting storage and query processing. We also compared WA-RDF with *S2RDF*, a recent Spark-based triplestore. WA-RDF was able to reduce the querying time by 200 ms without impacting the dataset size (WA-RDF dataset is slightly smaller than S2RDF). This result is due to the necessity of S2RDF to query *Hadoop Distributed File System (HDFS)* even when the response is an empty result, and the need for, especially for chain queries, multiple joins between their vertically partitioned tables.

Future works include improvement in our workload-awareness strategy by considering new measures and machine learning techniques. Moreover, we intend to support other SPARQL clauses, like OPTIONAL and LIMIT.

References

1. Aluç, G., Hartig, O., Özsu, M.T., Daudjee, K.: Diversified stress testing of RDF data management systems. In: Mika, P., et al. (eds.) ISWC 2014. LNCS, vol. 8796, pp. 197–212. Springer, Cham (2014). https://doi.org/10.1007/978-3-319-11964-9_13
2. Aluç, G., Özsu, M.T., Daudjee, K.: Workload matters: Why RDF databases need a new design. Proc. VLDB Endowment **7**(10), 837–840 (2014)
3. Chawla, T., Singh, G., Pilli, E.S.: A shortest path approach to SPARQL chain query optimisation. In: 2017 International Conference on Advances in Computing, Communications and Informatics (ICACCI), pp. 1778–1778. IEEE (2017)
4. Dobos, L., Pinczel, B., Kiss, A., Rácz, G., Eiler, T.: A comparative evaluation of NoSQL database systems. Anales Universitatis Scientiarum Budapestinensis de Rolando Eotvos Nominatae Sectio Computatorica **42**, 173–198 (2014)
5. Galárraga, L., Hose, K., Schenkel, R.: Partout: a distributed engine for efficient RDF processing. In: Proceedings of the 23rd International Conference on World Wide Web, pp. 267–268. ACM (2014)
6. Hose, K., Schenkel, R.: WARP: workload-aware replication and partitioning for RDF. In: 2013 IEEE 29th International Conference on Data Engineering Workshops (ICDEW), pp. 1–6. IEEE (2013)
7. Ilarri, S., Stojanovic, D., Ray, C.: Semantic management of moving objects: a vision towards smart mobility. Expert Syst. App. **42**(3), 1418–1435 (2015)
8. Kobashi, H., Carvalho, N., Hu, B., Saeki, T.: Cerise: an RDF store with adaptive data reallocation. In: Proceedings of the 13th Workshop on Adaptive and Reflective Middleware, p. 1. ACM (2014)
9. MahmoudiNasab, H., Sakr, S.: AdaptRDF: adaptive storage management for RDF databases. Int. J. Web Inf. Syst. **8**(2), 234–250 (2012)
10. Santana, M.: Workload-aware RDF partitioning and SPARQL query caching for massive RDF graphs stored in NoSQL databases. In: Brazilian Symposium on Databases (SBBD), pp. 1–7. SBC (2017)
11. Schätzle, A., Przyjaciel-Zablocki, M., Skilevic, S., Lausen, G.: S2RDF: RDF querying with SPARQL on Spark. Proc. VLDB Endowment **9**(10), 804–815 (2016)
12. Ullah, F., Habib, M.A., Farhan, M., Khalid, S., Durrani, M.Y., Jabbar, S.: Semantic interoperability for big-data in heterogeneous IoT infrastructure for healthcare. Sustain. Cities Soc. **34**, 90–96 (2017)

nativeNDP: Processing Big Data Analytics on Native Storage Nodes

Tobias Vinçon[1]($\boxtimes$), Sergey Hardock[1,2], Christian Riegger[1], Andreas Koch[3], and Ilia Petrov[1]

[1] Data Management Lab, Reutlingen University, Reutlingen, Germany
{tobias.vincon,sergey.hardock,christian.riegger,
ilia.petrov}@reutlingen-university.de
[2] Databases and Distributed Systems Group, TU Darmstadt, Darmstadt, Germany
sergey.hardock@dvs.tu-darmstadt.de
[3] Embedded Systems and Applications Group, TU Darmstadt, Darmstadt, Germany
andreas.koch@esa.informatik.tu-darmstadt.de

Abstract. Data analytics tasks on large datasets are computationally-intensive and often demand the compute power of cluster environments. Yet, data cleansing, preparation, dataset characterization and statistics or metrics computation steps are frequent. These are mostly performed ad hoc, in an explorative manner and mandate low response times. But, such steps are I/O intensive and typically very slow due to low data locality, inadequate interfaces and abstractions along the stack. These typically result in prohibitively expensive scans of the full dataset and transformations on interface boundaries.

In this paper, we examine R as analytical tool, managing large persistent datasets in *Ceph*, a wide-spread cluster file-system. We propose *nativeNDP* – a framework for *Near-Data Processing* that pushes down primitive R tasks and executes them in-situ, directly within the storage device of a cluster-node. Across a range of data sizes, we show that *nativeNDP* is more than an order of magnitude faster than other push-down alternatives.

Keywords: Near-Data Processing · In-storage processing · Cluster · Native storage

1 Introduction

Modern datasets are large, with near-linear growth, driven by developments in IoT, social media, cloud or mobile platforms. Analytical operations and ML workloads result therefore in massive and sometimes repetitive scans of the entire dataset. Furthermore, data preparation and cleansing cause expensive transformations, due to varying abstractions along the analytical stack. For example, our experiments show that computing a simple *sum* on a scientific dataset in R takes 1% of the total time, while the remaining 99% are spent for I/O and *CSV* format conversion.

© Springer Nature Switzerland AG 2019
T. Welzer et al. (Eds.): ADBIS 2019, LNCS 11695, pp. 139–150, 2019.
https://doi.org/10.1007/978-3-030-28730-6_9

Such data transfers, shuffling data across the memory hierarchy, have a negative impact on performance and scalability, and incur low resource efficiency and high energy consumption. The root cause for this phenomenon lies in the typically low data locality as well as in traditional system architectures and algorithms, designed according to the *data-to-code* principle. It requires data to be transferred to the computing units to be processed, which is inherently bounded by the *von Neumann bottleneck*. The negative impact is amplified by the slowdown of *Moore's Law* and the end of *Dennard Scaling*. The limited performance and scalability is especially painful for nodes of high-performance cluster environments with sufficient processing power to support computationally-intensive analytics.

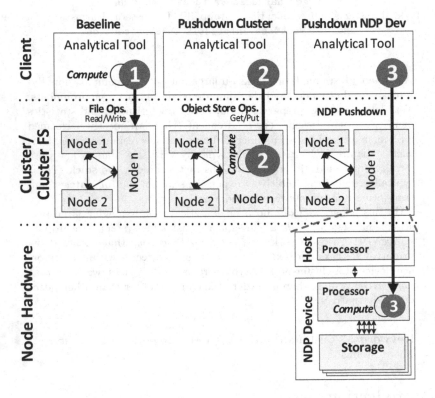

Fig. 1. Three different options to execute analytical operations on a cluster environment. (1) Baseline: Execute on the client; (2) Pushdown Cluster: Execute on a cluster's node; (3) Pushdown NDP Device: Execute on the NDP Device of a cluster's node

Luckily, recent technological developments help to counter these drawbacks. Firstly, hardware vendors can *fabricate combinations of storage and compute elements at reasonable costs*. Secondly, this trend covers virtually all levels of the memory hierarchy (e.g. IBM's AMC for Processing-in-Memory, or

Micron's HMC). Thirdly, the device-internal bandwidth and parallelism significantly exceed the external ones (Device-To-Host), for non-volatile semiconductor (NVM, Flash) storage devices.

Such *intelligent storage* allows for *Near-Data Processing (NDP)* of analytics operations, i.e. such operations are executed in-situ, close to where data is physically stored and transfer just the result sets, without moving the raw data. This results in a *code-to-data* architecture.

Analytical operations are diverse and range from complex algorithms to basic mathematical, statistical or algebraic operations. In this paper, we present execution options for basic operations in nodes of clustered environments as shown in Fig. 1: (1) The computation is within the client and the cluster node is used as part of a traditional distributed file system; (2) The operation is transmitted to the cluster and processed within the cluster node itself; (3) The operation is executed in-situ, within the NDP devices of the cluster's node. The investigated operations are simple, yet they clearly give evidence for the NDP effects on internal bandwidth and the ease of system and network buses. The execution of more extensive operations like betweenness centrality within graphs or clustering and k-nearest neighbor searches are planed for future work.
The main contributions of this paper are:

- End-to-end integration of NDP interfaces throughout the entire system stack
- The performance evaluation shows improvements of NDP operation pushdown of at least 10x
- Analysis of the impact of and necessity for NDP-based abstractions and interfaces.
- We identify the following aspects as the main drawbacks to implementing NDP: Interfaces; Abstractions; Result-Set consumption semantics; Data Layout and NDP Toolchain

The rest of the paper is structured as follows. Section 3 presents the architecture of *nativeNDP*. In Sect. 4 we discuss the experimental design and performance evaluation. We conclude in Sect. 5.

2 Related Work

The concept of *Near-Data Processing* is not new. Historically it is deeply rooted in *database machines* [3,6], developed in the 1970 and 1980s. [3] discuss approaches such as processor-per-track or processor-per-head as an early attempt to combine magneto-mechanical storage and simple computing elements to process data directly on mass storage and to reduce data transfers. Besides reliance on proprietary and costly hardware, the I/O bandwidth and parallelism are claimed to be the limiting factor to justify parallel DBMS [3]. While this conclusion is not surprising, given the characteristics of magnetic/mechanical storage combined with Amdahl's balanced systems law [8], it is revised with modern technologies. Modern semi-conductor storage technologies (NVM, Flash) are offering high raw bandwidth and high levels of parallelism. [3] also raises the issue of temporal locality in database applications, which has already been questioned earlier and is considered to be low in modern workloads, causing unnecessary data transfers. Near-Data Processing presents an opportunity to address it.

The concept of *Active Disk* emerged towards the end of the 1990s. It is most prominently represented by systems such as: Active Disk [2], IDISK [12], and Active storage/disk [15]. While database machines attempted to execute fixed primitive access operations, *Active Disk* targets executing application-specific code on the drive. Active storage [15] relies on processor-per-disk architecture. It yields significant performance benefits for I/O bound scans in terms of bandwidth, parallelism and reduction of data transfers. IDISK [12], assume a higher complexity of data processing operations compared to [15] and targets mainly analytical workloads and business intelligence and DSS systems. Active Disc [2] targets an architecture based on on-device processors and pushdown of custom data-processing operations. [2] focuses on programming models and explores a streaming-based programming model, expressing data intensive operations, as so called *disklets*, which are pushed down and executed on the disk processor.

With the latest trend of applying different compute units, besides CPUs, to accelerate database workloads, a more intelligent FPGA-based storage engine for databases has been demonstrated with Ibex [19]. It focuses mainly on the implementation of classical database operations on reprogrammable compute units to satisfy their characteristics, such as parallelism and bandwidth. A completely distributed storage layer, targeting NDP on DRAM over the network, is presented by Caribou [11]. Its shared-data model is replicated from the master to the respective replica nodes using Zookeeper's atomic broadcast. Utilizing bitmaps, Caribou is able to scan datasets with FPGAs only by the limiting factor of the selection itself (low selectivity) or the network (high selectivity). Moreover, [4,5,9,14] investigate further host-to-device interfaces for general-purpose applications or specific workloads.

However, previous research focused mainly either on the concrete implementation of the reconfigurable hardware, or on single device instances. In this paper, we attempt to combine both topics and focus on the abstraction and interfaces necessary to complete an efficient NDP pushdown.

3   nativeNDP Framework

The architecture shown in Fig. 2 presents a bird's eye view of the essential components, interfaces, and abstractions of the nativeNDP framework. An analytical client executes an R script, triggering an analytical operation (filtering, simple computation - SUM, AVG, STDDEV, or a clustering algorithm). It can be processed on different levels of the system stack:

- directly in R (Fig. 1–*baseline*). This is a classical approach, which can be done with out of the box software, requiring little overhead. The downside is that the complete dataset needs to be transferred through the stack causing excessive data transfers and posing significant memory pressure on the client.

- within a *Cluster node* (Fig. 1–*pushdown cluster*). The same function can be offloaded to the HPC cluster system and distributed across nodes. Hence the compute and data transfer load can be reduced, but not eliminated as such data transfers are performed locally on a node.
- on the *Storage Device* (Fig. 1–*pushdown NDP dev*). With NDP, the operations are offloaded directly on the device, utilizing the internal bandwidth, parallelism and compute resources to reduce data transfers and improve latency.

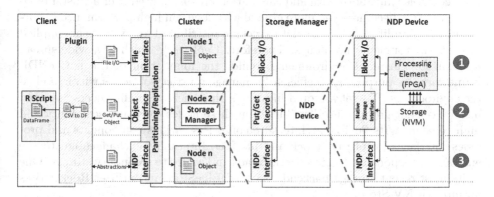

Fig. 2. The high-level architecture showing the applied interfaces and data abstractions along the access path for the three compared experiments: baseline, pushdown cluster node, and pushdown device

3.1 System Stack

In the following we describe the layers of the analytical stack in more detail.

Client: We utilize R as one of the most popular client software for analytical and statistical computation. To interact with the Ceph cluster and the underlaying layers, we designed a custom R plugin, RCeph. It uses the RADOS API [18] to connect to the cluster and is able to issue specific commands with following features:

Put/Get of Files/Objects: To facilitate the first scenario, presented in Fig. 1, the dataset file has to be retrievable from the cluster. Therefore, the standard file I/O API is reused. However, the transfer of results from the second and third scenario necessitates further interfaces such as RADOS's provided Object API as explained in Sect. 3.2.

Pushdown of Domain-specific Operations: This feature is mainly addressed with the second and third scenario, where domain-specific operations, usually executed within the client, are pushed down to either a cluster's node or even throughout the node's storage engine to the NDP Device. I.e. such domain-specific operations comprise R-native operations on their storage abstraction *DataFrame* or could even be extended to small algorithmic expressions.

Format Conversion: As interfaces and abstractions of lower levels often rely on backwards compatibility in nowadays complex systems, format conversions of the results or CSV-encoded files and objects into the R-specific abstraction *DataFrame* are necessary.

The RCeph is complied using Rcpp [7] to a plugin package and can be installed, loaded, and applied within the R runtime environment of the client.

Cluster: To process todays datasets with analytical or statistical workloads in an acceptable time, both data and calculation are distributed over a cluster environment. This becomes even more crucial with focus on high performance in particular. To simplify low latency data accesses, distributed file systems are applied in such environments nowadays. Therefore, Ceph [17], which is a wide-spread solution for clustered environments, builds the foundation of the nativeNDP framework. Its purpose is to efficiently manage a variety of nodes within a cluster environment. Thereby, stored files are striped across small objects, grouped into placement groups and distributed on these nodes to ensure scalability and high reliability. Its flexible architecture comprises various components and provides interfaces for object, block and file I/O. Internally, exchangeable storage engines are responsible to manage the reads and writes to secondary storage. One of its most recent storage backends is called BlueStore and utilizes RocksDB as an internal KV-Store.

Storage Manager: We replaced the internal KV-Store of BlueStore with our own native storage engine NoFTL-KV [16]. Hereby, hardware characteristics, like in-parallel accessible flash chips of the storage device, are known by NoFTL-KV, which in turn is able to efficiently leverage those. Consequently, the physical location of persisted data is defined by the KV-Store itself rather than any Flash Translation Layer (FTL) of a conventional stack. This opens the opportunity to issue commands directly on the physical locations throughout NoFTL-KV and to streamline low-level interfaces along the entire access path.

NDP Device: Devices are emulated by our own storage-type SCM Simulator, based on [10]. Running as a kernel module it provides the ability to delay read and write request depending on its emulated physical locations by utilizing the accurate kernel timer functions. As a consequence, reads or writes across physical page borders claim respectively multiple I/O latencies. For the experimental evaluation, the simulator is instrumented with realistic storage-type SCM latencies from [1]. Moreover, by its flexible design it allows us to extend it with the necessary NDP interface.

3.2   Interfaces and Abstractions

The first, most commonly applied interface is the traditional file I/O (Fig. 2.1). It abstracts the cluster as a large file system, storing its data distributed on multiple nodes. A partitioning and/or replication layer takes care of the internal data placement on various nodes. Instead of the KV-Store the conventional Block

I/O is used to issue reads and writes to the NDP Device. This also involves any kind of Flash-Translation-Layer on the device itself to reduce the wear on a single storage cell and consequently ensure longevity of the entire device.

Secondly, a modern object interface offered by RADOS [18] (Fig. 2.2) can be utilized to put/get objects on the cluster. This abstraction might comprise single or multiple records of a file, or the result set of a pushed down user defined function executed on the respective node. Since the cluster handles data placement, it can transparently execute such algorithms in parallel with the full processing power of the node's servers if the operations are data independent. Within the lower levels, depending on the storage manager, one can either exploit the conventional Block I/O to access the NDP Device or leverage NoFTL-KV's *Native Storage Interface*.

Thirdly, an NDP pushdown necessitates a different kind of interface definition (Fig. 2.3). The NDP execution of application-specific operations requires open interfaces. These should support NDP of application-specific abstractions such as *DataFrame* for R. Consequently, these interfaces and abstractions mandate flexibility, since various result types of the application logic on the device must be transferred back to the client. Expensive format conversion along the system stack can be avoided almost entirely. Yet, an extensive toolchain and NDP framework support is required, beginning from the analytical tool to the employed hardware devices in the cluster. Utilizing the processing elements near-storage (e.g. FPGA), the internal, on-device parallelism and bandwidth can be fully leveraged. For instance, [13] projects of up to 50 GB/s, while the workload on slower buses (e.g. PCIe 2.0 $\approx$ 6.4 GB/s) in the system is eased by reducing transfer volumes (i.e. $resultset \ll rawdata$).

4 Experimental Evaluation

To compare the different execution options on the presented system stack and evaluate their bottlenecks, we conduct three experiments aligned to the scenarios of Fig. 1.

4.1 Datasets and Operations

To ensure the comparability of the scenarios, datasets and operations are predefined. The datasets are created synthetically as CSV files with random numbers, with varying rows and columns from 1k to 10k. When stored in the KV-Store, each cell of the CSV file is identifiable by an auto-generated key with the structure:

$$[object\_name].[column\_index].[row\_index]$$

Inevitably, this is bloating out the raw file size by approximately 16x–17x but enables to access cells by this unique id. Alternatively, depending on the workload, an arrangement per row or per column is likewise feasible. Table 1 summarizes the properties of each dataset for the present experiments.

Table 1. Synthetically generated datasets for the experiments. The raw CSV file size is according the Key-Value format bloated out.

Dataset	KV Pairs	CSV Size [MB]	KV size [MB]	Bloating ratio
1k/1k	1000000	2.8	44	15.9
2k/2k	4000000	12	182	15.2
4k/4k	16000000	45	738	16.4
6k/6k	36000000	101	1668	16.5
8k/8k	64000000	178	2971	16.7
10k/10k	100000000	278	4649	16.7

The operations performed in all experiments is independent of the data distribution and constitutes a typical data science application - calculation of the sum or the average over a given column (Because of the marginal differences only sum is shown further on). The final result set comprises a 32-byte integer value and some additional status data. We leave the implementation of further analytical and/or statistical operations open for future work.

4.2 Experimental Setup

The server, *nativeNDP* is evaluated on, is equipped with four Intel Xeon x7560 8-core CPUs clocked at 2.26 GHz, 1TB DRAM running Debian 4.9, kernel 4.9.0. The NDP storage device is emulated by our real-time NVM Simulator, extended with an NDP interface and functionality. I/O and pushdown operations are handled internally with the storage-type SCM latencies [1].

Since the main target is to evaluate the streamlining of NDP interfaces and abstractions, interferences caused by data distribution or multi-node communication have to be avoided. Therefore, the Ceph cluster is set up with a single object store node. This allows conducting experiments along a clean stack and measuring execution and transfer size for each architectural layer individually.

4.3 Experiment 1 – Baseline

The first experiment utilizes the Ceph cluster in the most common and conventional way - as a file system (Fig. 2.1). Therefore, the file abstractions, interfaces, and subsequently Block I/O are used to retrieve the entire file. The sum over the 10th column is calculated in R by calling *readCSVDataFrame* of RCeph and caching the resulting DataFrame into the R runtime environment. Here, R's capabilities can be used to filter the DataFrame on the respective column and perform the arithmetic operation.

```
sum <- sum(RCeph::readCSVDataFrame(o_name)[col_id])
```

This experiment defines the baseline for any improvements of nativeNDP. However, it exemplifies multiple drawbacks yielding in a significant performance

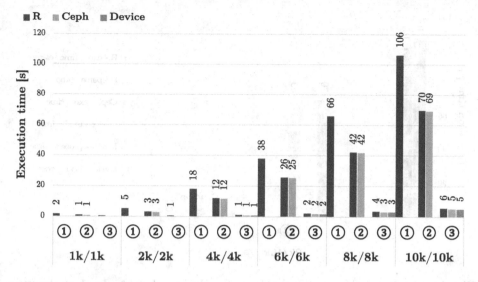

Fig. 3. Execution time for varying dataset sizes shows the performance impact of data transfers/volume, and the improvement through NDP.

degradation. Firstly, the entire file has to be read via block I/O, even though only a small portion of it, the 10th column, is necessary to be processed by the operation (Fig. 5). Secondly, the latency and bandwidth limitations of the network interconnect between the R host and the Ceph cluster, contribute to additional delays to the R processing. The significantly higher transfer size of Host-To-Client, illustrated in Fig. 5, leads inevitably to a slower request duration. Additionally, as R DataFrames do not support any streaming algorithmic, the processing has to idle until the entire dataset is retrieved from Ceph. Thirdly, additional compute-intensive format conversions along multiple interface boundaries are necessary to create R *DataFrames*, which increase delays even further. For example the "R - parse_time" is 95% of the total time as shown in Fig. 4. Moreover, such format conversions are directly depending on the data size, which is subsequently affected by the large Host-To-Client transfer size. Lastly, client systems often comprise limited hardware (e.g. notebook or workstation), while typical working sets can range from tens to hundreds of gigabytes. Thus, processing the whole dataset is not always possible without any performance degrading swapping to disk.

These drawbacks lead to a significantly higher total execution time for the calculation in general, as shown in Fig. 3 (at least 10x).

In total, the baseline experiment results in the lowest performance for all datasets, which is mainly caused by the time spent in transfer and conversion of the CSV object into the R-specific data type *DataFrame* ("R - parse_time" Fig. 4).

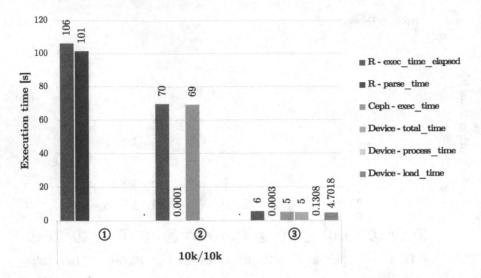

Fig. 4. A detailed execution time analysis shows the main bottlenecks along the analytical stack.

4.4 Experiment 2 – Pushdown Cluster

For the second experiment, Ceph's advanced object interface is extended to execute a user defined function. It queries the KV-Pairs of the respective dataset from NoFTL-KV of the Storage Manager by filtering on the 10th column. Thereby, the retrieved values are cumulated (Fig. 2.2). In a full-fledged cluster scenario, Ceph will automatically distribute this algorithm on the respective nodes within the cluster and aggregate their results afterwards. Obviously, the result size after the operation pushdown is dramatically smaller than the raw data, which relieves the network and accelerates subsequent expensive data format conversions. Hence, the almost non-existing "R - parse_time" (Fig. 4) and the respective transfer size from *Host-To-Client* (Fig. 5). Both result in an overall performance improvement of up to 30% in comparison to the baseline (Fig. 3).

```
sum <- RCeph::execCmd(o_name, "NDP_CEPH SELECT SUM COLUMN col_id")
```

Nonetheless, the I/O overhead of reading the entire data from the storage subsystem, as shown in Fig. 5 by *Device-To-Host*, represents a major bottleneck. Therefore, the time spent in format conversions within Ceph increases as well. For the largest dataset, it takes more than 99% of the time. However, it can be avoided by applying NDP.

4.5 Experiment 3 – Pushdown NDP Device

Our last experiment relies on Near-Data Processing (Fig. 2.3). Abstractions and interfaces are statically created for the purpose of filtering on a given column and computing sums to enable a device pushdown.

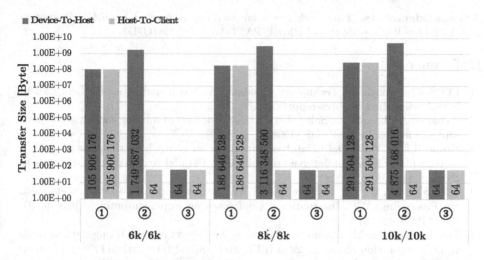

Fig. 5. Transfer sizes from Device-To-Host and Host-To-Client of varying datasets shows the counteraction of NDP to the *von Neumann bottleneck*

```
sum <- RCeph::execCmd(obj_name,"NDP_DEV SELECT SUM COLUMN col_id")
```

The NDP pushdown leverages the much higher levels of compute and I/O parallelism supported by the on-device processing elements (FPGA, GPU) to compute the sum an order of magnitude faster (Fig. 3). Thereby, transferring data from the storage chips takes most of the time (Fig. 4 "Device - load_time"), while the processing is only about 3% of the total time (Fig. 4 "Device - process_time"). Not only is the network relieved by this early reduction of volume, but also the system-wide number of data transfers is significantly reduced. This is mainly driven by the on-device computation and result size reduction as shown in Fig. 5. As this is only possible with the application-specific abstractions, a push down command must compulsorily comprise those to apply computation on the device, *in-situ*. In R, for instance, *DataFrame* may be a suitable application-specific abstraction.

5   Conclusion

We present *nativeNDP*, a NDP approach to effectively pushdown analytical operations to a native storage node of a clustered environment. The evaluation shows improvements of at least 10x over the baseline. Besides the known issues with todays computer architectures, we identify ill-suited interfaces and abstractions along the analytical stack as major drawbacks of current solutions. Moreover, the necessity to push down application-specific abstractions, and data layouts interpretable by the NDP Device is considered a key aspect for a true *in-situ* processing in complex system stacks. To mitigate format conversions along interface boundaries of such stacks, a comprehensive but flexible NDP toolchain is required.

150   T. Vinçon et al.

Acknowledgements. This work has been partially supported by *HAW Promotion MWK, Baden-Würrtemberg* and BMBF PANDAS 01IS18081C/D.

References

1. ITRS - International Technology Roadmap for Semiconductors Reports (2014). http://www.itrs2.net/itrs-reports.html
2. Acharya, A., Uysal, M., Saltz, J.H.: Active disks: programming model, algorithms and evaluation. In: ASPLOS (1998)
3. Boral, H., De Witt, D.J.: Database machines: an idea whose time has passed? A critique of the future of database machines. In: Parallel Architectures for Database Systems (1989)
4. Cho, S., Park, C., Oh, H., Kim, S., Yi, Y., Ganger, G.R.: Active disk meets flash. In: Proceedings 27th International Conference on Supercomputing - ICS, p. 91. ACM Press (2013)
5. De, A., Gokhale, M., Gupta, R., Swanson, S.: Minerva: accelerating data analysis in next-generation SSDs. In: 2013 IEEE 21st Annual International Symposium on Field-Programmable Custom Computing Machines, pp. 9–16. IEEE, April 2013
6. DeWitt, D., Gray, J.: Parallel database systems: the future of high performance database systems. Commun. ACM **35**, 85–98 (1992)
7. Eddelbuettel, D.: Seamless R and C++ integration with Rcpp. Springer, New York (2013). https://doi.org/10.1007/978-1-4614-6868-4
8. Gray, J., Shenoy, P.J.: Rules of thumb in data engineering. In: Proceedings ICDE, p. 3 (2000)
9. Gu, B., et al.: Biscuit: a framework for near-data processing of big data workloads. In: ACM/IEEE 43rd Annual International Symposium on Computer Architecture, vol. 8, pp. 153–165. IEEE, June 2016
10. Hardock, S., Petrov, I., Gottstein, R., Buchmann, A.: NoFTL: database systems on FTL-less flash storage. Proc. VLDB Endow. (2013)
11. István, Z., Sidler, D., Alonso, G.: Caribou. Proc. VLDB Endow. **10**(11), 1202–1213 (2017)
12. Keeton, K., Patterson, D.A., Hellerstein, J.M.: A case for intelligent disks (IDISKS). SIGMOD Rec. **27**(3), 42–52 (1998)
13. Kim, S., Oh, H., Park, C., Cho, S., Lee, S.W., Moon, B.: In-storage processing of database scans and joins. Inf. Sci. (Ny) **327**, 183–200 (2016)
14. Minutoli, M., Kuntz, S.K., Tumeo, A., Kogge, P.M.: Implementing Radix Sort on Emu 1. Work. Near-Data Process, pp. 1–6 (2015)
15. Riedel, E., Gibson, G.A., Faloutsos, C.: Active storage for large-scale data mining and multimedia. In: Proceedings of the 24th International Conference on Very Large Data Bases, pp. 62–73. VLDB, Morgan Kaufmann Publishers Inc., San Francisco (1998)
16. Vinçon, T., Hardock, S., Riegger, C., Oppermann, J., Koch, A., Petrov, I.: NoFTL-KV: Tacklingwrite-amplification on KV-stores with native storage management. In: EDBT (2018)
17. Weil, S.A., Brandt, S.A., Miller, E.L., Long, D.D.E., Maltzahn, C.: Ceph: a scalable, high-performance distributed file system. In: OSDI (2006)
18. Weil, S.A., Leung, A.W., Brandt, S.A., Maltzahn, C.: RADOS: a scalable, reliable storage service for petabyte-scale storage clusters. In: PDSW (2007)
19. Woods, L., Teubner, J., Alonso, G.: Less watts, more performance. In: Proceedings 2013 Int. Conference Management of Data - SIGMOD, p. 1073. ACM Press, New York (2013)

Calculating Fourier Transforms in SQL

Dennis Marten, Holger Meyer$^{(\boxtimes)}$, and Andreas Heuer

Institute of Computer Science, Rostock University,
Albert-Einstein-Strasse 22, 18059 Rostock, Germany
{dm,hme,ah}@informatik.uni-rostock.de

Abstract. The Fourier transform is an important tool for analyzing, transforming and searching multi-media content in databases. SQL is the *lingua franca* for querying structured data. Implementing the Discrete Fourier Transform (DFT) in SQL itself has several benefits. The DFT can directly be executed in the database system. It can be reused for several, different content processing steps from feature extraction to query transformation and evaluation.

We not only discuss different algorithmic aspects but also do a performance evaluation on top of different database systems of different architectures, i.e. row and column stores. The SQL-based implementation is also compared to a Python-based implementation on the client side. There is no variant that always performs best.

Keywords: Fourier transform · SQL · Databases · Multi-media · Performance evaluation

1 Introduction

This research is part of the PArADISE framework (**P**rivacy **A**ware **A**ssistive **D**istributed **I**nformation **S**ystem **E**nvironment) [15]. The project aims at supporting developers of assistive systems in the development and usage phase. Therefore, machine learning algorithms and more generally methods of scientific computations (mainly based on linear algebra) are transparently translated into SQL. In the development phase, depending on data size, these statements will be decomposed into sub-queries for an efficient, horizontally distributed calculation on a parallel database system [16].

In the usage phase, the SQL statements will be further decomposed in order to push the sub-queries vertically down a hierarchy of database systems with decreasing functionality, but increasing proximity to the data source (sensors) [7]. This supports privacy aware computation as sensitive data will not be sent and processed in the cloud. Even without privacy as a requirement for vertical push-down of queries, there are many other reasons to calculate such algorithms on database systems using SQL. As an example, one can benefit of established database techniques as transparent logical and physical optimization of data heavy (distributed) operations, fast access on highly selective algorithms (sparse data) due to index structures, and concurrency control of transactions.

© Springer Nature Switzerland AG 2019
T. Welzer et al. (Eds.): ADBIS 2019, LNCS 11695, pp. 151–166, 2019.
https://doi.org/10.1007/978-3-030-28730-6_10

Furthermore, as SQL is a standardized language, developers can easily change their backend database system without the time-demanding (and expensive) need of reimplementation. Additionally, standardization does also provide persistence of code as SQL-89-queries will run without any problem on modern database systems, even in the following decades.

After investigating Machine Learning (ML) algorithms in [15], we have decided to focus on Fourier transforms which are not only used in big data analytics but also have a wide area of applications in content-based retrieval and multi-media database systems.

The Fourier transform is a fundamental method for analyzing periodicity, e.g. in media types. Periodicity is essential for describing texture in images or the chroma feature of audio signals, especially in music. Due to the complexity of such data, the extraction of interpretable parameters, e.g. from the waveform of an audio signal, is a difficult problem. The Fourier transform is the most important tool for the analysis of such data, which breaks up an audio signal into its constituting frequencies. It is also an important technique not only for feature extraction but also for transformation, selection, reduction of data, and for querying and retrieval.

Feature extraction and transformation are typical preprocessing steps in indexing multi-media content in database systems. Content-based retrieval often exploits special low- or high-dimensional index structures that are built upon these features. The index structures are needed for efficient similarity search in databases.

The Discrete Fourier Transform (DFT) is also a data-independent transform for feature selection and reduction by concentrating the information, and thereby produce fewer coefficients containing most of the information [1].

The Fast Fourier Transform (FFT) or DFT are therefore essential for content based search application across one or several multi-media contents in databases. Nevertheless, there is a much broader range of applications that can benefit from executing the Fourier transform directly on the database management system.

In this paper we present a feasibility study on calculating Fourier Transforms in relational database systems via SQL-statements. We outline a concise state of the art analysis in Sect. 2, followed by a theoretical overview of Fourier transforms, the DFT and the FFT, as well as a discussion on how to translate these methods efficiently into SQL in Sect. 3. The following Sect. 4 presents an evaluation of the corresponding in-database calculation method and a small discussion and future work discussion is provided in the final section.

2   State of the Art

Several research projects exists that investigate a push down of ML algorithms or scientific computations into the database system. A detailed discussion is given in [15,16]. To sum it up, one can distinguish between three project types in this research area.

– The first group integrates in-memory environments like R into the database system, e.g. MonetDBs R-Integration [11]. This allows users to call efficiently implemented algorithms and directly reuse their results in SQL.

– The second family is extending the SQL interface with linear algebra data types and operations, like MADLib [8], SciDB [3] or [12]. Both of these groups aim at improving the database system by internally implementing new features and extending the SQL-interface with non-standard operations.

– The third group, as presented here, concentrates on the efficient implementation of (parallel) algorithms of scientific computations and Machine Learning in ISO-SQL. This also includes investigations on which kind of database systems perform more efficiently and what level of ISO-SQL standard is needed and supported by most of the common systems. This research area is therefore not focused on the optimization of one database system, but rather on a long-term implementation with an effortlessly interchangeable backend database system.

Before our research started, very few ML projects tested pure SQL implementations on suitable database systems. For instance, a Principal Component Analysis (PCA) has been tested in SQL Server in [17] and a simple matrix multiplication in MySQL in [26]. Both projects neglected further investigations, due to somewhat unsatisfying performances. These were most likely caused by the choice of database systems, its tuning, the SQL statements used and (in the case of PCA) the kind of algorithm tested. As we have shown in [15], SQL implementations can perform very good if used for algorithms of the Hidden Markov Model, even in comparison to in-memory environments for scientific calculations.

While the most recent Big Data Analytics research is promoting to push analysis into the database, multi-media database applications have a long tradition of using the Fourier transform. Without doubt, the application fields of the Fourier transform in content-based retrieval and multi-media databases is manifold and essential. Applications include similarity search, feature-based content indexing, feature extraction, feature selection, dimensionality reduction and transformation on different and combined media types, to mention only some.

The FFT is used for efficient similarity search in sequence databases [2] as well as for high dimensional indexing [1] or dimension reduction for distance-based indexing in [13,14]. It is also used for indexing the different media types like audio [21], music [25], or for audio classification and segmentation [10]. Index building and query transformation can directly benefit from in-database FFTs. Image signature [4] and feature generation [9] as well as indexing [20,22] for image retrieval is another scenario of using built-in database FFT functionality. As one last example, integrated image and audio analysis for content-based video indexing [5] uses FFTs for feature extraction and combination.

All these aforementioned techniques would profit from FFT built into database updates or queries. Since SQL-operations are directly evaluated within the database systems, large amounts of multi-media content has not to be transferred between client and server during feature extraction, transformation, combination, and feature-based retrieval in multi-media database applications.

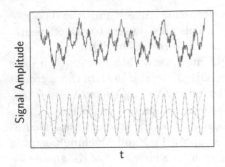

Fig. 1. Graphical illustration of the sine and cosine waves from Eq. 9 (bottom) and their superposition (top) with noise added.

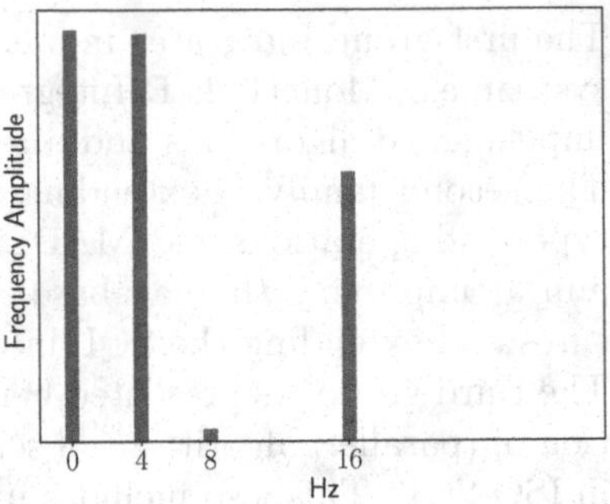

Fig. 2. Section of the spectrum from Fig. 1 calculated via Fourier transformations with peaks at 0 (noise), 4, 8 and 16 Hz.

3   Fourier Transform in SQL

In this section we will briefly discuss different implementations of the Fourier transform in SQL. We start by giving a brief introduction of the underlying theory. For a more precise and elaborate discussion of the theory of Fourier transforms, we would like to refer to [19].

3.1   Theory of Fourier Transforms

The Fourier transform of a square-integrable function $f \in L_2(\mathbb{R})$ is a function $\hat{f} : \mathbb{R} \mapsto \mathbb{C}$ defined by

$$\hat{f}(\omega) := \int_{\mathbb{R}} f(t)\, e^{-2\pi i t \omega}\, dt, \tag{1}$$

with

$$L_2(\mathbb{R}) = \left\{ f : \mathbb{R} \mapsto \mathbb{R} \mid f \text{ measureable}, \int_{\mathbb{R}} |f|^2\, dx < \infty \right\}, \tag{2}$$

usually used to calculate the frequency distribution of a function of time (see Fig. 1). The value of $|\hat{f}(\omega)|$ from (1) can be interpreted as the amplitude of sine waves with frequency ω (see Fig. 2). The superposition of all these waves compose the original function f.

As measured time series are discrete and not continuous, it is necessary to adjust the transform from (1).

Discrete Fourier Transform (DFT). Therefore, consider a discrete time series $\mathbf{x} = \left(x_0 \ldots x_{n-1} \right)^T \in \mathbb{R}^n$. Due to the discretization, the integral in (1) converts into a sum and thus the discrete Fourier transform (DFT) is defined as

$$\hat{\mathbf{x}} = (\hat{x}_j)_{j=0,\ldots,n-1} := \left(\frac{1}{n} \sum_{k=0}^{n-1} e^{-2\pi i j k/n}\, x_k \right)_{j=0,\ldots,n-1}. \tag{3}$$

At this point, one can easily verify that this definition can be expressed as a simple matrix-vector-multiplication

$$\hat{\mathbf{x}} = F\mathbf{x} \tag{4}$$

where

$$F = \begin{pmatrix} 1 & 1 & 1 & \dots & 1 \\ 1 & w_n & w_n^2 & \dots & w_n^{n-1} \\ 1 & w_n^2 & w_n^4 & & w_n^{n-2} \\ \vdots & \vdots & & \ddots & \vdots \\ 1 & w_n^{n-1} & w_n^{n-2} & \dots & w_n \end{pmatrix}$$

and

$$w_n = e^{-2\pi i/n} .$$

The variables w_n^j ($j = 0, \dots, n-1$) are called twiddle factors and only depend on the length of the time series. This is why these factors are usually stored in lookup tables in order to save computation time. The symmetrical matrix F is usually referred to as DFT matrix and has, due to its structure, several useful properties. If F is already computed, a discrete Fourier transform as described in (4) needs $\mathcal{O}(n^2)$ floating point operations. However, the fast Fourier transform uses the special structure of the DFT matrix or (3), respectively, to significantly optimize the calculation.

Fast Fourier Transform (FFT). The fast Fourier transform (FFT) implements a divide and conquer strategy in order to efficiently calculate the DFT of a discrete signal $\mathbf{x} \in \mathbb{C}^n$. More accurately, the FFT is a family of methods, that share a similar form, mainly depending on the size of $\mathbf{x}$. Here, we will only discuss the most common implementation: the Radix-2-DIT-algorithm [19]. For this, the length of the time series has to be a power of 2: $n = 2^p$ ($p \in \mathbb{N}^{\geq 1}$), or at least be divisible by 2; hence the name. The method divides the original sum into two sub-DFTs as

$$\hat{x}_k = \underbrace{\sum_{m=0}^{n/2-1} x_{2m} e^{-\frac{2\pi i}{n/2} mk}}_{\text{DFT of even indices} =: E_k} + e^{-\frac{2\pi i}{n} k} \underbrace{\sum_{m=0}^{n/2-1} x_{2m+1} e^{-\frac{2\pi i}{n/2} mk}}_{\text{DFT of odd indices} =: O_k} = E_k + w_n^k O_k. \tag{5}$$

Because of periodicity, one can show that

$$\hat{x}_{k+\frac{n}{2}} = E_k - w_n^k O_k \tag{6}$$

and therefore

$$\hat{x}_k = x_k + w_n^k O_k \qquad\qquad \hat{x}_{k+\frac{n}{2}} = E_k - w_n^k O_k. \tag{7}$$

Algorithm 1 Recursive Radix-2-DIT algorithm ($s = 1$, $n = 2^p$) for the calculation of the DFT of a time series **x**.

1: $(\hat{x}_k)_k = \text{FFT}(x, n, s)$
2: **if** $n = 1$ **then**
3:     $\hat{x}_0 = x_0$
4: **else**
5:     $(\hat{x}_k)_{k=0,\ldots,n/2-1} =$ $\text{FFT}(x, n/2, 2s)$
6:     $(\hat{x}_k)_{k=n/2,\ldots,n-1} =$ $\text{FFT}(x_{.+s}, n/2, 2s)$
7:     **for** $k = 0, \ldots, n/2 - 1$ **do**
8:         $t = \hat{x}_k$
9:         $\hat{x}_k = t + w_n^k \hat{x}_{k+n/2}$
10:        $\hat{x}_{k+n/2} = t - w_n^k \hat{x}_{k+n/2}$
11:    **end for**
12: **end if**

Algorithm 2 Iterative Radix-2-DIT algorithm ($n = 2^p$) for the calculation of the DFT of a time series **x**.

1: $h = 1$
2: $x = x[\text{reverse\_bit\_order}(1 : n)]$
3: **for** $s = 1, 2, \ldots, \log_2(n)$ **do**
4:     **for** $j = 0, 2^s, 2 \cdot 2^s, 3 \cdot 2^s, \ldots, n-1$ **do**
5:         **for** $t = 0, 1, \ldots, h - 1$ **do**
6:             $p = t + j + 1$
7:             $r = w_n^{t 2^{\log_2(n)-s}}$
8:             $z = x_{p+h} r$
9:             $x_{p+h} = x_p - z$
10:            $x_p = x_p + z$
11:        **end for**
12:    **end for**
13:    $h = 2h$
14: **end for**

Since $E_k, O_k \in \mathbb{C}^{2^{p-1}}$ the procedure can be repeated until $E_k, O_k \in \mathbb{C}$. This leads to the recursive Radix-2-DIT method presented in Algorithm 1. In contrast to the quadratic costs of the matrix-vector-multiplication, this method does only take $\mathcal{O}(n \log n)$ floating point operations. Anyhow, as deep recursion levels on comparatively big data sets can be very demanding on memory management, this method is often implemented as an iterative version shown in Algorithm 2.

3.2   Translation into SQL

With the theory explained before, it is now possible to discuss a variety of SQL implementations of the DFT. All of these use a subset of the following database schema:

```
x ( i int, re double, im double )
f ( i int, re double, im double )
w ( i int, re double, im double )
fmat ( i int, j int, re double, im double )
bro ( i int, v int )
```

Here, x stores the time series, f is the transform to be calculated, w the twiddle factors, fmat the DFT matrix and bro the bit reverse order (needed for FFT; see Algorithm 2). Furthermore, the attribute i represents row indices, j column indices, re the real and im the imaginary parts of the associated complex numbers of vectors x and f or the F. The suitability of this schema has been discussed in detail in [16].

DFT in SQL. The first thing to notice for matrix-based DFT as well as FFT implementations is that the twiddle factors only depend on the sample size $n = 2^p$ and can therefore be stored independently in `w`. This allows the reuse of these factors for equally sized transforms, which can be considered standard in real applications. From this point on, it can be decided if it is beneficial to physically store the DFT matrix in `fmat`, which can be done in SQL via

```
insert into fmat
select t.i, t.j, w.re, w.im
from w join ( select w1.i as i,w2.i as j from w w1, w w2 ) t
    on mod(t.i*t.j,$n)=w.i
```

or be computed online in a `with`-clause of the corresponding query:

```
insert into f
select fmat.i, sum(fmat.re*x.re-fmat.im*x.im),
    sum(fmat.im*x.re+fmat.re*x.im)
from fmat join x on fmat.j=x.i
group by fmat.i.
```

Naturally, online calculation of the DFT matrix does slow down the performance, especially if multiple transforms have to be computed. Besides the quadratic complexity, one of the main downsides of this approach is that the physical storage of the matrix can become very disk space demanding as it is quadratically increasing with problem size. For instance: at n samples the relation `x` stores

$$\underbrace{4+4+8 \mid 8}_{\text{2 integers and 2 doubles}} \cdot \underbrace{n^2}_{\text{matrix elements}} / \underbrace{2^{30}}_{\text{byte to GB}} \text{ GB} \tag{8}$$

of data (neglecting internal meta data at this point). This means that at $2^{15} = 32768$ samples the fourier matrix relation would take up 24 GB and for 2^{16} samples already 96 GB disk space. Since the matrix vector multiplication does usually include a hash join between matrix and vector, this will ultimately lead to a large amount of disk operations. Unfortunately, without the usage of recursive queries (see FFT) it seems not to be possible to calculate the DFT without some kind of cartesian product of the twiddle factor relation as one needs to satisfy a join condition like `x.i = mod (w1.i * w2.i, $n)`. Anyhow, one can use compression techniques to reduce the amount of the data processed. One of such techniques we evaluated uses the symmetrical shape of the matrix, storing only upper triangle and diagonal elements in a relation `sfmat( i int, j int, re double, im double)`. This does only require around half of the data size of `fmat`, but also requires adjustment of the DFT query to

```
insert into f
select i, sum(re), sum(im)
from (
    select sfmat.i as i, x.re * sfmat.re -x.im * sfmat.im as re,
        x.re * sfmat.im + x.im * sfmat.re as im
    from sfmat join x on x.i = sfmat.j
    union all
select sfmat.j as i, x.re * sfmat.re -x.im * sfmat.im as re,
        x.re * sfmat.im + x.im * sfmat.re as im
    from sfmat join x on x.i = sfmat.i
    where sfmat.j <> sfmat.i
) t
group by i.
```

Here, instead of one big join with $\mathcal{O}(2^{2p} \log 2^{2p})$ operations needed ($n = 2^p$), this approach is aggregating over two joins with approximately half the size and costs of $\mathcal{O}(2^{2p-1} \log 2^{2p-1})$ operations each. Both of these DFT matrix based queries have been evaluated in Sect. 4. It is worth mentioning that both approaches are not selective and can therefore not benefit from classical index structures, like B-trees. This changes if one is calculating a series of transforms on sub-signals as also described in Sect. 4.

FFT in SQL. We have implemented an iterative Radix-2-DIT-method in SQL based on Algorithm 2. In contrast to the DFT implementations, this SQL statement is comparatively space demanding and can therefore not be fully displayed here. The main challenge for a SQL translation is that the algorithm is highly iterative. As described in [16] such algorithms can perform very poorly in SQL, especially if it is necessary to realize them via several successive queries. One common example for this is the bulk insertion of external data into the database system. Usually each system provides a special method instead of the consecutive insertion via insert into. In this case, due to the special structure of the FFT, it is possible to realize the method with a reasonable amount of SQL statements. Hereby, the structure is as follows: The two inner loops ($j = 0, 2^s, 2 \cdot 2^s, 3 \cdot 2^s, \ldots, n-1$ and $t = 0, 1, \ldots, h-1$) are translated into one recursive query. With regards to the variables from Algorithm 2, one can describe the structure of the recursive step of the query in a SQL-like pseudocode as

```
with z = g₁ ( (x ⋈ xrec) ⋈ w ), xjz = g₂ ( x ⋈ z )
select xₚ + z
from xjz
union all
select xₚ₊ₕ − z
from xjz
where p < h.
```

The **xrec** relation presents the name of the recursive relation and is used to define the quasi-recursion step and therefore the size of z and **xjz**. The outer

Table 1. Hardware setup of notebook used for experiments.

Component	Value
Processor	Intel Core i7-4600U CPU @ 2.10 GHz × 4
L1 Cache	32 KB
L2 Cache	256 KB
L3 Cache	4 MB
RAM	12 GB DDR3
Operating System	Ubuntu 18.04.2 (64-bit)
Secondary Storage (SSD)	250 GB

loop $(s = 1, 2, \ldots, \log_2(n))$ is represented by successive execution of the respective recursive queries of the inner loops. Here, we have tested three different strategies:

- inserting the result of every loop into an extended relation with an additional level attribute x (i **int**, level **int**, re **double**, im **double**),
- storing every loop result consecutively into a common table expression of one big query, and
- **update** the x relation on every loop.

Based on test runs not presented here, the update-based implementation with a B-tree index on x (i) has shown to be superior by a large margin (speed-up factors up to 3) on any dimension $(n = 2^k, k = 5, \ldots, 15)$ tested and is therefore our method of choice for the evaluation in the upcoming section.

4 Experimental Evaluation

In this section we present the results of two experiments we have conducted in order to get an overview of the performance of the SQL statements from Sect. 3.2 in comparison to implementations to more traditional (imperative) languages. Both of these have been run on a notebook with hardware specifications listed in Table 1. As a basis for further considerations, the first scenario tested the performance of Fourier transformation with varying sample sizes and different implementations and database systems. The second experiment investigates the performance on short time Fourier transform (STFT), in which Fourier transforms are calculated on small multiple overlapping windows of an input signal (see Fig. 4). For both experiments we have created signals as a superposition of 4 and 16 Hz cosine waves, as well as a 8 Hz sine wave and uniformly distributed random noise

$$S(t) = \cos(2\pi f_1 t) + 0.2\sin(2\pi f_2 t) + 0.8\cos(2\pi f_3 t) + \varepsilon(t). \tag{9}$$

This signal and its sub-signals are depicted in Fig. 1. The sample (and problem) size has been de- or increased by adjusting the sampling rate or the signal length. As representatives of relational database systems, PostgreSQL 11 [18]

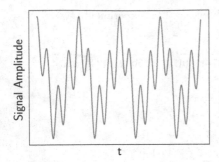
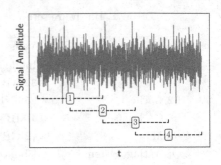

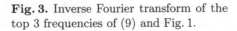

Fig. 3. Inverse Fourier transform of the top 3 frequencies of (9) and Fig. 1.

Fig. 4. Exemplary distribution of windows in a short-time Fourier transform (STFT). In the presented experiments, consecutive windows overlap on 50% of their size.

and Actian's Vector 5.1 [27] have been chosen. PostgreSQL is a widely known open-source relational database system that is built on a row-wise storage model and provides a rich set of database functionality and extensions. Furthermore, it is released under a BSD-like license, which is especially interesting for industrial usage. In contrast to this, we chose the commercial relational database system Actian's Vector. This column store has shown great potential for similar use cases in [16], due to its newly implemented kernel and features like clustered index structures. Furthermore, we have established in [15], that column stores do usually perform better on this kind of operations (Fig. 3).

As a representative of traditional environments for scientific computing, we have implemented the iterative version of the FFT according to Algorithm 2 in Python3/NumPy [23]. It is worth noting that there exist more powerful native functions even in NumPy itself, but these implementations are far more complex and are not comparable with the FFT implementations we have done in SQL so far.

In addition to the pure processing time in Python ('Py Pure Calc' in Figs. 5, 6, and 7, data already in main memory) we also have provided the overall time needed for the calculation, reading the original signal from and inserting the results into PostgreSQL ('Py IO PSQL') via the Python-DB connector Psycopg2 [6]. The latter method has been added to offer a more real and fair comparison to the persistent database solutions. Here, the choice of database system is mainly founded on ease-of-use and is unlikely to have significant performance influence, due to comparatively small data sizes and simple IO operations.

4.1 Calculating Fourier Transforms

The first experiment compares the calculation (and storage) of the Fourier transform in each of the mentioned system. Hereby, the FFT as well as the implementations with the DFT matrix and the symmetrical DFT matrix from Sect. 3.2

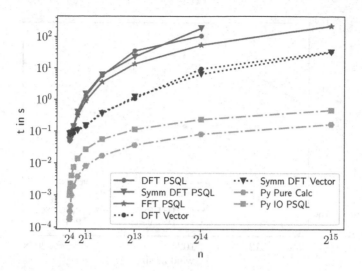

Fig. 5. Performance comparison of different implementations of Fourier transformation in PostgreSQL 11, Actian's Vector 5.1 and Python3/NumPy.

have been tested. We did not implement the FFT in Actian Vector, as the system does not support recursive queries. The computation of the twiddle factors, DFT matrices and bit reverse orders are not included in the results, as they are serving as look-up tables. In this scenario the sample size from signal (9) has been altered by adjusting the sampling rate, while maintaining the duration of one second. The results are depicted in Fig. 5 and represent the best out of 5 separate runs. It can be seen that Python is clearly the fastest system (as expected) as it outperforms Actian's Vector with an increasing factor around 10 to 50 and PostgreSQL with at least another order of magnitude. As the latter gap might be to wide for most of the common use-case-scenarios of Fourier transformations, the gap between Vector and Python is not as high and therefore one might consider Vector as a viable option for moderate size signals, especially when taken into account the numerous advantages of in-database calculation discussed in the Introduction. Another interesting aspect that can be seen is that the symmetrical DFT consistently does better than the classical DFT in both database systems up to around 2^{13} to 2^{14} samples, but becomes slower after this point. We suspect that this might be due to disk or memory limits as the performance of PostgreSQL on 2^{15} has been abandoned because of insufficient disk space. Anyhow, one of the most important findings of this experiment is that the FFT implementation in PostgreSQL gain a significant performance advantage after an initial transient phase with factors around 2 to 3 in comparison to the established DFT implementations. This outcome was not clearly predictable, since highly iterative algorithms can perform very poorly in SQL [16], even if theoretically superior. Also, this might be an indication that

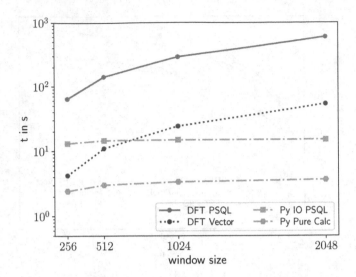

Fig. 6. Depiction of experimental results of the STFT with variable window size on a 10 s signal in PostgreSQL, Actian's Vector and Python.

systems like Actian's Vector might even further close the gap between database solutions and environments for scientific computing, once recursive queries are supported.

4.2 Short-Time Fourier Transform

As the pure computation of Fourier transforms turned out to be comparatively slow, especially in PostgreSQL, we suspected that testing the SQL based approach on a common real application that would include a significant amount of data handling (selection and grouping) might give a better understanding on whether this approach is feasible. Therefore, we implemented a short-time Fourier transform (STFT), which does calculate Fourier transforms on small consecutive windows (see Fig. 4). One of the main objectives of STFT is to find significant changes of dominating frequencies of signals over time. As the analysis and detection of music is a common field of STFT applications, we simulated digital audio signals by creating random integer sequences with a sampling rate of 44.1 kHz, which is probably the most used rate in audio applications. The window sizes chosen are commonly used in real applications (up to 2048 samples), consecutive windows overlap on 50% of their samples [24] and signal durations vary from 10 to 60 s. With this specification, we investigated the performance of the aforementioned systems again. Hereby, we tested calculations with the DFT matrix on sliding windows as both, the classical and the symmetrical DFT matrix approach have performed fairly similar on the given window sizes in SQL. Furthermore, we neglected the window-wise calculation of FFTs in PostgreSQL as on the one hand we have repeatedly observed that looping through

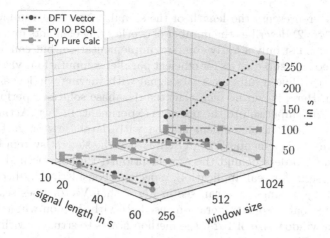

Fig. 7. Depiction of experimental results of the STFT with variable window size and signal length in Actian's Vector and Python.

many low-cost-queries does scale poorly in SQL [16] and on the other hand it is remarkably easy to transform the DFT query from Sect. 3.2 into a corresponding STFT query. For the latter, one needs to combine two sub-queries: one for the calculation of the Fourier transform of all even-numbered windows and one for all odd-numbered windows. The sub-queries can be created from the SQL statements with DFT matrices from Sect. 3.2 by adding an attribute for the window size, adjusting the join condition with a modulo operation in order to assign the elements of the Fourier matrix to the respective ones of each relevant window and finally sum up for any row-window-pair:

```
insert into stft
select i, window_even, sum(re), sum(im)
from (
    select fmatrix.i as i, 2*( x.i / $ws ) as window_even,
        x.re*fmatrix.re as Re, x.re * fmatrix.Im as im
    from x join fmatrix on mod(x.i,$ws)=fmatrix.j
    where x.i < $n-mod($n,$ws)
) ttt
group by i, window_even
union all
select i, window_odd, sum(re), sum(im)
from (
    select fmatrix.i as i, 1+2*( (x.i-$shift) / $ws ) as window_odd,
        x.Re*fmatrix.re as re, x.re * fmatrix.im as im
from x join fmatrix on mod(x.i-$shift,$ws)=fmatrix.j
where x.i>=$shift and x.i < ($n - mod($n-$shift,$ws) )
) ttt
group by i, window_odd
```

Here, $n represents the length of the signal, $ws the window size and $shift $= (\$ws/2)$ describes the number of overlapping elements for consecutive windows. This is not only a fairly compact implementation, but can also be easily decomposed into sub-queries for efficient parallel computations via techniques presented in [16]. The results for a 10 s signal with varying window sizes can be seen in Fig. 6. Here, it can be seen that the database solutions perform significantly better in comparison to the previous experiment. In fact, Actian's Vector even surpasses the Python implementation with database storage for window sizes of 256 and 512 elements. PostgreSQL is still the slowest system but is only between 1 and 2 orders behind the aforementioned Python method, making it a potential system for real application, especially considering its license. Additionally to these findings, one can see in Fig. 7 that Vector does scale linearly (as well as the floating point operations needed) with time on window sizes 256 and 512. At a window size of 1024, the method starts to struggle with increasing signal lengths. This is most likely again due to more intensive main memory usage in the database implementation. Anyhow, this problem should be easily worked around by repeatedly calculating the STFT on a smaller subset of the whole signal. Another lowpoint of the presented implementation in SQL is that it scales comparatively badly regarding the window size. This is not surprising when one considers the results of the first experiment and notices that Python computes FFTs that only take $\mathcal{O}(\$ws \log(\$ws))$ operations per window in comparison to the quadratic costs of the DFT. Anyhow, usually STFTs are meant to find changes on relatively wide frequency bands, which does support the overall positive results of this experiment.

5   Conclusion

In this paper we have presented a study on the feasibility of calculating Fourier transforms in SQL on top of relational database systems. Many reasons for in-database SQL-based calculation, like long-term implementations, data security, and system interchangeability have been discussed and motivated an investigation of the Fourier transformation, being one of the most common tools for data analysis.

It has been shown that discrete Fourier transforms and even fast Fourier transforms (if recursive queries are supported) can be efficiently calculated using SQL statements. Furthermore, we compared these implementations in PostgreSQL, an open source row store, and Actian's Vector, a commercially available column store, with Python as a more traditional environment. The results of this experiment have shown that Vector could perform within a range of one to two orders of magnitude better compared to Python, if results need to be stored. Furthermore, while overall the slowest system, it has be shown that the FFT in PostgreSQL performed significantly better than the presented DFT matrix based solutions. This is especially interesting as it might be worth revisiting this research, if faster systems like Vector will eventually support recursive queries.

Last but not least, we have shown that real applications, like the short-time Fourier transform, can perform very good and even faster than a Python implementation doing the transform on the client side. Especially with the aforementioned additional advantages of in-database calculation this might be considered as an implementation strategy for applications in the near future.

References

1. Agrawal, R., Equitz, W.R., Faloutsos, C., Flickner, M.D., Swami, A.N.: Method for high-dimensionality indexing in a multi-media database, US Patent 5,647,058, July 1997
2. Agrawal, R., Faloutsos, C., Swami, A.: Efficient similarity search in sequence databases. In: Lomet, D.B. (ed.) FODO 1993. LNCS, vol. 730, pp. 69–84. Springer, Heidelberg (1993). https://doi.org/10.1007/3-540-57301-1_5
3. Brown, P.G.: Overview of SciDB: large scale array storage, processing and analysis. In: Proceedings of the ACM SIGMOD International Conference on Management of Data, SIGMOD 2010, Indianapolis, Indiana, USA, 6–10 June 2010, pp. 963–968 (2010). https://doi.org/10.1145/1807167.1807271
4. Celentano, A., Di Lecce, V.: FFT-based technique for image-signature generation. In: Storage and Retrieval for Image and Video Databases V, vol. 3022, pp. 457–467. International Society for Optics and Photonics (1997)
5. Chang, Y., Zeng, W., Kamel, I., Alonso, R.: Integrated image and speech analysis for content-based video indexing. In: Proceedings of the IEEE International Conference on Multimedia Computing and Systems, ICMCS 1996, Hiroshima, Japan, 17–23 June 1996, pp. 306–313. IEEE (1996)
6. Di Gregorio, F., Varrazzo, D.: Psycopg – PostgreSQL database adapter for Python. http://initd.org/psycopg/docs/
7. Grunert, H., Heuer, A.: Query rewriting by contract under privacy constraints. OJIOT 4(1), 54–69 (2018)
8. Hellerstein, J.M., et al.: The MADlib analytics library or MAD skills, the SQL. Technical report, UCB/EECS-2012-38, EECS Department, University of California, Berkeley, April 2012
9. Kekre, H., Mishra, D.: CBIR using upper six FFT sectors of color images for feature vector generation. Int. J. Eng. Technol. 2(2), 49–54 (2010)
10. Kiranyaz, S., Qureshi, A.F., Gabbouj, M.: A generic audio classification and segmentation approach for multimedia indexing and retrieval. IEEE Trans. Audio Speech Lang. Process. 14(3), 1062–1081 (2006)
11. Lajus, J., Mühleisen, H.: Efficient data management and statistics with zero-copy integration. In: Proceedings of the 26th International Conference on Scientific and Statistical Database Management, SSDBM 2014, pp. 12:1–12:10. ACM, New York (2014). https://doi.org/10.1145/2618243.2618265
12. Luo, S., Gao, Z.J., Gubanov, M., Perez, L.L., Jermaine, C.: Scalable linear algebra on a relational database system. In: 2017 IEEE 33rd International Conference on Data Engineering (ICDE), pp. 523–534, April 2017. https://doi.org/10.1109/ICDE.2017.108
13. Mao, R., Miranker, W.L., Miranker, D.P.: Dimension reduction for distance-based indexing. In: Proceedings of the Third International Conference on SImilarity Search and APplications, pp. 25–32. ACM (2010)

14. Mao, R., Miranker, W.L., Miranker, D.P.: Pivot selection: dimension reduction for distance-based indexing. J. Discrete Algorithms **13**, 32–46 (2012)
15. Marten, D., Heuer, A.: Machine learning on large databases: transforming hidden Markov models to SQL statements. Open J. Databases (OJDB) **4**(1), 22–42 (2017)
16. Marten, D., Meyer, H., Dietrich, D., Heuer, A.: Sparse and dense linear algebra for machine learning on parallel-RDBMS using SQL. OJBD **5**(1), 1–34 (2019)
17. Navas, M., Ordonez, C.: Efficient computation of PCA with SVD in SQL. In: Proceedings of the 2nd ACM SIGKDD Workshop on Data Mining using Matrices and Tensors, Paris, France, 28 June 2009 (2009). https://doi.org/10.1145/1581114.1581119
18. Obe, R., Hsu, L.: PostgreSQL: Up and Running. O'Reilly Media, Inc. (2012)
19. Rao, K.R., Kim, D.N., Hwang, J.J.: Fast Fourier Transform - Algorithms and Applications, 1st edn. Springer, Dordrecht (2010). https://doi.org/10.1007/978-1-4020-6629-0
20. Sabharwal, C.L., Subramanya, S.R.: Indexing image databases using wavelet and discrete Fourier transform. In: Proceedings of the 2001 ACM Symposium on Applied Computing (SAC), 11–14 March 2001, Las Vegas, NV, USA, pp. 434–439 (2001). https://doi.org/10.1145/372202.372395
21. Subramanya, S., Simha, R., Narahari, B., Youssef, A.: Transform-based indexing of audio data for multimedia databases. In: Proceedings of IEEE International Conference on Multimedia Computing and Systems, pp. 211–218. IEEE (1997)
22. Tsapatsoulis, N., Avrithis, Y.S., Kollias, S.D.: Facial image indexing in multimedia databases. Pattern Anal. Appl. **4**(2–3), 93–107 (2001)
23. van der Walt, S., Colbert, S.C., Varoquaux, G.: The NumPy Array: a structure for efficient numerical computation. Comput. Sci. Eng. **13**(2), 22–30 (2011). https://doi.org/10.1109/MCSE.2011.37
24. Weihs, C., Ligges, U., Mörchen, F., Müllensiefen, D.: Classification in music research. Adv. Data Anal. Classif. **1**(3), 255–291 (2007). https://doi.org/10.1007/s11634-007-0016-x
25. Yang, C.: MACS: music audio characteristic sequence indexing for similarity retrieval. In: Proceedings of the 2001 IEEE Workshop on the Applications of Signal Processing to Audio and Acoustics (Cat. No. 01TH8575), pp. 123–126. IEEE (2001)
26. Zhang, Y., Herodotou, H., Yang, J.: RIOT: I/O-Efficient Numerical Computing without SQL. CoRR abs/0909.1766 (2009)
27. Zukowski, M., Boncz, P.: From x100 to Vectorwise: opportunities, challenges and things most researchers do not think about. In: Proceedings of the 2012 ACM SIGMOD International Conference on Management of Data, SIGMOD 2012, pp. 861–862. ACM, New York (2012). https://doi.org/10.1145/2213836.2213967

Novel Applications

Finding Synonymous Attributes
in Evolving Wikipedia Infoboxes

Paolo Sottovia[1], Matteo Paganelli[2], Francesco Guerra[2(✉)],
and Yannis Velegrakis[3]

[1] University of Trento, Trento, Italy
`paolo.sottovia@unitn.it`
[2] Università di Modena e Reggio Emilia, Modena, Italy
`{matteo.paganelli,francesco.guerra}@unimore.it`
[3] Utrecht University, Utrecht, The Netherlands
`i.velegrakis@uu.nl`

Abstract. Wikipedia Infoboxes are semi-structured data structures organized in an attribute-value fashion. Policies establish for each type of entity represented in Wikipedia the attribute names that the Infobox should contain in the form of a template. However, these requirements change over time and often users choose not to strictly obey them. As a result, it is hard to treat in an integrated way the history of the Wikipedia pages, making it difficult to analyze the temporal evolution of Wikipedia entities through their Infobox and impossible to perform direct comparison of entities of the same type. To address this challenge, we propose an approach to deal with the misalignment of the attribute names and identify clusters of synonymous Infobox attributes. Elements in the same cluster are considered as a temporal evolution of the same attribute. To identify the clusters we use two different distance metrics. The first is the co-occurrence degree that is treated as a negative distance, and the second is the co-occurrence of similar values in the attributes that are treated as a positive evidence of synonymy. We formalize the problem as a correlation clustering problem over a weighted graph constructed with attributes as nodes and positive and negative evidence as edges. We solve it with a linear programming model that shows a good approximation. Our experiments over a collection of Infoboxes of the last 13 years shows the potential of our approach.

Keywords: Temporal schema matching · Evolving data · Wikipedia

1 Introduction

Wikipedia, with its more than 5.8 million entries[1] is one of the largest human curated sources of knowledge. A Wikipedia entry provides information about some real world entity, of a specific type, like an event, a person, an organization,

---

[1] https://en.wikipedia.org/wiki/Wikipedia:Statistics updated on 24 March 2019.

© Springer Nature Switzerland AG 2019
T. Welzer et al. (Eds.): ADBIS 2019, LNCS 11695, pp. 169–185, 2019.
https://doi.org/10.1007/978-3-030-28730-6_11

a product, etc. It consists of two parts: the unstructured part, which is free text, and the structured part, that is known as the *infobox* and is a set of attribute-value pairs. These pairs describe the main characteristics of the entity that the entry describes. The importance of the Infoboxes is significant. They may contain information that is also found in the text of the of Wikipedia entry, yet, they are highly more structured. This means that the semantics of the information they contain is much easier to interpret, queried, analyzed, combined and explored in general. The attributes (i.e., the names of the attributes) to be present in an Infobox of an entity depend on the type of the entity and are dictated by the Wikipedia policies.

The Wikipedia entries are highly dynamic data. Real world entities evolve over time, and so does the knowledge that we have about them. This real world evolution is reflected into the Wikipedia entries. Users are continuously updating the Wikipedia entries in order to always contain in the best possible way the knowledge we have about an entity. This means that by studying the evolution of the Wikipedia pages, it is possible to understand the evolution of the entities through time. To do so, a fundamental task is to be able to identify and link, across different versions in time of the same Wikipedia page, the parts that model the same kind of information. This is typically done using the schema information, i.e., the attribute names.

Unfortunately, the evolution of the Wikipedia entries is not only on the content but also on the attribute names, making the required linking a challenging task. Attribute names are often changed to more accurately or completely represent the semantics of the attribute values in the Infobox. As a result, different attribute names in different versions of the same Wikipedia entry may be used to represent the same semantic information, and the same attribute name in two different versions may be used to model semantically different pieces of information.

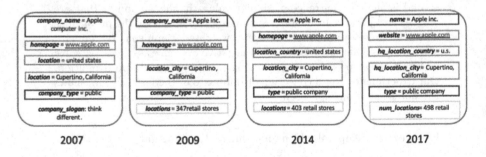

Fig. 1. Evolution of schema and values of the entity Apple Inc.

As an illustration of that situation, consider Fig. 1 that contains four Infoboxes from different versions in time of the entity "Apple Inc.". Note the attribute that describes the location of the company. Initially (in 2007), the attribute name *location* was used to specify the country and other geographical data. In 2009 it disappeared. In 2014 two new attributes were introduced to indicate the location: the *location_country* and *location_city*. Finally, in 2017, the attributes were renamed to *hq_location_country* and *hq_location_city* to more accurately specify that the location is the location of the headquarters.

The aforementioned example is also indicating another situation. The fact the same can happen to the attribute values. For instance, it can be noticed that the value indicating the country USA was originally "united states" and later changed to "u.s.".

In this work we deal with the problem of attribute name alignment in Wikipedia pages across time. We want to analyse and identify sets of attribute names that, across the evolution history of the pages of a specific type in Wikipedia, have been used to represent the same semantic concept. At the same time, we want to identify cases in which the same name has been used in attributes that model semantically different information.

Attribute alignment is a well-known problem that has been studied extensively in the past, mainly in the case of schema matching and in ontology alignment. The straightforward approach would be to look for synonym words in attribute names. This is the approach that has been followed in the area of Natural Language Processing. The techniques that have been developed there can be classified in two main categories. The first are those that exploit dictionaries, i.e., being based on the semantics of the attribute names. They are however, limited.

Their limitation lies mainly in the number of synonyms that can be encoded in a knowledge base. Furthermore, they are context-independence, i.e., are not able to differentiate synonyms according to the context in which they are used. Another approach that has been studied in the context of identifying correspondences in Web form schema matching [12] is one that exploits correlations. The idea behind these techniques is that the search for synonyms is implemented by discovering attributes that correlate negatively or that do not co-occur. Unfortunately, this concept cannot be directly applied in the context of the Wikipedia Infobox attributes. This, because the Wikipedia content is so rich that using only co-occurrence information results in high false positives and also false negatives.

In an effort to overcome the limitations of the co-occurrence approach we have developed an extension of it that exploits the attribute values. In particular, the occurrence of same values between two different attributes in Infoboxes of different versions of the same type is treated as a positive indication that the two attribute names model the same real world concept. Furthermore, we treat co-occurrences as a negative indication. In particular, high degree of correlation (co-occurrence) in Infoboxes between two different attributes is an indication that these two attribute are referring to different concepts, thus, the high correlation is a negative indication. We turn the above two indicators into two different metrics, and create a network of attribute names where the values of the different metrics are used as a distance function. Then we apply clustering [10] to identify those sets that form mutually close names. Such sets are those we consider as semantically related.

Our contributions are specifically as follows: (i) We provide a novel approach to the problem of attribute name alignment in Wikipedia Infoboxes that exploits co-occurrence information as a negative evidence and common attribute values as a positive evidence; (ii) We turn the evidences into metrics and treat the problem as a clustering problem, providing an efficient implementation of it that is based on lineal programming that provides a good approximation; (iii) We apply the technique on the set of Wikipedia entries of 13 years and report our findings on how effective such approach is indeed.

The remainder of the paper is structured as follows. Section 2 defines formally the problem with which we deal and Sect. 3 introduces our approach. Section 4 provides an extensive evaluation of our technique and reports our findings. Related works are presented in Sect. 5 alongside details on how our approach differs from these works.

2   Problem Statement

The paper deals with entities described in Wikipedia articles. We assume that an article describes only an entity, identified by an identifier id (e.g., the page title). The type T specifies the subject of the entity, e.g, an event, person, organization, product, etc. Wikipedia articles consist of two components: an unstructured textual component and a list of attribute-value pairs called *infobox*.

Definition 1 (Infobox and infobox schema). *We define the entity infobox* $I = \{\langle a_1, v_1 \rangle, ..., \langle a_n, v_n \rangle\}$, *where* $\langle a_i, v_i \rangle$ *are attribute-value pairs. We denote with* S_I *the infobox schema, that is the set of attributes included within it, and with* V_I *its values.*

For each type of entity T, *Wikipedia policies specify a template for the infobox schema, i.e. the list of attributes that should describe that type of entity.*

The data shown in an infobox may change over time. This happens mainly for two reasons: (1) the referred entity changes, i.e, the infobox values change and/or (2) Wikipedia releases new policies defining the infobox schema associated to a type of entity.

We define I_t the infobox at time t and E_t the entity at time t it is describing.

Definition 2 (Entity). *An entity at time t, denoted as* E_t, *is a triple* $\langle id, T, I_t \rangle$ *where id is the entity identifier,* T *is the entity type and* I_t *the associated infobox.*

The set of all the changes occurred to the entity can be collected from all the infoboxes and constitutes the *entity evolution*.

Definition 3 (Entity evolution). *Assuming the existence of a set of times values* T *that correspond to all possible times instances* t_i, *we define entity evolution* E *as the triple* $\langle id, T, I_T \rangle$ *where id is the entity identifier,* T *is the entity type and* I_T *the set of all infoboxes* I_{t_i} *describing the entity over time. We identify with* S_{I_T} *the schema of* I_T, *that is the union of the schemas of all the infoboxes contained within it. Similarly, we define* V_{I_T} *as the set of values in all the included infoboxes.*

The problem we want to address is to find, for each entity type, lists of synonymous attributes, i.e., attributes that are used over the time to describe the same property of an entity. The set of attributes used in the infoboxes of a specific entity type is called *set of entity type attributes*.

Definition 4 (Set of entity type attributes). *For each entity type T, A^T is the set of entity type attributes and includes all attributes used in at least an infobox schema at any time for describing an entity of type T.*

Synonymous attributes are clusters of entity type attributes that describe the same real-world entity property.

Problem 1 (Finding synonymous attributes). *Given a set of entity type attributes A^T, we want to find a disjoint partitioning of A^T, denoted as $S = \{S_1, ..., S_m\}$, where the attributes $a_j \in S_i$ are used to describe the same real-world entity property.*

Furthermore, in the following, we denote:

- $V_I(a)$ as the set of values assumed over time by the attribute a within all the infoboxes I_T associated with an entity;
- $\Delta t_I(a)$ as the time interval (i.e., a list, even if not contiguous, of time instants) in which attribute a is *valid*, i.e. it appears in some infoboxes I_T associated with an entity;
- $\mathcal{I}$ as the set of the infoboxes I_T collected over time for a collection of entities.

3   The Approach

In this section, we present our proposal for finding synonymous attributes in Wikipedia entities having the same type. For each pair of attributes, two measures are computed, assessing the extent in which the attribute represent (and do not represent) the same entity property, respectively. In this way, they provide a positive and a negative evidence of the synonymy. The measures are presented in Sect. 3.1. Then Sect. 3.2 shows how to use the knowledge provided by these measures to generate clusters of synonymous attributes. For this purpose, we reduce our problem to the one addressed by correlation clustering [4], where data points are partitioned into groups based on their similarity. A linear-programming approach has been adapted for this purpose. The work has been inspired from [12], where a similar technique has been adopted in the context of web search engines.

3.1   Positive and Negative Evidence for Synonymy

We can model the synonymy relationship between attributes by analyzing their co-occurrences in the same infobox. In this perspective, we assume that synonymous attributes cannot appear simultaneously in the same infobox otherwise there would be information redundancy. In other words, we leverage the co-occurrence of two attributes in the same infobox as a negative evidence for their synonymy.

Example 1. Consider for example the attributes *name* and *type* which definitely describe different aspects of an entity. They are very common attributes: in a random sample of 60,760 infoboxes describing companies collected over the last 13 years, they appeared, together or separately, in 74.61% of the cases (i.e. for describing 45,332 entities). Within these entities the attributes coexist in the same infobox in 99.89% of the cases (i.e., 45,281 times), and they do not co-occur 51 times. According to our idea, they cannot be considered as synonyms. Conversely, the *name* and *company_name* attributes, that instead can be used to describe the same characteristic of an entity, show an inverse co-occurrence pattern: in 0.08% of the cases are present simultaneously in the same infobox and in 99.92% of the cases do not co-exist.

More formally, given two attributes a_i, a_j belonging to infoboxes describing the same entity type, Eq. 1 provides a measure of their "negative" co-occurrence.

$$NegCoocc(a_i, a_j) = \frac{|\{I_T \in \mathcal{I} | \Delta t_I(a_i) \cap \Delta t_I(a_j) \neq \emptyset\}|}{|\{I_T \in \mathcal{I} | a_i, a_j \in S_{I_T}\}|} \tag{1}$$

To compute this measure, all infoboxes in the collection are evaluated. For each of them, the presence of both input attributes (i.e., a_i and a_j) in a time frame is verified. This check is carried out by identifying whether there are overlaps in their validity time interval (i.e., $\Delta t(a_i)$ and $\Delta t(a_j)$ respectively). The number of entities for which there is overlap, normalized by the overall number of entities in $\mathcal{I}$, provides the correlation value that we consider as negative evidence for their synonymy.

As the experimental evaluation shows, the adoption of this measure only is not enough to accurately identify the synonymous attributes.

Example 2. The highest values obtained by the application of Eq. 1 to the attribute *company_logo*, are with the attributes *logo*, *name* and *type*. The results, in our collection are respectively 0.9988, 0.965 and 0.94. Obviously, only the first pair of attributes are synonyms. The other pairs are attributes representing very different information. After a careful analysis of the temporal evolution of the infobox schemas we noticed that the attributes with the "company" prefix have been introduced with an old Wikipedia policy to identify all attributes describing "company" type entities. Today, this policy is no longer adopted, in favor of more concise and direct attributes (such as *type* and *name* instead of *company_type* and *company_name* respectively). However, a delay in the application of the new policy produces misalignments in the infoboxes and make Eq. 1 not enough accurate.

To produce more accurate results, we introduce a measure for positive synonymy evidence. In particular, we measure the values shared between attributes as the indication that they are really synonyms. We analyze the different value representations of the attributes throughout the entire history of Wikipedia and we calculate their fraction of overlap through the Jaccard similarity. We do not deliberately consider other string similarity techniques to have a more general approach, which does not rely on specific domain knowledge. In more detail, for each pair of attributes we select the values that generate the maximum fraction of overlap within the data collection. Equation 2 provides the formulation of the measure we adopt.

$$PosOverlap(a_i, a_j) = \frac{\sum\limits_{I_T \in \mathcal{I}|a_i,a_j \in S_{I_T}} max\left(\sum\limits_{v_i \in V_I(a_i), v_j \in V_I(a_j)} Jaccard(v_i, v_j)\right)}{|\{I_T \in \mathcal{I}|a_i, a_j \in S_{I_T}w\}|}$$

(2)

where, with reference to the notation introduced in Sect. 2, $V_I(a_i)$ and $V_I(a_j)$ represent respectively all values assumed over time by the attributes a_i and a_j for all infoboxes in I_T where they are valid.

Example 3. Let us consider the application of Eq. 2 with the same input as in Example 2. We obtain the following results: $PosOverlap(company\_logo, logo) = 0.8899$; $(company\_logo, name) = 0.003$; and $(company\_logo, type) = 0.007$. We can observe that the high value computed for the pair $(company\_logo, logo)$ confirms the previous evidence of synonymy. The very low values for the other pairs do not confirm the evidence of synonymy resulting from Eq. 1.

3.2 Holistic Approach for Synonym Discovery

The measures of synonymy between pairs of attributes are used to compute clusters of synonymous attributes which constitute the result of our work. Our idea is to model the synonymy relations between the attributes by means of a graph and to apply a clustering algorithm over the graph to extract groups of synonymous attributes.

Given some positive and negative evidence for attributes synonymy, we model attributes and their synonymy relationship as an *attribute-synonymy graph*, where the nodes correspond to the attributes and the edges to the synonymy relations between the attributes. The edges are labeled according to whether the measure associated with them should be interpreted as positive or negative evidence of synonymy.

Definition 5 (*attribute-synonymy graph*). *An* attribute-synonymy graph *is a graph $G = (V, E)$ with vertices representing the attributes of the infoboxes we want to analyze. The edges associate to each pair of vertices provide a measure of their synonymy through a weight $w_{i,j} \geq 0$. Let $L_{i,j}$ be the label associated to each edge (i, j). L can assume the value $+$ or $-$ according to whether the edge is representing the measure of the negative or the positive evidence for their synonymy expressed by Eq. 1 and Eq. 2, respectively. Let E^+ be the set of edges identified by a label of value $+$: $E^+ = \{(i,j)|L_{i,j} = +\}$, and, analogously, E^- (i.e., $E^- = \{(i,j)|L_{i,j} = -\}$) the set of edges identified by a label of value $-$. A representation of this graph is provided in Fig. 2, where solid edges indicate edges with positive weights and edges with crosses the negative ones.*

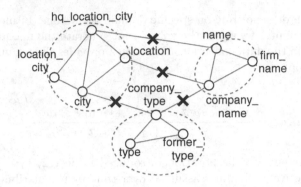

Fig. 2. Attribute-synonym graph for "company" type entities

Our goal now is to apply a clustering strategy that partitions the nodes of the *attribute-synonymy graph* so that each attribute is associated with a single cluster with its synonyms (see the dashed blue circles of Fig. 2). To obtain this result, we adopt a correlation clustering algorithm [4] which provides a method for clustering data points into the optimum number of clusters based on their similarity without specifying that number in advance. In our implementation, the aim is to identify the partitioning of the infobox attributes that best respects the positive and negative evidence of synonymy provided as input.

Problem 2 (discovery of synonymous infobox attributes). *Given an attribute-synonymy graph $G = (V, E)$, we want to find a disjoint partitioning of V, denoted as $S = \{S_1, ..., S_m\}$, that agrees as much as possible with the labels L associated to the edges E of the attribute-synonymy graph. More precisely, we want a clustering that maximizes the weight of agreements: the weight of + edges within clusters plus the weight of – edges between clusters.*

The resolution of this problem exploits a heuristic procedure already proposed in the literature [10] for solving the correlation clustering problem. This technique is divided into two steps. First a linear programming approach is used to provide an approximate solution to the problem. The results produced by this model are fractional values that correspond to scores of synonymy between attributes. In a second step a technique called region-growing is applied to group attributes with a high synonymy level within the same cluster and remove the attributes that describe different information about the referred entity.

Linear-Programming Approach. In the first phase of the approach the following linear model has to be solved.

$$\text{minimize} \quad \sum_{(i.j)\in E^-} w_{i,j}(1 - x_{i,j}) + \sum_{(i.j)\in E^+} w_{i,j}x_{i,j}$$
$$\text{subject to} \quad x_{i,j} \in [0, 1s], \qquad x_{i,j} + x_{j,k} \geq x_{i,k}, \qquad x_{i,j} = x_{j,i}.$$

The goal of this model is to identify a valid assignment of the variable $x_{i,j}$ that minimizes the sum of the negative edges included in a cluster and maximizes the sum of positive edges. Intuitively, this variable provides an indication of the collocation of the nodes in the clusters (i.e., it assumes, in the borderline cases, the value 0 when two attributes are included in the same cluster and 1 in the opposite case). An assignment of $x_{i,j}$ is considered valid if $x_{i,j} \in [0,1]$ and $x_{i,j}$ satisfies the triangular inequality. This motivates the inclusion of the constraints in the problem formulation. The adaptation of this linear model to our problem requires the addition of a further constraint, which requires that the negative weights (i.e., $w_{i,j}$ in the first sum) are defined according to Eq. 1, and the positive ones (i.e., $w_{i,j}$ in the second sum) according to Eq. 2.

Region Growing. Once a first approximated solution to the problem is obtained, we apply the region growing technique. Its objective is to convert the approximate cluster membership indication of the attributes provided by this first solution, into an exact distribution of the attributes in the different clusters. More precisely, this technique is used to convert the fractional solution x in an integral solution which identifies if two attributes belong to the same cluster. Since this technique represents a classical clustering strategy, below we provide only an insight into its operation. More details instead can be found in [10]. The intuition behind this technique is to construct, in an iterative way and starting from randomly selected seed nodes, some balls (i.e., groups of graph nodes) modifying, step by step, their coverage radius on the graph. The growth of these balls is determined by the weights associated with the graph edges: a ball will continue to grow as the sum of the positive weights included inside the subgraph identified by the ball is advantageous. On the contrary, the ball will stop growing, causing the creation of a new ball (or a new cluster), when its growth would incorporate dissimilar nodes compared to those already included in the cluster. The arrangement of these balls within the graph determines its final partitioning.

4 Experimental Evaluation

In this section, we firstly provide a description of the dataset used for the experimental evaluation (Sect. 4.1) and then we qualitatively (Sect. 4.2) and quantitatively (Sect. 4.3) evaluate the effectiveness of the approach. Finally, a case study is presented (Sect. 4.4) to show how Wikipedia synonymous attributes can be used in a real scenario.

4.1 Dataset Description

The dataset used in the experimental evaluation is a collection of infoboxes of entities having type associated to the "concept of company" (i.e., we consider entities having type company, organization, dot-com company, etc.). This collection includes, for each entity, its complete history between August 2004 to August 2017, i.e. all updates in the infobox schemas that have been introduced by Wikipedia users. The result is 60, 760 entities and around 1, 861, 252 changes.

Table 1. Number of attributes and values per entity

	Avg	Std	Max	Min
# attribute per entity	12.84	5.64	253	1
# value per entity	25.35	23.67	503	1

The number of attributes used in the infoboxes varies: it is not fixed per entity type and in the time. Table 1 provides some statistics about attributes and values. The average number of attributes and values per entity are 12.84 and 25.35 respectively. Moreover, the maximum number of attributes associated to an entity, in the considered period of time, is equal to 253, and the maximum number of different values is 503.

In Tables 2a and b the top 10 most frequent attributes and values are reported. The attribute "name" is the most used: it appears in 97% of the collected entities, while "company_name", appearing in 47% of the entities, is the 10th most used attribute. Concerning the values, "united states", "privately held companies" and "public companies" appear respectively in 31.78%, 31.63% and 23.47% of the entities and are the most frequently values used in the collection.

Table 2. Frequencies of attributes and values

(a) Top 10 most frequent attributes

attr	freq	freq (%)
name	59017	97.13%
industry	51845	85.33%
foundation	49033	80.70%
homepage	47076	77.48%
type	46015	75.73%
logo	40102	66.00%
key_people	36176	59.54%
products	33388	54.95%
location	32490	53.47%
company_name	28565	47.01%

(b) Top 10 most frequent values

value	freq	freq (%)
united states	19310	31.78%
privately held company	19221	31.63%
public company	14259	23.47%
united states dollar	9692	15.95%
private	7983	13.14%
subsidiary	7144	11.76%
united kingdom	5986	9.85%
worldwide	5973	9.83%
yes	5289	8.70%
chief executive officer	4793	7.89%

Tables 3 provides an insight on the evolution of the attributes and values in the considered period of time. In particular, Table 3a shows the attributes whose values were most frequently subject to change, and Table 3b the top 10 entities affected by the greatest number of changes over time. Note that 10% of all the infobox updates involves the "key_people" attribute, and the most modified entity is "Eurosport".

A more detailed analysis of the evolution is shown in Fig. 3 and Table 4, where the top 5 most updated types of entity are analyzed. Table 4 shows the number of entities collected per type and the total number of changes. Figure 3 plots some statistics about the number of updates per entity. Although the total number of

Table 3. Updates in Wikipedia entries

(a) Top 10 most changed attributes

attr	freq	freq (%)
key_people	162465	10.07%
products	112440	6.97%
location	94936	5.89%
foundation	87828	5.45%
industry	86707	5.38%
homepage	84182	5.22%
name	77894	4.83%
logo	62653	3.88%
type	61655	3.82%
revenue	59206	3.67%

(b) Top 10 most changed entities

entity title	freq	freq (%)
Eurosport	1924	0.10%
National Geographic (TV channel)	708	0.04%
Canada	672	0.04%
Apple Inc.	594	0.03%
Nintendo	580	0.03%
HBO	538	0.03%
Cuba	527	0.03%
Animax Asia	526	0.03%
General Motors	525	0.03%
Amazon.com	509	0.03%

updates in the "company" category is the highest, a particularly high number of average updates has been applied to entities belonging to the "television" type. The other categories of entities, on the other hand, present an average number of updates which is approximately the same (i.e., the range varies between 20 and 40 updates).

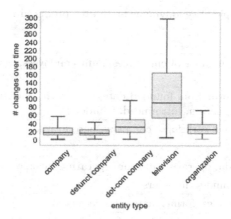

Fig. 3. Number of changes of the top 5 most updated entity types.

4.2 Qualitative Evaluation of the Effectiveness

In this section we qualitatively evaluate the effectiveness of our approach by analyzing a sample of its results. Table 5 shows 10 clusters of synonymous attributes generated by our approach. We can observe that our approach is able to identify interesting and non-trivial synonymy relations between attributes. For example,

Table 4. Entities and updates for the top 5 most updated entity types.

Entity type	# entities	# total changes
Company	57,553	1,494,245
Defunct company	664	15,304
Dot-com company	127	6,340
Television	40	4,953
Organization	155	4,819

it is able to find the correspondences between attributes like "established" and "founded" or "predecessor" and "former_name" which would not be identifiable by a string similarity technique. Furthermore, we match attributes expressed in different languages, such as "employees" with "mitarbeiterzahl" and "city" with "sitz". Analyzing these results, we can observe the various textual forms used over time by the Wikipedia community to indicate the same characteristic of an entity. This variety of forms presumably derives from the adoption of different schema guidelines/policies imposed by Wikipedia[2]. The attributes "name", "company_name" and "type", "company_type" are examples of this situation. The inclusion of the prefix "company" has been introduced by a policy to make more explicit the type of entity described by the attributes.

Table 5. Example of synonymous attributes produced by our approach

Cluster
num_staff, employees, number of employees, num_employees, numemployees, mitarbeiterzahl
established, opened, formation, founded_date, start_year, date_founded, foundation, gründungsdatum, introduced, founded
logo, non-profit_logo, network_logo, company_logo, firm_logo
web, url, website
operating_profit, ebitda, operating income, operating_income
creator, founder(s), founder, founders
predecessor, former_names, former_name, predecessors
company_type, type, unternehmensform, former type, former_type, company type, non-profit_type
headquarters, headquaters, hq_city, location_city, city, sitz, residence, hq_location_city, location, place, hq_location
agency_name, network_name, group_name, name, non-profit_name, company_name, firm_name, company name

---

[2] Note a cleaning procedure has been applied to the input infoboxes to remove the "noise" generated by human mistakes.

4.3   Quantitative Evaluation of the Effectiveness

The effectiveness of the proposed approach is assessed in quantitative terms. The main goal of this analysis is to empirically demonstrate that both the measures contribute in identifying synonymous attributes. To perform this evaluation, firstly a ground truth has been manually created. We have exploited public attribute mappings directly provided by the Wikipedia Template pages to obtain a first minimal set of attribute matches. This basic information has been then extended with new manually inserted attribute correspondences. The generated ground truth includes about 2, 000 attributes clustered in 454 groups of synonyms. Once an exact set of attribute correspondences has been generated, we evaluated our approach on a sample of the entire collection of Wikipedia infoboxes. In more detail, we tested our approach on an *attribute-synonymy graph*, generated starting from the input dataset, consisting of 6, 854 attributes and 52, 707 synonymy relations. The results provided by our approach were finally compared with the ground truth.

To provide a measure of the quality of the clusters generated by our approach with respect the ground truth, we adopted four measures: precision, recall, f1 score and rand index. We calculate precision as $\frac{\#\ true\ synonyms\ in\ cluster}{\#\ total\ attributes\ in\ estimated\ cluster}$, and recall as $\frac{\#\ true\ synonyms\ in\ cluster}{\#\ total\ attributes\ in\ real\ cluster}$. The f1 score is a combination of precision and recall defined as $2 * \frac{precision*recall}{precision+recall}$. Finally, the rand index [14] was used to evaluate the similarity between the clustering solution produced by our approach and that provided by the ground truth.

The experiment aims, in particular, to evaluate the contribution of each measure in obtaining the final result. To force such behavior, a linear combination of the two measures has been introduced. Its formulation is proposed in Eq. 3, where the α parameter is used to weight the contribution of the measures.

$$SynonymyScore(a_i, a_j) = \alpha*PosOverlap(a_i, a_j)+(1-\alpha)NegCoocc(a_i, a_j) \quad (3)$$

The results of this experimentation are given in Table 6. The results show that, with reference to the company entity type, linear combinations that assign more importance to the positive evidence of synonymy produce better results. Table 6 shows only α ranging from 0.6 to 1, however with lower α results follow a similar trend: precision decreases and instead recall increases. The best configuration is $\alpha = 0.8$, that obtains the highest values in all evaluation measures. We observe that the configuration with $\alpha = 1$, where there is no contribution from the negative evidence of synonymy, is the one that obtain the highest precision level. Nevertheless, in that configuration, the recall, rand and f1 score levels decrease considerably.

4.4   Case Study

In this section we provide a small case study to show that synonymous attributes can support the extraction of high quality and accurate information from Wikipedia. Table 7 introduces 5 information needs a user would like to satisfy against the collection of infoboxes described in Sect. 4.1. Each information

Table 6. Effectiveness evaluation with different positive contributions

α parameter (positive contribution)	Precision	Recall	f1 score	Rand index
0.6	0.375	0.764	0.372	0.146
0.7	0.817	0.759	0.757	0.886
0.8	0.797	0.767	0.760	0.947
0.9	0.791	0.754	0.752	0.942
1	0.831	0.668	0.723	0.858

need has been transformed into 2 structured queries: one with the original filtering condition formulated by the user and the second where the attributes have been substituted with a number of disjunctive clauses, each one expressing the same information need but by using synonymous attributes. Table 7 shows the number of entities retrieved when both the queries are executed and the number of results it is expected to be retrieved. Last two columns show the same information in percentage. We observe that synonymous attributes largely support the retrieval of all results. The maximum improvement is obtained with the last query (i.e., `type="public company"`, `num_employees>10,000`, `year=2010`) where the application of synonymous attributes allows us to retrieve all results, instead of 19.15% of them, as we obtain with the original formulation.

Table 7. Case study

Query	# query results				
	Original clauses	Using synonymous attributes	Ground truth	Original clauses (%)	Using synonymous attributes (%)
location_city="tokyo"	268	293	298	89.93	98.32
founded<1900	681	705	705	96.59	100
location="USA", year=2014	381	531	531	71.75	100
location="united states", net_income>1 billion	371	429	429	86.48	100
type="public company", num_employees>10,000, year=2010	221	1154	1154	19.15	100

5   Related Work

Importance and Usage of Wikipedia Infoboxes. Wikipedia infoboxes have been used in a large number of research projects. The most significant works include techniques for building structured knowledge bases [3,18], for analyzing the evolution of specific kinds of entities [13] and for applying structured queries on the Wikipedia content [2]. Although several works support the formulation of structured queries, no previous effort has considered the evolutionary nature of Wikipedia. All previous approaches consider only a static snapshot of the infoboxes as input.

Schema Matching. Schema matching is one of the most studied topics in the database community. Books [5,11] and surveys [6,16] introduce the existing approaches in the literature. According to the categorization proposed by [16], schema matching approaches can be classified into schema-only matchers and instance-based matchers. Our proposal follows a hybrid approach since we incorporate holistic correspondence refinement that belongs to the category of collective matching approaches [6].

Our approach implements a strategy similar to the one proposed by He et al. [12] that consider as negative evidence the co-occurence of attribute names in the same schema. Their approach focus on the discovery of synonyms supporting a web search engine and is based on the combined use of web tables and query logs. The intuition is that users who are looking for the same results provide different synonyms as query terms on a search engine. This is used as positive evidence for attribute synonymy. Instead, attributes within the same web tables are not likely to be synonyms, thus providing a negative evidence. Our approach adapts that idea to work with Wikipedia infoboxes by extracting positive and negative evidence of synonymy analyzing co-occurrences of attributes and values.

Schema Matching and Alignment Over Infoboxes. Schema matching techniques have been applied against Wikipedia infoboxes in the context of finding correspondence between schemas in different languages. Adar et al. [1] propose a framework called Ziggurat that creates a supervised classifier based on features that are learned from a set of positive and negative examples extracted from data with heuristics. Bouma et al. [9] match attributes based on the equality of their values. Two values are equal if they have the same cross-language link or exactly the same literals. A different approach [17] exploits value similarity over infobox templates, where first an entity matching process is done, then templates are matched to obtain inter-language mappings between templates and finally attribute matching is done by means of similarity metrics. These approaches rely on similarity metrics that are sensitive to the syntax of the underline languages: they work well if the compared languages have the same root. To overcome this limitation [15] exploits different evidences for similarity and combines them in a systematic manner.

Exploration of Schema and Value Changes. In the context of data exploration, a recent line of research focuses on the exploration of changes over time. [7] is vision paper that introduces innovative concepts related to understanding changes that happen in the data over time. Furthermore, [8] designs a set of primitives supporting the exploration over schema and data of evolving datasets.

6   Conclusion

In this paper, we introduced an approach that automatically defines clusters of synonymous temporal-evolving infobox attributes. The approach is mainly based on two kinds of knowledge: a negative evidence of synonymy provided by co-occurrences of the attributes in the same infobox in a given time instance, and a positive evidence of synonymy generated by co-occurrences of similar values for the attributes in different time instances. We formalized this issue as a correlation clustering problem over a weighted graph and we used a linear programming model to solve it. Our experiments, over the last 13 years infoboxes history, shows that our approach is effective in discovering synonymous attributes.

References

1. Adar, E., Skinner, M., Weld, D.S.: Information arbitrage across multi-lingual Wikipedia. In: Proceedings of WSDM, pp. 94–103 (2009)
2. Agarwal, P., Strötgen, J.: Tiwiki: searching Wikipedia with temporal constraints. In: Proceedings of WWW, pp. 1595–1600 (2017)
3. Auer, S., Bizer, C., Kobilarov, G., Lehmann, J., Cyganiak, R., Ives, Z.G.: DBpedia: a nucleus for a web of open data. In: ISWC, pp. 722–735 (2007)
4. Bansal, N., Blum, A., Chawla, S.: Correlation clustering. Mach. Learn. **56**(1–3), 89–113 (2004)
5. Bellahsene, Z., Bonifati, A., Rahm, E. (eds.): Schema Matching and Mapping. Data-Centric Systems and Applications. Springer, Heidelberg (2011). https://doi.org/10.1007/978-3-642-16518-4
6. Bernstein, P.A., Madhavan, J., Rahm, E.: Generic schema matching, ten years later. PVLDB **4**(11), 695–701 (2011)
7. Bleifuß, T., Bornemann, L., Johnson, T., Kalashnikov, D.V., Naumann, F., Srivastava, D.: Exploring change - a new dimension of data analytics. PVLDB **12**(2), 85–98 (2018)
8. Bleifuß, T., Bornemann, L., Kalashnikov, D.V., Naumann, F., Srivastava, D.: DBChEx: interactive exploration of data and schema change. In: Proceedings of CIDR (2019)
9. Bouma, G., Duarte, S., Islam, Z.: Cross-lingual alignment and completion of Wikipedia templates. In: Proceedings of the Workshop on Cross Lingual Information Access, pp. 21–29. Association for Computational Linguistics (2009)
10. Demaine, E.D., Emanuel, D., Fiat, A., Immorlica, N.: Correlation clustering in general weighted graphs. Theor. Comput. Sci. **361**(2–3), 172–187 (2006)
11. Euzenat, J., Shvaiko, P.: Ontology Matching. Springer, Heidelberg (2007). https://doi.org/10.1007/978-3-540-49612-0
12. He, Y., Chakrabarti, K., Cheng, T., Tylenda, T.: Automatic discovery of attribute synonyms using query logs and table corpora. In: Proceedings of WWW, pp. 1429–1439 (2016)
13. Hoffart, J., Suchanek, F.M., Berberich, K., Weikum, G.: YAGO2: a spatially and temporally enhanced knowledge base from Wikipedia. Artif. Intell. **194**, 28–61 (2013)
14. Rand, W.M.: Objective criteria for the evaluation of clustering methods. J. Am. Stat. Assoc. **66**, 846–850 (1971)

15. Nguyen, T., Moreira, V., Nguyen, H., Nguyen, H., Freire, J.: Multilingual schema matching for wikipedia infoboxes. PVLDB 5(2), 133–144 (2011)
16. Rahm, E., Bernstein, P.A.: A survey of approaches to automatic schema matching. VLDB J. 10(4), 334–350 (2001)
17. Rinser, D., Lange, D., Naumann, F.: Cross-lingual entity matching and infobox alignment in Wikipedia. Inf. Syst. 38(6), 887–907 (2013)
18. Suchanek, F.M., Kasneci, G., Weikum, G.: YAGO: a large ontology from wikipedia and wordnet. J. Web Semant. 6(3), 203–217 (2008)

Web-Navigation Skill Assessment Through Eye-Tracking Data

Patrik Hlavac$^{(\boxtimes)}$ (iD), Jakub Simko (iD), and Maria Bielikova (iD)

Faculty of Informatics and Information Technologies, Slovak Technical University,
811 07 Bratislava, Slovakia
{patrik.hlavac,jakub.simko,maria.bielikova}@stuba.sk

Abstract. Eye-tracking data provide many new options in domain of user modeling. In our work we focus on the automatic detection of web-navigation skill from eye-tracking data. We strive to gain a comprehensive view on the impact of navigation skills on addressing specific user studies and overall interaction on the Web. We proposed an approach for estimating the web navigation skill, with support of self-evaluation questionnaire. We have conducted eye-tracking study with 123 participants. Dataset from this study serves as a base for exploration analysis. We pair different web-navigation behavior metrics with result score from our questionnaire in order to find differences between participant groups. The results of the classification show that some stimuli are more appropriate than others.

Keywords: Web navigation · Eye-tracking · Web literacy

1 Introduction

Aim of our research is to show that analyzing user's gaze can bring valuable information into user classification. In our paper we describe its relation to web-navigation skill.

The main goal is to automatically distinguish between two groups based on inferred gaze metrics. Groups differ in efficiency when performing navigational tasks.

We estimate one specific aspect of web-navigation skill, the web-navigation efficiency. Our proposed questionnaire focuses on user's success and efficiency during navigational tasks on the Web.

Short and automatic web-navigation skill assessment would be helpful UX tool with perspective usage during user studies but also in general user modeling. Our main motivation for starting to research a topic of web-navigation impact on user behavior was very limited existing knowledge about gaze relations in the field of User studies. Nowadays, User Experience (UX) research become an interesting domain because of growing popularity of UX testing worldwide. Thanks to the power of UX testing, it quickly became a massively used approach in industry. More and more companies started to use UX methods to improve their

© Springer Nature Switzerland AG 2019
T. Welzer et al. (Eds.): ADBIS 2019, LNCS 11695, pp. 186–197, 2019.
https://doi.org/10.1007/978-3-030-28730-6_12

products and services. More and more companies that support UX testing and consulting were established. Even general public is aware of these possibilities. This area deserves more exploration and deeper understanding.

Whilst qualitative studies generally consist of the interaction of a participant with an environment with presence of moderator as an important mediator, quantitative studies are carried mostly without the moderator and thus without further analysis of a perception of the given participant. Given that in this approach we usually gather a large amounts of data from logs or questionnaires, the evaluation of quantitative studies is inferred from methods of mathematical analysis [14].

Quantitative studies allow to generalize results for a greater population. However, this group could be so diversified, that we could not compare or evaluate the results. Broad study was performed with 445 participants [15] using an application to record data and session by cameras. There was an indication that each user has different characteristics and properties that enter the process and therefore that detecting of individual impacts could be more important than quantity of participants.

Our work is structured as follows. Related work section contains different views on the Web Navigation definition and skill assessment. Our study goal is explained in Sect. 3. Experimental methodology is described by scenario, participants, devices and acquired data, followed by results. We summarized the observed results in section Conclusions.

2   Related Work

2.1   Web Navigation Definition

Web navigation includes many areas that could be inspected as well as there exist many factors that should be taken into account while investigating.

The navigation term can describe user activity and also website element (that usually contains hyperlinks). When we explore the user navigation skill, we have to consider also navigational environment that includes different elements. And vice versa, when we measure the quality of website navigation, we should consider the navigational ability of each user.

Website navigation is typically based on visual perception, accompanied by the goal to get somewhere (most likely to meet information needs) and exploring ways of doing this.

Since most of the information regarding the various products, actions and situations from daily life has gradually shifted to a web-based platform, there is a need for definition of new concept.

Nowadays, we consider digital literacy as a person's ability to perform tasks effectively in the digital environment, with emphasis on the representation of information in a form usable for computers [13]. It covers a wide range of qualities, abilities and skills, such as reading and interpretation of the media, replicating and modifying data (from documents to pictures and music). It also deals with the electronic equipment such as computers, mobile phones and portable devices and software cooperating with them. It is very connected to the Information literacy (search, examination and handling of information) and media literacy (competences allowing to work with the media in various formats and genres) [1].

Information literacy, as well as the Digital literacy, has ambiguous and broadly defined the problem area. Thus, a new, unique concept of Web literacy [3] has been earmarked.

Unlike computer literacy or digital literacy, Web literacy includes more detailed aspects of Web interaction. Many definitions divide Web literacy into several areas, including navigation. Some standards describe it and provide specific examples; other standards rely on a detailed and unambiguous definition that does not require examples.

We consider the web-navigation skill as the ability to effectively move across the Web.

2.2   Web Navigation as an Ability

We expect the navigation behavior (the user's ability to intentionally move across the Web) is influenced mostly by two factors - the web layout and the user web-navigational skill. Navigation layout may also influence interaction and overall user satisfaction [9]. Navigation behavior is influenced by familiarity [6] (navigational expectations can help or distract if website uses elements of unexpected functionality) and the ability of keeping the path to correct destination.

Web navigation is a path traversed with or without main intention (goal). The interaction path then consists of all visited web-pages. When we look to a lower level, we can say the navigation even consists of all website elements and related interaction on each visited web-page element [11].

Web navigation without specific aim, but with intention could be depicted by typical task: *Get understand what is the site about.* In this case we are measuring the ability to traverse the most important parts of website by the most efficient way. Thus, the real ability is that we know where to get crucial information and how to navigate to the next information point.

Web navigation with intention should evoke the need of choosing the most correct and efficient way to reach the final website (destination): *Find specific product.*

2.3   Web Navigation as a Layout

Web navigation could be divided according the size of working environment into: wearable, mobile device (smartphones, tablets), standard screen (PC, notebook), we should consider even voice form of navigation for blind users. We can navigate through icons, texts (and text abbreviations), shortcuts, sounds (read links for sightless persons).

Navigation design includes specific differences on the level of context domains [7] (e.g. e-shop, e-banking, e-government), or even on the level of individual websites of the same context domain. Navigation layout is influenced by individual design approaches, which are based on constantly evolving technologies. Secondly, it is also influenced by the graphic style. In some practical cases, the navigation could be influenced by a framework or brand that serves the software (e.g. specific color, icons, actions).

2.4 Web Navigation as a Process

Web navigation has been evolving over the years as well as technical aspects of the Internet. At present, we can see some of the well-established standards of creating web navigation by navigational elements that will be familiar to users even when they first arrive at an unknown page. Effective web browsing is the foundation for good usability, and so ongoing approaches for creating navigation are constantly evolving and improving [12].

The way to the navigation goal could lead through many different paths. The chosen way can lead to assessment of user's navigation skill. This could be a very powerful method in the topic of skill estimation. As we discovered during work on skill estimation [8], users can be clustered by the path they chose during navigation to the target.

Every path has its own reasoning. Whether it is web-page influenced by website owner, that tried to lead customers to specific information (or additional advertisement), or whether it is well known shopping cart process, we can describe the navigational path.

2.5 Web Literacy Estimating

Most current assessments test users for overall computer usage, however they test web interaction only marginally and even less they test web-navigation behavior. Naturally, research experiments are specially designed for testing specific issues [7], therefore they are not applicable to various aspects. Despite the expectations, we did not find a great variety of proven and established Web literacy assessments. Existing tests for Information, Computer or Internet literacy go only partially into details that we can observe in Web literacy domain.

The results of well-known worldwide skills measurements are based mainly on the subjective opinions of the participants. They are calculated from the scale of how often student do some activity [10], but not how successful or effective he actually is. PISA 2012 tests were accompanied by computer-based tests focused on reading literacy [2]. Interaction was logged during work in specialized framework. Raw files were then treated to extract the navigation sequence and standard metrics (e.g. number of steps) was inferred.

Verification of literacy is significant for companies that provide education. In some cases, the assessment is just the last part of online educational process. Microsoft Digital Literacy[1] provides courses based on testing practical skills

---

[1] https://www.microsoft.com/en-us/DigitalLiteracy.

(Build Your First App; Creating an Internet Email Account; Writing a Great Resume; or newer Digital Literacy: Get Connected, Browse the Web, Search the Web etc.). Assessments contain from 30 to 90 questions according to the level of tested literacy. The user usually has to choose some of prepared options. Northstar[2] provides much sophisticated tool, which is another assessment service, but provides a more attractive and verbally moderated tests. It tests the user from the basics of text processing, presentations, web or information literacy in a nine assessments. Its environment simulates a computer desktop, interactive captcha verification, completing a web form or searching the Web. Questions are answered by click or by explicit "I don't know" button. iSkills[TM] Assessment[3] contains many types of environment for Defining, Accessing (filtering with select boxes), Evaluating, Managing (sorting content in e-mail client), Integrating, Creating (statistical environment with graphs) or Communicating (composing e-mail).

Our initial effort is to automatize estimation of user's web-navigation skill. There are many theoretically defined standards which help us to understand the issue in depth. However, there are almost no practical tests or questionnaires. While at the beginning of the millennium several attempts were made to estimate digital and Web literacy through questionnaires, they are no longer available today. In addition, with constant development and additional discoveries, what does the interaction on Web mean and what does it include - the number of such instruments to cover each area should have been created along with their definition. But specialized tools missed arise.

User experience research is full of different methods and modern approaches. On the one hand, in the UX industry, the time-limited website testing is the tool to understand the perception of the user. One of the pertaining method is the Three second rule, where after three seconds, user should be able to answer what he can expect from the website. This method could be also found as Five second rule[4] (with little adjustments).

On the other hand, UX research[5] shows that first few seconds of interaction with website are critical in terms of deciding to leave or to stay. There are even hints [5] that claim, that critical time that is needed to leave the website is decreasing, being now about 8 s.

How is it, when user has to orientate in 8 s? We assume that they will focus on the elements that they consider most important to gain most of information fast. Do experts choose the important elements better than novices?

3   Experiment: Influence of Web-Navigation Skill on User Behavior

The way that user perceive the website through exploring multiple elements allows us to model his navigation. We have chosen the scenario tasks that help

---

[2] https://www.digitalliteracyassessment.org/.

[3] https://forms.ets.org/sf/iskills/rfi/.

[4] https://fivesecondtest.com/.

[5] https://www.nngroup.com/articles/how-long-do-users-stay-on-web-pages/.

us to motivate the user to search for key information on the web through multiple elements (thus navigate). Key information would be those, which user consider to be helpful for orientation on current homepage.

Fig. 1. Sample of four homepage websites among all 32. Website differs in domain, content and elements.

Through exploratory analysis we strive to gain deeper insight into human behavior during navigational tasks on the Web. We chose 32 screenshots[6] from different website homepages that are provided as testing stimuli for each participant with simple task "to understand what is the website about". We record gaze data during sessions that are further processed to transition matrices. Our approach consists of three main parts:

– collection of eye-tracking data from the first encounter with the homepage,
– collection of a web-navigation skill questionnaire and calculating results,
– classification of quantified data on participant behavior.

We expect that the division of Web literacy by other aspects will be the step to modelling the Web literacy more transparently and automatically. Thus, maybe this will be the "missing link" in modern user modelling and a primal approach to fast and easy web-navigation skill detection.

We approach with analysis of user orientation on the website. Dataset contains behavior from the orientation on the homepages of different websites (see Fig. 1). We expect that users with similar skill would have similar behavior in terms of specific gaze metrics. We analyze only first 7 s of participant interaction to obtain the information about what elements considers participant as the most important when perceiving new website.

---

[6] http://eyetracking.hlavac.sk/web-navigation/.

3.1   Experiment Goals

The main goal is to find differences in eye-tracking data that allow us to classify users according their web-navigation skill. We have to go through the following steps:

- **To verify whether the questionnaire has a distinctive ability to estimate web-navigation skill.**
  The experiment was designed to compare the results of the proposed questionnaire with real behavior - more precisely, to find out whether the questionnaire can in specific cases significantly demonstrate a difference in metrics (duration of the task execution, number of tasks, SPI, gaze metrics).
- **To compare the behavior on different groups of participants on modern websites.**
  All participants perform four tasks on modern single-page websites. All four websites are of single-page structure, therefore links in the navigation only scroll the content and do not open new pages.
- **To analyze the user orientation on the homepages among groups.**
  Contains collecting the dataset from the orientation on the home page. These websites come from different domains (education, stores, services).

3.2   Questionnaire

In order to partially abolish the subjective assessment effect, we did not ask directly for web-navigation skill estimate. We rather asked about the situations associated with it. Using examples from existing questionnaires, we have formulated 19 statements, where the participant indicates on Likert scale 1–5 how much they agree with each (from 1 - definitely disagree, to 5 - definitely agree).

The questionnaire contains a participant identifier (usually 6 characters), filling time time-stamp, and integer values for 19 statement questions[7]. For example: "I often do not finish my online purchase because the page is too chaotic", "I often get to a page where I can not orientate myself", "It often happens that I do not find what I'm looking for at all."

Questionnaire is provided as online form. It is anonymous, individual codes are used to pair with gaze data. Questionnaire is not limited by the time or requirements to be met. Questions are provided in random order to each participant.

3.3   Apparatus

The testing room was equipped by 20 computers with eye-trackers and headphones, which enables multiple studies to be carried out at the same time with identical conditions. Participants were recorded by standalone Tobii X2-60 eye-tracker with 60 Hz sampling rate. 20 simultaneously running computers with OS Windows 8 was used in our setup [4]. The participants' display were 24" LCD. Referenced freedom of head movement was 30 × 15 cm at the distance of 70 cm.

---

[7] http://eyetracking.hlavac.sk/web-navigation/q_description.pdf.

3.4 Session Description

After welcoming and introducing study we performed eye-tracker calibration. Calibration of each participant was checked individually by a team member, only then the participants could start their session or re-calibrate. Recording part started by unrelated study with video stimuli, followed by our experiment: navigation tasks on live websites, getting familiar with static homepages and filling the questionnaire.

Scenario had the following purpose:

1. to collect reference eye-tracking data from the use of live websites (4 stimuli),
   - Task: to accomplish given task (find information, order product etc.). Task is finished by clicking on the correct link or choosing button "I do not know".
2. to collect eye-tracking data from the first encounter with the website homepage for short time (32 stimuli),
   - Task: familiarize with current website. 32 different website screenshots stimuli were presented in random order for 7 s. After each stimuli, participants were provided with form containing two questions ("what is the site about" and "what can you expect of it"). The main role of two tasks after each screenshot was to motivate all participants to achieve a conscientious interaction. We limited the answers by the size of text input field. Only small number of responds contained information that participant was unable to answer (2,9% of responds).
3. to complete a web-navigation skill questionnaire (19 questions).

3.5 Participants

123 participants, took part in this exploratory study. Very few participants were not able to pass the calibration process or we were not able to pair questionnaire score with few gaze recordings. The average age in the sample was 25.29 (SD = 7.16, min = 19, max = 55). Most of the sample were females (n = 95),

Participant groups came from different environments. Different groups participated:

- students of pedagogy on bachelor grade (N = 45),
- students of pedagogy on master grade (N = 35),
- high school teachers (N = 28),
- IT faculty doctoral students and researchers (N = 15).

3.6 Data

Eyetracking Data. The recorded gaze interaction duration on each stimuli across all participants was lower than expected 7 s, but stable (M = 5.66 s, SD = 0.18 s). The duration of participant's gaze interaction across all stimuli varies much more (M = 6.30 s, SD = 1.62 s).

Three unexpected situations happened when stimuli was displayed considerably more than 7 s (12 s, 25 s, 30 s) due to software error. These recordings were cropped to standard 7 s in further analysis.

For analysis were only used cleaned gaze data where eye-tracker was able to estimate the Gaze Point position on the screen.

We defined 54 different areas of interest (AOI), among all stimuli, from which we picked 4 (logo, site options, footer, top navigation) with the highest occurrence among stimuli. Their intersection resulted in 21 websites, that we further used for classification.

We worked with gaze data processed into AOI transition matrix. We generated unique matrix for every participant on each stimuli. We use their fixation points to calculate transitions between AOI. As the result we got relatively sparse matrices with count of transitions in each cell, see Fig. 2.

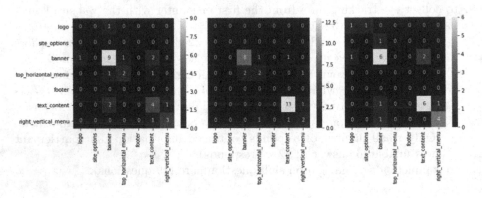

Fig. 2. Sample of three transitional matrices, for three different participants on the same stimuli. Number of transitions will become a classification feature.

Questionnaire Data. Questions are not strongly correlating, as can be seen on Fig. 3, so we should get comprehensive information about participant.

Although, overall success on four tasks was relatively high (M = 3.44, SD = 0.93). Success shown on participants ordered according questionnaire score differ for top 20% participants (M = 3.85, SD = 0.37) and bottom 20% (M = 3.5, SD = 0.95).

3.7 Normalization

Each cell on Fig. 2 represents behavior, whether participant has transition between two specific areas. Each matrix was converted to 1-dimensional array and used for Support Vector Classification, therefore each matrix cell became a classification feature. These data are referred as *Raw* in Table 1.

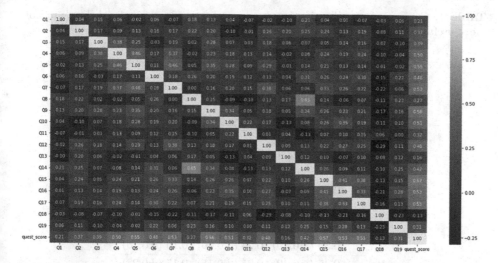

Fig. 3. Correlation matrix of 19 tasks based on 158 respondents

In order to determine further relations, we used data normalization in two more testings. *Norm1* is used for labeling arrays where values are scaled to 0–1 but in each array solely - so this normalization is related for single participant. *Norm2* is used for labeling arrays where values are scaled to 0–1, but maximum value is calculated among all participants.

For classification was used SVC with k-fold cross-validation with 20 splits.

3.8   Results

From the wide range of participants (117) we picked extreme samples (20 best and 20 worst according to their web-navigation score). These were labeled by two classes for binary classification ("less skillful", "more skillful" participant). For every stimuli we got evenly balanced 40 instances.

Results of the classification for each stimuli separately can be seen in Table 1. Each row contains classification accuracy with standard deviation. Only few websites (stimuli 6, 8, 11, 12, 14) support the classification to our two classes. Almost in all cases, standard normalization approaches were not useful for classification.

For further analysis we see possible improvements in understanding the differences between stimuli that are useful for classification and that which are not useful, and therefore not suitable. Because even though they have similar features and website elements, they have various usage and purpose. Possible solution for improving results would be to infer more general metrics, that could be used across all stimuli and will enlarge the testing sample.

Table 1. Comparing classification results on each stimuli. Normalized data are not more useful for this type of task.

Stimuli	Raw (std)	Norm1 (std)	Norm2 (std)
0	0.42 (±0.40)	0.72 (±0.33)	0.42 (±0.43)
1	0.47 (±0.37)	0.03 (±0.11)	0.15 (±0.28)
2	0.28 (±0.37)	0.15 (±0.28)	0.03 (±0.11)
3	0.45 (±0.42)	0.47 (±0.37)	0.42 (±0.43)
4	0.25 (±0.34)	0.17 (±0.33)	0.25 (±0.34)
5	0.47 (±0.37)	0.35 (±0.36)	0.15 (±0.28)
6	0.60 (±0.41)	0.42 (±0.40)	0.20 (±0.37)
7	0.50 (±0.39)	0.55 (±0.35)	0.17 (±0.29)
8	0.70 (±0.33)	0.68 (±0.36)	0.30 (±0.43)
9	0.30 (±0.40)	0.28 (±0.40)	0.10 (±0.25)
10	0.45 (±0.42)	0.05 (±0.15)	0.15 (±0.28)
11	0.72 (±0.37)	0.60 (±0.41)	0.15 (±0.28)
12	0.60 (±0.41)	0.30 (±0.43)	0.25 (±0.40)
13	0.38 (±0.38)	0.45 (±0.42)	0.38 (±0.47)
14	0.72 (±0.33)	0.47 (±0.40)	0.10 (±0.20)
15	0.28 (±0.29)	0.38 (±0.38)	0.17 (±0.33)
16	0.55 (±0.38)	0.20 (±0.24)	0.10 (±0.20)
17	0.45 (±0.44)	0.35 (±0.39)	0.33 (±0.43)
18	0.50 (±0.42)	0.38 (±0.41)	0.17 (±0.29)
19	0.35 (±0.39)	0.35 (±0.42)	0.15 (±0.28)
20	0.40 (±0.34)	0.07 (±0.18)	0.17 (±0.33)

4   Conclusions

In this paper, we provide new approach for estimating user web-navigation skill. We strive to determine user web-navigation skill from his interaction represented by eye-tracking metrics. We needed to take several steps to conduct this analysis. We performed exhaustive study on variety of participants in order to compare their web-navigation skill. Skill estimation was assessed by specialized self-evaluation questionnaire. Distribution of the questionnaire results suggests a realistic representation, additionally confirmed by the success in control tasks.

The overall results show that it is necessary to infer more user-related features to obtain better classification accuracy. The results of the transition matrix analysis revealed a need for sophisticated normalization of features among users that enter the classification. However, as we conducted the user study on multiple website homepages, we can see the results vary with each tested stimuli. For our classification we chose stimuli with four most common features to maintain a large number of stimuli. Equally important is to understand all the website elements relations and differences among stimuli.

Acknowledgement. This work was partially supported by the Slovak Research and Development Agency under the contracts No. APVV-15-0508, APVV SK-IL-RD-18-0004, grants No. VG 1/0725/19, VG 1/0409/17. The authors would like to thank for financial contribution from the STU Grant scheme for Support of Young Researchers.

References

1. Ala-Mutka, K.: Mapping digital competence: towards a conceptual understanding. Institute for Prospective Technological Studies, Sevilla (2011)
2. Bardini, C.: Computer-based assessment of mathematics in PISA 2012. In: Stacey, K., Turner, R. (eds.) Assessing Mathematical Literacy, pp. 173–188. Springer, Cham (2015). https://doi.org/10.1007/978-3-319-10121-7_8
3. Belshaw, D.E.A.: Working towards a framework to understand the skills, competencies and literacies necessary to be a webmaker (2013). https://wiki.mozilla.org/Learning/WebLiteracyStandard/Legacy/WebLiteraciesWhitePaper
4. Bielikova, M., et al.: Eye-tracking en masse: group user studies, lab infrastructure, and practices. J. Eye Mov. Res. **11**(3), 6 (2018)
5. Canada Microsoft: Attention spans. Technical report, Microsoft (2015). http://dl.motamem.org/microsoft-attention-spans-research-report.pdf
6. Eraslan, S., Yesilada, Y.: Patterns in eyetracking scanpaths and the affecting factors. J. Web Eng. **14**(5&6), 363–385 (2015)
7. Juvina, I., Van Oostendorp, H.: Individual differences and behavioral metrics involved in modeling web navigation. Univ. Access Inf. Soc. **4**(3), 258–269 (2006)
8. Kubanyi, J., Hlavac, P., Simko, J., Bielikova, M.: Towards automated web navigation and search skill assessment: an eye-tracking study on the skill differences. In: 2018 13th International Workshop on Semantic and Social Media Adaptation and Personalization (SMAP), pp. 49–54. IEEE (2018)
9. Murano, P., Lomas, T.J.: Menu positioning on web pages. Does it matter? Technical report, 4 (2015). www.ijacsa.thesai.org
10. OECD: Educational Research and Innovation Innovating Education and Educating for Innovation The Power of Digital Technologies and Skills: The Power of Digital Technologies and Skills. OECD Publishing (2016). https://books.google.sk/books?id=ylYhDQAAQBAJ
11. van Oostendorp, H., Aggarwal, S.: Modeling and supporting web-navigation. J. Interact. Sci. **3**(1), 3 (2015)
12. Pilgrim, C.J.: Website navigation tools: a decade of design trends 2002 to 2011. In: Proceedings of the Thirteenth Australasian User Interface Conference, vol. 126, pp. 3–10. Australian Computer Society, Inc. (2012)
13. Roberts, N.J.: Toward a literate future: pairing graphic novels and traditional texts in the high school classroom. Ph.D. thesis, Colorado State University, Libraries (2012)
14. Rohrer, C.: When to use which user-experience research methods. Nielsen Norman Group (2014)
15. Şengel, E.: Usability level of a university web site. Proc. Soc. Behav. Sci. **106**, 3246–3252 (2013)

Ontologies and Knowledge Management

Updating Ontology Alignment on the Concept Level Based on Ontology Evolution

Adrianna Kozierkiewicz[ID] and Marcin Pietranik[✉][ID]

Faculty of Computer Science and Management, Wroclaw University of Science
and Technology, Wybrzeze Wyspianskiego 27, 50-370 Wroclaw, Poland
{adrianna.kozierkiewicz,marcin.pietranik}@pwr.edu.pl

Abstract. According to one of the base definitions of an ontology, this representation of knowledge can be understood as a formal specification of conceptualization. In other words - they can be treated as a set of well-defined concepts, which represent classes of objects from the real world, along with relationships that hold between them. In the context of distributed information systems, it cannot be expected that all of the interacting systems can use one, shared ontology. It entails a plethora of difficulties related to maintaining such a large knowledge structure. A solution for this problem is called an ontology alignment, sometimes it is also referred to as an ontology mapping. It is a task of designating similar fragments of ontologies, that represent the same elements of their domain. This allows different components of a distributed infrastructure to preserve its own independent ontology while asserting mutual interoperability. However, when one of the participating ontologies change over time, the designated alignment may become stale and invalid. As easily seen in a plethora of methods found in the literature, aligning ontologies is a complex task. It may become very demanding not only in terms of its computational complexity. Thus relaunching it from the beginning may not be acceptable. In this paper, we propose a set of algorithms capable of updating a pre-designated alignment of ontologies based solely on the analysis of changes applied during their evolution, without the necessity of relaunching the mapping algorithms from scratch.

Keywords: Ontology alignment · Ontology evolution ·
Knowledge management

1 Introduction

In recent years knowledge bases incorporated in modern information systems must adapt to rapidly changing requirements originating from the real world. A modeled universe of discourse may change, and therefore, its formal representation must follow these changes. Such a situation entails difficulties that concern asserting that components interacting with the evolving knowledge base will remain up to date.

© Springer Nature Switzerland AG 2019
T. Welzer et al. (Eds.): ADBIS 2019, LNCS 11695, pp. 201–214, 2019.
https://doi.org/10.1007/978-3-030-28730-6_13

One of the tools which allow expressing knowledge in information systems is using ontologies. In literature, they are frequently described as a formal specification of conceptualization and are one of the building blocks of modern semantic applications. Informally speaking, they can be treated as a set of well-defined concepts, which represent classes of objects from the real world, along with relationships that hold between them.

In the context of distributed, heterogeneous information systems, one cannot expect that all of them will utilize one, shared ontology as a foundation of their knowledge bases. Different systems have different requirements and enforcing a common ontology would eventually make it impossible to maintain. To allow different systems to incorporate different ontologies, while asserting consistent communication, a bridge between ontologies is needed. In the literature, the topic of creating such a bridge is referred to as ontology alignment. In essence, it can be described as a task of selecting parts of two (or more) ontologies that express the same parts of a modeled universe of discourse, although differently ([5]).

There are a variety of approaches and solutions to this task. None of which is computationally inexpensive - processing large, expressive knowledge structures and finding similarities between them is obviously a very complex task. Therefore, a designated alignment between ontologies is very valuable. However, what if one of the aligned ontologies needs to be altered due to unforeseen, new requirements? Obviously, the procedure of designating the alignment may be executed from scratch, but this solution requires to incur all computational costs once more.

In our earlier publications ([19] and [13]) we created a mathematical foundation for managing changes that may be applied to ontologies during their lifespan. Based on these notions, a criterion for updating ontology alignment was developed. It can be used to answer a question concerning when changes applied to ontologies are significant enough to potentially invalidate the alignment between ontologies.

The main goal of the following paper is focused on a question - is it possible to update the existing alignment between two ontologies, if one of them changes in time, based only on the analysis of these changes. Formally, it can be defined as: *For a given source ontology $O^{(m)}$ in a moment in time denoted as m, a target ontology $O'^{(n)}$ in a moment in time denoted as n, and an alignment between these two ontologies denoted as $Align(O^{(m)}, O'^{(n)})$, one should provide algorithms which can update this alignment if the source ontology evolves from the state $O^{(m)}$ to the state $O^{(m+1)}$ according to applied changes.* Due to the limited space, we will focus only on a concept level available in ontologies.

The article is structured as follows. In the very next section, a selection of related works is presented. Section 3 provides basic notions and definitions that will be further used throughout the paper. In Sect. 4 algorithms for updating the ontology alignment on the concept level can be found. These algorithms have been tested using an experimental procedure described in Sect. 5. A brief summary and an overview of our upcoming research are given in Sect. 6.

2   Related Works

An ontology is a structure that allows to store and process knowledge. It allows reflecting the complexity of the knowledge, relations, hierarchies and other dependencies between objects or concepts. However, knowledge stored in ontologies could be out of date and an update process needs to be conducted to allow reliable reasoning based on a current ontology. Many authors notice the need for the management of an ontology evolution process. In the literature [1,6,10,17,20] it is possible to find some systems and methods devoted to detecting changes in the ontology and management procedures.

The ontology evolution management becomes more complicated if the knowledge is distributed. The carrying out the ontology integration (also referred to as merging) requires an actual alignment between input ontologies. However, any modification of ontologies may entail changes in the existing alignment. Thus, the ontology alignment evolution is a real problem and to the best of our knowledge, it has not been well investigated so far.

In [19] authors proposed a preliminary algorithm for revalidation and preserving the correctness of alignment of two ontologies. An original method for identifying the most relevant subset of the concept's attributes, which is useful for interpreting the evolution of mappings under evolving ontologies, was presented in [4]. The proposed solution aims at facilitating maintenance of mappings based on the detected attributes. COnto-Diff [7] is a rule-based approach that detects high-level changes according to a dedicated language of changes. COnto-Diff can also manage the alignment evolution process. Spotted ontology changes are coupled with a mapping between the elements (concepts, properties) of two ontology versions. The application of found modifications (and their inverses) is also considered in this work.

Euzenat and others ([5,21]) noticed that if ontologies evolve, then new alignment has to be produced following the ontology evolution. Authors claimed that this can be achieved by recording the changes made to ontologies and transforming those changes into an alignment (from one ontology version to the next one). This can be used for computing new alignments that will update the previous ones. In this case, previously existing alignments can be replaced by their composition with the ontology update alignment. However, the alignment evolution problem is raised only from a theoretical point of view. Thus, the Alignment API and Alignment Server tool [3] do not provide any service related with changing of alignment.

Authors [11] distinguished two approaches to the problem of mapping evolution and claimed that both solutions proposed so far have several drawbacks and cannot constitute a generic solution. The first approach, called mapping composition, relies on designating direct mappings between data sources S and T' if two mappings between S and T also T and T' are given. This method is used for maintaining semantic mappings between relational schemas and ontologies in [2]. The second approach, mapping adaptation, focus on incrementally adapting mappings following schema changes. Thus, after every, even small, change (like adding, moving or deleting elements from ontologies) the mappings are updated. This approach is more often used in the literature.

In [23] authors presented a semi-automatic approach to update the adapting mapping document. For an existing ontology change, pattern mapping updating rules are directly applied to remove the staleness. When new changes arrive an ontology engineer computes the rules and updates their repository.

In [14] Authors defined two operators for updating an ontology-based data access specification when it's TBox and/or source schema change. Each such operator is based on a specific notion of mapping repair. The deletion-based mapping repair approach looks for repairs among subsets of the original mappings. Then, entailment-based mappings' repair procedure, aimed at preserving as much as possible of mapping assertions, is implied by the original specification.

The [15] proposes two approaches for managing of the ontology mapping evolution. The first one is a user-centric approach, in which the user defines the mapping evolution strategies to be applied automatically by the system. The second is a semantic-based approach, in which the ontology's evolution logs are exploited to capture the semantics of changes and then adapted to (and applied on at) the ontology mapping evolution process. The authors noticed, that changes to ontologies affect the ontology mapping in two ways depending on whether it is an addition or removal. In the case of removal, existing semantic relationships are affected and must evolve. Additionally, if adding new ontology entities, new semantic relations might be necessary. This remark has been applied for managing of the ontology mapping evolution. In the first approach, the user gets only the list of possible corrective changes. The proposed semantic-based approach automatically suggests the best corrective strategy, based on the information manage provided by the ontology evolution process.

Khattak and others [9] also used a change history log. These logged changes are later used in a change history log to store the ontology change information, which helps to drastically reduce the reconciliation time of the mappings between dynamic ontologies. The authors conducted a very comprehensive evaluation of the mapping reconciliation procedure which confirmed its effectiveness.

Despite the fact that in the literature it is possible to find some methods for detecting changes in ontologies the problem of managing the mapping evolution is not widely examined. There is a lack of tools dedicated to the mentioned problems. Indeed, PrompDiff [16] is particularly good at finding alignments between versions of ontologies. When such an alignment is made available, it is possible to provide a composition of new versions of the alignment tied to the previous version and eventually, to perform a migration. However, there is still a lack of tools dedicated directly to updating an ontology alignment when ontologies evolve.

3   Basic Notions

In the following article, we focus on the level concepts. Therefore, provided definitions concern only elements related to this level. For a broader overview of the formal foundations of our research, please refer to one of our previous publications, for example, [12].

A pair *(A, V)*, where *A* is a set of attributes and *V* is a set of attributes domains is called "a real world". An *(A, V)*-based ontology is defined as a quintuple:

$$O = (C, H, R^C, I, R^I) \qquad (1)$$

where *C* is a set of concepts; *H* is a concepts' hierarchy; R^C is a set of relations between concepts $R^C = \{r_1^C, r_2^C, ..., r_n^C\}$, $n \in N$, such that $r_i^C \in R^C$ ($i \in [1, n]$) is a subset of $C \times C$; *I* is a set of instances' identifiers; $R^I = \{r_1^I, r_2^I, ..., r_n^I\}$ is a set of relations between concepts' instances.

A concept's $c \in C$ structure from the *(A, V)*-based ontology is defined as:

$$c = (id^c, A^c, V^c, I^c) \qquad (2)$$

where: id^c is its identifier, A^c is a set of concept's *c* attributes ($A^c \subseteq A$) and their domains V^c ($V^c = \bigcup_{a \in A^c} V_a$ where V_a is a domain of an attribute *a* taken from the set *V*), and I^c is a set of concept's *c* instances.

To give attributes from the set A^c meanings, we assume an existence L_s^A, which is a sub-language of the sentence calculus, and a function $S_A : A \times C \rightarrow L_s^A$, which assigns logic sentences to attributes taken from a concept *c*. For example, an attribute *BirthDate* within a concept *Human* may be assigned with the following semantics: $S_A(BirthDate, Human) : birthYear \wedge birthMonth \wedge birthDay \wedge age$.

A concept's overall meaning (its context) can be defined as a conjunction of its attributes' semantics. Formally, for a concept *c*, such that $A^c = \{a_1, a_2, ..., a_n\}$, its context is as follows $ctx(c) = S_A(a_1, c) \wedge S_A(a_2, c) \wedge ... \wedge S_A(a_n, c)$.

Our approach to the ontology evolution is based on a notion of a timeline, which can be treated as an ordered set of discrete moments in time. Formally, it is defined as $\overline{TL} = \{t_n | n \in N\}$. $TL(O)$ is a subset of this timeline, containing only those moments from $\overline{TL}$ during which the ontology *O* has been modified changed. A superscript $O^{(m)} = (C^{(m)}, H^{(m)}, R^{C(m)}, I^{(m)}, R^{I(m)})$ is used to denote the ontology *O* in a given moment in time $t_m \in TL(O)$. A symbol $\prec$ is used to denote a fact that $O^{(m-1)}$ is an earlier version of *O* than $O^{(m)}$ ($O^{(m-1)} \prec O^{(m)}$). This notation is further used for particular elements of the given ontology, e.g. $c^{(m-1)} \prec c^{(m)}$ represents a fact that a concept *c* has at least two versions, and $c^{(m-1)}$ is an earlier one. A repository of an ontology *O*, is an ordered set of its successive versions, It is defined as $Rep(O) = \{O^{(m)} | \forall m \in TL(O)\}$.

A function $diff_C$ is defined to allow comparing different versions of the same ontology. Its input are two successive states of a single ontology *O*, $O^{(m-1)}$ and $O^{(m)}$ (such that $O^{(m-1)} \prec O^{(m)}$). Its output is three sets containing concepts added, deleted and modified. Formally:

$$diff_C(O^{(m-1)}, O^{(m)}) = \Big\langle new_C(C^{(m-1)}, C^{(m)}),$$
$$del_C(C^{(m-1)}, C^{(m)}), \qquad (3)$$
$$alt_C(C^{(m-1)}, C^{(m)}) \Big\rangle$$

where:

1. $new_C(C^{(m-1)}, C^{(m)}) = \left\{ c \middle| c \in C^{(m)} \wedge c \notin C^{(m-1)} \right\}$

2. $del_C(C^{(m-1)}, C^{(m)}) = \left\{ c \middle| c \in C^{(m-1)} \wedge c \notin C^{(m)} \right\}$

3. $alt_C(C^{(m-1)}, C^{(m)}) = \left\{ (c^{(m-1)}, c^{(m)}) | c^{(m-1)} \in C^{(m-1)} \wedge c^{(m)} \in C^{(m)} \wedge \right.$
   $c^{(m-1)} \prec c^{(m)} \wedge (A^{c^{(m-1)}} \neq A^{c^{(m)}} \vee V^{c^{(m-1)}} \neq V^{c^{(m)}} \vee I^{c^{(m-1)}} \neq I^{c^{(m)}}) \vee$
   $\left. ctx(c^{(m-1)}) \neq ctx(c^{(m)}) \right\}$

The first two descriptors are self-explanatory. The last one represents a modification of concepts from $O^{(m-1)}$, as a set of pairs of concept's versions, that are neither new nor removed, but internally different.

Assuming that there are two independent, (A, V)-based ontologies, O and O', their alignment on a concept level can be defined as a set $Align(O, O')$ containing tuples (further referred to as mappings) of the form:

$$(c, c', \lambda_C(c, c'), r) \tag{4}$$

where: c and c' are concepts from O and O' respectively, $\lambda_C(c, c')$ is a real value representing a degree to which the concept c can be aligned into the concept c', and r is a relation's type connecting c and c' (equivalency, generalisation or contradiction). $\lambda_C(c, c')$ can be designated using one of the matching methods described in a literature related to ontology alignment ([22]). In our research we use an approach used in [8], which has been proved useful in a variety of applications and evaluation procedures.

In order to extend the definition above to time-tracked ontologies, we use the aforementioned superscript notation. For example, $Align(O^{(m)}, O'^{(n)})$ represents an alignment of the ontologies O and O' in states in a moments m and n respectively (where both $m, n \in \overline{TL}$).

In our work we assume that the alignment $Align(O^{(m)}, O'^{(n)})$ fulfils a *taxonomic completeness postulate*, defined formally below:

$$\forall (c, c', \lambda_C(c, c'), r) \in Align(O^{(m)}, O'^{(n)}) \bigg((b, c) \in H \implies$$
$$\exists (b, c', \lambda_C(b, c'), r') \in Align(O^{(m)}, O'^{(n)}) \bigg) \tag{5}$$

The equation above states that if some concept c taken from the first ontology O has a predecessor b within the concepts' hierarchy H, then they can be both mapped to a concept c' from the second ontology.

4   Updating Ontology Alignment on a Concept Level

In this section of the paper, a set of three algorithms for updating the ontology alignment on the concept level will be presented. Every algorithm takes as an input a result of a comparison of two states of the source ontology $diff_C(O^{(m-1)}, O^{(m)})$ and an alignment between the source ontology (before the applied changes) and a target ontology $Align(O^{(m-1)}, O'^{(n)})$. As a result, algorithms will return the updated alignment $Align(O^{(m)}, O'^{(n)})$, valid for the most recent versions of both ontologies.

As defined on Eq. 3, the ontology evolution on the concept may entail appearing new concepts within an ontology, modifying some of them and deleting them. These three aspects may have a big influence on an established alignment between the tracked source ontology (that will be further denoted as O) and some other target ontology (further denoted as O'). Therefore, revalidating the alignment $Align(O^{(m-1)}, O'^{(n)})$ established in the past and asserting that it is valid in the current moment in time $Align(O^{(m)}, O'^{(n)})$ requires to analyse elements of sets from Eq. 3. This task can be divided into three, simultaneously run sub-procedures.

The first one (defined on Algorithm 1) is a fairly simple procedure, due to the fact that the task at hand is very straightforward. Removing concepts from the ontology O must be followed by removing all of the mappings that involved these concepts. The algorithm at first creates an auxiliary set of mappings that needs to be removed (Line 2) and eventually deletes them from the alignment $Align(O^{(m-1)}, O'^{(n)})$ (Line 3).

Algorithm 2 addresses a more complex issue concerning new concepts added to the ontology O. In such a situation all new concepts must be confronted with the target ontology O' in case they can be mapped with any of them. In this particular algorithm (for illustrative reasons) we use the most basic approach (Lines 3–4) which incorporates the concept alignment degree function $\lambda_C(c, c')$ (Line 5) defined in our previous research ([18]). This method may be easily swapped with a more refined ontology alignment tool (that can be found in the literature, e.g. [22]). However, despite the complexity of the approach presented on Algorithm 2, its output fulfills Eq. 5, which may not be met by other alignment methods out of the box.

Algorithm 3 targets the most complex issue concerning concepts which structures have been updated. In the new state, the applied updates may entail three subproblems:

Algorithm 1: Removing stale mappings of deleted concepts from the existing alignment

Input : $diff_C(O^{(m-1)}, O^{(m)})$, $Align(O^{(m-1)}, O'^{(n)})$
Output: $Align(O^{(m)}, O'^{(n)})$
1 **begin**
2     $del := \{(c, c', \lambda_C(c, c'), r) \in Align(O^{(m-1)}, O'^{(n)}) | c \in del_C(C^{(m-1)}, C^{(m)})\}$;
3     $Align(O^{(m)}, O'^{(n)}) := Align(O^{(m-1)}, O'^{(n)}) \setminus del$;
4   **return** $Align(O^{(m)}, O'^{(n)})$;
5 **end**

---

1. mappings connecting a particular, changed concept may no longer be valid and must be removed
2. the taxonomic completeness may no longer be fulfilled (due to its incompleteness or mappings that are no longer valid) and needs to be re-asserted
3. an updated concept may be mapped into concepts from the target ontology, which was not possible in its previous state

Algorithm 3 is, therefore, build around three sections, each dealing with specific situations described above. At first (Line 3) the algorithm generated a set of mappings connecting changed concepts. It is denoted as $\widetilde{alt}$ and contains concepts' mappings from the source ontology and the target ontology in a state from before the applied modification. This is achieved using the alt_C component of $diff_C$ function from the Eq. 3. Then, throughout Lines 5–9, each of the mappings from $\widetilde{alt}$ is re-checked for its correctness using current states of concepts. The validity is checked (Line 6) based on a difference between the degree to which two concepts in a state before the modification can be aligned and the degree they can be aligned after the alteration. If this difference is higher than some assumed threshold, than the validity no longer holds (Line 6). In such a situation a particular, invalid mapping is removed (Lines 7–8).

After revalidating existing alignments, the algorithm begins (Line 11) to process a set of changed concepts in order to re-assert the taxonomic completeness. It is based on iterating over the set of mappings of modified concepts using the alt_C component of $diff_C$ function from the Eq. 3. At first (Line 12), an auxiliary set C' is created - it is a copy of concepts from the target ontology. For each pair it is checked (Line 13) if a particular concept from the source ontology in the new state m acquired any new super-concepts within a taxonomy $H^{(m)}$ (Line 14). If that is the case, then the alignment is updated with appropriate mappings of ancestors in order to fulfill the taxonomic completeness postulate from Eq. 5 (Line 15).

Algorithm 2: Updating the existing alignment with new mappings

Input : $diff_C(O^{(m-1)}, O^{(m)}), Align(O^{(m-1)}, O'^{(n)}), \tau$
Output: $Align(O^{(m)}, O'^{(n)})$

1 **begin**
2    $Align(O^{(m)}, O'^{(n)}) := Align(O^{(m-1)}, O'^{(n)})$;
3   **for** $c^{(m)} \in new_C(C^{(m-1)}, C^{(m)})$ **do**
4   **for** $c' \in C'^{(n)}$ **do**
5   **if** $\lambda_C(c^{(m)}, c') \geq \tau$ **then**
6          $Align(O^{(m)}, O'^{(n)}) :=$
7            $Align(O^{(m)}, O'^{(n)}) \cup \{(c^{(m)}, c', \lambda_C(c^{(m)}, c'), r)\}$
8   **end**
9   **end**
10   **end**
11   **return** $Align(O^{(m)}, O'^{(n)})$;
12 **end**

The opposite situation is checked in lines 18–20. If some concepts have lost some of its super-concepts (Line 18) - the corresponding mappings are removed (Line 19) in order to meet the same taxonomic completeness postulate.

Checking both situations concerning changes in the taxonomy of the source ontology if followed by potential removing target concepts from the aforementioned auxiliary set $\widetilde{C'}$ (Line 15 and 19). This allows the algorithm to end with checking if any new mappings may by included into the final alignment. It iterates over the remaining, "untouched" concepts from the target ontology (Line 21) and performs in Lines 22–24 an analogical procedure to the one presented in Algorithm 2 Obviously, this part of the presented algorithm may be performed along with the aforementioned Algorithm 2 However, all three presented algorithms may be conducted simultaneously.

The next section of the paper contains a description of an experimental procedure we have conducted, along with an analysis of collected results. The experiment was designed to verify the presented algorithms in terms of correctness and completeness of their output in comparison with alignments obtained by re-launching the mappings procedures for whole ontologies.

5   Experimental Verification

Our updating procedure has been experimentally verified using benchmark datasets provided by the Ontology Alignment Evaluation Initiative (OAEI). We have chosen "a Conference Track" consisting of 16 ontologies describing the domain of organizing conferences, that was used in the OAEI'2018 campaign. The experiment has been divided into two parts which are described in detail below.

Our approach for updating ontology alignment on the concept level has been compared with the determination a mapping between two ontologies from the beginning by using LogMap ([8]). LogMap is an ontology alignment and alignment repair system which earned high positions in subsequent OAEI campaigns. For the comparison of both tested methods, we have used an accuracy measure. It is calculated as the number of common links (concepts' mappings) between two ontologies divided by the number of all connections found by both methods.

Formally, let us denote by $Align_{LogMap}$ a set of mappings (defined according to Eq. 4) created by LogMap, and by $Align$ a set being a result of the update procedures described in Sect. 4. The accuracy can be therefore measured using the equation below:

$$Accuracy = \frac{|Align_{LogMap}(O^{(m+1)}, O') \cap Align(O^{(m+1)}, O')|}{|Align_{LogMap}(O^{(m+1)}, O') \cup Align(O^{(m+1)}, O')|} \quad (6)$$

The aim of the first part of the experiment was to show how changes in base ontologies influence changes of alignments. For this purpose, we have chosen a source ontology (called *edas*) and a target ontology (called *ekaw*). The source ontology has been modified in a random way. In the subsequent steps, we have applied more changes by adding new concepts. In the beginning, only a 10%

Algorithm 3: Updating modified alignments of modified concepts

    Input : $diff_C(O^{(m-1)}, O^{(m)}), Align(O^{(m-1)}, O'^{(n)}), \epsilon$
    Output: $Align(O^{(m)}, O'^{(n)})$

1 **begin**
2    $Align(O^{(m)}, O'^{(n)}) := Align(O^{(m-1)}, O'^{(n)})$;
3    $\widetilde{alt} := \{(c^{(m-1)}, c', \lambda_C(c^{(m-1)}, c'), r) \in Align(O^{(m-1)}, O'^{(n)})|$
4      $c^{(m-1)} \in alt_C(C^{(m-1)}, C^{(m)})$;
5   **for** $(c^{(m-1)}, c', \lambda_C(c^{(m-1)}, c'), r) \in \widetilde{alt}$ **do**
6   **if** $|\lambda_C(c^{(m-1)}, c') - \lambda_C(c^{(m)}, c')| \geq \epsilon$ **then**
7        $Align(O^{(m)}, O'^{(n)}) := Align(O^{(m)}, O'^{(n)}) \backslash$
8        $\{(c^{(m-1)}, c', \lambda_C(c^{(m-1)}, c'), r)\}$;
9   **end**
10   **end**
11   **for** $(c^{(m-1)}, c^{(m)}) \in alt_C(C^{(m-1)}, C^{(m)})$ **do**
12      $\widetilde{C'} := C'^{(n)}$
13   **if** $\exists(b, c^{(m)}) \in H^{(m)} \wedge \exists(c^{(m)}, c') \in Align(O^{(m)}, O'^{(n)})$ **then**
14   **if** $\neg\exists(b, c^{(m-1)}) \in H^{(m-1)}$ **then**
15          $Align(O^{(m)}, O'^{(n)}) := Align(O^{(m)}, O'^{(n)}) \cup \{(b, c', \lambda_C(b, c'), r)\}$
           $\widetilde{C'} := \widetilde{C'} \backslash \{c'\}$
16   **end**
17   **end**
18   **if** $\exists(b, c^{(m-1)} \in H^{(m-1)}) \wedge \neg\exists(b, c^{(m)}) \in H^{(m)}$ **then**
19        $Align(O^{(m)}, O'^{(n)}) := Align(O^{(m)}, O'^{(n)}) \backslash \{(b, c', \lambda_C(b, c'), r)\}$
         $\widetilde{C'} := \widetilde{C'} \backslash \{c'\}$
20   **end**
21   **for** $c' \in C'^{(n)}$ **do**
22   **if** $\lambda_C(c^{(m)}, c') \geq \tau$ **then**
23          $Align(O^{(m)}, O'^{(n)}) :=$
24            $Align(O^{(m)}, O'^{(n)}) \cup \{(c^{(m)}, c', \lambda_C(c^{(m)}, c'), r)\}$
25   **end**
26   **end**
27   **end**
28   **return** $Align(O^{(m)}, O'^{(n)})$;
29 **end**

---

change has been applied. Eventually, the number of source ontology concepts has been doubled. Therefore, the described experimental verification has been repeated ten times for different amounts of changes applied to the source ontology. It covers all the situations in real world ontology application - from minor updates to complete rebuild of some maintained ontology.

The obtained results are presented in Table 1. The conducted experiment allows us to draw a conclusion that our approach and LogMap give similar alignments (in terms of the accuracy measure defined in Eq. 6) of two ontologies.

Table 1. Different number of modifications in the source ontology *edas*

No. of changes	Number of maps found by our approach	Number of maps finded by LogMap	Accuracy
10%	16	15	1
20%	17	17	1
30%	18	18	1
40%	19	20	0.95
50%	21	21	0.91
60%	21	21	0.91
70%	23	23	0.92
80%	26	24	0.93
90%	26	24	0.93
100%	25	24	0.96

However, considering the computational complexity, updating an existing alignment is obviously less expensive than building new mappings from the beginning. Additionally, our procedure is able to find more correct connections between concepts - a bigger alignment simplifies and reduces the costs of the further integration of two ontologies.

In the second part of our experiment, we have verified how the same changes affect the alignment if different ontologies are chosen as a source. As a target ontology, we have chosen *confOf*. During the procedure, the same amount of modifications have been applied to the remaining ontologies from the conference track taken from the OAEI dataset. Eventually, a comparison analogous to the first part of the experiment has been conducted for every ontology pair. We have confronted alignments updated by algorithms presented in the following paper with alignments designated from scratch by LogMap for modified source ontologies and the preselected target ontology. Thus, the procedure has been repeated fourteen times, for different ontologies, which verifies proposed methods in different situations.

The obtained results are presented in Table 2. It is shown, that the second experiment also proved the utility of our approach. However, the alignments determined by tested methods are sometimes not consistent. We can draw a conclusion that the bigger the source ontology, the bigger a difference between alignments. Additionally, LogMap found more links between concepts only in 2 cases from overall 13. In terms of accuracy, our approach is more precise. It allows finding more mappings in an easier way due to the fact that only modified parts of source ontologies are analyzed. It is obvious that it will always be faster than processing the whole ontologies.

Table 2. Different source ontologies

Source ontology	Number of concepts in source ontology	Number of maps finded by our approach	Number of maps finded by LogMap	Accuracy
CMT	30	14	14	0.875
Conference	60	21	20	0.77
Confious	57	12	12	0.648
Crs	14	15	15	0.875
Edas	104	19	17	0.552
Ekaw	74	24	27	0.7
Iasted	141	13	13	0.53
Linklings	37	15	12	0.5
Micro	32	16	15	0.84
MyReview	39	14	15	0.83
OpenConf	62	13	12	0.55
Parperdyne	46	18	15	0.71
PCS	24	14	13	0.81
Sigkdd	50	13	13	0.86

6   Future Works and Summary

In this paper, a set of algorithms capable of updating a pre-existing ontology alignment on the concept level is given. All of the presented procedures are based solely on changes applied to the evolving ontologies. The experimental verification was conducted using broadly accepted datasets provided by the Ontology Alignment Evaluation Initiative.

Results collected during the experiment showed the usefulness of our ideas. Due to the fact that our algorithms do not process whole ontologies, but only their alterations, the times required by our algorithms were shorter than re-launching mapping algorithms for whole ontologies. Therefore, during the verification procedure, we investigated the accuracy of the proposed algorithms.

The conducted experiments showed that LogMap and the proposed algorithms for ontology alignment update give similar alignments of two ontologies. Having in mind the computational complexity of designating mappings between two ontologies, updating an existing alignment is obviously more cost effective. Moreover, our procedure can find more correct connections between concepts which may be useful during the eventual integration of ontologies.

In the future, we will extend the created framework for different levels of ontologies, namely a level of relations and a level of instances. We will also perform more extensive experiments that will involve larger ontologies.

Acknowledgement. This research project was supported by grant No. 2017/26/D/ST6/00251 from the National Science Centre, Poland.

References

1. Allocca, C., d'Aquin, M., Motta, E.: Detecting different versions of ontologies in large ontology repositories. In: Proceedings of IWOD 2009, Washington, D.C., USA (2009)
2. An, Y., Topaloglou, T.: Maintaining semantic mappings between database schemas and ontologies. In: Christophides, V., Collard, M., Gutierrez, C. (eds.) ODBIS/SWDB -2007. LNCS, vol. 5005, pp. 138–152. Springer, Heidelberg (2008). https://doi.org/10.1007/978-3-540-70960-2_8
3. David, J., Euzenat, J., Scharffe, F., Trojahn dos Santos, C.: The alignment API 4.0. Semant. Web **2**((1), 3–10 (2011)
4. Dinh, D., Dos Reis, J.C., Pruski, C., Da Silveira, M., Reynaud-Delaître, C.: Identifying relevant concept attributes to support mapping maintenance under ontology evolution. Web Semant. Sci. Serv. Agents World Wide Web **29**, 53–66 (2014)
5. Euzenat, J., Mocan, A., Scharffe, F.: Ontology alignments. In: Hepp, M., De Leenheer, P., De Moor, A., Sure, Y. (eds.) Ontology Management. Computing for Human Experience, vol. 7, pp. 177–206. Springer, Boston (2008). https://doi.org/10.1007/978-0-387-69900-4_6
6. Grandi, F.: Multi-temporal RDF ontology versioning. In: Proceedings of IWOD 2009, Washington, D.C., USA (2009)
7. Hartung, M., Groß, A., Rahm, E.: Conto-diff: generation of complex evolution mappings for life science ontologies. J. Biomed. Inform. **46**(1), 15–32 (2013)
8. Jiménez-Ruiz, E., Cuenca Grau, B.: LogMap: logic-based and scalable ontology matching. In: Aroyo, L., et al. (eds.) ISWC 2011. LNCS, vol. 7031, pp. 273–288. Springer, Heidelberg (2011). https://doi.org/10.1007/978-3-642-25073-6_18
9. Khattak, A.M., et al.: Mapping evolution of dynamic web ontologies. Inform. Sci. **303**, 101–119 (2015). https://doi.org/10.1016/j.ins.2014.12.040
10. Klein, M., Fensel, D., Kiryakov, A., Ognyanov, D.: Ontology versioning and change detection on the web. In: Gómez-Pérez, A., Benjamins, V.R. (eds.) EKAW 2002. LNCS (LNAI), vol. 2473, pp. 197–212. Springer, Heidelberg (2002). https://doi.org/10.1007/3-540-45810-7_20
11. Kondylakis, H., Flouris, G., Plexousakis, D.: Ontology and schema evolution in data integration: review and assessment. In: Meersman, R., Dillon, T., Herrero, P. (eds.) OTM 2009. LNCS, vol. 5871, pp. 932–947. Springer, Heidelberg (2009). https://doi.org/10.1007/978-3-642-05151-7_14
12. Kozierkiewicz, A., Pietranik, M.: The knowledge increase estimation framework for integration of ontology instances' relations. In: Lupeikiene, A., Vasilecas, O., Dzemyda, G. (eds.) DB&IS 2018. Communications in Computer and Information Science, vol. 838, pp. 172–186. Springer, Cham (2018). https://doi.org/10.1007/978-3-319-97571-9_15
13. Kozierkiewicz, A., Pietranik, M.: A formal framework for the ontology evolution. In: Nguyen, N.T., Gaol, F.L., Hong, T.-P., Trawiński, B. (eds.) ACIIDS 2019. LNCS (LNAI), vol. 11431, pp. 16–27. Springer, Cham (2019). https://doi.org/10.1007/978-3-030-14799-0_2
14. Lembo, D., Rosati, R., Santarelli, V., Savo, D.F., Thorstensen, E.: Mapping repair in ontology-based data access evolving systems. In: IJCAI International Joint Conference on Artificial Intelligence, pp. 1160–1166 (2017)

15. Martins, H., Silva, N.: A user-driven and a semantic-based ontology mapping evolution approach. In: ICEIS, vol. 1, pp. 214–221 (2009)
16. Noy, N.F., Musen, M.A.: The PROMPT suite: interactive tools for ontology merging and mapping. Int. J. Hum.-Comput. Stud. **6**(59), 983–1024 (2003)
17. Papavassiliou V., Flouris G., Fundulaki I., Kotzinos D., Christophides V.: High-level change detection in RDF(S) KBs. ACM Trans. Database Syst. **38**(1), 1–42 (2013)
18. Pietranik, M., Nguyen, N.T.: A Multi-atrribute based framework for ontology aligning. Neurocomputing **146**, 276–290 (2014). https://doi.org/10.1016/j.neucom.2014.03.067
19. Pietranik, M., Nguyen, N.T.: Framework for ontology evolution based on a multi-attribute alignment method. In: CYBCONF 2015, pp. 108–112 (2015). https://doi.org/10.1109/CYBConf.2015.7175915
20. Sassi, N., Jaziri, W., Gargouri, F.: Z-based formalization of kits of changes to maintain ontology consistency. In: Proceedings of KEOD 2009, pp. 388–391 (2009)
21. Shvaiko, P., Euzenat, J.: Ten challenges for ontology matching. In: Meersman, R., Tari, Z. (eds.) OTM 2008. LNCS, vol. 5332, pp. 1164–1182. Springer, Heidelberg (2008). https://doi.org/10.1007/978-3-540-88873-4_18
22. Shvaiko, P., Euzenat, J., Jiménez-Ruiz, E., Cheatham, M., Hassanzadeh, O.: Proceedings of the 13th International Workshop on Ontology Matching Co-located with the 17th International Semantic Web Conference, OM@ISWC 2018, Monterey, CA, USA, 8 October 2018, CEUR Workshop Proceedings, vol. 2288 (2018). CEUR-WS.org
23. Thenmozhi, M., Vivekanandan, K.: A semi-automatic approach to update mapping for ontology evolution. In: Proceedings of International Conference on Computational Intelligence and Information Technology, CIIT, pp. 319–324 (2012)

On the Application of Ontological Patterns for Conceptual Modeling in Multidimensional Models

Glenda Amaral$^{(\boxtimes)}$ and Giancarlo Guizzardi

Free University of Bozen-Bolzano, 39100 Bolzano, Italy
{gmouraamaral,giancarlo.guizzardi}@unibz.it

Abstract. Data warehouses (DW) play a decisive role in providing analytical information for decision making. Multidimensional modeling is a special approach to modeling data, considered the foundation for building data warehouses. With the explosive growth in the amount of heterogeneous data (most of which external to the organization) in the latest years, the DW has been impacted by the need to interoperate and deal with the complexity of this new type of information, such as big data, data lakes and cognitive computing platforms, becoming evident the need to improve the semantic expressiveness of the DW. Research has shown that ontological theories can play a fundamental role in improving the quality of conceptual models, reinforcing their potential to support semantic interoperability in its various manifestations. In this paper we propose the application of ontological patterns, grounded in the Unified Foundational Ontology (UFO), for conceptual modeling in multidimensional models, in order to improve the semantic expressiveness of the models used to represent analytical data in a DW.

Keywords: Multidimensional modeling · Data warehouse ·
Conceptual modeling · Ontological patterns

1 Introduction

Multidimensional modeling is the foundation for building data warehouses (DW). Data warehouses were initially designed to support business intelligence applications in the internal context of the organization. In the latest years, the explosion in the volume of data on the web and in social networks, together with the accumulation of data generated by mobile devices, sensors and other semi-structured and unstructured data sources brought a challenge to the traditional analysis model, based on the DW, giving rise to the need of an approach that is suited to deal with the complexity of this new type of information, such as big data, data lakes and cognitive computing platforms. In this scenario, the necessity of integrating the Data Warehouse with new heterogeneous sources of information (most of which external to the organization) emerges. In addition, with the open data phenomena, the data stored in the DW was made available outside the

© Springer Nature Switzerland AG 2019
T. Welzer et al. (Eds.): ADBIS 2019, LNCS 11695, pp. 215–231, 2019.
https://doi.org/10.1007/978-3-030-28730-6_14

organization, becoming evident the need to make explicit the meaning of the information disclosed. In the light of the above, there is the need to improve the semantic expressiveness of the multidimensional models used to represent DW analytical data, making explicit the worldview to which they are committing (i.e., their ontological commitments), thus providing *intra-worldview consistency* and *inter-worldview interoperability*. In this paper, we move towards addressing this issue by means of the application of *ontological patterns* for conceptual modeling in the design of multidimensional models.

Conceptual modeling is the activity of formally describing some aspects of the physical and social world for the purposes of understanding and communication [23]. It plays a fundamental role, helping us to understand, elaborate, negotiate and precisely represent subtle distinctions in our multiple conceptualizations of reality. The discipline of conceptual modeling is supported by a wide range of methods and tools for representing the conceptualization of subject domains of interest. In this paper, we focus on a set of conceptual modeling techniques, which can be applied to address recurrent multidimensional modeling issues and to improve the semantic expressiveness of multidimensional models. This set includes three techniques related to the notion of ontological patterns that are grounded in the Unified Foundational Ontology (UFO) [11], namely, *Foundational Ontology Patterns* [29], *Reification and Truthmaking Patterns* [10] and the *Powertype Pattern* [3].

UFO is an axiomatic formal theory based on theories from Formal Ontology in Philosophy, Philosophical Logics, Cognitive Psychology and Linguistics. For an in-depth discussion, empirical support and formalization see [11,15]. UFO is the theoretical basis of OntoUML, a language for ontology-driven conceptual modeling that has been successfully employed in several projects in different domains [14]. A recent study shows that UFO is the second-most used foundational ontology in conceptual modeling and the one with the fastest adoption rate [33].

Several approaches have been proposed to multidimensional modeling in the conceptual level, either as extensions to the Entity-Relationship model [6,30], as extensions to UML [1,21], or ad hoc models [8,17]. The past decade has seen an increasing interest in ontology-driven approaches for multidimensional modeling, which led to a number of research initiatives in this area, most of them using domain ontologies for representing shared conceptualizations [16,18,25,27,28,31, 32]. Different from other approaches that use domain ontologies to provide more semantics to the information stored in the data warehouse, we have focused in this paper on improving the semantic expressiveness of multidimensional models by applying ontological patterns in their design.

The remainder of this paper is organized as follows. In Sect. 2, we give a brief review on multidimensional modeling and introduce the reader to the main notions on ontological patterns. Section 3 presents our approach for applying ontological patterns in the design of multidimensional models. In Sect. 4, to validate and demonstrate the contribution of our approach, we apply it to model a case study on education, extracted from [20]. We finalize the paper in Sect. 5 with some final considerations and directions.

2   Multidimensional Modeling and Ontology Patterns

2.1   Multidimensional Modeling

Multidimensional modeling is the process of modeling data in a universe of discourse, under the multidimensional paradigm. This is widely accepted as the preferred technique for modeling analytic data [20].

Multidimensional models categorize data either as facts with associated measures, which correspond to events occurred in the business domain, or as dimensions that characterize the facts and are mostly textual [26]. For example, in financial sector payment systems, money is transferred between financial institutions in certain amounts and at certain times. A typical fact would be a payment. Typical measures would be the debited and the credited amounts. Typical dimensions would be the debited financial institution, the credited financial institution, the currency and the time of the money transfer. Queries aggregate measure values over ranges of dimension values to produce results, such as the total value credited per financial institution, per month.

Traditionally, a cube metaphor is used to represent the multidimensional data view. The cells of the data cube contain the measures describing the fact. The axes of the cube, called dimensions, represent different ways of analyzing the data [2]. Classification hierarchies containing levels are used for the structuring of dimensions. A hierarchy level contains a distinct set of members and different levels correspond to different data granularities. Another orthogonal way of structuring dimensions from a user's point of view is the use of dimension level attributes. These attributes describe dimension level members but do not define hierarchies (e.g. the name and address of a financial institution).

Multidimensional models implemented in relational databases are referred to as star schemas because of their resemblance to a star-like structure [19]. Basically, the star schema represents each dimension as a dimension table and each fact as a fact table with a many-to-many relationship with all the dimensions. Figure 1 shows an example of a star schema. In this particular schema, the fact is the PAYMENT table. Measures are the non-foreign keys in the PAYMENT fact table (e.g. amount). Dimensions (TIME, CREDITED FINANCIAL INSTITUTION, DEBITED FINANCIAL INSTITUTION and CURRENCY) are all the tables connected to the fact table in a one-to-many relation-ship. Note that in this example the FINANCIAL INSTITUTION is referenced multiple times in the fact table, with each reference linking to a logically distinct role for this dimension (CREDITED and DEBITED FINANCIAL INSTITUTION), what is commonly referred to as "role-playing dimension" [19].

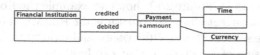

Fig. 1. Star schema payments

Although these schemas provide some level of modeling abstraction that is understandable to the user, they are not proper conceptual models in the sense of [13], given that they assume an underlying relational model implementation choice and contain further decisions that are proper of a physical design phase.

2.2 Ontological Patterns as Tools for Conceptual Modeling

Foundational Ontology Patterns

Foundational Ontology Patterns are reusable fragments of foundational ontologies. As foundational ontologies span across many fields and model the very basic and general concepts and relations that make up the world, Foundational Ontology Patterns can be applied in any domain [2]. They are reused *by analogy*, i.e., by establishing a structural correspondence (or structural transfer) between the structure of the pattern and the one of the problem at hand. In this article, we focus on the use of some of the Foundational Ontology Patterns that constitute the OntoUML Pattern Grammar [29].

Over the past decade, a number of Foundation Ontology Patterns have been derived from UFO, using OntoUML as a pattern language. Given the objectives of this paper, we focus here on four examples extracted from [29], selected for their applicability in the scope of multidimensional modeling: the *RoleMixin*, the *Phase*, the *Role* and the *Collective* Patterns. For a detailed description of these and other OntoUML Patterns, one should refer to [29].

The RoleMixin Pattern has been extracted from UFO's theory of sortal universals and addresses the problem of specifying *roles with multiple disjoint allowed types* [11].

UFO makes a fundamental distinction between Sortal and Non-Sortal types. A sortal is a type that either provides or carries a uniform principle of identity for its instances. A principle of identity supports the judgment whether two individuals are the same or, as a special case, what changes an individual can undergo and still be the same. A Kind is a sortal that is rigid, meaning that all its instances cannot cease to be so without ceasing to exist. In contrast with rigidity is the notion of anti-rigidity that characterizes a type whose instances can move in and out of its extension without altering their identity. A Role is a sortal, anti-rigid and relationally dependent type. Therefore, every Role in UFO must be connected to an association representing this relational dependence condition. Moreover, the association end connected to the depended type in this relation must have a minimum cardinality ≥ 1.

A RoleMixin is an anti-rigid and relationally dependent non-sortal that aggregates properties that are common to different Roles. Different from Roles, RoleMixins classify entities that instantiate different kinds (and that obey different principles of identity). Figure 2(a) shows an example the RoleMixin Pattern. In this picture, the abstract class CUSTOMER is the RoleMixin that covers different Role types. Classes PERSONAL CUSTOMER and CORPORATE CUSTOMER are the disjoint subclasses of CUSTOMER that can have direct instances, representing the Roles (i.e., sortal, anti-rigid and relationally dependent types) that carry

the principles of identity that govern the individuals that fall in their extension. Classes PERSON and ORGANIZATION are the ultimate Kinds that supply
the principles of identity carried by PERSONAL CUSTOMER and CORPORATE
CUSTOMER, respectively.

The Phase Pattern consists of a *phase partition*, i.e., a disjoint and complete
set of two or more complementary phases that specialize the same sortal and
that are associated with the same *dividing principle* (e.g., gender, life status,
developmental state). Phases in UFO are relationally independent, anti-rigid
types, defined as a partition of a sortal. This partition is derived based on an
intrinsic property of that sortal (e.g., Child is a phase of Person, instantiated by
instances of person who are less than 12 years). Figure 2(b) presents an instance
of the Phase Pattern. In this picture, class PERSON is the sortal and classes
CHILD, ADOLESCENT and ADULT represent the different phases that specialize
this sortal. The sortal instances can move in and out of the extension of the
phases, due to a change in the intrinsic properties of these instances. Analogous,
in the Role Pattern we have one or more roles that specialize a sortal (Fig. 2(c)).

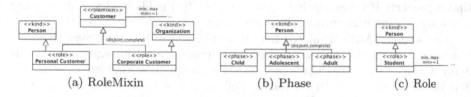

(a) RoleMixin   (b) Phase   (c) Role

Fig. 2. Foundational ontology patterns

The Collective Pattern, exemplified in Fig. 3, describes a Collective Universal
and the universals whose instances are members of these collectives. The unity
principle of collectives is a uniform relationship (i.e., a relation instance) that
holds between all parts and only those parts [12]. Because of the uniformity of
this relationship, the collective has a uniform structure, i.e., all its members are
undifferentiated with respect to (w.r.t.) the whole. In other words, they can be
said to play the same role w.r.t. the whole. Take for example collectives such as
a crowd or a forest with their corresponding instances of the *member of* relation
(i.e., person-crowd, tree–forest). In all of these cases, the wholes have a uniform
structure provided by a uniform unity principle (e.g., a crowd is a collective
of persons all which are positioned in a particular topologically self-connected
spatial location) and their parts are all considered to play the same role w.r.t.
the whole (e.g., all persons are equally considered to be *members Of* the crowd).

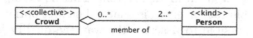

Fig. 3. Collective pattern

Reification and Truthmaking Patterns

Reification is a standard technique in conceptual modeling, which consists of including in the domain of discourse entities that may otherwise be hidden or implicit [10]. Classic examples are the reification of relationships [4,9,10,24] and events [5,7]. Recent work on formal ontology suggests that entities that should be put in the domain of discourse are those responsible for the (alleged) truth of our propositions. These are called truthmakers [22].

In [10], the authors propose a systematic analysis of truthmaking patterns (TMP) for properties and relations, based on the ontological nature of their truthmakers (TM) and present a number of Truthmaking Patterns for properties and relations at different levels of expressivity. In this paper we focus on two Truthmaking Patterns proposed in [24], which are more relevant in the context of multidimensional modeling.

The first one is the TMP proposed for intrinsic descriptive properties. Regarding the concept of intrinsic property, [10] states that a property holding for x is extrinsic iff it requires the existence of something else external to x in order to hold, and intrinsic otherwise. As for descriptive property, [10] defines that a property P is descriptive iff, for every x, P(x) holds in virtue of (at least) a quality q being existentially dependent on x.

The second TMP considered here was proposed in [10] for descriptive relations. Analogously to the case of descriptive properties, a descriptive relation is defined as a relation that holds in virtue of some qualities that are existentially dependent on one or both its relata. Following is a brief description of these two TMP, extracted from [10]. For a formal definition of them, as well as for additional TMP not mentioned here, the reader should refer to [10].

Before proceeding, there is an important notion that should be defined, namely the distinction between strong and weak truthmakers. In the strong version of truthmakers t is a truthmaker of the sentence ϕ if the existence of t is sufficient to make ϕ true. By contrast, t is a weak truthmaker of ϕ if it makes the proposition true not just because of its existence, but because of the way t contingently is.

Intrinsic Descriptive Properties. Intrinsic descriptive properties rarely correspond to classes, because they do not carry a principle of identity [11]. So, for example, the property of being red for a rose is typically expressed as an attribute-value pair within the class Rose (Fig. (4a)), where the attribute name implicitly denotes the color quality [3]. We have three reification options, corresponding to different Truthmaking Patterns. A weak TMP emerges when the quality is reified as a separate class (Fig. 4(b)). Note the 1-1 cardinality constraint, showing that a quality inheres in exactly one object, and an object has exactly one quality of a given kind. A strong TMP is exemplified in Fig. 4(c), where an event of "color occurrence" is reified. The first option is generally more flexible, making it possible to describe the way the quality interacts with the world (Mary likes the color of this rose), or further information about the quality itself (the color of a rose is located in its corolla). The second option is however necessary when we need to account for temporal information (e.g., how long the redness lasted), or for the spatiotemporal context (what happened meanwhile and where...). To achieve the maximum expressivity, a third option is that of a full TMP, including both strong and weak TMs plus the relationship among them (Fig. 4(d)). Concerning the latter, note that there is a formal ontological connection between qualities and events, discussed in [9]: events can be seen as manifestations of qualities, and qualities as the focus of events.

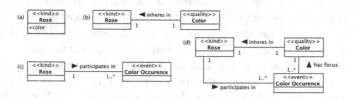

Fig. 4. Truthmaking patterns for an intrinsic descriptive property [10]

External Descriptive Relations. External descriptive relations hold in virtue of at least one relational quality inhering in at least one relatum. We distinguish two main cases: single-sided relations holding in virtue of one or more qualities inhering in just one relatum, and multi-sided relations holding in virtue of at least two qualities, each inhering in a different relatum. The reification of multi-sided relations is often necessary to model social and legal relationships, such as marriages, economic contracts, employment relationships, and so on. An example of the first kind is an attitudinal relation such as desires, represented in Fig. 5(a). A weak TMP is shown in Fig. 5(b), where a desire quality inhering in an agent and depending on some resources is reified. Note that we have represented it as a quality, but it could be seen as well as a relator consisting of just one quality. The addition of a strong TM, resulting in a full TMP, is shown in Fig. 5(c). The event labeled DesireEvolution' describes whatever happens in reality whose focus is that particular desire, such as the arising of the desire and its satisfaction.

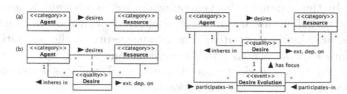

Fig. 5. Weak and full truthmaking patterns for a single-sided relation [10]

The Powertype Pattern
In several subject domains there is the need to deal with multiple classification levels. In such domains, the occurrence of situations in which instances of a type are specializations of another type is recurrent [3]. This phenomenon is known in the conceptual modeling community as the Powertype Pattern [3].

The Powertype Pattern is an example of an early approach for multi-level modeling in software engineering. This approach is used to model situations in which the instances of a type (the power type) are specializations of a lower-level type (the base type), and both power types and base types appear as regular classes in the model.

In [3], the authors address multi-level modeling from the perspective of the Powertype Pattern. They propose an axiomatic well-founded theory called MLT (for Multi-Level Theory) and apply it to revise the powertype support in UML. In their approach, they propose to mark the association between the base type and the higher order type with the ≪instantiation≫ stereotype, in order to distinguish it from other domain relations that do not have an instantiation semantics. An association stereotyped ≪instantiation≫ represents that instances of the target type are instantiated by instances of the source type and, thus, denote that there is a *characterization relation* (in the technical sense of [3]) between the involved types (regardless of possible generalization sets). The multiplicities of the "target" side of an ≪instantiation≫ association can be used to distinguish between the different variations of characterization. Whenever the lower bound multiplicity of the target association end is set to one, each instance of the base type is instance of, at least one instance of the powertype (e.g., every instances of person is necessarily either a living person or a deceased person). Thus, the higher order type *completely characterizes* the base type. In contrast, if the lower bound multiplicity of the target association end is set to zero, the inferred characterization relation is not a complete characterization. Analogously, if the upper bound multiplicity of the target association end is set to one, each instance of the base type is instance of, at most one instance of the higher order type. Thus, in this case, the higher order type *disjointly characterizes* the base type (again, no person can be both an instance of living person and of deceased person). In contrast, if the upper bound multiplicity of the target association end is set to many (*), the inferred characterization relation is not a disjoint characterization. Figure 6 shows the application of the Powertype Pattern proposed in [3]. As the authors show, there are non-trivial interactions between the semantics of the ≪instantiation≫ relation and the meta-properties of a given generalization set. In this example the generalization set is *incomplete* and *disjoint* meaning that: (i) there are instances of "Employee" which are not instances of any instance of "Management Role" (as a consequence of the semantics of the ≪instantiation≫association); and (ii) there are instances of "Employee" which are neither "Organization President" nor "Department Dean" (as a consequence of the semantics of incomplete generalization sets).

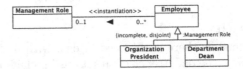

Fig. 6. Using ≪instantiation≫ [3]

The UML extensions proposed in [3] go beyond the ≪instantiation≫ stereo-type and the lower/upper bound multiplicities. Further details of their approach fall outside the scope of this paper and are not presented here. For a complete description of the approach just described the reader should refer to [3].

3   Piecing It All Together

3.1   Applying to Dimensions

Foundational Ontology Patterns (FOP) can be used to improve the expressive-ness of multidimensional models, thus, facilitating activities, such as communica-tion and meaning negotiation, as well as the semantic interoperability regarding the domains represented therein. The application of FOPs in the modeling of dimensions provide more semantics for the concepts represented.

For example, the modeling of role-playing dimensions can benefit from the use of the Role Pattern, as it can be used to represent the different roles played by a dimension, at the same time that it makes it explicit that the same entity plays different roles in that specific context. Figure 7 illustrates the application of the Role Pattern in the dimension FINANCIAL INSTITUTION of the star-schema illustrated in Fig. 1. In a payment event (fact), FINANCIAL INSTITUTION (dimen-sion) plays two different roles: CREDITED FINANCIAL INSTITUTION (the Finan-cial Institution whose account should be credited) and DEBITED FINANCIAL INSTITUTION (the Financial Institution whose account should be debited).

Fig. 7. Application of the role pattern in the modeling of role-playing dimensions

When the role played by a dimension aggregates properties that are common to different Roles, the RoleMixin Pattern can be applied. Again, at the same time that the pattern reinforces the truthfulness of the concepts represented, it makes explicit the nature and the restrictions applicable to the entity represented by the dimension. The OntoUML model presented in Fig. 8 illustrates the application of the RoleMixin Pattern in the modeling of dimensions that represent borrowers, in the context of Finance. In this case, a borrower may be defined as a person or an organization that obtains a loan from a Financial Institution. In the figure, LOAN represents the fact table about the loans.

In the figure, BORROWER is the RoleMixin that covers different role types. CORPORATE BORROWER and PERSONAL BORROWER are the disjoint subclasses of BORROWER that can have direct instances, representing the sortal roles that carry the principles of identity that govern the individuals that fall in their

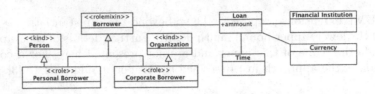

Fig. 8. Application of the RoleMixin pattern

extension. Dimensions ORGANIZATION and PERSON are the ultimate substance sortals (kinds) that supply the principles of identity carried by CORPORATE BORROWER and PERSONAL BORROWER, respectively. The application of the RoleMixin Pattern preserves the unity of the concept borrower at the same time that clarifies the distinction between different types of borrowers (personal borrower and corporate borrower), satisfying both the modeling of facts related to all types of borrowers and the modeling of facts related only to a specific type of borrower (person or organization).

Analogously, when it is necessary to relate a dimension to a fact whose instances apply only to a subset of the dimension instances (corresponding to a phase partition) the Phase Pattern may be applied. Figure 9 depicts an example of the Phase Pattern applied to the modeling of a fact representing exams taken by applicants for a driver's license. As only persons over 18 years are eligible to a driver's license, the PERSON dimension related to the fact DRIVER LICENSE EXAM should be restricted to people meeting the minimum age requirement. The Phase Pattern was applied to create three phase-partitions specializing the dimension PERSON (CHILD, ADOLESCENT and ADULT). Then it was possible to relate the fact DRIVER LICENSE EXAM to a subset of the PERSON dimension representing only adults (Phase ADULT).

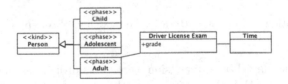

Fig. 9. Application of the Phase Pattern

Finally, the Collective Pattern is applicable to dimensions that represent entities as integral wholes, composed by members that play the same role in the collective. In many cases, in multidimensional models, it is important to distinguish the conceptualization of the whole from the conceptualization of the parts, because it is necessary to relate the whole to a fact that applies to the collective and the parts to a fact applicable only to the individuals. At the same time, it is important to make explicit the existence of a uniform relationship that holds between all parts (and only those parts). Figure 10 presents an example of the Collective Pattern applied to a multidimensional model in the context of product manufacturing systems, which work with the concepts of Lot and Item. In this case, a Lot is defined as a group composed of a definite quantity of some product, manufactured under conditions of production that are considered uniform, while the Item corresponds to each product in the Lot. In the model, the dimension LOT represents the collection, while ITEM represents its members. In this approach, it is possible to relate the dimension LOT to the fact DELIVERY containing information applicable to the collective (for example, the lot weight), as well as to relate the dimension ITEM to the fact SELL whose granularity is the individual product (for example, unit price).

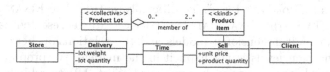

Fig. 10. Application of the Collective Pattern

Turning now to Truthmaking Patterns, this technique can be applied in the modeling of dimensions to improve the expressivity of attributes describing dimension level members. These attributes are mostly intrinsic descriptive properties that can be reified as previously discussed.

Take as example the dimension HOTEL illustrated in Fig. 11(a). The property "star rating", used to classify hotels according to their quality, is typically expressed as an attribute-value pair within the dimension HOTEL (Fig. 11(a)), where the attribute name implicitly denotes the hotel star rating quality. The first option is to reify this quality (weak truthmaker) as separate class (Fig. 11(b)), making it possible to describe the ways the quality interacts with the world (e.g., people prefer hotels rated from four to five stars), or further information about the quality itself (e.g., the hotel star rating is reviewed annually). The second option is to reify the event of "star rating occurrence" (strong truthmaker), which allows to account for temporal information (e.g., how long the hotel has been rated as five stars), or for the spatiotemporal context (what happened when the rating changed from five to four stars). The third option, which gives maximum expressivity, is that of a full TMP, including both strong and weak TMs plus the relationship among them (Fig. 11(d)).

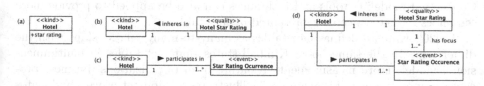

Fig. 11. Hotel dimension with a "star rating" attribute

Finally, there is another modeling issue that, despite being often neglected in the design of multidimensional models, should be addressed in the models to reinforce truthfulness to the reality. This is the case of dimensions that represent entities of different classification levels. For example, let us take the case of Financial Institutions and their types. Consider that FINANCIAL INSTITUTION can be specialized in BANK, INSURANCE COMPANY, INVESTMENT COMPANY and BROKERAGE FIRMS. In this case, "Bank A" and "Bank B" are particular BANKS, both instances of FINANCIAL INSTITUTION. Data analysis under the perspective of the type of FINANCIAL INSTITUTION are particularly common in this context, then the TYPE OF FINANCIAL INSTITUTION should also be considered as an entity, whose instances are "Bank", "Insurance Company", "Investment Company" and "Brokerage Firm". Traditionally, entities like FINANCIAL INSTITUTION and TYPE OF FINANCIAL INSTITUTION are represented as unrelated dimensions in multidimensional models and the relationship between the different classification levels is not explicit in the models. We propose the use of the Powertype Pattern previously mentioned (Sect. 2.2) to address this issue. Figure 12 presents an example of the application of the Powertype Pattern to the scenario of Financial Institutions and their types. In the example, the association stereotyped ≪instantiation≫ has both the lower and the upper bound multiplicity set to one, meaning that the target dimension (TYPE OF FINANCIAL INSTITUTION) disjointly and completely characterize the source dimension (FINANCIAL INSTITUTION). Thus, the model in Fig. 14 represents that: (i) every instance of FINANCIAL INSTITUTION must be either an instance of BANK, an instance of INSURANCE COMPANY, an instance of INVESTMENT COMPANY, or an instance of BROKERAGE FIRM and that (ii) "Bank" and "Insurance Company", "Investment Company" and "Brokerage Firm" are the only admissible instances of TYPE OF FINANCIAL INSTITUTION.

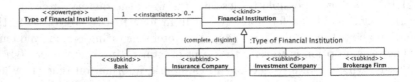

Fig. 12. Application of the Powertype Pattern

3.2 Applying to Facts

Conceptual modeling tools and techniques can also be applied to provide more semantics for the concepts represented by fact tables.

In [20], Kimball defines fact tables as many-to-many relationships with the dimensions. In the same book, Kimball states that fact tables in multidimensional models store measurements resulting from organizations' business processes events. In one of his examples he illustrates a shipment process and states that each movement of product onto an outbound truck generates performance measures or facts, such as the shipment quantity. In this way, it seems that

Kimball is committed to the view that fact tables are relationships, but he also admits that a fact table corresponds to a physical observable event.

In [9], the authors propose a view in which events emerge from scenes as a result of a cognitive process that focuses on relationships: relationships are therefore the focus of events, which in turn can be seen as manifestations of relationships. Further in the paper, they state that referring to the relationship (which maintains its identity during the event) is unavoidable when we need to describe what changes in time, while referring to the event is unavoidable when we need to describe contextual aspects that go beyond the relationship itself.

In the light of what has been discussed in [20] and [9] regarding relationships and events, a reasonable approach would be to consider two elements w.r.t. fact tables: the fact as a relationship involving multiple participants (dimensions) and, on the other hand, the event that is the sum of the manifestations of the qualities constituting this relationship (measures). According to [26], not only the relationships should be reified but also the events.

Following the terminology for kinds of relationships defined in [10], we may classify fact tables as external descriptive relations, as they hold in virtue of relational qualities (measures) inhering in their relata (dimensions). Thus, both the relationship and the event (whose focus is the relationship) can be reified by applying the TMP for external descriptive relations previously mentioned.

An example of the application of the full TMP is presented in Fig. 13, where the TMP was applied to the fact table LOAN represented in the star schema of Fig. 13(a). The example describes a loan relation holding between a FINANCIAL INSTITUTION and a BORROWER. The relator is shown as a LOAN RELATIONSHIP composed of the amount, which has a value in a CURRENCY conceptual space, and of the loan interest rate. Because the amount was reified as a «quality» (whose instances inhere in the loan), it is possible to express further information about it, for instance: (1) "This was the highest loan amount so far" or (2) "The amount borrowed did not reach the credit limit. It is still possible to grant new loans". In addition, the application of the TMP allows to explicitly represent other relevant information regarding the LOAN RELATIONSHIP, such as the reciprocal commitments and claims inhering in the financial institution or the borrower (and externally dependent on each other). The event labeled LOAN EVENT describes the loan date as well as whatever happens in reality whose focus is that particular loan, such as the occurrence of loan disbursements, repayments and credit risk assessments.

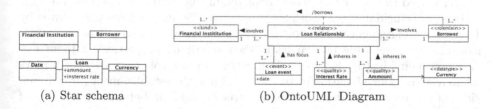

(a) Star schema   (b) OntoUML Diagram

Fig. 13. Application of Truthmaking Patterns to fact tables

The reification of measures as individual qualities represents an interesting improvement in the semantic expressiveness of measures in multidimensional models. It allows to express the correlated units of measures, magnitudes, and scales, which are generally overlooked in multidimensional approaches. This empowers multidimensional models because each scale type defines a mathematical structure on which the permissible statistics and scale transformations are allowed. It also provides a better understanding about the nature of additivity constraints, as many statistic functions may be used to aggregate data cells of measures, though their use depends on which sort of measure and aggregation criteria are involved. Identifying these concepts in the multidimensional models, based on their ontological foundations, enables designers to describe properly what is being modeled, and therefore, to elucidate how data should be analyzed.

4   Case Illustration on Education: Student Attendance

To validate and demonstrate the contribution of our proposition to the multidimensional modeling practice, we have applied it to model a tangible example: a case study on education, extracted from [20], designed to track student attendance in a course. In this model, the grain is each occurrence of a student walking through the course classroom door each day, considering multi-instructor courses (that is, co-taught courses). The original multidimensional model, depicted in Fig. 14(a) [20], allows business users to answer questions concerning student attendance at courses, including: which courses were the most heavily attended, which students attended which courses, and which faculty member taught the most students.

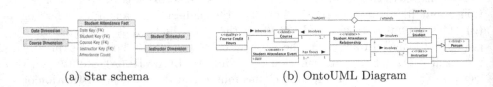

(a) Star schema   (b) OntoUML Diagram

Fig. 14. Applying Ontology Patterns to the multidimensional model

We applied ontology patterns to reengineer the original model (Fig. 14(a)) and produced the OntoUML model depicted in Fig. 14(b). By applying the Role Pattern, we elucidated that both STUDENT and INSTRUCTOR are roles played by PERSONS. Consequently, PERSON instances can move in and out of the extension of these roles (due to changes in their relational properties), without any effect on their identity. For example, a STUDENT is a role that a PERSON plays when related to an education institution, and it is the establishment (or termination) of this relation that alters the instantiation relation between an instance of PERSON and the type STUDENT. The application of the ontological pattern not only provides more clarity and expressiveness to the model, but also favors the reuse of encoded experiences and good practices. Considering the existence of a property COURSE CREDITS HOURS in the dimension Course, we have applied the TMP for Intrinsic Properties to reify the COURSE CREDITS HOURS as separate class, thus making it possible to describe the ways in which this quality interacts with the world (e.g., this amount of credit hours can also be earned by taking part on a summer school), or further information about the quality itself (e.g., course credit hours are specified in the course regulation). In this case, the TMP contributes to enrich the expressivity of the model. Finally, we applied a full TMP for External Descriptive Relations on the original table STUDENT ATTENDANT FACT to reify the truthmaker of the STUDENT ATTENDANT FACT RELATIONSHIP (between student, course and instructor) by means of the STUDENT ATTENDANT EVENT. The application of the TMP allows to explicitly represent relevant information regarding the STUDENT ATTENDANT RELATIONSHIP, as well as to explicit represent the STUDENT ATTENDANT EVENT, which describes the date of the student attendance, as well as whatever happens in reality, whose focus is that particular student attendance, such as late arrivals or early leaves from the classes. By applying the TMP we improve the model expressivity, conceptual clarity as well as its truthfulness to reality. The aforementioned benefits seem to corroborate the fact that the use of ontological patterns in multidimensional modeling helps domain experts to externalize the knowledge about the domain, making the ontological commitments explicit and the models more truthful to the domain being represented.

5   Conclusions

This paper described our approach to systematically apply ontological patterns in the design of multidimensional models. We have discussed how conceptual modeling techniques can be applied in combination for building consistent multidimensional models. In our approach we focused on the application of Foundational Ontology Patterns, Reification and Truthmaking Patterns and the Powertype Pattern to improve the semantic expressiveness of multidimensional models. The case illustration in the area of Education exemplified how our propositions contribute to improve the quality of multidimensional models, enhancing their quality as artifacts to support communication, problem solving, meaning negotiation and, principally, semantic interoperability in its various manifestations.

Conceptual modeling is a fundamental discipline to several communities in computer science. In the future we plan to extend our work by conducting an analysis of the role played by conceptual models and philosophically grounded foundational ontologies in the scope of other technologies used for data analytics.

Acknowledgment. CAPES (PhD grant# 88881.173022/2018-01) and OCEAN project (UNIBZ).

References

1. Abelló, A., Samos, J., Saltor, F.: YAM^2: a multidimensional conceptual model extending UML. Inf. Syst. **31**(6), 541–567 (2006)
2. de Almeida Falbo, R., Guizzardi, G., Gangemi, A., Presutti, V.: Ontology patterns: clarifying concepts and terminology. In: WOP (2013)
3. Carvalho, V.A., Almeida, J.P.A., Guizzardi, G.: Using a well-founded multi-level theory to support the analysis and representation of the powertype pattern in conceptual modeling. In: CAISE (2016)
4. Dahchour, M., Pirotte, A.: The semantics of reifying N-ary relationships as classes. In: ICEIS (2002)
5. Davidson, D.: The individuation of events. In: Rescher, N. (ed.) Essays in Honor of Carl G. Hempel, vol. 24, pp. 216–234. Springer, Dordrecht (1969)
6. Franconi, E., Kamblet, A.: A data warehouse conceptual data model. In: SSDBM (2004)
7. Galton, A.: Reified temporal theories and how to unreify them. In: IJCAI, pp. 1177–1183. Citeseer (1991)
8. Golfarelli, M., Maio, D., Rizzi, S.: The dimensional fact model: a conceptual model for data warehouses. Int. J. Coop. Inf. Syst. **7**(02n03), 215–247 (1998)
9. Guarino, N., Guizzardi, G.: Relationships and events: towards a general theory of reification and truthmaking. In: AI*IA (2016)
10. Guarino, N., Sales, T.P., Guizzardi, G.: Reification and truthmaking patterns. In: ER (2018)
11. Guizzardi, G.: Ontological Foundations for Structural Conceptual Models. CTIT, Centre for Telematics and Information Technology, Trento (2005)
12. Guizzardi, G.: Ontological foundations for conceptual part-whole relations: the case of collectives and their parts. In: CAiSE (2011)
13. Guizzardi, G., Halpin, T.: Ontological foundations for conceptual modelling. Appl. Ontol. **3**(1–2), 1–12 (2008)
14. Guizzardi, G., Wagner, G., Almeida, J.P.A., Guizzardi, R.S.S.: Towards ontological foundations for conceptual modeling: the unified foundational ontology (UFO) story. Appl. Ontol. **10**(3–4), 259–271 (2015)
15. Guizzardi, G., et al.: Towards ontological foundations for the conceptual modeling of events. In: ER (2013)
16. He, L., Chen, Y., Meng, N., Liu, L.Y.: An ontology-based conceptual modeling method for data warehouse. In: ICM (2011)
17. Hüsemann, B., Lechtenbörger, J., Vossen, G.: Conceptual Data Warehouse Design. Universität Münster, Angewandte Mathematik und Informatik (2000)
18. Khouri, S., Ladjel, B.: A methodology and tool for conceptual designing a data warehouse from ontology-based sources. In: Proceedings of the ACM 13th International Workshop on Data Warehousing and OLAP, pp. 19–24. ACM (2010)
19. Kimball, R., Ross, M.: The Data Warehouse Toolkit: The Complete Guide to Dimensional Modeling. Wiley, Hawaii (2011)
20. Kimball, R., Ross, M., Thornthwaite, W., Mundy, J., Becker, B.: The Data Warehouse Lifecycle Toolkit. Wiley, Hawaii (2008)
21. Luján-Mora, S., Trujillo, J., Song, I.Y.: A UML profile for multidimensional modeling in data warehouses. Data Knowl. Eng. **59**(3), 725–769 (2006)
22. MacBride, F.: Truthmakers. In: Zalta, E.N. (ed.) The Stanford Encyclopedia of Philosophy. Stanford University, Metaphysics Research Lab (2019)
23. Mylopoulos, J.: Conceptual modeling and telos. In: ER (1992)

24. Olivé, A.: Relationship reification: a temporal view. In: CAiSE (1999)
25. Pardillo, J., Mazón, J.N.: Using ontologies for the design of data warehouses. arXiv preprint. arXiv:1106.0304 (2011)
26. Pedersen, T.B.: Multidimensional modeling. In: Liu, L., Özsu, M.T. (eds.) Encyclopedia of Database Systems, pp. 1777–1784. Springer, Boston (2009)
27. Romero, O., Abelló, A.: Automating multidimensional design from ontologies. In: Proceedings of the ACM Tenth International Workshop on Data Warehousing and OLAP, pp. 1–8. ACM (2007)
28. Romero, O., Abelló, A.: A framework for multidimensional design of data warehouses from ontologies. Data Knowl. Eng. **69**(11), 1138–1157 (2010)
29. Ruy, F.B., Guizzardi, G., Falbo, R.A., Reginato, C.C., Santos, V.A.: From reference ontologies to ontology patterns and back. Data Knowl. Eng. **109**, 41–69 (2017)
30. Sapia, C., Blaschka, M., Höfling, G., Dinter, B.: Extending the E/R model for the multidimensional paradigm. In: ER (1998)
31. Selma, K., Ilyès, B., Ladjel, B., Eric, S., Stéphane, J., Michael, B.: Ontology-based structured web data warehouses for sustainable interoperability: requirement modeling, design methodology and tool. Comput. Ind. **63**(8), 799–812 (2012)
32. Thenmozhi, M., Vivekanandan, K.: A tool for data warehouse multidimensional schema design using ontology. Int. J. Comput. Sci. Issues (IJCSI) **10**(2), 161 (2013)
33. Verdonck, M., Gailly, F.: Insights on the use and application of ontology and conceptual modeling languages in ontology-driven conceptual modeling. In: ER (2016)

Process Mining and Stream Processing

Accurate and Transparent Path Prediction Using Process Mining

Gaël Bernard[1]([envelope]) and Periklis Andritsos[2]

[1] Faculty of Business and Economics (HEC),
University of Lausanne, Lausanne, Switzerland
`gael.bernard@unil.ch`
[2] Faculty of Information, University of Toronto, Toronto, Canada
`periklis.andritsos@utoronto.ca`

Abstract. Anticipating the next events of an ongoing series of activities has many compelling applications in various industries. It can be used to improve customer satisfaction, to enhance operational efficiency, and to streamline health-care services, to name a few. In this work, we propose an algorithm that predicts the next events by leveraging business process models obtained using process mining techniques. Because we are using business process models to build the predictions, it allows business analysts to interpret and alter the predictions. We tested our approach with more than 30 synthetic datasets as well as 6 real datasets. The results have superior accuracy compared to using neural networks while being orders of magnitude faster.

Keywords: Process mining · Predictive process monitoring · Predictive analytics · Path prediction · Trace clustering

1 Introduction

After observing a few events of an incomplete sequence of activities, we can predict the next events until process completion by learning from historical event logs, an activity coined path prediction [1]. Anticipating the next events is valuable in a wide range of scenarios. For instance, when a service desk team predicts the paths taken by open tickets, the results can be used in many different ways. One proposition is to cut the number of predicted complaints due to delays by changing the priority of tickets. Another is to reduce the negative impact on customer satisfaction by preemptively informing them about a delay. One more is to align the expertise of service desk agents with the events predicted for a ticket. The predictions could also be used by inexperienced agents to anticipate the next events better, allowing them to communicate more accurate information to the customers. Overall, predicting paths can help to improve worker and customer satisfaction, as well as improve operational efficiency.

There are two main approaches to making predictions for a series of events. The first uses process mining while the second relies on neural networks. Both

© Springer Nature Switzerland AG 2019
T. Welzer et al. (Eds.): ADBIS 2019, LNCS 11695, pp. 235–250, 2019.
https://doi.org/10.1007/978-3-030-28730-6_15

approaches have their strengths and limitations. Process mining is more transparent because it relies on models that can be inspected by business analysts. This is important, as business analysts may have hidden knowledge that will influence their confidence in the prediction. Furthermore, "business stakeholders are not data scientists [...] they are more likely to trust and use these models if they have a high-level understanding of the data that was used to train these models" [2]. In contrast, reasoning about predictions made by artificial neural networks is complex, if not impossible. Furthermore, a neural network requires a long training time [1]. However, in terms of performance, the most recent research shows that predictions using long short-term memory (LSTM) in a neural network achieves high accuracy [3].

We address the research gap that exists between accurate, but black-box techniques and transparent, but less accurate process mining approaches. Indeed, we aim to make predictions that are accurate, fast, and interpretable by business analysts. We propose a matrix named the loop-aware footprint matrix (LaFM), which captures the behaviors of event logs when replayed on a business process model obtained automatically using process mining techniques. The captured behaviors are then retrieved from LaFM to make predictions about uncompleted traces. We also propose a clustered version of LaFM (c-LaFM) that can cope with the inherent complexity of real datasets. We evaluate the prediction accuracy of LaFM with 30 synthetic datasets and the accuracy of c-LaFM with 6 real datasets. We show that our technique outperforms the LSTM approach introduced in [3].

The paper is organized as follows. In Sect. 2, we introduce the main definitions and discuss process mining. Section 3 provides an overview of existing works. Section 4 presents the main idea behind LaFM. In Sect. 5, we present the evaluation procedure. Section 6 evaluates and compares the accuracy of the method using synthetic datasets. Section 8 introduces the clustered version of LaFM, coined c-LaFM, which is evaluated in Sect. 8. The paper ends in Sect. 9 with a conclusion.

2 Preliminaries

In this section, we lay out the main definitions and concepts of our approach. They are part of the well-established process mining discipline. In this paper, we consider only the sequence of events, disregarding the timestamps or any other contextual information in the data. By doing so, we present a simplified view of process mining, to be complemented with the foundational book about process mining [4].

Events. An event is a discrete type of data representing the activities executed in a process. For instance, 'transferring a ticket' is an event in a ticket's lifecycle. Let e be an event and E be the set of all distinct events; i.e., $e \in E$.

Trace. A trace is an instance of a process execution. In a service desk context, a trace is a ticket. Let $t = \{e_1, e_2, ...; e \in E\}$ be a trace: a list of events. For instance $\langle ABBC \rangle$ is a trace with three distinct events of length 4 ($|t| = 4$).

Prefix. Let a prefix $p_n = \{e_1, e_2, ..., e_n; e \in t\}$ be the first n events of a trace. Typically, if $t = \langle \text{ABBC} \rangle$, then $p_3 = \langle \text{ABB} \rangle$. A prefix represents the few events observed from an uncompleted trace that we use to make a prediction.

Suffix. A suffix represents the n last events of a trace. Formally, $s_n = \{e_{|t|-n}, ..., e_{|t|-1}, e_{|t|}; e \in t; e \notin p_n; |p_n| + |s_n| = |t|\}$, i.e., the suffix is the complement of the prefix. The suffix is the set of events that we are trying to predict.

Event logs. An event log $L = \{t_1, t_2, ...;\}$ is a collection of traces.

By looking only at the event log, process discovery techniques allow us to infer the business process model that describes well the behavior of the traces. This is a challenging task because the algorithm should be able to generalize behaviors even if only a subset of them is observed, to exclude noise and outliers, and to discover a model that is simple enough that it can be analyzed by a business analyst but also precise enough to reflect the behaviors of the event logs. Several techniques and approaches have been proposed to tackle this task. In this work, we use the inductive miner [5].

The inductive miner works by finding the best split in an event log and seeing how the two parts are related. It does this recursively on both parts. The output is a process tree (Fig. 1), which is a representation of a process model that was introduced in [6]. A process tree uses four operators: (1) the exclusive choice operator, xor, expresses that only one of the branches is executed; (2) the parallel operator, and, indicates that all the branches should be executed, in any order; and (3) a sequence, seq, forces the execution of the branches from left to right. Finally, (4) a loop has a more complex execution scheme: the

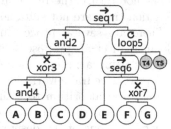

Fig. 1. Process tree obtained by the inductive miner with the traces: $\{\langle \text{ABDEF} \rangle, \langle \text{BDAEGEF} \rangle, \langle \text{DCEFEG} \rangle, \langle \text{CDEG} \rangle\}$.

first branch is executed at least once. Then, either we enter the loop by executing the second branch and the first branch again (which can be done once or multiple times), or we execute the third branch to exit the loop. As can be seen in Fig. 1, except for the leaves, these four operators fill the whole tree. The leaves of the tree are composed of the events E as well as silent activities. Silent activities, τ, can be executed like any other events in the model, but they will not be seen in the traces.

We have now introduced the main terminology, the inductive miner, and the process tree. Path prediction is concerned with predicting the suffix for a given prefix by learning from event logs. It differs from process model discovery in which the goal is to discover a process model from event logs. While the output is different, both methods are about understanding the control flow of traces. We leverage this by using the inductive miner as a first step in making predictions.

3   Related Work

The area of predictive analytics is wide as trace predictions can be time-related (e.g., predicting the remaining time), outcome-oriented (e.g., success vs. failure), or control-flow oriented (e.g., next event(s) prediction). In this work, we specifically focus on the latter type of prediction.

A widely adopted approach to prediction is to build a Markov chain that describes the transition probabilities between events. These transition probabilities are used to make predictions. A prediction depends only on the previously observed event. In the all-K-order Markov model, [7], the number of levels in the Markov chain is increased, but this increases the execution time. While the accuracy of the prediction increases, it suffers from rigidness in terms of the "patterns that it can learn" [8]. As another approach, Gueniche et al., propose the compact prediction tree [8]. It uses three data structures that can be used efficiently to retrieve the most probable event that might occur after having observed a prefix. While it predicts with high accuracy which events might occur in the suffix, it does not return the order in which they will be executed. Hence, compact prediction trees are not suitable for predicting paths.

There are several process mining approaches for predicting paths. In [9], Lakshmanan et al. propose a method that estimate the likelihood of the next activities using a process model and Markov chain. Breuker et al. propose in [10] a predictive framework that uses grammatical inference and an expectation-maximization algorithm to estimate the model parameters. Among its predictions, it can predict the next event. Improving the comprehensibility of the predictions is one of the design goals of their approach, so that "users without deep technical knowledge can interpret and understand" [10]. In [1], Polato et al. propose a labeled transition system and methods for several predictive analytic tasks. Path prediction can be done by finding a path in the transition system that minimizes the sum of the weights between the edges.

Recently, neural networks have been studied for predicting the next events. To the best of our knowledge, Evermann et al. were the first to use a LSTM neural network approach to predict the next event of an ongoing trace [11]. LSTM, [12], is a special type of neural network for sequential inputs. It can learn from long-term dependencies using a sophisticated memory system. The sophisticated memory system is a double-edged sword: it achieves high accuracy; however, its inherent complexity prevents any inspection of the reasoning behind the predictions. In [3], Tax et al. generalize the approach of [11]. They evaluate–amongst other methods–the performance of the algorithm in path prediction and show that it is more accurate than [1,10,11]. Because it achieves the best accuracy, we use it as a baseline when evaluating the accuracy of LaFM.

Overall, two streams of research dominate path prediction. On one hand, using process mining techniques, we can make predictions using models that can be inspected by business analysts. On the other hand, neural networks attain better performance in terms of accuracy. Our contribution is an algorithm that utilizes the best aspects of both methods.

4   LaFM: Loop-Aware Footprint Matrix

We designed LaFM to store the behavior of traces efficiently when replayed on
business process models. The aim is that the behaviors can be retrieved when
predicting a suffix of events. First, we present the LaFM data structure. Next,
we explain how to build it. Finally, we detail how to use it to make predictions.

4.1   LaFM Data Structure

LaFM records the behavior of traces when replayed on top of a business process
model. An illustration of LaFM is shown in Fig. 2. Each row corresponds to a
trace and each column describes the behavior of an operator. LaFM captures the
execution orders of parallel branches, the exclusive choices, and the number of
iterations of each loop. We next describe in more detail the information recorded
by LaFM as well as the used terminology.

Traces	*Terminology* and2(1)|	and2(2)|	and2(3)|	and4(1)|	and4(2)|	loop5|	xor7|loop5{1}	xor7|loop5{2}	xor3|
ABDEF	1	1	2	1	2	1	F	∅	and4
BDAEGEF	1	2	1	2	1	2	G	F	and4
DCEFEG	2	1	∅	∅	∅	2	F	G	C
CDEG	1	2	∅	∅	∅	1	G	∅	C

Fig. 2. Result of LaFM when the traces ⟨ABDEF⟩, ⟨BDAEGEF⟩, ⟨DCEFEG⟩, and ⟨CDEG⟩
are replayed on top of the process tree of Fig. 1.

Parallel Branches. LaFM stores the order in which parallel branches are executed. An incremental index is assigned to each outgoing branch of the **and**
operators and then propagated to the events and silent activities underneath.
For instance, **and2** in Fig. 1 has two outgoing branches. The index 1 is assigned
to the first branch, which is propagated to the events below, i.e., 1 is assigned
to A, B, and C. Similarly, task D has index 2. The index is recorded in LaFM for
each **and** operator.

Exclusive Choices. The decision made for each exclusive choice is recorded
in LaFM. For example, at **xor3** in Fig. 1, a choice must be made between **and4**
and C. For the trace ⟨CDEG⟩, the choice is C. Hence, C is recorded in LaFM.

Loops. LaFM stores the number of times loops are executed. In Fig. 1 for the trace ⟨CDEG⟩, the value recorded for loop5 is 1 because it was executed once.

Terminology. An operator might be executed multiple times during a single process execution. For instance, when the trace ⟨BDAEGEF⟩ is replayed on the process tree in Fig. 1, we execute the operator xor7 twice because loop5 above it is also executed twice. The name 'loop-aware footprint matrix reflects that the matrix can store all behaviors, regardless of the number of times a loop is executed. The terminology used for columns in LaFM allows us to retrieve the behaviors of an operator using a standardized name: operator|loop. Each operator is assigned a unique name. For example, in Fig. 1, loop5 is an operator. For parallel gateways, we also append the execution order inside parentheses. For instance, the second execution of and4 is and4(2). If there are loops, a single operator can be executed many times, resulting in multiple pieces of information that must be recorded. Adding the loop position to the terminology allows us to distinguish this information. Let L be a list of loops that are in the path starting from but excluding the operator itself to the root of the process tree. L can be empty if an operator is not contained in a loop. Then, we concatenate $\forall l \in L$ the following strings: $l_{name}(l_{index})$, i.e., for each loop above an operator, we include its name. In parentheses, we add the index of the loop. As an example, xor7|loop5{2} points to the column returning the decisions that are made when the operator xor7 is executed for the second time.

Three behaviors are captured in the LaFM in Fig. 2. Columns 1 to 5 retain the execution order of parallel gateways; column 6 records the number of times a loop was taken, and columns 7 to 9 store the decisions made at exclusive choice gateways.

4.2   Training Phase: Building LaFM

To record the decisions made for each operator in the discovered process tree, we replay the traces we want to learn from a Petri net version of the process tree. Petri nets can easily be derived from process trees using simple transformation rules [5]. Petri nets have a strong and executable formalism, which means we can replay a trace on a Petri net by playing the token game [13]. The token game takes as input a trace and a Petri net. Then, using a particular set of rules (see Chapter '3.2.2 Petri Nets' in [4]), the game indicates if the trace fits into the process model (i.e., the Petri net). Algorithm 1 defines few extra operations that are performed during the token game to build LaFM. The next section explains how predictions can be made from LaFM.

```
  /* Map the parallel operators above the events using a list of tuples (andOperator,
     branchIndex). Return an empty list if the event is not included in a parallel
     operators.                                                                    */
  /* e.g.,: {a: [(and4,0), (and2,0)], b: [(and4,1), (and2,0)], c: [(and2,0)]...}   */
1 tsToAnds = getTransitionToAnds(processTree)

  /* Map the transitions that occur right after an exclusive gateway.              */
  /* e.g.,: {and4: Xor3, c: Xor3, F: Xor7, G: Xor7 }                               */
2 tsToXors = getTransitionToXor(processTree)

  /* Map the second branch of loops to tsIncrementLoops and the third one to
     tsLeavingLoops
  /* e.g., tsIncrementLoops: {τ4: loop5}; tsLeavingLoops: {τ5: loop5}              */
3 tsIncrementLoops = getTransitionToIncrementLoop(processTree)
4 tsLeavingLoops = getTsToLeaveLoop(processTree)
5 laFM = Matrix[]
```

```
 6 foreach trace in logs do
 7     counter = initializeCounters()
 8     foreach tsFired in tokenGame do
 9         manageCounter(tsFired)
10         foreach andOperators in tsToAnds[tsFired] do
11             foreach andOperator, branchIndex in andOperators do
12                 record(trace, andOperator, branchIndex)

13         if tsFired in tsToXors then
14             record(trace, tsToAnds[tsFired], tsFired)

15         if tsFired in tsToLeaveLoop then
16             record(trace, tsLeavingLoops[tsFired], counter[tsFired])

17 function manageCounter(tsFired):
18     if tsFired in tsToAnds then
19         foreach andOperator in tsToAnds[tsFired] do
20             counter[andOperator].increment()

21     if tsFired in tsIncrementLoops then
22         counter[tsFired].increment()
23         foreach dependentTransition in dependentTransitions[tsFired] do
24             counter[tsFired].reset()

25 function record(trace, transition, value):
26     laFM[trace][getTerminology(transition)] = value
```

Algorithm 1: Set of extra operations performed during the token game to build LaFM.

4.3 Prediction Phase: Using LaFM

Making predictions using LaFM is a five step recursive process, illustrated in Fig. 3.

Step 1. We play the token game with the prefix to get a list of active tokens.

Step 2. From the tokens, we get the list of active transitions, i.e., the activities that are currently allowed by the business process model. If only one transition is active, we skip steps 3 and 4 to fire the transition (step 5). Otherwise, we recursively eliminate transitions that are less likely (steps 3 and 4).

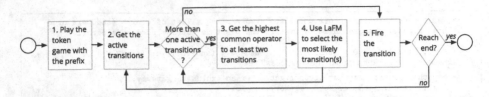

Fig. 3. Five steps in making prediction using LaFM

Step 3. We find the highest (closest to the root) operator in the process tree common to at least two transitions. For example, in Fig. 1, if the active transitions are A, B, and D, the highest common operator is **and2**.

Step 4. We make a decision about the operator selected in step 3. Depending on the operator type, we select the branch to execute next, what decision to make at an exclusive gateway, or whether to stay in or leave a loop. Figure 4 details how we retrieve the information in LaFM. In Fig. 2, in order to know which one of F and G is the transition most likely to be chosen the first time we are at xor7, we look at LaFM for xor7|loop5{1} and observe that F occurs more often (three times out of four). When a tie occurs, we pick the first one. The number of loops in the prefix might exceed the number of loops that were observed in the data. Alternatively, we might have a particular order in the prefix that was never observed in the event logs. We define three levels of abstraction that we apply consecutively when the previous abstraction fails. The first level of abstraction is to use LaFM as is. The second level of abstraction is to drop the loop part of the terminology and stack the columns for the same operator. For example, if xor7|loop5{3} does not exist in LaFM, we stack the two columns starting with xor7|. If there is still not enough information, the third abstraction is to make a decision by looking only at the Petri net. For parallel and exclusive choice transitions, we pick the first branches with active transitions. For a loop, the decision is always to leave the loop. Using these three abstractions, we can always make a prediction. If the list of potential transitions has been reduced to 1, we go to step 5. Otherwise, we recursively go back to step 3 where the highest common operator will inevitably be lower.

Step 5. We fire the transition. If it is a task ∈ E, we add it to the suffix. Then, we check to see if we have reached the end of the Petri net. If yes, we return the suffix. If not, we go back to step 3.

Having defined how to build and use LaFM, we detail in the next section the evaluation procedure used to assess the quality of the predictions.

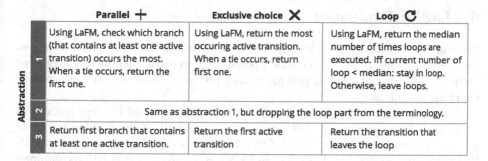

		Parallel **+**	Exclusive choice **✕**	Loop **C**
Abstraction	**1**	Using LaFM, check which branch (that contains at least one active transition) occurs the most. When a tie occurs, return the first one.	Using LaFM, return the most occuring active transition. When a tie occurs, return first one.	Using LaFM, return the median number of times loops are executed. Iff current number of loop < median: stay in loop. Otherwise, leave loops.
	2	Same as abstraction 1, but dropping the loop part from the terminology.		
	3	Return first branch that contains at least one active transition.	Return the first active transition	Return the transition that leaves the loop

Fig. 4. Decisions for each operator type at three level of abstractions.

5 Evaluation Procedure

The evaluation procedure is the same as that described by Tax et al. in [3]. Two-thirds of the traces in the event logs are added to the training set. Each trace in the evaluation is tested from a prefix length of 2 to a prefix length of $l - 1$, l being the length of the trace. For instance, the trace $\langle ABCD \rangle$ is decomposed into: prefix:$\langle AB \rangle$, suffix:$\langle CD \rangle$ and prefix:$\langle ABC \rangle$, suffix:$\langle D \rangle$. The extracted prefix is added to the evaluation set and the suffix is added to the ground truth set. After learning from the training set, we use the evaluation set to make predictions about the prefix. The accuracy is obtained by measuring the Damerau-Levenshtein similarity between the predicted suffix and the ground truth set. The Damerau-Levenshtein distance, [14], is an edit-distance-based metric that minimizes the number of substitutions, deletions, or additions that are needed to align two sequences. In contrast with the Levenshtein distance, the Damerau-Levenshtein distance allows us to swap two adjacent activities. Let e be the evaluation set, p_i the ith predicted suffix, and t_i the ith ground truth suffix. We evaluate a whole evaluation set using the following formula:

$$DamerauSimilarity(e) = 1 - \frac{\sum_{i=1}^{|e|} \frac{DamerauDistance(p_i, t_i)}{max(length(p_i), length(t_i))}}{|e|} \qquad (1)$$

A Damerau similarity of 1 means that the predicted suffix is identical to the ground truth. We use the evaluation procedure in the next section to evaluate the performance of LaFM on synthetic datasets as well as in Sect. 8 where the performance of c-LaFM is tested on real datasets.

All evaluations were processed on a Mac Pro with the following configuration: 3.5 GHz 6-Core Intel Xeon E5, 64 GB 1866 MHz DDR3. We slightly updated LSTM[1] so that it does not predict the time remaining. We confirmed that this change does not impact the accuracy of the next event predictions and slightly reduces the execution time. LaFM and c-LaFM, as used in the evaluations, are available at: http://customer-journey.unil.ch/lafm.

---

[1] available here: https://verenich.github.io/ProcessSequencePrediction/.

6   LaFM: Evaluation

To evaluate LaFM, we used a collection of 30 synthetic datasets[2] that were created from process trees of varying shapes and complexities. These datasets were initially created and used in [13] for testing process discovery and conformance checking techniques.

There are three rounds of evaluation. In each round, 10 process trees were generated. The complexity of the process trees as well as the number of traces generated increase with the round. Overall, 64 traces were generated in round 3, 256 traces in round 4, and 1025 in round 5. We compared the predictions obtained using LaFM, Markov chains, and LSTM. We ran the evaluation five times. The arithmetic means of these five runs is shown in Fig. 5. LaFM is deterministic, therefore, its variance is null. The predictions made using LaFM are closest to the ground truth (21 times), followed by LSTM (8 times), and Markov chains (4 times).

There are important differences in the execution times of the three techniques (Fig. 6). Because its predictions rely only on the previous observed event, it is not surprising that the fastest predictions are made using Markov chains, followed by LaFM. To put the duration into perspective, the average execution time per dataset is approximately 111 times slower for LaFM compared to a Markov chain, and 6140 times slower for LSTM compared to a Markov chain.

	treeSeed	1	2	3	4	5*	6	7	8	9	10
Round 3	LaFM	**1.00**	**1.00**	**0.58**	0.31	n/a*	**0.70**	**0.57**	**0.46**	**0.66**	0.91
	lstm	**1.00**	**1.00**	0.50	0.29	n/a*	0.60	0.42	0.44	0.50	**0.92**
	markov	0.60	**1.00**	0.20	**0.37**	n/a*	0.60	0.15	0.33	0.46	**0.92**
	treeSeed	1	2	3	4	5	6	7	8	9	10
Round 4	LaFM	**0.84**	**0.90**	0.39	0.30	0.66	**0.54**	**0.39**	0.34	**0.51**	**0.52**
	lstm	0.81	0.85	**0.43**	**0.35**	**0.83**	0.51	0.35	**0.36**	0.50	0.24
	markov	0.55	0.89	0.26	0.31	0.72	0.43	0.17	0.32	0.45	0.50
	treeSeed	1	2	3	4	5	6	7	8	9	10
Round 5	LaFM	0.51	**0.48**	**0.24**	**0.50**	**0.86**	**0.63**	**0.36**	**0.21**	0.48	**0.54**
	lstm	**0.56**	0.36	0.21	0.42	0.85	0.62	0.30	0.14	0.22	0.29
	markov	0.28	0.42	0.13	0.41	0.45	0.56	0.17	0.17	**0.50**	0.47

		Max	Arithmetic mean	Median
Variance	LaFM	0.0000	0.0000	0.0000
	lstm	0.0146	0.0018	0.0002
	markov	0.0120	0.0009	0.0000

*not enough data for the evaluation because 84% of the traces have a length of 1.

Fig. 5. Comparing LaFM, LSTM and Markov Chains using the Damerau similarity metric. The closer to 1, the closer the predictions are to the ground truth.

---

[2] https://data.4tu.nl/repository/uuid:745584e7-8cc0-45b8-8a89-93e9c9dfab05,   sets '1 - scalability', 'round 3 to 5'.

	round	3	4	5
Training	LaFM	~24 sec	~28 sec	~2.5 min
	lstm	~2.5 min	~18 min	~6 hours
	markov	< 1 sec	< 1 sec	< 1 sec

	round	3	4	5
Prediction	LaFM	~3 sec	~1 min	~27 min
	lstm	~1 min	~5 min	~22 hours
	markov	< 1 sec	< 1 sec	~17 sec

Fig. 6. Performance comparison of the training and predictions times.

7 c-LaFM: Clustered Loop-Aware Footprint Matrix

The accuracy of the predictions made using LaFM is dependent on the quality of the discovered process tree. While the previous section showed that LaFM performs well with synthetic datasets generated from well-structured process trees, the accuracy will drop with real datasets, which often have very complex behaviors and noise that cannot be described well using a single model. Our intuition is that we should group similar traces using clustering techniques and, for each group, discover a process tree that well describes a subset of similar traces. Hence, we propose an updated version of LaFM with a clustering step, coined c-LaFM for clustered LaFM.

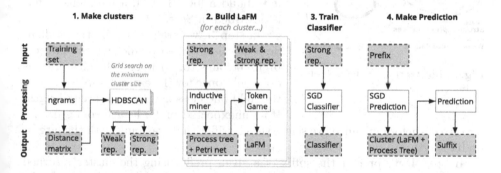

Fig. 7. Overview of the 4 steps approach of c-LaFM.

We propose a four-step clustering method, as shown in Fig. 7. In step 1, we extract the features that will be used to group similar traces. Thus, we count the number of ngrams ranging in size from 1 to 2. For instance, the trace ⟨ABA⟩ becomes: {A:2, B:1, AB:1, BA:1}. Then, we cluster the traces using HDBSCAN[3], which has the advantage of having only one intelligible parameter to set, the minimum number of traces per cluster. According to our experiment, from 2 to 10 traces per cluster yields the best results. However, it is difficult to anticipate the best minimum cluster size. Hence, we perform a hyperparameter optimization of a type grid search by using 10% of the training data set to evaluate the accuracy of the minimum cluster size and retain the best-performing one. Instead

---

[3] https://github.com/scikit-learn-contrib/hdbscan.

of attributing each trace to a single cluster, we rely on a soft clustering approach, which returns, for each trace, the probability of it belonging to all the clusters.

Figure 8 illustrates the soft clustering approach. Each point represents a trace. The closer two traces are, the more ngrams they share. The strong representatives are used to discover the process tree, while the weak and the strong representatives will be replayed over the process tree and are available in LaFM. The strong representatives are the traces that have a probability higher than 80% of belonging to a cluster and the weak representatives have a probability higher than 20% but less than 80%. Using a soft clustering approach has two main advantages. First, the inductive miner is sensitive to noise. Hence, we want to learn only from the strong representatives (i.e., with a high probability of belonging to the clusters) with the aim of capturing only the core behaviors. Second, although we do not use them to build the process trees, borderline traces might contain interesting behaviors for several clusters. By using a soft clustering approach, we can assign these single traces to several clusters.

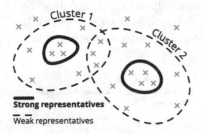

Fig. 8. Illustration of the soft clustering concept.

In step 2, the strong representatives are used to build the process tree. Then, the process tree is transformed to a Petri net so that the weak representatives can be replayed on it to build a local LaFM, a mechanism that is described in Sect. 4.2.

In step 3, we train a stochastic gradient descent classifier[4] to predict which cluster a prefix belongs to. Although the clustering is done only once for the entire complete traces, we build one classifier for each prefix length. If an unexpected prefix length comes from a never-seen-before instance, we select the classifier that was built with the largest prefix length.

In step 4, we predict the suffix of a given prefix using the cluster returned by the classifier. Altogether, these four steps allow us to make predictions in the presence of noise and outliers, which are often found in real datasets. This is what we evaluate in the next section.

8   c-LaFM: Evaluation

To test our approach, we used six publicly available event logs, as described in Table 1. Because the event logs reflect activities performed in real life, making predictions is a complex task. Typically, for the event logs describing the execution of a building permit application (envPermit), "almost every case follows a unique path" [3].

---

[4] http://scikit-learn.org/stable/modules/sgd.html.

Table 1. Datasets used for the evaluation.

	Name (doi)	Description	#traces	#events
1	helpdesk (10.17632/39bp3vv62t.1)	Events from a ticketing system	3'804	13'710
2	bpi12 (10.4121/uuid:3926db30-f712-4394-aebc-75976070e91f)	Loan process for a financial industry. Note: keeping only manual task and lifecycle: complete as described in [3]	9'658	72'413
3	bpi13 closeP (10.4121/c2c3b154-ab26-4b31-a0e8-8f2350ddac11)	Closed problem - management system from Volvo IT Belgium	6'660	1'487
4	bpi13 incidents (10.4121/3537c19d-6c64-4b1d-815d-915ab0e479da)	Incidents - management system from Volvo IT Belgium	7'554	65'533
5	bpi13 openP (10.4121/500573e6-accc-4b0c-9576-aa5468b10cee)	Open problems - management system from Volvo IT Belgium	819	2'351
6	envPermit (10.4121/uuid:26aba40d-8b2d-435b-b5af-6d4bfbd7a270)	Execution of a building permit application process. Note: we pick the Municipality 1	38'944	937

In contrast to LaFM, c-LaFM is non-deterministic due to the clustering step. Hence, we ran the experiment 10 times with c-LaFM and LSTM using the procedure described in Sect. 5. Figure 9 compares the accuracy of LSTM and c-LaFM. c-LaFM is more accurate for five datasets out of six. We compare the execution times in Fig. 10. On average, c-LaFM is 9 times faster than LSTM. Overall, we have shown that the clustered version of LaFM is accurate and fast.

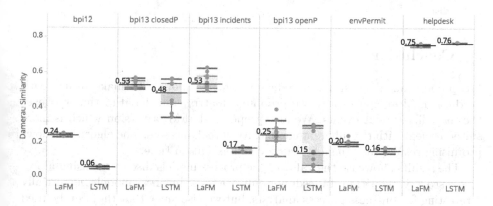

Fig. 9. Comparing c-LaFM to LSTM using real datasets. Each datasets was evaluated 10 times.

Figure 11 shows one of the predictions for the execution of a building permit using a business process model, which was derived from the process tree that was used to make the prediction. This is an illustration of how we can provide, not only the predictions itself, but a way to express the reasoning behind the prediction. For instance, a business worker could–after investigating traces like those used to make the prediction–decide not to trust the prediction because they have knowledge about the context that is not available in the event logs.

Dataset	bpi12	bpi13 closedP	bpi13 incidents	bpi13 openP	envPermit	helpdesk
LaFM	~45 min	~2 min	~47 min	~1 min	~3.1 hours	~2 min
LSTM	~15.4 hours	~35 min	~20.6 hours	~18 min	~5.6 hours	~41 min

Fig. 10. Comparing the total execution time to obtain predictions using c-LaFM and LSTM. The value reported is the average of 10 executions.

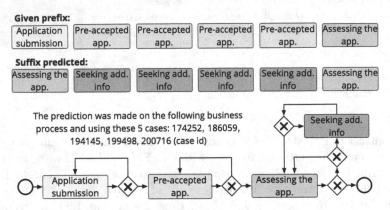

Fig. 11. Displaying an actual prediction from the dataset envPermit next to the business process model that was used to make the prediction. The labels has been translated in English.

9   Conclusion

We propose an algorithm that relies on process models to make future path prediction. More specifically, we propose a matrix coined LaFM that retrieves the most likely next events. We also propose c-LaFM, a version which is more suited to deal with the inherent complexity of real datasets. The algorithm shows promising results in terms of accuracy and execution time.

The results showcase the value of the process models discovered using a process discovery algorithm. Indeed, not only are these business models intrinsically interesting for business process analysts, but we also show that they can be used to make predictions. A limitation of this work is that we choose to rely on the inductive miner. In our future work, we plan to measure how the use of different process discovery techniques may impact the accuracy of the predictions. We anticipate that mining hidden rules between LaFM columns will yield interesting results, especially if we consider extending LaFM with contextual information. This would allow us to detect long-term dependencies that could be used to improve the accuracy further.

Business analysts can be reluctant to trust predictions they do not understand [10]. Because in our work the predictions are made with business process models, the predictions can be manually inspected by business analysts. Currently, our algorithm returns only the predictions, limiting the explainability of the results. However, we envision a framework that includes an advanced visualization system that explains how the predictions are made and allows business analysts to alter the predictions if they have knowledge that is not in the data. This type of system would display the process model, the traces on which the predictions were made, and the reasoning behind the predictions. Gartner has urged us to move toward explainable artificial intelligence that gives visibility to business stakeholders "by leveraging historical data, explaining model inputs, simplifying results or exposing underlying data in human understandable ways" [2]. Our work contributes by providing the foundation on which a fully comprehensible prediction system can be built. Interestingly, in the same report, [2], Gartner states that there is a trade-off between explainability and accuracy. Our results highlight that this trade-off does not necessarily hold here as we can provide results that are both transparent and more accurate than state-of-the-art neural network approaches.

References

1. Polato, M., Sperduti, A., Burattin, A., De Leoni, M.: Time and activity sequence prediction of business process instances. Computing **100**, 1005–1031 (2018)
2. Alaybeyi, S., Baker, V., Clark, W.: Build trust with business users by moving toward explainable AI. Technical report, Gartner, October 2018. https://www.gartner.com/doc/3891245/build-trust-business-users-moving
3. Tax, N., Verenich, I., La Rosa, M., Dumas, M.: Predictive business process monitoring with LSTM neural networks. In: Dubois, E., Pohl, K. (eds.) CAiSE 2017. LNCS, vol. 10253, pp. 477–492. Springer, Cham (2017). https://doi.org/10.1007/978-3-319-59536-8_30
4. van der Aalst, W.: Process Mining: Data Science in Action. Springer, Heidelberg (2016). https://doi.org/10.1007/978-3-662-49851-4
5. Leemans, S.J.J., Fahland, D., van der Aalst, W.M.P.: Discovering block-structured process models from event logs - a constructive approach. In: Colom, J.-M., Desel, J. (eds.) PETRI NETS 2013. LNCS, vol. 7927, pp. 311–329. Springer, Heidelberg (2013). https://doi.org/10.1007/978-3-642-38697-8_17
6. Vanhatalo, J., Völzer, H., Koehler, J.: The refined process structure tree. In: Dumas, M., Reichert, M., Shan, M.-C. (eds.) BPM 2008. LNCS, vol. 5240, pp. 100–115. Springer, Heidelberg (2008). https://doi.org/10.1007/978-3-540-85758-7_10
7. Pitkow, J., Pirolli, P.: Mining longest repeating subsequences to predict world wide web surfing. In: Proceedings of the UsENIX Symposium on Internet Technologies and Systems, p. 1 (1999)
8. Gueniche, T., Fournier-Viger, P., Tseng, V.S.: Compact prediction tree: a lossless model for accurate sequence prediction. In: Motoda, H., Wu, Z., Cao, L., Zaiane, O., Yao, M., Wang, W. (eds.) ADMA 2013. LNCS (LNAI), vol. 8347, pp. 177–188. Springer, Heidelberg (2013). https://doi.org/10.1007/978-3-642-53917-6_16
9. Lakshmanan, G.T., Shamsi, D., Doganata, Y.N., Unuvar, M., Khalaf, R.: A Markov prediction model for data-driven semi-structured business processes. Knowl. Inf. Syst. **42**(1), 97–126 (2015)
10. Breuker, D., Matzner, M., Delfmann, P., Becker, J.: Comprehensible predictive models for business processes. MIS Q. **40**(4), 1009–1034 (2016)

11. Evermann, J., Rehse, J.-R., Fettke, P.: A deep learning approach for predicting process behaviour at runtime. In: Dumas, M., Fantinato, M. (eds.) BPM 2016. LNBIP, vol. 281, pp. 327–338. Springer, Cham (2017). https://doi.org/10.1007/978-3-319-58457-7_24
12. Hochreiter, S., Schmidhuber, J.: Long short-term memory. Neural Comput. **9**(8), 1735–1780 (1997)
13. Leemans, S.: Robust process mining with guarantees. Ph.D. thesis, Eindhoven University of Technology (2017)
14. Damerau, F.J.: A technique for computer detection and correction of spelling errors. Commun. ACM **7**(3), 171–176 (1964)

Contextual and Behavioral Customer Journey Discovery Using a Genetic Approach

Gaël Bernard[1]([✉]) and Periklis Andritsos[2]

[1] Faculty of Business and Economics (HEC), University of Lausanne,
Lausanne, Switzerland
gael.bernard@unil.ch
[2] Faculty of Information, University of Toronto, Toronto, Canada
periklis.andritsos@utoronto.ca

Abstract. With the advent of new technologies and the increase in customers' expectations, services are becoming more complex. This complexity calls for new methods to understand, analyze, and improve service delivery. Summarizing customers' experience using representative journeys that are displayed on a Customer Journey Map (CJM) is one of these techniques. We propose a genetic algorithm that automatically builds a CJM from raw customer experience recorded in a database. Mining representative journeys can be seen a clustering task where both the sequence of activities and some contextual data (e.g., demographics) are considered when measuring the similarity between journeys. We show that our genetic approach outperforms traditional ways of handling this clustering task. Moreover, we apply our algorithm on a real dataset to highlight the benefit of using a genetic approach.

Keywords: Customer journey mapping · Process mining ·
Customer journey analytics · Genetic algorithms

1 Introduction

A customer experience can be defined as a customer's journey with an organization over time across multiple interactions called touchpoints [1]. Recent studies show that customer interactions are increasing [2], services are becoming more complex, and customers are often unpredictable [3]. In this context, understanding the main trajectories that were followed by customers to consume a service is a complex task. According to Verhoef et al., a strategy based on customer experience may provide a superior competitive advantage [1]. It is, therefore, not surprising that "Characterizing the Customer Journey along the Purchase Funnel and Strategies to Influence the Journey" has been ranked as one of the most important research priorities for the coming years by the Marketing Science Institute [4]. A challenge faced by many practitioners is that of understanding

© Springer Nature Switzerland AG 2019
T. Welzer et al. (Eds.): ADBIS 2019, LNCS 11695, pp. 251–266, 2019.
https://doi.org/10.1007/978-3-030-28730-6_16

the large number of combinations of activities that may exist when consuming a service. As a result, new methods employed to design, analyze, and understand customer journeys are emerging from the industry and are becoming popular among researchers. One of these conceptual methods that will be the focus of this work, is called the Customer Journey Map (CJM). By showing typical journeys experienced by customers across several touchpoints, a CJM helps to better understand customers' trajectories [5].

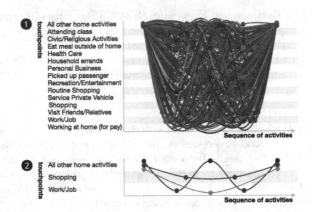

Fig. 1. ❶ Less than 0.01% of the entire dataset on a CJM, and ❷, a summary of the dataset using 3 representatives.

Figure 1 shows CJMs derived from a real dataset[1]. In this dataset, a journey is all the activities that are performed by a citizen throughout the day. For instance being at home, attending class and going back home is one of the potential journeys. As can be seen in ❶ of Fig. 1, displaying such actual journeys on the CJM without preprocessing the data results in an overwhelming chart. It becomes clear that when a company deals with very large numbers of actual journeys, it is necessary to reduce the complexity and to look at these journeys at a higher level of abstraction. Specifically, representative journeys address this issue, [6], by summarizing the dataset (using three journeys visible in ❷ of Fig. 1).

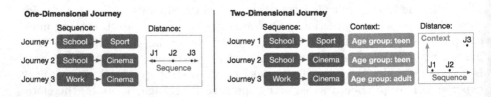

Fig. 2. Measuring the distance among three journeys with and without the context.

[1] http://www.cmap.illinois.gov/data/transportation/travel-survey.

The existing solutions to summarize collections of journeys [6,7] consider only the sequence of touchpoints when measuring the distance between journeys. Figure 2 illustrates the process with 3 short journeys. Using a basic distance measure between sequences (e.g., edit-distance), we cannot say which one of 'Journey 1' or 'Journey 3' is closer to 'Journey 2' (upper part of Fig. 2). We suggest that demographics and other contextual information might be equally important to measure the distance between journeys. Hence, in this paper, we propose to integrate such information when mining journeys. The bottom part of Fig. 2 shows that when we also consider the age group, it becomes clearer that the closest journey to 'Journey 2' is 'Journey 1'.

We propose an algorithm summarizing a customer experience using both the sequence of activities as well as the contextual information. Our genetic approach uses only three intuitive parameters: (1) the approximate number of representative journeys to use, (2) the weight of the sequence of activities, and (3) the weight of the contextual data. In the evaluation section, we demonstrate that we outperform existing techniques. Finally, we highlight the impact of the three parameters using a real dataset and illustrate the results with CJMs.

The paper is organized as follows. Section 2 discusses the discovery of customer journeys. In Sect. 3, we outline the existing techniques. Section 4 depicts our genetic algorithm. In Sect. 5, we evaluate our approach using internal and external evaluation metrics. Section 6 illustrates CJMs produced by our algorithm. Finally, in Sect. 7 we conclude the paper.

2 Customer Journey Discovery

The goal of a customer journey discovery algorithm is to find a reasonable amount of representative journeys that summarize well the observed journeys.

Definition 1 (Touchpoint): We define a touchpoint as the interaction between a company's products or services and a customer [5]. 'Buying a product' or 'complaining about a product' are two examples of touchpoints in an online retail context. We define t as the touchpoint while T is the collection of all touchpoints. The touchpoints are visible in the y-axis of the CJMs (Fig. 1).

Definition 2 (Actual Journey): An actual journey J_a is a sequence of touchpoints observed from customers. To improve readability, we refer to touchpoints using alphabetical characters (e.g., J becomes $\langle ABC \rangle$). The order in which the activities are executed is represented by the x-axis of the CJMs visible in Fig. 1.

Definition 3 (Representative Journey): A representative journey, J_r, is a journey that summarizes a subset of actual journeys. In Fig. 1, ❶, shows how a CJM would look like when we display actual journeys, while the bottom part, ❷, uses representative journeys. Clearly, as can be seen in Fig. 1, the use of representative journeys increases the readability of the CJM.

Definition 4 (Event Logs): An event log is denoted by $J_\mathcal{A}$, which is the list of all journeys observed by customers.

Definition 5 (Customer Journey Map): By using representative journeys, a CJM summarizes customer trajectories. Let a customer journey map J_R be the set of all the J_rs summarizing J_A. k_R denotes the total number of journeys. Typically, the part ❷ of Fig. 1 is a CJM, J_R, containing three representative journeys summarizing an event log.

We define the discovery of customer journeys as a function that maps all members of J_A to a member of J_R; i.e., that maps all the actual journeys to representative journeys ultimately displayed on a CJM. Discovering customer journeys from event logs can be seen as an unsupervised clustering task. This task has interesting challenges. First, choosing the number of representatives is difficult. When the goal is to have a general overview about a particular dataset, it seems reasonable to display only few journeys so the CJM is readable. However, discovering a few dozens of representative journeys might also be a relevant choice if the goal is to catch complex and less generic patterns. Finally, the sequence that best summarizes its assigned actual journeys needs to be found. It might be the case that an ideal representative journey was never observed but still summarizes the actual journeys well. These phenomena were observed by Gabadinho et al., and illustrated as follows: "We could imagine synthetic – not observed – typical sequences, in the same way as the mean of a series of numbers that is generally not an observable individual value" [8]. Before presenting our solution, the next section describes related work.

3   Related Work

There is a body of work in social sciences that is relevant to the summarization of customer journeys. Typically, in [7,8], Gabadinho et al. are summarizing observed sequences with representatives. They define a representative as *"a set of non-redundant 'typical' sequences that largely, though not necessarily exhaustively, cover the spectrum of observed sequences"* [7]. The authors propose four ways to choose a representative. *'Frequency'*, (1), considers the most frequent sequence as the representative. *'Neighborhood density'*, (2), selects the sequence that has the most neighbors in a defined diameter. *'Centrality'*, (3), picks the most central object, i.e., the one having the minimal sum of distances from all other objects. Finally, *'sequence likelihood'* considers a sequence derived from the first-order Markov model.

Since Process Mining operates in a bottom-up fashion, from data all the way to the discovery of conceptual models, it is another discipline closely related to the topic of customer journey discovery. The link between customer journey maps and process mining was highlighted in [5]. However, business process models and CJMs are not built for the same purpose. While a business process model captures how a process was or should be orchestrated, a CJM is built for the purpose of better understanding what customers have experienced.

In [9], we propose CJM-ex, an online tool to explore CJMs. Because it uses a hierarchical structure, it allows to efficiently navigate the space of journeys in

CJMs. In [10], it was shown that customer journey maps can be discovered using Markov models. In [6], we suggested a genetic approach to discover representative journeys that uses only the sequence of touchpoints to measure the distance between journeys. Hence, this current work can be seen as an extension of [6] to allow taking both the sequence of touchpoints and the contextual information into account when build CJMs.

4 A Genetic Algorithm for Customer Journey Discovery

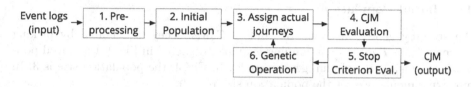

Fig. 3. Our genetic approach

Our work is inspired by the genetic approaches proposed in [11–13] to discover business process models from event logs. However, we tailored it towards CJMs by introducing specific evaluation metrics suited for them. Figure 3 depicts the main phases: (1) a preprocessing phase, (2) a phase for the generation of the initial population, (3) the assignment of each actual journey to its closest representative, (4) the evaluation of the quality of the CJMs, (5) the stopping criterion evaluation, and (6) the creation of new CJMs by applying some genetic operations. We introduce these phases in details while the Fig. 4 illustrates how it works.

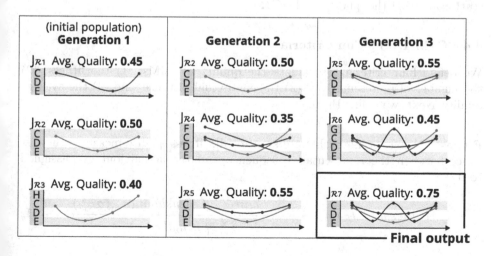

Fig. 4. Illustration of the genetic process for the discovery of the best CJMs

4.1 Preprocessing

We assume that the representative journeys will be similar to the journeys with the most frequent patterns of activities. Hence, to reduce computation time, we extract the most frequent patterns that we use to create new journeys and generate the initial population. Let Top_{ℓ_n} be the n most occurring patterns of activities of length ℓ and $Top_n \supseteq Top_{\ell_{[2,m]}}$ be the list of all the most occurring patterns of lengths 2 to m. By using Top_n, we reduce the execution time by two without impairing the quality of the final output.

4.2 Initial Population

We start by generating a set of random CJMs. They are created by picking journeys from Top_n. In our running example, depicted in Fig. 4, the initial population is visible in column 'generation 1'. In Fig. 4, the population size is 3. In our experiments, we set the population size to 100.

4.3 Assignment of Actual Journeys

In order to evaluate the quality of the generated CJMs, it is required to assign each actual journey to its closest representative. The closeness between J_a and J_r is measured using the Levenshtein distance [14]. This metric counts the number of edit operations (i.e., deletions, insertions, and substitutions) required to match two sequences. Typically, the distance between $\langle \text{XYZ} \rangle$ and $\langle \text{XYYW} \rangle$ is 2. The closest representative is the one having the smallest Levenshtein distance with the actual journeys. If a tie occurs, we assign the actual journey to the representative having the less journeys already assigned to it. When the actual journeys have been assigned to their respective closest representative, we can start evaluating the quality of the CJMs.

4.4 CJM Evaluation Criteria

We define four criteria to evaluate the quality of CJMs: (1) the fitness, (2) the number of representatives, (3) the contextual distance, and (4) the average quality. Next, we define them.

Fitness. Using the Levenshtein distance [14], fitness quality measures the distance between the representative sequence and the actual journeys assigned to it.

$$Fitness(J_a, J_{\mathcal{R}}) = 1 - \frac{\sum_{i=1}^{|J_a|} min_{j=1}^{|J_{\mathcal{R}}|}(Levenshtein(\sigma_{\mathcal{A}_i}; \sigma_{\mathcal{R}_j}))}{\sum_{i=1}^{|J_a|} Length(\sigma_{\mathcal{A}_i})} \tag{1}$$

where

$\sigma_{\mathcal{A}_i}$: i^{th} actual (observed) sequence in $J_\mathcal{A}$
$\sigma_{\mathcal{R}_j}$: j^{th} representative contained in $J_\mathcal{R}$
$Length(x)$: Number of touchpoints in the sequence x

When an actual journey is strictly identical to its representative journey, the fitness measure is equal to 1.

Number of Representatives. The more representative journeys we use, the more likely the fitness will be high. Hence, without a metric that allows a low number of representatives, we would obtain a final CJM with several thousands of representative journeys. Therefore, the goal of this metric is to keep a low number of representatives. To guide the algorithm towards an 'ideal' number of representatives, we employ a clustering technique that helps in choosing the number of clusters. More specifically, we used the Calinski-Harabasz index [15]. Let k_h be the optimal number of clusters returned by the Calinski-Harabasz index. To evaluate the quality, we measure the distance between $k_\mathcal{R}$ and k_h using the following distribution function:

$$NumberOfRepresentatives(k_\mathcal{R}, k_h, x_0) = \frac{1}{1 + (\frac{|k_\mathcal{R} - k_h|}{x_0})^2} \qquad (2)$$

where

$k_\mathcal{R}$: Number of J_r journeys on $J_\mathcal{R}$ (i.e., $|J_\mathcal{R}|$)
k_h : ideal number of journeys according to the Calinski-Harabasz index
x_0 : x value of the midpoint

We set the value of the midpoint, x_0, to 5 for all our experiments. The intuition behind this parameter is the following: if we have 11 representative journeys on a CJM and the ideal number of journeys is 6, we would have a quality of 0.5 (midpoint) because the absolute distance between 11 and 6 is 5. Often, the final output will have a number of representative journeys that differs from k_h. This is due to the fact that there are other evaluation criteria.

Contextual Distance. The contextual distance allows us to consider the set of contextual data C when grouping similar journeys. The more distant the set of contextual data is between J_a that are represented by distinct J_r, the better the quality is. To measure the distance, we first build a value frequency table which counts all the values per representative (v_i is the value frequency counter for J_{r_i}). Then, for each pair of clusters, we calculate the cosine similarity, which is defined as:

$$ContextualDistance(v_1, v_2) = \frac{v_1 \cdot v_2}{||v_1|| \cdot ||v_2||} \qquad (3)$$

Finally, the cosine distances are averaged to get the overall contextual distance. A short overall distance indicates that the contextual data of J_a that are assigned to distinct J_r are similar. In other words, the contextual data does not help in classifying J_a between several J_r.

Average Quality. We get the average weighted quality by getting the arithmetic mean of: the fitness, the number of representatives, and the contextual-distance.

4.5 Stopping Criterion

Once we assess the quality of generated CJMs, we assess the stopping criterion. Inspired by the process mining genetic algorithms, [11,13], we found three stopping criteria: (1) a certain amount of generations has been reached, (2) the quality does not improve anymore, or, (3), a quality threshold has been reached. Predicting the quality that will be reached by a CJM is difficult. Hence, we believe that the latter stopping criterion is not advisable. If a stopping criterion is met, the algorithm stops, returning the best $J_\mathcal{R}$. If none of the stopping criteria is met, we generate new candidates by recursively calling a function that generates the next population, described in the next section.

4.6 Genetic Operations

Before transforming the CJMs, we evaluate and rank them by average quality. We copy a fraction (i.e., e) of the best CJMs in a set named *elite*. In Fig. 4, the elite size is 1. In our experiments, we set the elite size to 5.

By keeping the best CJMs as-is, we ensure that the quality will increase or stay unchanged. We also generate $p - e$ new CJMs using the following operators. (1) Addition of a random journey (mutation): A sequence from Top_n is added to $J_\mathcal{R}$. (2) Addition of an existing journey (crossover): A journey from the elite population is added to $J_\mathcal{R}$. (3) Deletion of a journey (mutation): A journey is removed from $J_\mathcal{R}$. Nothing happens if $J_\mathcal{R}$ contains only one journey. (4) Addition of a touchpoint (mutation): A touchpoint is inserted in one of the existing journeys. (5) Deletion of a touchpoint (mutation): A touchpoint is removed from $J_\mathcal{R}$.

We loop over each of these 5 types of transformations three times. Each time, the probability of applying the transformation is 10%, which means that more than one transformation is applied. It also means that the same transformation might be applied up to three times (with a probability of 0.1%). At the very least, one transformation has to be applied. If it is not the case, we loop over each transformation three times again until at least a transformation is performed.

In Fig. 4, $J_{\mathcal{R}5}$ have been produced by taking $J_{\mathcal{R}2}$ and adding a journey picked from Top_n (defined in Sect. 4.1). Once new $J_\mathcal{R}$s have been created, we return to the evaluation phase as shown in Fig. 3.

5 Evaluation Using Synthetic Datasets

In order to evaluate the quality of our approach to return the best set of representative journeys in $J_\mathcal{R}$, we evaluate the results using a collection of synthetic customer journeys that includes some contextual data. We first describe how we generated the dataset. Then, using this synthetic dataset, we evaluate and compare our algorithm with existing techniques.

5.1   Datasets

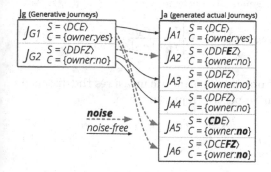

Fig. 5. Dataset with 50% of noise.

In order to evaluate the results of our algorithm, we generated synthetic event logs that simulate journeys using generative journeys. A generative journey is a known sequence of activities with a known set of characteristics from which we generate the event logs. These generative journeys represent the ground truth. If we used only those known generative journeys to produce the dataset, we would get only k_G distinct journeys. From a business point of view, this would describe an ideal situation where each group of customers behaves in an homogeneous way. However, we know that this is not the case. Having group of similar journeys that slightly differ from a representative is a more realistic setting. To achieve this, we add some noise to the generated journeys. Typically, when the noise level is set to 50%, $J_a = J_g$ is true for half of the data. Figure 5 illustrates how six journeys are generated from two generative journeys. If we assume that the noise level is defined to be 50%, three actual journeys in the event logs deviate from the original generative journeys. The goal of our experiments is to retrieve the set of generative journeys, as representatives, from the produced actual journeys. The 40 generated datasets as well as details on how we produced them are made publicly available[2].

5.2   Metrics

To evaluate and compare the quality of representative journeys, we rely both on external and internal evaluations. The former evaluates the results by using the generative journeys. Since we add some random noise, it might be the case that the ground truth is not the best solution. For this reason, we also use internal evaluation measures that rely on cluster analysis techniques. These metrics are described in [8].

External Evaluation - Distance in the Number of Journeys. Measures the distance between the number of generative journeys and the number of representative journeys. We evaluate this metric using the following equation:

$$NbJourneysDistance(k_G, k_R) = abs(k_G - k_R) \qquad (4)$$

---

[2] http://customer-journey.unil.ch/datasets/.

External Evaluation - Jaccard Distance. We use the Jaccard distance to evaluate how well we can retrieve the generative journeys.

$$JaccardDistance(\sigma_\mathcal{R}, \sigma_\mathcal{G}) = 1 - \frac{|\sigma_\mathcal{R} \cap \sigma_\mathcal{G}|}{|\sigma_\mathcal{R} \cup \sigma_\mathcal{G}|} \tag{5}$$

Internal Evaluation - Mean Distance [8]. This metric measures the distance between the actual journeys and their respective representative.

$$MeanDistanceScore_i = \frac{\sum_{j=1}^{k_i} D(S_i, S_{ij})}{k_i} \tag{6}$$

where

$D(x_1, x_2)$: Levenshtein distance between two sequences
k_i : Number of journeys attached to the representative i
S_i : Representative sequence i
S_{ij} : Sequence of journeys j attached to the representative i

Internal Evaluation - Coverage [8]. This metric represents the density of journeys in the neighborhood n of a representative.

$$Coverage_i = \frac{\sum_{j=1}^{k_i} (D(S_i, S_{ij}) < n)}{k_i} \tag{7}$$

where

$D(x_1, x_2)$: Levenshtein distance between two sequences
k_i : Number of journeys attached to the representative i
S_i : Representative sequence i
S_{ij} : Sequence of journeys j attached to the representative i

Internal Evaluation - Distance Gain [8]. This metric quantifies the gain in using representative journeys rather than the medoid of the dataset.

$$DistGain_i = \frac{\sum_{j=1}^{k_i} D(C(\sigma_\mathcal{A}), S_{ij}) - \sum_{j=1}^{k_i} D(S_i, S_{ij})}{\sum_{j=1}^{k_i} D(C(\sigma_\mathcal{A}), S_{ij})} \tag{8}$$

where

$D(x_1, x_2)$: Levenshtein distance between two sequences
k_i : Number of journeys attached to the representative i
S_i : Representative sequence i
S_{ij} : Sequence of journeys j attached to the representative i
$C(x)$: True center of the set

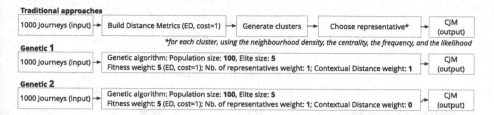

Fig. 6. Approach used to evaluate our clustering algorithm from traditional approaches.

5.3 Settings

We test two settings of the algorithm against traditional approaches. The traditional approaches are state-of-the-art techniques that are used to cluster and summarize sets of sequential and categorical data. Figure 6 depicts the approach at a high-level. As can be seen, with traditional approaches, we first build a distance metric. We use the edit distance with a constant cost operation set to 1. Once the distance matrix is built, we create k clusters. Because we do not know the number of representative journeys to be found, we test using from 2 to 12 clusters and use the squared Calinski-Harabasz index described in [15] to return the most statistically relevant. Next, we get the best representatives using the neighborhood density, the centrality, the frequency, or the likelihood using Traminer [16]. These techniques do not use the contextual data. Hence, to allow for a fair comparison, we compare these techniques with a version of our genetic algorithm that does not use contextual data and which was presented in [6]. We call this version $Genetic_1$. We also test our genetic algorithm with a version that considers the contextual data, called $Genetic_2$. Note that both the traditional and genetic approaches use the same techniques to find k_h and the distance is measured using the edit distance with a constant cost operation. To account for the fact that the genetic algorithm is non-deterministic, we run the algorithm ten times for each setting.

5.4 Results

Figure 7 shows the external evaluation metrics. It can be seen that the best solution is the $Genetic_2$, highlighting that considering the contextual information when grouping journeys improves the quality. Next, the best solution that does not use contextual data is $Genetic_1$ proposed in [6].

The internal evaluation of Fig. 8 shows that not only does the genetic algorithm outperforms the traditional approaches, it also proposes a better solution than the ground truth. This can be explained by the fact that when we inject noise, we potentially change the optimal solution.

The execution time for one thousand journeys is improved using Traminer [16] compared to our genetic approach. We compare how the different algorithms scale when the number of journeys increases. Hence, we ran each configuration five times with the 40 different datasets. Figure 9 summarizes the results. As can

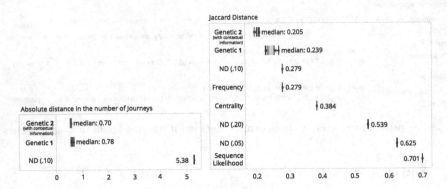

Fig. 7. External evaluation. The genetic algorithm that uses the contextual information (i.e., *Genetic₂*) performs best.

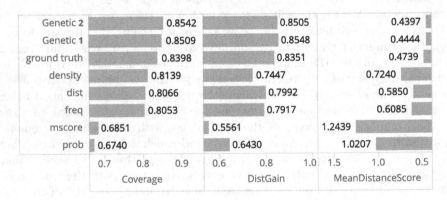

Fig. 8. Internal evaluation. The *Genetic₂* has the best coverage and mean distance while *Genetic₁* has the best distance gain.

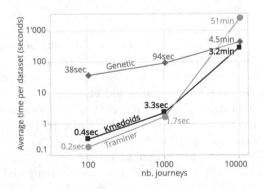

Fig. 9. Execution time for 100, 1'000, and 10'000 journeys.

be seen, the algorithms implemented in Traminer are orders of magnitude faster than our approach when dealing with 100 or 1,000 journeys. However, note that our algorithm has a better scaling potential when the number of journeys grows. All the algorithms tested tend to be slow and will not scale when dealing with several thousand journeys.

6   Experiments Using Real Datasets

This section reports on the experiments with a real dataset, the goal being to illustrate how a change in the settings impacts the results. We used a publicly available dataset[3] describing the activities performed throughout the day by Chicago's citizens. There are 15 types of activities, such as, 'being at home', 'attending class', 'going shopping', or 'doing households errands'. In the context of this dataset, a journey is the sequence of activities starting from the morning until the night. Typically, 'being at home' → 'attending class' → 'being at home'

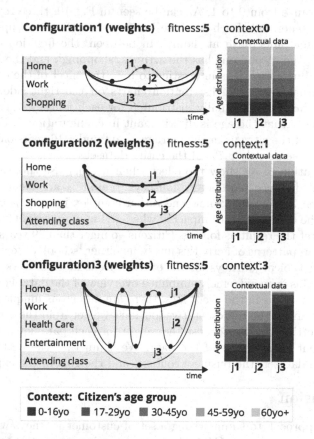

Fig. 10. Results with real dataset using three configurations

<hr/>

[3] http://www.cmap.illinois.gov/data/transportation/travel-survey.

is a journey consisting of three activities. The total number of journeys is 29,541 and there are 123,706 activities (with an average of 4.817 activities per journey). This dataset is interesting not only for the relatively large number of data points describing life trajectories, but also because of the available detailed contextual data, such as information on the citizens' demographics.

The goal of this experiment is to show the influence of taking the citizen's age in consideration when measuring the distance between journeys. Figure 10 shows the results using three different configurations. In configuration 1, we did not leverage the contextual data (i.e., the contextual distance weight is set to 0). We interpret the resulting CJM as follows. The first journey represents people going to 'work', going back 'home' at noon, and returning to 'work' in the afternoon. The second journey is close to the first one, the main difference being that people do not seem to go back 'home' at noon. The third journey shows citizens being at 'home', going 'shopping' twice in the afternoon, and going back 'home'.

In configuration 2, we test the effect on the resulting CJM when considering the ages of the customers. Therefore, we changed the weight assigned to the contextual distance from 0 to 1. As can be seen in Fig. 10, three representative journeys were generated. Each of these journeys has three touchpoints. They start from 'home' and finish at 'home'. In between, the first journey has the activity 'work', the second one has the activity 'shopping', and the last one the activity 'attending class'. It is interesting to note the effect of the configuration on the contextual data (the distribution charts on the right side of Fig. 10). Indeed, while the age was equally distributed for each journey in configuration 1, we can observe that the age is discriminant in configuration 2. For instance, more than half of the citizens in the journey j_3 are under 16 years old, while this population represents only 8.7% of the entire dataset.

In configuration 3, we show the effect when we increase the weight put on the contextual distance parameter. Journeys j_1 and j_3 are identical to those in configuration 2. However, a new and rather complex journey j_2 emerges. We observe that the distribution is impacted when giving more weight to homogeneity. We interpret the result as follows: Citizens younger than 29 years old tend to have two typical patterns of activities involving either 'school' or 'entertainment' while the most typical journeys for the other citizens involve 'work'.

Of course, this is an extremely simplified overview of the data. For the almost 30,000 actual journeys in the event logs, there are numerous unique actual journeys that differ from the representative journeys we get from these three configurations. By letting the user choose the weight for each parameter, we let them explore different perspectives of the data. We claim that the best parameters depend on the dataset, the business context, and the goal of the exploration.

7 Conclusion

Our genetic approach to summarizing a set of customer journeys with the purpose of displaying them on a CJM offers an interesting alternative to approaches used in social sciences for three reasons. First, the quality of the results is better, which is true using both internal and external evaluation metrics. Second, the weights of the three quality criteria are a flexible way to analyze a dataset under different perspectives. All the other parameters, such as the number of representative journeys to display or the length of the representative journeys are left entirely to the genetic algorithm. Third, in addition to the sequence of activities, our genetic algorithm can leverage contextual data to group similar journeys. By doing so, we provide a way to summarize insights from customers that are hidden in the data.

We tackle the challenging task of building a CJM from event logs as a single-objective optimization problem so that a single 'best' CJM is returned. Due to the inherent conflicting objectives of the quality criteria, we acknowledge that a multi-objective approach might be a relevant choice that we did not investigate.

References

1. Lemon, K.N., Verhoef, P.C.: Understanding customer experience throughout the customer journey. J. Mark. **80**, 69–96 (2016)
2. Gürvardar, İ., Rızvanoğlu, K., Öztürk, Ö., Yavuz, Ö.: How to improve the overall pre-purchase experience through a new category structure based on a compatible database: Gittigidiyor (Ebay Turkey) case. In: Marcus, A. (ed.) DUXU 2016. LNCS, vol. 9747, pp. 366–376. Springer, Cham (2016). https://doi.org/10.1007/978-3-319-40355-7_35
3. Peltola, S., Vainio, H., Nieminen, M.: Key factors in developing omnichannel customer experience with finnish retailers. In: Nah, F.F.-H., Tan, C.-H. (eds.) HCIB 2015. LNCS, vol. 9191, pp. 335–346. Springer, Cham (2015). https://doi.org/10.1007/978-3-319-20895-4_31
4. Research priorities 2018–2020. Technical report, Marketing Science Institute (2018). https://www.msi.org/research/2018-2020-research-priorities/cultivating-the-customer-asset/1.1.-characterizing-the-customer-journey-along-the-purchase-funnel-and-strategies-to-influence-the-journey/
5. Bernard, G., Andritsos, P.: A process mining based model for customer journey mapping. In: Proceedings of the Forum and Doctoral Consortium Papers Presented at the 29th International Conference on Advanced Information Systems Engineering (CAiSE 2017) (2017)
6. Bernard, G., Andritsos, P.: Discovering customer journeys from evidence: agenetic approach inspired by process mining. In: Cappiello, C., Ruiz, M. (eds.) CAiSE 2019. LNBIP, vol. 350, pp. 36–47. Springer, Cham (2019). https://doi.org/10.1007/978-3-030-21297-1_4
7. Gabadinho, A., Ritschard, G., Studer, M., Müller, N.S.: Summarizing sets of categorical sequences: selecting and visualizing representative sequences, pp. 94–106, October 2009
8. Gabadinho, A., Ritschard, G., Studer, M., Müller, N.S.: Extracting and rendering representative sequences. In: Fred, A., Dietz, J.L.G., Liu, K., Filipe, J. (eds.) IC3K 2009. CCIS, vol. 128, pp. 94–106. Springer, Heidelberg (2011). https://doi.org/10.1007/978-3-642-19032-2_7
9. Bernard, G., Andritsos, P.: CJM-ex: goal-oriented exploration of customer journey maps using event logs and data analytics. In: 15th International Conference on Business Process Management (BPM 2017) (2017)
10. Harbich, M., Bernard, G., Berkes, P., Garbinato, B., Andritsos, P.: Discovering customer journey maps using a mixture of Markov models, December 2017

11. Buijs, J.C., van Dongen, B.F., van der Aalst, W.M.: A genetic algorithm for discovering process trees. In: 2012 IEEE Congress on Evolutionary Computation (CEC), pp. 1–8. IEEE (2012)
12. Vázquez-Barreiros, B., Mucientes, M., Lama, M.: ProDiGen: mining complete, precise and minimal structure process models with a genetic algorithm. Inform. Sci. **294**, 315–333 (2015)
13. van der Aalst, W.M.P., de Medeiros, A.K.A., Weijters, A.J.M.M.: Genetic process mining. In: Ciardo, G., Darondeau, P. (eds.) ICATPN 2005. LNCS, vol. 3536, pp. 48–69. Springer, Heidelberg (2005). https://doi.org/10.1007/11494744_5
14. Levenshtein, V.I.: Binary codes capable of correcting deletions, insertions, and reversals. Sov. phys. dokl. **10**, 707–710 (1966)
15. Caliński, T., Harabasz, J.: A dendrite method for cluster analysis. Commun. Stat. Theory Methods **3**(1), 1–27 (1974)
16. Gabadinho, A., Ritschard, G.: Searching for typical life trajectories applied to childbirth histories. Gendered life courses-between individualization and standardization. A European approach applied to Switzerland (2013), pp. 287–312 (2013)

Adaptive Partitioning and Order-Preserved Merging of Data Streams

Constantin Pohl[✉] and Kai-Uwe Sattler

Databases and Information Systems Group, TU Ilmenau, Ilmenau, Germany
{constantin.pohl,kus}@tu-ilmenau.de

Abstract. Partitioning is a key concept for utilizing modern hardware, especially to exploit parallelism opportunities from many-core CPUs. In data streaming scenarios where parameters like tuple arrival rates can vary, adaptive strategies for partitioning solve the problem of over-estimating or underestimating query workloads. While there are many possibilities to partition the data flow, threads running partitions independently from each other lead to unordered output inevitably. This is a considerable difficulty for applications where tuple order matters, like in stream reasoning or complex event processing scenarios.

In this paper, we address this problem by combining an adaptive partitioning approach with an order-preserving merge algorithm. Since reordering output tuples can only worsen latency, we mainly focus on the throughput of queries while keeping the delay on individual tuples minimal. We run micro-benchmarks as well as the Linear Road benchmark, demonstrating correctness and effectiveness of our approach while scaling out on a single Xeon Phi many-core CPU up to 256 partitions.

Keywords: Adaptive partitioning · Order preservation ·
Stream processing · Parallelism · Many-core · Xeon Phi

1 Introduction

Recent trends in hardware have lead to a continuously increasing core count in processors, following Moore's law, stating that the number of transistors in a chip doubles every two years. With higher core counts and thus more available threads executing in parallel, it becomes more and more crucial to spread out algorithms and applications to run partly independent from each other. For database systems, user queries can benefit from multithreading by intra-query or intra-operator parallelism, e.g., by scanning different regions of data simultaneously. This means that DBMS operators can be parallelized mostly individually, processing input data at once or in batches, producing results afterwards in one single output step.

C. Pohl—This work was partially funded by the German Research Foundation (DFG) within the SPP2037 under grant no. SA 782/28.

© Springer Nature Switzerland AG 2019
T. Welzer et al. (Eds.): ADBIS 2019, LNCS 11695, pp. 267–282, 2019.
https://doi.org/10.1007/978-3-030-28730-6_17

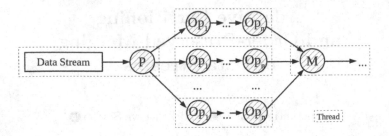

Fig. 1. Partitioning and merge schema

However, data stream management system (DSMS) operators have additional constraints. Since most of the DSMS queries are long-running, producing results continuously over time, a well-known concept is the partitioning of the data flow to benefit from multiple cores or CPUs through data parallelism. Multiple instances of an operator are running concurrently with the same functionality but only on a fraction of input data, increasing overall query throughput. This is achieved by a partitioner P in front of the parallel region, being responsible for splitting data for partitions under the goal of an even load balancing to minimize the further delay. Results from the partitions are finally gathered by a so-called merge operator M, which combines the individual output into a single consistent stream again (visualized in Fig. 1).

Recent research, especially in the field of machine learning, focuses on adaptive partitioning, where the number of partitions increases or decreases during query execution dynamically, depending on the input data stream behaviour [14]. An immediate challenge arising from this is how costly state migrations are handled, presuming stateful operations within the partitions. To avoid these costs, other recent work concentrates on optimizing the partitioning algorithm to reduce repartitioning events [6].

On top of that, an additional restriction on partitioning is the need of an order-preserved output for any subsequent operators that depend on tuple-wise order. Because of partitions running independently from each other to maximize parallelism, some partitions can process tuples faster than others. To point an example, the same selection predicate in one partition could drop more tuples than in another partition or a join operator could find a varying amount of matches. Therefore it might be necessary to reorder the output for detecting patterns or avoiding priority inversion. Since a data stream can run for a long time, this reordering step has to minimize blocking to keep latency of individual tuples low.

Another challenge is given directly from modern hardware, like the Xeon Phi *Knights Landing (KNL)* many-core CPU. The huge amount of cores on a single chip allows to scale out partitions into high numbers, prompting the question how good the partition-merge schema can be utilized with the restrictions of adaptation and order preservation. A solution to minimize thread synchronization for exchanging tuples are micro-batches being applied at the expense of individual latency. However, this leads to additional parameters to optimize (e.g., different batch sizes), increasing the complexity of optimization.

In this paper, we address these challenges and problems by the following contributions:

- We propose a classification of related work, combining state migration handling and determination of the ideal partition number.
- We show how adaptive partitioning is realized in our stream processing engine (SPE) *PipeFabric*[1],
- optimizing the main bottleneck of data exchange for a many-core CPU with micro-batching strategies.
- We describe how a minimal blocking merge operator can be realized when the partition count is dynamically changing,
- using micro-benchmarks as well as the well-known Linear Road benchmark [1] to demonstrate correctness and performance of our solution.

2   Related Work

Research on partitioning in data stream processing has lead to many approaches on how to optimize load balancing while keeping aggregation costs low. We suggest the following classification of the partitioning strategies (see Fig. 2), taking account of how the number of partitions is determined as well as state migration handling is realized.

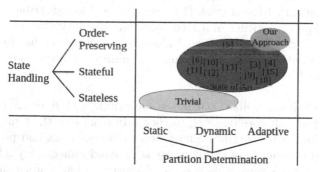

Fig. 2. Classification of partitioning strategies

2.1   Determining the Number of Partitions

The number of partitions depends mainly on the available hardware and use case. While it is often a good idea to fully utilize resources to speed up query execution and hiding latencies by oversubscription (i.e., using more partitions than CPU cores), it is not necessary and reasonable in all cases, e.g., for monetary reasons (cloud computing). Processing of slow-paced data streams or simple stateless queries like notifications from anomaly detection needs no scale out, wasting processing power and memory not available for other queries. However, there are different ways to determine the ideal number of partitions of a query.

---

[1] Open Source, https://github.com/dbis-ilm/pipefabric.

Static. In a static case, the number of partitions and also the partitioning function is determined **once at the beginning of a query**, staying unchanged afterwards. This works well without any overhead of recalculating in cases when tuple delivery rate of data streams does not change a lot or when there are no concurrent queries in a system. To point an example, querying multiple sensors that emit a signal every second each, followed by a calculation if the measurement is above a certain threshold is such a case. The partition count can then be determined mathematically, per calibration, or simply by experience.

Dynamic. For a dynamic environment with varying data streams and multiple queries competing for resources, a static approach is not applicable. Especially since queries can be long-running, it is necessary to adjust the partitioning approach continuously. The main goal of a dynamic strategy is to achieve good load balancing and throughput for the partitions. There are two diametral parameters that determine the performance of a query regarding partitions - tuple imbalance on the partitioner side and aggregation cost to merge results. A dynamic approach **changes the partitioning function during runtime** often based on statistics, determining tuple routing to partitions.

Katsipoulakis et al. [6] provide an algorithm to weigh up tuple imbalance and aggregation cost that can be parametrized by the partitioner. They compared their work to the *Partial Key Grouping* algorithm of Nasir et al. [10], which was the state of the art algorithm up to that time, later refined by heavy hitter detection [11]. Rivetti et al. [13] also analyzed key distribution for partitioning techniques, resulting in a Distribution-aware Key Grouping algorithm. Pacaci et al. [12] published additional algorithms focused on distributed stream processing on multiple machines, trying to minimize memory overhead and key splitting to partitions.

Adaptive. While it is possible to achieve good results with dynamic partitioning, an adaptive approach goes one step further, adjusting also the number of partitions. This allows a fine-grained adaptation to skewed data and peaks in tuple rates. The main challenge of this strategy is to avoid unnecessary additions and removals of partitions since it adds noticeable overhead and processing delays to the system. The usage of statistics is also a common solution for deciding when partitions are changed.

Shah et al. [15] proposed the Flux operator that handles repartitioning with state movements on-the-fly, focused on query execution in a cluster with multiple machines. Zhu et al. [18] further analyzed state migration strategies, leading to two different algorithms named *moving state* and *parallel track*, underlined with cost models. Gedik [3] investigated partitioning functions in general, breaking them down into three main properties. This allows adapting the used function to the individual use case while keeping migration costs low. In addition, Gedik et al. [4] provided an elastic auto-parallelization approach being able to adapt to changing workloads with state migration on runtime. De Matteis et al. [9] revisited the adaptive partitioning approach by the idea of prediction instead of reaction, using online optimization with queue theory models to allow reconfigurations to be done just in time.

2.2   State Handling

Partitioning approaches can also be classified by regarding state handling of operations, like hash tables of a *join* or sum of a *group by*.

Stateless. Operators that are not dependent on previously seen tuples, like selection predicates or projection of attributes, can easily be parallelized. With a dynamic partitioning function, the input rate of single partitions can vary, but that does not require any changes inside of a partition. Changing the number of partitions is also intuitive since stateless operations, in general, have the same memory footprint that does not change over time.

Stateful. Aggregations, joins, or windows are examples for stateful operations. The partitioner is responsible for a suitable distribution of tuples in such a way that the partitions do not miss results because a key is forwarded into the wrong partition. If the number of partitions is reduced, state migration is unavoidable. The migration mechanism depends on the type of state - a sum could easily be added to another sum of a different partition, for example.

Order-Preserving. Another layer of restriction on top of stateful operations is the preservation of correct tuple order after merging partition results. Since some partitions can run longer than others and partitions are generally not synchronized to maximize parallelism, the order of tuples after the merge step cannot be guaranteed automatically.

In the past, a solution to this problem in the setting of distributed computing was provided by Gulisano et al. [5]. They sorted the output of partitions by timestamp, using dummy elements to avoid long blocking periods if a partition produces no output tuples for a time interval that must be specified by the user.

3   Data Stream Processing

More and more applications require a non-blocking processing schema, where results are continually produced. Prominent examples for such applications are social networks like Facebook or Twitter, leading to own SPEs and real-time analytic platforms like Apache Storm [16] or its successor, Apache Heron [7].

The general focus of stream processing lies on high throughput of queries while keeping latencies low. The already mentioned partition-merge strategy can be found commonly in systems today, being capable of parallelizing query operators for throughput increase without sacrificing too much latency.

However, when scaling out partitions to maximize throughput, the right number of partitions fitting to problem size is necessary to not waste hardware resources. For a DBMS on regular servers, a simple strategy is used most of the time: Utilize every core and memory controller that is available to the query. Using more cores means more processing power and bandwidth, leading to (hopefully) faster execution time overall. Streaming queries, on the other hand, can get limited by the rate of data arrival. If this rate is low, it does not make any sense to increase the degree of parallelism by adding more partitions, wasting memory for local data structures and increasing contention to the partitioner and merge operator. Underestimating the rate of incoming tuples instead leads to tuples buffered or enqueued waiting to be processed, raising latency as well as leading to wrong results inevitably when the buffer or queue is full and tuples get discarded.

Because of the dynamic behavior of long-running streams, even an initially ideal configuration can become very bad, violating guarantees like latency constraints. To fully adapt to stream behavior, a DSMS must be able to support *scale out* as well as *scale in* techniques on partitioning. Increasing the number of partitions requires an initialization process of operators and states obviously. Reducing the number instead means that states of the partition have to be migrated into the other partitions to not lose information. For both cases, the partitioning function, as well as the merging step, must be adapted accordingly.

3.1 Linear Road Streaming Benchmark

The Linear Road benchmark [1] simulates a configurable number of vehicle expressways on which cars enter and leave different segments of the road continuously. Each added expressway increases the number of concurrent cars and thus the number of calculations necessary to correctly answer the benchmark query. The long-running query calculates tolls raised dynamically and tracks accidents, which lead to higher congestion on segments accordingly.

The challenge of this benchmark is to respond to each input tuple within five seconds while keeping track of the different vehicles. Since a single thread has limited computing power, over time it is inevitable to increase parallel computations. Since the tuple arrival rate increases vastly within the first hours because of more and more cars entering the expressways, this benchmark is a good candidate for adaptive partitioning. With a fixed partition size, a high amount of partitions must be initialized right from the start to keep up with later arrival rates, blocking hardware resources unnecessarily.

3.2 Adaptive Partitioning

The goal of our partitioning approach is twofold: (1) Minimize partition change events to avoid costly state migrations and (2) Utilize as many partitions as necessary to avoid over-occupation of computing resources as well as underestimation of query workload. This is achieved in general by a suitable partitioning function as well as a strategy when adding/removing partitions, further improved by micro-batching.

Partitioning Function. The partitioning function is used to determine which partition is responsible for processing a batch or a single tuple, used by the partitioning operator. In general, it has to balance the load in such a way that all partitions got input to process without overburdening individual partitions.

In addition, when stateful operations are involved, the same keys should not end in too many partitions to avoid costly aggregation steps in the merge operation.

The function can also be dependent on the use case. For the Linear Road benchmark, information about individual cars is requested as well as information about the different road segments. Therefore, cars of the same segment should not become distributed over multiple partitions. Since cars change segments over time, it is still necessary for account balance queries that the same car is only processed by a single partition, which leads to a conflict on how to partition the incoming tuples.

A good partitioning function can improve the overall throughput of a query by lowering aggregation costs and balance load. However, our work is one step further, providing an adaptive scaling of partition numbers which can use *almost any partitioning function* like current state of art from Katsipoulakis et al. [6]. For our micro-benchmarks, we simply use a round-robin-based partitioning function while the Linear Road benchmark partitions data according to pairs of car and segment ID: Since cars travel linearly through segments until they leave, their combination is a good candidate for partitioning overall.

Scaling Strategy. We base our decision when to add partitions on an equation for fulfilling general latency constraints. Whenever the equation returns false, new partitions get added to the current processing:

$$\forall i \in partitions : size(q_i) + tp\_ins(q_i) < tps(p_i) \tag{1}$$

$size(q_i)$ expresses the number of tuples the input queue of partition i holds since we use queues to exchange tuples between threads. $tp\_ins(q_i)$ is the amount of tuples that were added in the last time frame (usually within a second) to partition i and $tps(p_i)$ is the tuple per second processing rate of the partition i. The idea behind this formula is that when a latency constraint is given (like answering within 5 s for the Linear Road benchmark), the partitions should not store input tuples for later processing when they cannot catch up current input rates. This relationship is realized and supervised by an optimizer which also polls the necessary information from the partitions within a specified time interval (e.g., one second for tight latency constraints or once in a minute for more relaxed adaptation). Whenever a partition has to be added, the optimizer initializes the new partition and adapts the partitioning function afterwards. The reduction of partitions needs to find a balance between two aspects: state migration and resource allocation. If the partition size is often reduced, the overall throughput and latency are harmed by many state migrations from the removed partitions. On the other hand, if more partitions than necessary are kept, resources are wasted and merging efforts are needlessly high. We, therefore, propose an intensity-based approach (see Algorithm 1), being able to solve this tradeoff problem very well for at least our chosen experiments.

Whenever the input queue of a partition is fully empty, the intensity of reduction is increased. If the partitions got tuples again at the next timestep, the intensity is reset, which is also done if the amount of partitions is reduced

```
    Input: Initial configuration, Threshold t, Partitions ps
 1  Intensity_old=Intensity_new=0;
 2  while query is running do
 3  |   Intensity_old = Intensity_new;
 4  |   for each p in ps do
 5  |   |   if q_inp(p) == empty then  //a partition has no input tuples
 6  |   |   |   Intensity_new++;
 7  |   |   |   mark_partition(p);  //for later removal
 8  |   |   end
 9  |   end
10  |   if Intensity_old == Intensity_new then  //all partition queues filled
11  |   |   Intensity_new=0;
12  |   end
13  |   if Intensity_new >= t then  //reduction can be applied
14  |   |   reduce_partition();  //based on marked partitions
15  |   |   Intensity_new=0;
16  |   end
17  end
```

Algorithm 1: Intensity Algorithm

successfully. After reaching a parametrizable amount of intensity (expressed in a threshold), the partition is marked for reduction, not getting tuples anymore. When it has finished current processing, it is removed by the optimizer, integrating its state atomically into another partition. This procedure depends on the complexity of the states as well as the use case. It is also possible to reduce multiple partitions at once by storing references to marked partitions in a list, forwarding the list to the reducing function of the optimizer. However, it is not recommendable to reduce multiple partitions in short optimizing time intervals, since it can easily lead to partition *ping pong* by over- and underestimation of necessary resources.

Micro-batching. A notable bottleneck in query performance is the data exchange between the partitioner and the threads of partitions when scaling out to tens and hundreds of partitions on a single many-core CPU. We, therefore, added a batching operation, which gathers tuples at the expense of individual tuple latency. The batch is then forwarded by the partitioner to the corresponding partition at once, reducing the amount of thread synchronization, shown later in this paper. The partition extracts the individual tuples from the batch afterwards, continuing with further processing. The general concept of batching in stream processing applications is not new, Apache Flink [2] is a well-known example combining both strategies.

Batching up tuples can follow different approaches, providing a tradeoff between latency and aggregation overhead in the merge operator. On the one hand, if tuples are gathered into a single batch, forwarded when full, the latency delay can be kept low (depending on batch size). However, this leads to a key distribution where multiple partitions hold the same key, increasing aggregation efforts afterwards. On the other hand, if the batch size is kept small and tuples are batched together according to their key, it is possible not to distribute the same key to many partitions at the expense of latency.

For our experiments, we use batching only for the micro-benchmarks, since Linear Road has high latency constraints that cannot be fulfilled otherwise.

3.3   Order-Preserving Merge

The goal of an order-preserving merge operation is again twofold: (1) All output tuples from partitions are forwarded in a single stream, ordered accordingly and (2) Latency delay by keeping arrived tuples for ordering is minimized, holding back tuples only as long as necessary. To reach this goal, the architecture of merging is described first. Afterwards, the order-preserving step is briefly explained.

Merge Concepts. There are mainly two ways of exchanging tuples between partitions and a merge operator: a single queue for all partitions or one queue per partition (see Fig. 3).

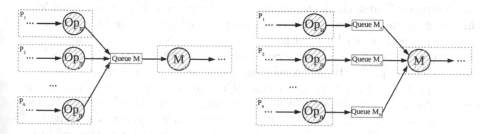

Fig. 3. Merge with one queue (left) and multiple queues (right)

Regarding performance, only one queue works well for small numbers of partitions, when the overall contention is low. Ordering the output, however, is difficult to realize without costly copy operations or iterations through the tuples inside of the queue. Regarding that fact as well as increased contention on a single queue for high amounts of partitions, we optimized the merge step by providing individual queues for all partition outputs. Therefore, the *single producer single consumer* (SPSC) paradigm can be applied, which is a well-known concept from lock-free data structures shared between threads.

Ordered Merge. We made the assumption that the order of tuples does not change inside of a partition. When tuples are batched together, forwarding the batch at once, the partition extracts the tuples in the same way as they are inserted, leading to the same order. This means that the order of tuples inside of an output queue is correct if there is no other partition involved. Even when multiple partition-merge schemes are applied to a query, the assumption holds if each merge step achieves the right order for the tuples. The general concept of order preservation for the merge operation is shown in Fig. 4.

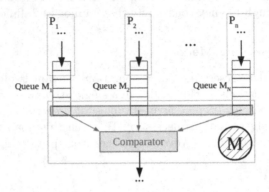

Fig. 4. Order-preserving merge

When all partitions publish results in their own queues, the merge operator can compare all first elements among each other. At a high conceptual level, this merging step is comparable to merging sorted arrays like in a sort-merge join. Any function can be used to produce the correct order, like ordering by timestamp, value or even lexicographically. Since the comparison has only to check one value per partition, good scaling can be achieved overall.

The problem of a partition producing no output tuples for a longer time period, delaying any output of the merge step, can be solved differently. One way to avoid blocking is to add dummy elements [5]. The K-Slack algorithm [8] can also be applied here, guaranteeing a time frame in which tuples are in the right order. The end of the time frame can be observed by comparison of the timestamps of tuples at the end of the queues.

4   Experimental Analysis

We expect that our approach scales the number of partitions directly with the tuple input rate of the data stream, increasing and decreasing accordingly. With batching of tuples, reducing communication between threads, the overall throughput in terms of tuples processed per second should rise at the expense of individual latency. Regarding the merge operator ordering output tuples, each added partition leads to additional comparisons worsening throughput. This raises the expectation that the gap in performance compared to regular merging gets bigger the more partitions are involved.

For the experiments we built our SPE *PipeFabric* with the Intel compiler version 17.0.6. and AVX512 enabled. The KNL 7210 processor runs in SNC4 clustering mode, while the MCDRAM is used as last-level cache.

4.1   Micro-benchmarks

Since our approach contains different aspects like adaptive partitioning, batching, or order-preserving merge, we run individual benchmarks for each of them. The general query we use for benchmarking can be written in SQL in the following way:

> **SELECT** key, **SUM**(payload)
> **FROM** stream
> **GROUP BY** key

Tuples from the datasets we choose (described further below) consist of three attributes each: a timestamp, a 4 byte key and 4 byte payload. The query executes the grouped aggregation *GBy* with a varying number of instances, while the projection is applied afterwards by a single projection operator *Proj* as shown in Fig. 5, since a stateless projection, in general, can be executed very fast.

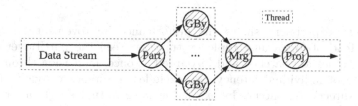

Fig. 5. Micro-benchmark query

We use two different synthetical datasets to simulate data stream behavior. The first dataset follows a sinus curve over time, where the amount of tuples produced per second varies between zero and 100.000 tuples per second. The second dataset simulates a burst, where 10 million tuples are available to process right from the beginning, allowing the query to finish as fast as it is able to process them.

Figure 6 shows the distribution of the sinus dataset over time and the adaptive scaling of partition numbers. Each point of the dataset reflects the number of tuples arrived in that second. Not all points are shown in the Figure to keep track of the plot, though. Since the aggregation operator instances have comparable processing rates, we directly convert the **partition number** to **tuples processed per second overall**.

It can be seen that the adaptive approach scales correctly with the sinus curve, guaranteeing the latency restrictions. The intensity-based (delayed) reduction of partitions leads to some overextension of resources which is intentional. As many real data streams follow a more skewed and noisy behavior with the fact that removing partitions with stateful operations is costly, a *ping-pong* of partition sizes through adding and removing partitions within a second must be avoided.

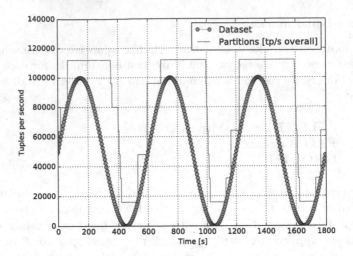

Fig. 6. Adaptive partitioning with sinus dataset

The next experiment shows the performance improvement for batching tuples. Instead of forwarding tuple by tuple between the partitioner and the partitions, tuples are batched and forwarded at once, increasing throughput at the expense of individual latency. We exclude advantages through SIMD for operations directly on batches by extracting them to tuples right after the partitions received them. The query stays the same as in the sinus dataset, but the second dataset is used to create pressure on throughput. Figure 7(left) shows the result on different numbers of partitions and varying batch sizes.

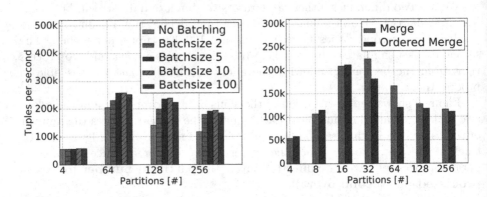

Fig. 7. Benchmark for batches (left) and merges (right)

Mainly two observations can be made. First, the overall throughput degrades after 64 partitions. This is attributable to the fact that a single partitioner forwarding data to other threads is limited in its processing rate. Additionally, the aggregation is a cheap operation in general, especially when compared to the necessary computations per tuple of the Linear Road benchmark. Second, there is an ideal batch size for throughput. With only four partitions it simply does not matter if batching is applied, because thread synchronization is not the limiting factor. Instead, the speed of aggregation is limited by the processing power of the partitions. When the number of partitions grows, the throughput with batching increases by up to 70% when compared with a non-batching strategy.

The last micro-benchmark applies to the order-preserving merge operator. The same setup like in the batching experiments before is used, excluding batching obviously. Results can be seen in Fig. 7(right).

Throughput of both merge strategies is compared directly, changing the number of partitions over time. With only a few partitions up to 16, the ordering does not hurt throughput overall. Because each partition has its own output queue following the SPSC concept, even a speedup can be achieved. But when the number of partitions further increases, the order-preserving merge operator has higher efforts of checking queues and comparing tuples, leading to a reduction of throughput over time up to 28% for 64 threads. Due to hyper-threading for 128 and 256 partitions, both merge steps decrease in performance. Since the regular merge operator only forwards incoming tuples, it gets more penalized than the order-preserving merge by high contention.

4.2   Linear Road

The dataset for Linear Road can be generated with a variable number of expressways. Each expressway adds more vehicles and thus more tuples per second that have to be processed, leading to the *L-rating* of a solution. To the best of our knowledge, the highest L-rating with 512 expressways was achieved by [17], using a distributed setting with 70 nodes, two quad-core CPUs per node. With a single many-core CPU, it is obviously not possible to reach 512 or even more expressways - our goal is still to achieve a high L-rating for one processor while scaling out our partitioning approach. In addition, we demonstrate that our algorithm is able to adapt to the increasing data arrival rate over three hours of traffic simulation. We use two Linear Road datasets for evaluation with one (L1) and two (L2) expressways generated, looking forward to increasing the number of expressways even further since not all partitions are used up to that point.

The implementation of the Linear Road query synchronizes the input tuples according to the data driver provided by the original implementation[2]. For each incoming tuple, a preprocessing step is performed, filtering the types of tuples and applying a five-minute window semantic for the average speed within a segment. The core of the query, storing the tuple information, processing tolls, and checking for accidents, is realized by *user defined functions* (UDFs) in a specialized operator. The UDFs follow the notations provided in the original paper [1]. This operator is scaled out by the partition-merge schema described above.

---

[2] www.cs.brandeis.edu/~linearroad/datadriverinstall.html.

Since the order of output tuples, in general, does not matter (they should only be processed within 5 s), the merge operation simply skips the ordering step, directly forwarding any results from the partitions. Figure 8 illustrates the datasets and adaptive partitioning results.

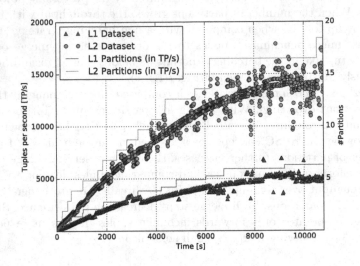

Fig. 8. Linear road benchmark

It can be seen that the number of tuples that arrive per second is much higher on the L2 dataset. Not all points for the datasets are plotted - just like the sinus micro-benchmark, to keep the overview. The results show that our approach is viable, although some skew in the datasets requires an over-provisioning of partitions to keep up with the 5 s latency constraint.

If we weight the latency constraint against wasting computing resources, it is obvious that the latency is much more important, failing the benchmark otherwise. Based on our intensity approach to scale down the partition number, the threshold necessary to reduce the number of partitions is therefore increased to avoid oscillating by adding and removing partitions caused by skew.

5   Conclusion

Partitioning of a data stream is a necessity to improve throughput and latency of queries since single threads have limited computational power. This is even more important for many-core CPUs or GPUs, which provide an intense amount of parallelism at the expense of serial performance. The partition-merge schema is a common solution to parallelize stream processing, allowing to scale out operators to multiple instances sharing their work.

For related work done in the field of adaptive partitioning as well as order preservation, we propose a classification regarding state handling and partition determination with their characteristics in combination of both. Order preservation is motivated by different use cases like complex event processing or pattern detection that require correct order of tuples delivered. In addition, stream behavior often changes over time, leading to inefficient queries when the partitioning strategy is static and does not adapt.

In this paper, we combined an adaptive partitioning approach with order-preserving merge to utilize the parallelism provided by the KNL processor. The partitioning aspect covers the partitioning function and scaling strategy, complemented with batching to overcome the data exchange bottleneck between partitioner and partitions. For an efficient order preserving output, the merge operator has to avoid blocking as much as possible, which can be realized by individual SPSC queues. The function that guarantees the correct order is configurable, allowing the merge operation to handle different order properties, like common timestamps. Finally, we demonstrated performance and correctness of our approach in microbenchmarks as well as the Linear Road use case, executed on a KNL many-core processor.

References

1. Arasu, A., et al.: Linear road: a stream data management benchmark. In: (e)Proceedings of the Thirtieth International Conference on Very Large Data Bases, Toronto, Canada, pp. 480–491, 31 August–3 September 2004
2. Carbone, P., et al.: Apache Flink^{TM}: stream and batch processing in a single engine. IEEE Data Eng. Bull. **38**(4), 28–38 (2015)
3. Gedik, B.: Partitioning functions for stateful data parallelism in stream processing. VLDB J. **23**(4), 517–539 (2014). https://doi.org/10.1007/s00778-013-0335-9
4. Gedik, B., et al.: Elastic scaling for data stream processing. IEEE Trans. Parallel Distrib. Syst. **25**(6), 1447–1463 (2014). https://doi.org/10.1109/TPDS.2013.295
5. Gulisano, V., et al.: StreamCloud: a large scale data streaming system. In: 2010 International Conference on Distributed Computing Systems, ICDCS 2010, Genova, Italy, pp. 126–137, 21–25 June 2010. https://doi.org/10.1109/ICDCS. 2010.72
6. Katsipoulakis, N.R., et al.: A holistic view of stream partitioning costs. PVLDB **10**(11), 1286–1297 (2017). https://doi.org/10.14778/3137628.3137639
7. Kulkarni, S., et al.: Twitter heron: stream processing at scale. In: SIGMOD, Melbourne, Victoria, Australia, pp. 239–250, 31 May–4 June 2015. https://doi.org/10. 1145/2723372.2742788
8. Li, M., et al.: Event stream processing with out-of-order data arrival. In: ICDCS Workshops, Toronto, Ontario, Canada, p. 67, 25–29 June 2007. https://doi.org/ 10.1109/ICDCSW.2007.35
9. Matteis, T.D., et al.: Keep calm and react with foresight: strategies for low-latency and energy-efficient elastic data stream processing. In: Proceedings of the 21st ACM SIGPLAN Symposium on Principles and Practice of Parallel Programming, PPoPP 2016, Barcelona, Spain, pp. 13:1–13:12, 12–16 March 2016. https://doi. org/10.1145/2851141.2851148
10. Nasir, M.A.U., et al.: The power of both choices: practical load balancing for distributed stream processing engines. In: 31st IEEE International Conference on Data Engineering, ICDE 2015, Seoul, South Korea, pp. 137–148, 3–17 April 2015. https://doi.org/10.1109/ICDE.2015.7113279

11. Nasir, M.A.U., et al.: When two choices are not enough: balancing at scale in distributed stream processing. In: 32nd IEEE International Conference on Data Engineering, ICDE 2016, Helsinki, Finland, pp. 589–600, 16–20 May 2016. https://doi.org/10.1109/ICDE.2016.7498273

12. Pacaci, A., et al.: Distribution-aware stream partitioning for distributed stream processing systems. In: Proceedings of the 5th ACM SIGMOD Workshop on Algorithms and Systems for MapReduce and Beyond, BeyondMR@SIGMOD 2018, Houston, TX, USA, pp. 6:1–6:10, 15 June 2018. https://doi.org/10.1145/3206333.3206338

13. Rivetti, N., et al.: Efficient key grouping for near-optimal load balancing in stream processing systems. In: Proceedings of the 9th ACM International Conference on Distributed Event-Based Systems, DEBS 2015, Oslo, Norway, pp. 80–91, 29 June–3 July 2015. https://doi.org/10.1145/2675743.2771827

14. Russo, G.R., et al.: Multi-level elasticity for wide-area data streaming systems: a reinforcement learning approach. Algorithms 11(9), 134 (2018). https://doi.org/10.3390/a11090134

15. Shah, M.A., et al.: Flux: an adaptive partitioning operator for continuous query systems. In: Proceedings of the 19th International Conference on Data Engineering, Bangalore, India, pp. 25–36, 5–8 March 2003. https://doi.org/10.1109/ICDE.2003.1260779

16. Toshniwal, A., et al.: Storm @Twitter. In: SIGMOD, Snowbird, UT, USA, pp. 147–156, 22–27 June 2014. https://doi.org/10.1145/2588555.2595641

17. Zeitler, E., et al.: Massive scale-out of expensive continuous queries. PVLDB 4(11), 1181–1188 (2011)

18. Zhu, Y., et al.: Dynamic plan migration for continuous queries over data streams. In: Proceedings of the ACM SIGMOD International Conference on Management of Data, Paris, France, pp. 431–442, 13–18 June 2004. https://doi.org/10.1145/1007568.1007617

Data Quality

CrowdED and CREX: Towards Easy Crowdsourcing Quality Control Evaluation

Tarek Awwad[1,3]([✉]), Nadia Bennani[1], Veronika Rehn-Sonigo[2], Lionel Brunie[1], and Harald Kosch[3]

[1] Université de Lyon, CNRS INSA-Lyon, LIRIS, UMR5205, Lyon, France
{tarek.awwad,nadia.bennani,lionel.brunie}@insa-lyon.fr
[2] FEMTO-ST Institute, Université Bourgogne Franche-Comté/CNRS, Besançon, France
veronika.sonigo@univ-fcomte.fr
[3] Department of Distributed and Multimedia Information Systems, University of Passau, Passau, Germany
{tarek.awwad,harald.kosch}@uni-passau.de

Abstract. Crowdsourcing is a time- and cost-efficient web-based technique for labeling large datasets like those used in Machine Learning. Controlling the output quality in crowdsourcing is an active research domain which has yielded a fair number of methods and approaches. Due to the quantitative and qualitative limitations of the existing evaluation datasets, comparing and evaluating these methods have been very limited. In this paper, we present CrowdED (Crowdsourcing Evaluation Dataset), a rich dataset for evaluating a wide range of quality control methods alongside with CREX (CReate Enrich eXtend), a framework that facilitates the creation of such datasets and guarantees their future-proofing and re-usability through customizable extension and enrichment.

Keywords: Crowdsourcing · Quality control · Dataset · Generic platform · Extendable campaign

1 Introduction

In the era where Artificial Intelligence is emerging at a steady fast pace through its underlying concepts such as Machine Learning and Data Mining, the quest for collecting labeled data like labeled images or annotated metadata is a persistent and fundamental task for researchers in these domains. In the last decade, crowdsourcing has proved its ability to address this challenge by providing a mean to collect labeled data of various types, at a low cost and short time as compared to expert labeling. However, the quality of the data produced through crowdsourcing is still questionable, especially when the labeling task shows a fair amount of subjectivity or ambiguity or requires some domain expertise [37].

© Springer Nature Switzerland AG 2019
T. Welzer et al. (Eds.): ADBIS 2019, LNCS 11695, pp. 285–301, 2019.
https://doi.org/10.1007/978-3-030-28730-6_18

Tackling this quality issue is, consequently, an active research domain that has yielded a large number of quality control (QC) methods ranging from optimizing the contribution aggregation process [10,14,40] and the worker selection step [3,23] to modeling context-specific reputation systems [24,27] and controlled crowdsourcing workflows [11]. Indeed, validating and comparing these methods raise the need for evaluation datasets which are sufficiently representative, information rich and easily extensible. Existing datasets [8,17,38,39,43] do not fulfill those requirements because they are tailored, form-wise, to evaluate one method or in the best cases, one category of approaches. This renders the cross-category comparison - like comparing aggregation approaches to selection approaches - unfeasible through sound scientific workflows. To address this challenge we designed and collected CrowdED (Crowdsourcing Evaluation Dataset), a publicly available information-rich evaluation dataset. In this paper, we detail and motivate the creation of CrowdED and describe CREX (CReate Enrich eXtend), an open platform that facilitates the collaborative extension and enrichment of CrowdED. The contributions of this paper can be summarized as follows:

- We provide a comprehensive specification sheet for a generic and future proof evaluation dataset, provide a comparative review of the existing datasets and discuss their compliance with those specifications.
- We propose CrowdED, a rich evaluation dataset of which we present the design and the contribution collection steps as well as the statistical and structural properties. We assess the ability of CrowdED in plugging the dataset gap through a qualitative study.
- We present the design of CREX and show how it facilitates the creation of crowdsourcing campaigns to extend and enrich evaluation datasets similar to CrowdED.

This paper is structured as follows: In Sect. 2, the state of the art of QC methods is briefly reviewed. In Sect. 3, the specifications of a suitable evaluation dataset are set. In Sect. 4, the state of the art crowdsourcing evaluation datasets are discussed w.r.t. the requirements stated earlier. Then, in Sect. 5, we describe the creation process of CrowdED as well as its structural and statistical characteristics. Finally, we present CREX in Sect. 6 and discuss its re-usability in Sect. 7, before concluding this paper in Sect. 8.

2   Crowdsourcing Quality Control

Many methods have been proposed to perform QC in crowdsourcing systems [3,5,15,22–24,31,35]. Most works in this domain have focused on optimizing the contribution aggregation process which consists in inferring the correct answer of a task using the collected contributions for this task. Early works used majority voting (MV) with multiple assignments to infer the correct answer to a given task. Giving different weights to the different votes improves the quality of the aggregation by penalizing less reliable answers. Those weights can be computed as graded and binary accuracy measures [15], credibility scores [24] or overall approval rates which are widely used in commercial crowdsourcing platforms e.g., Figure-Eight and AMT. More generic and widely used techniques [4,23,40] rely on probabilistic data completion methods like the expectation maximization algorithm (EM) [6,7]. In the latter, the weights and the correct answers are simultaneously inferred by maximizing a likelihood model. Li et al. [23] use, in their model, the worker accuracy and inaccuracy as weights for correct and wrong answers (respectively), while in [40], a Generative model of Labels, Abilities, and Difficulties (GLAD) is proposed; GLAD uses both the worker ability and the task difficulty as weights for the contributions in the aggregation process. In [35], the worker's reliability score is estimated using her participation behavior e.g., time for completing a task, the number of clicks, mouse travel, etc. Some methods propose to add more knowledge to the aggregation process using multiple stage crowdsourcing such as the produce/review workflow in [4].

Another way of controlling the quality consists in allowing only reliable workers to participate to the task completion. This can be done through pre-assignment qualification tests. Platforms like Figure-Eight use a gold-based quality assurance [19] which consists in continuously measuring the accuracy of the worker, using test tasks - with known answers - randomly injected in the workflow. A high error rate causes the rejection of the worker from the current campaign. Programmatic gold [28] is an extension of the gold-based QC where test tasks with incorrect answers are also used to train workers against common errors. Li et al. [23] propose a probing-based selection method. They describe an algorithm that finds, for each incoming task, a group of reliable online workers for this particular task. This is done by assigning, during the so called *probing stage*, a part of the tasks to the whole crowd in order to sample it and identify the reliable group for the remaining part. Awwad et al. [3] substitute the probing stage by an offline learning phase to learn the reliable group from previously completed tasks with a lower cost. Roy et al. [34] characterize in the same feature space the tasks by the skills they require and the workers with their skills, and then match workers and tasks according to their skills.

Moreover, some approaches in the literature leverage the worker incentive and preference aspects of the crowdsourcing process. For instance, in [1,2], the authors argue that proposing a personalized (based on the preferences) list of tasks for a given worker improves her throughput in terms of quality. Kamar et al. [16] propose incentive mechanisms that promote truthful reporting in crowdsourcing and discourage manipulation by workers and task owners.

3 Specifications

In this section, we analyze the requirements of the aforementioned QC approaches and deduce four specifications of a suitable evaluation dataset.

Specification 1: Data richness (S1)

Table 1 summarizes the requirements of a representative set of QC methods. The majority of classical methods such as aggregation techniques [6,15] do not

Table 1. The needs of selected QC methods in terms of dataset content.

Methods	Workers	Tasks	Contributions	Ref	Methods	Workers	Tasks	Contributions	Ref
Optimize design	ID	Content	Yes	[30]	Profile selection	Declar. profile	Content	Yes	[3, 23]
Priming	ID	Content	Yes	[26]	Reviewing/Editing	ID	Content	Yes	[4]
Train workers	ID	Content	Yes	[4]	Test questions	ID	Content	Yes	[19]
Reputation	ID	ID	Yes	[27]	Optimize pay	ID	Content	Yes	[9]
OAR	ID	ID	Yes	[42]	Fingerprinting	ID	ID	Yes	[35]
Skill matching	Skill profile	Skill set	Yes	[25]	Task modeling	ID	ID	Yes	[40]
Recommender	Preferences	Content	Yes	[1]	Worker model	ID	ID	Yes	[14]

require any specific features to be present in the dataset aside from the set of contributions, i.e., a set of labels indexed by (ID_{worker}, ID_{task}) keys. Those are indeed required by all the existing methods. Other methods, such as profile-based worker selection [3,14,18,23] necessitate the presence of the worker profiles[1] in the dataset. Methods which take into account the type of the task when selecting/screening workers - and which we refer to as contextual methods - necessitate either the existence of a category-labeled task or the content of the task from which the task type can be derived [34]. Finally, some methods [3] can require information on both the workers and the tasks to be present in the dataset at the same time. We distinguish two specifications related to the richness of a suitable evaluation dataset:

S1.1 The dataset must provide information about workers, that is, the worker declarative profiles.

S1.2 The dataset must provide information about tasks, that is, their full content, i.e., description, questions and answer options.

Specification 2: Data diversity (S2)

Crowdsourcing tasks cover a wide range of types [37]. Similarly, workers in a crowdsourcing system fall into multiple profile groups [13]. In order to allow assessing the genericity of the compared methods, it is crucial that the evaluation dataset reflects - to a sufficient extent - this type and profile diversity. Accordingly, we set two specifications related to the data diversity:

S2.1 The dataset must reflect the diversity of the profile features characterizing the workers of a real crowdsourcing system.

S2.2 The dataset must reflect the diversity of the task types. This includes the generic asked action e.g. labeling an image, judging relevance, etc., and the actual knowledge domain of the task e.g. sport, economy, etc.

Logically, **S.2** tightens **S.1** or equivalently **S.2** contains **S.1**. however to allow a more fine-grained comparison of existing datasets we keep both specifications.

---

[1] E.g., demographics and self-evaluation profiles.

Specification 3: Contribution abundance (S3)

To control the quality, one might need to estimate the global [42] or the contextual [3] reliability of the worker from his previous or current contributions, to compute the difficulty of the task using the workers' agreement on its answer [40], to assess the accuracy and the convergence ability of a proposed aggregation method [15,38], to compute the correlation between worker's reliability (computed using his contributions) and his declarative profile [23] etc. All this requires the dataset to provide sufficient contributions per worker and per task while ensuring that these contributions provide information about workers' reliability and the tasks' difficulty. We formulate this by the following specifications:

S3.1 The dataset must contain a large number of contributions. That is, both the tasks and the workers present in the dataset must have a reasonable number of contributions.

S3.2 The dataset must contain non-random contributions for tasks and for workers. We show later how this can be achieved during the campaign design and the data preprocessing steps.

Specification 4: Extensibility (S4)

The creation of a generic and information rich dataset should always be open to new contributors, so that absent and new features can be proposed and collected based on uncovered and new QC needs. Moreover, creating a realistic evaluation dataset for crowdsourcing QC necessarily passes by a crowdsourced data collection step, which is obviously a paid process. This makes the creation of a large enough dataset very costly, hence not achievable by only one entity (research laboratory, company, ...). Therefore, we add to the qualitative specifications **S.1**, **S.2** and **S.3** detailed earlier in this section a fourth specification as follows:

S4.1 The dataset must be collaboratively extensible both in terms of tasks, workers and contributions and in terms of worker features and task types.

In the remainder of this paper, we show how we design and build CrowdED and how CREX guarantees its extensibility to fulfill Specifications **S.1**, **S.2** and **S.3** and **S.4**.

4   State-of-the-Art of Crowdsourcing Evaluation Datasets

Table 2 details the characteristics of the evaluation datasets available in the crowdsourcing literature. For the sake of completeness, both publicly available and non-publicly available datasets are reported even though the latter ones are not accessible and thus cannot be used as benchmarking dataset. The table also shows the compliance of these datasets with Specifications **S.1**, **S.2** and **S.3**. As none of these datasets is compliant with S4, this specification is not shown in the table. The compliance with those specifications is judged based on a set of observed characteristics in the dataset which we enumerate as follows:

- The worker features (*Feat.*): is the number of worker profile features found in the dataset (related to **S.1.1** and **S.2.1**).
- The task content (*Cont.*): shows whether the dataset contains information about task content or not (related to **S.1.2**).
- The task diversity (*Div.*): shows whether the dataset contains more than one type of tasks or not (related to **S.2.2**).
- The contribution density (*Den.*): shows whether the set of contributions is Dense (D), i.e., all of the tasks were solved by all of the workers, Semi-Dense (DS), i.e., the sets of workers who answered different tasks overlap or Sparse (S), i.e., the workers who answered one task are different from those who answered another task (related to **S.3.1**).

Table 2. A comparison of a sample of dataset used in the literature to evaluate crowd-sourcing quality control.

Ref	Dataset	Characteristics							PA	RD	Compliance with S1−S3					
		Worker		Tasks			Contrib.				S1.1	S1.2	S2.1	S2.2	S3.1	S3.2
		#	Feat.	#	Cont.	Div.	#	Den.								
[14]	Stack overflow	505	8	14021	Yes	No	42063	S	−	+	+	+	+	−	+	+
	Evergreen webpage	434	9	7,336	Yes	No	22,008	S	−	+	+	+	+	−	+	+
	TREC 2011	160	9	1826	Yes	No	5478	S	−	+	+	+	−	−	+	+
[19]	Product search	255	0	256	No	No	NA	S	−	+	−	−	−	−	NA	+
[21]	Synthetic	11	0	300	No	No	3300	D	−	−	−	−	−	−	−	−
[23]	Knowledge dataset	100	5	75	Yes	No	7500	D	−	+	+	+	−	−	−	+
	RTE	NA	5	80	Yes	No	NA	D	−	+	+	+	−	−	−	+
	Disambiguation data	277	5	50	Yes	No	13850	D	−	+	+	+	−	−	−	+
[38]	Aff. text analysis	10	0	700	Yes	No	7000	D	+	+	−	+	−	−	−	+
	RTE	10	0	800	Yes	No	8000	D	+	+	−	+	−	−	−	+
	Word Similarity	10	0	30	Yes	No	3000	D	+	+	−	+	−	−	−	+
[39]	Image annot. Synth.	12*	0	500	No	No	NA	NA	−	−	−	−	−	−	−	−
	Image annot. Real	40*	0	100	No	No	4000	NA	−	+	−	−	−	−	+	+
[43]	Image labeling	109	0	807	No	No	NA	SD	−	+	−	−	−	−	−	+
	Relevance judgment	6*	0	2665	No	No	16000	S	−	+	−	−	−	−	−	+

Feat.: worker Features, **Cont.**: task Content, **Div.**: task Diversity, **Den.**: contrib. Density
D: Dense contrib., **DS**: Semi-Dense contrib., **S**: Sparse contrib., **n/a**: not available
PA: Public availability, **RD**: Real Dataset, −: Un-fulfilled, +: Fulfilled, *: per task

In the literature, many datasets have been used to evaluate crowdsourcing QC techniques. Only a few among those provide information about the declarative profile of the workers [14,23] which is in line with the low number of QC methods leveraging this aspect. The same observation was made by Ye et al. in [42]. The previous reasoning also applies on the content of the tasks which is not always present in the datasets [14,38]. On the opposite side, the contribution abundance requirement is almost met by all of the datasets [8,14,17,19,23,38,39]. This might be due to the fact that aggregation methods, which constitute a large part of the crowdsourcing related literature as shown in Sect. 2, usually require this requirement to be met.

The *Data For Everyone (DFE)*[2] corpus from Figure-Eight provides a large number of real task sets for which many contributions have been collected. While these sets are varied enough in the task types, they suffer from at least one of the following limitations: First, the majority of them provide aggregated contributions instead of individual contributions, which violate Specification **S 3.1**. Second, to the best of our knowledge, none of these datasets provide the profiles of the workers which violates Specification **S 1.1**. Third, the content of the task is not always present which does not meet Specification **S 1.2**. One can argue that it is possible, through some data engineering effort, like transferring missing data like profiles from one set to the other, to combine a number of these sets into a larger specification-fulfilling dataset. However, the datasets found in the DFE corpus are designed and generated independently by different requesters. Hence, the intersection between the workers and tasks of different datasets, when computable e.g., for unaggregated or un-anonymized datasets, might be empty or sparse which hinders any "match and transfer" step.

The aforementioned datasets are all *real crowdsourcing datasets*. That is, datasets generated through an actual crowdsourced data collection step. Alternatively, *Synthetic datasets* have been also used in the literature. Roy et al. [34] and Rahman et al. [29] generated a set of workers and tasks distributed over a set of skills found in a multilayer skill taxonomy in order to test the efficiency of their skill matching approaches. Others, such as Welinder et al. [39] and Hung et al. [12], generated synthetic datasets to evaluate the performance of their aggregation algorithms. While generating synthetic evaluation datasets for aggregation and skill matching optimization approaches is relatively an easy and scientifically valid approach, generating synthetic datasets to evaluate approaches that leverage worker's behavior (e.g., fingerprinting [35]) and profile (e.g., declarative profile based worker selection [3,23]) is unfeasible. That is because, on the one hand, ignoring the uncertainty and noise resulting from the subjectivity of the human being in generating the data, produces a dataset which does not reflect the real crowdsourcing context. And, on the other hand, modeling the uncertainty and noise is impossible due to the lack of behavioral studies of the crowd in crowdsourcing systems. Hence, a synthetic dataset could, theoretically, fulfill all the specifications except Specification **S 3.2**.

5 CrowdED: Crowdsourcing Evaluation Dataset

In this section, the process used to create CrowdED is described in detail and its statistical and structural characteristics are presented. This process is divided into three steps: First, the data preparation during which the raw resources such as the task input are collected and preprocessed. Second, the data collection step during which the actual contributions and profiles crowdsourcing occurred. Finally, the data formatting step during which the collected contributions and profiles are cleaned and restructured.

---

[2] https://www.figure-eight.com/data-for-everyone/.

5.1   Raw Data Preparation

We built our task corpus by collecting publicly available task sets from the Data
For Everyone datasets provided freely by Figure Eight[3] (FE). The main motivation behind choosing the DFE datasets is to use tasks that have served real world
applications. In fact, it is possible to generate random labeling and knowledge
related tasks from scratch and to use them in the dataset generation process.
However, those will not be as significant as real world tasks. Furthermore, DFE
is a sustainable source[4] of task sets for future extension of CrowdED (Specification **S 4.1**). Our initial task pool consisted of 280K+ tasks, originally belonging
to 11 different task sets. The task content was distributed over various domains
such as sport (2), fashion (1), politics (2), economy (1), disaster relief (2), technology (1) and natural sciences (2) and over different action types like relevance
judgment, image labeling, tweet categorization etc. The task questions consisted
mainly multiple choice questions. In some cases free text answers where also
possible. All this helps fulfilling Specification **S 2.2**. The tasks were unevenly
distributed over the various task sets. For instance, one set constitutes 67% of
the entire corpus. That is why, in order to balance our task corpus we sampled
4000 tasks out of each set (i.e., the size of the smallest set). The set of 44k
resulting tasks constitutes our task corpus. In the next step, a random sample
of 525 tasks (Limited by our crowdsourcing budget) within the task corpus was
published for crowdsourcing.

5.2   Data Collection

We designed a crowdsourcing job and submitted it to FE. Workers who selected
the job were asked to read a detailed description of the task solving process
and conditions and to fill their contributor IDs. Those who decided to proceed
with the job completion were redirected to an external web page on which the
data collection took place. In the first stage of the task, we asked workers to fill
their contributor IDs again (for an easier matching and control) and to answer
a set of profile related and self-evaluation questions (Specification **S 1.2**) (see
Sect. 5.3). Once done, workers proceeded in the actual task solving. For each job
instance, tasks were randomly distributed over 11 pages in order to prevent the
concentration of the negative impact of weariness on one subset of tasks. After
completing the whole task set, a unique submission code was provided to each
worker allowing her to receive her reward on FE.

Workers were rewarded a base pay equal to 1$ US. Additionally, a bonus of 2$
US was awarded (manually) to workers whose answers and declared profiles were
of a good quality and high consistency (see Profile Rating in Sect. 5.3). Moreover,
we estimated the job completion time by 45 min, thus workers who finished the
job in a very short time (i.e., less than 40 min) were automatically eliminated and
did not receive any reward. Finally, we only accepted workers of at least level

---

[3] https://www.figure-eight.com. Formerly named CrowdFlower.
[4] Yet, it is not the only one since any other task corpus can be used.

2 in the FE worker classification[5]. On the one hand, these three parameters (bonus, contribution duration and minimum worker level) were strict enough to ensure that malicious workers (i.e., workers who intentionally fill random or wrong answers) are eliminated (Specification S 3.2). On the other hand, they are loose enough to allow a real representation of the quality issue in crowdsourcing. Contributions were collected during 3 months over all week days and covering all times of the day. This is to eliminate the bias related to the time zones, holidays and working hours during the data collection, e.g., workers representing few countries, limited educational and work profiles, etc.

5.3 Data Structure and Statistics

Figure 1 shows the structural characteristics of CrowdED as well as the features of tasks and workers that it contains. In total, we collected 280K+ contributions for 525 tasks from 450 workers among which 200 completed the entire set of tasks. We call the set of contributions given by those 200 workers a "dense set". Structurally, CrowdED consists of 4 files: *contributions.csv* which contains the worker contributions, *workers.csv* which contains the worker profiles, *rating.csv* where profile ratings are stored and finally task.zip where the tasks content and description are stored in JSON format. CrowdED have been made public on Figshare and on Github.

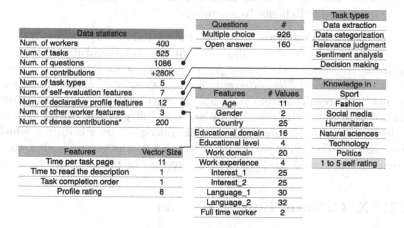

Fig. 1. An overview of the structural characteristics of CrowdED. (*) a dense contribution is a set of answers given by a single worker to the entire task set.

Some of the 525 tasks in CrowdED contain up to three independent questions. The total number of answered questions is 1086. The majority of these questions (926) are multiple choice questions and the remaining part consists of open answer questions. The input of the tasks are tweets, images, scientific article

---

[5] FE levels range from 1 to 3 where level 3 represents the most experienced and reliable workers and 1 represents all qualified workers.

quotes or news articles and headlines. Their action types fall into five categories: *data extraction, data categorization, relevance judgment, sentiment analysis, and decision making.* While the gold answers for these question are not available, it is safe to say that the large number of answers collected per tasks allows to estimate these gold answers through any aggregation technique with a high confidence.

For each worker, we collected a profile consisting of 23 features divided into three categories:

Declarative Profile. We collected 12 features consisting of the following demographical, education and interest related information about the user: *age, gender, country, education domain, education level, work domain, work experience, interests (two features), native language, other spoken language and full time worker (i.e., whether the worker is a full time or occasional crowdsourcing worker).* We observed that these numbers are, for their majority, compliant with the numbers reported in previous studies found in the literature such as the study of the Mechanical Turk marketplace [13].

Self-evaluation Features. We collected 7 features consisting of a 5-star self rating for 7 knowledge domains: *sport, fashion, technology, natural sciences, humanitarian work, politics, and social media.* We observed that in average, female workers seemed more confident in their knowledge in fashion and Humanitarian work, while male workers, rated themselves higher for sport. For the remaining domains, i.e., technology, natural sciences, politics and social media, both female and male workers rated themselves similarly.

Behavior-Related Features. Four features related to the behavior of the workers during the campaign were collected. Three of these features were collected automatically in the interface:*time for completing a task page, time for reading the description and filling the profile and the order of task completion.* The fourth, however, resulted from a complementary crowdsourcing campaign; in fact, in order to judge the consistency and reliability of the worker declarative and self-evaluation profiles, we ran a profile rating job on FE during which the profile of each worker who participated to our job was rated (from 1 to 4) for consistency by at least 7 workers (with an average of 11 workers).

6   CREX: CReate, Enrich, eXtend

Generating **and extending** the data described earlier is a technically tedious and time consuming task. In this section, we present CREX (CReate, Enrich, eXtend), a framework that allows the generation and extension of such data (CREX has been used to create CrowdED).

CREX uses a two-component architecture. This architecture is shown in Fig. 2. The first component, CREX-D, allows a configurable task data selection while the second, CREX-C, provides tools to automatically generate crowdsourcing campaigns from the output of CREX-D. The computational modules of CREX are developed with Python3. CREX uses well established and sustainable natural language processing and machine learning libraries such as *scikit-learn, nltk, gensim,* etc. The web user interface uses a combination of *Bootstrap, JavaScript* and *PHP* and the used database technology is $MySQL^6$.

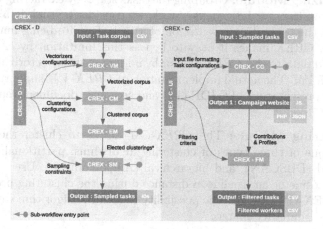

Fig. 2. An overview of the CREX framework that combines two main component; **CREX-D** for data selection and **CREX-C** for campaign generation and data collection.

6.1 Data Preparation Component (CREX-D)

A typical crowdsourcing workflow consists of 3 steps: first, designing the task, second, crowdsourcing the task and last, collecting the results. Indeed, this typical workflow is suitable for classical crowdsourcing where the aim of the requester is to exploit the results in a limited application-centric way, e.g., label multimedia data to facilitate indexing, translate a given corpus, etc. In other words, it suits applications where the input data are fixed and limited in size. When it comes to research-related crowdsourcing, e.g., building evaluation, validation or training datasets where the usage of the collected data goes beyond the limited exploitation, the input data space is usually huge and more complex. Therefore, an upstream input data selection effort is needed. A more suitable workflow is then a four step process that adds an input data selection step at the beginning of the aforementioned workflow. We propose a data selection step encapsulated in the data preparation component CREX-D that allows the requester to group his tasks according to their types through clustering, and then, to reduce their number according to his budget through sampling.

Figure 2 depicts the structure of the CREX-D component. It comprises four modules: the *vectorizing module (CREX-VM),* the *clustering module (CREX-CM),* the *sampling module (CREX-SM)* as well as the *evaluation module (CREX-EM).* Those modules are available and **inter-operable** yet **independent**.

---

[6] A demo of CREX's user interface and a real world use scenario can be found on https://project-crowd.eu/.

That is, each module can be used separately or as an entry point for the remaining steps or substituted by another module of equivalent role. This allows a more flexible usage and thus a wider cross-domains utility of CREX.

The Vectorizing Module: Grouping the tasks starts by extracting the features of interest from the raw data. In this work, we consider textual data where each data point is the textual representation of a task. Despite being limited to this type of data, CREX makes it easy to bypass this limitation by either feeding pre-vectorized data to the *CREX-CM* or by adding custom vectorizing functions to the *CREX-VM*. The actual implementation of *CREX-VM* supports frequency based text representation (TF-IDF [36]) and semantic document representation (Doc2vec [20]).

The Clustering Module: The *CREX-CM* allows to cluster the vectorized tasks using one of three types of clustering algorithms: partitional (K-means), density-based (DBSCAN), and hierarchical (Agglomerative). User can natively use either a cosine or an Euclidean distance during the clustering process. However, the CREX-CM provides the possibility to feed the algorithm with a custom pre-computed similarity matrix.

The Sampling Module: This module allows to sample from an input task corpus a smaller set of tasks that can be crowdsourced while respecting the budget constraints of the requester. This module implements a basic stratified sampling algorithm and a type-aware constrained sampling process which is out of the scope of this paper.

The Evaluation Module: The *CREX-EM* module allows to evaluate the clustering process using internal and external validity measures such as silhouette [33], homogeneity, completeness and V-measure [32] as well as a custom validity measure consisting of a similarity to co-occurrence correlation matrix.

6.2 Campaign Management Component (CREX-C)

From a requester perspective, a mandatory step of the crowdsourcing workflow is the task design and generation. This step is tedious and time consuming due to two factors: first, the interest and use of crowdsourcing is growing to reach a wider sphere of scientific and social domains. Thus, the range of task forms and content is getting larger. Second, a crowdsourcing task, itself, might be dynamic, i.e., it may require conditional or real-time computed components. Therefore, it becomes harder for commercial crowdsourcing platforms to quickly adapt their design tools, preset templates and real-time computational means[7]. A common way of dealing with these limitations is to build campaign sites with dedicated databases and back-end computations and to make them accessible through a common crowdsourcing platform to provide reward payment and worker management (for security and trust). The campaign management component of CREX,

---

[7] e.g., requester accessible back-end services or API to **dynamically** modify tasks and assignments.

CREX-C, provides an easy-to-use tool for generating campaign sites from the sampled tasks using the *campaign generator module* (CG).

The Campaign Generator Module: CREX-CG takes two inputs: the set of tasks to be published on the campaign site and the requester input consisting of the task descriptions, examples and instructions. It parses these inputs to intermediate JSON files and uses them to generate the campaign pages. The campaign site communicates directly with the database where the contributions and the worker profiles are stored. Contributions in the database are stored using a JSON format which allows a straightforward use of CREX-C for different task structures/types without the need for a new database model and query rewriting.

The Filtering Module: For a set of workers, tasks and contributions collected after publishing the campaign generated by the CREX-CG module, the filtering module allows to select a subset of these data based on qualitative and quantitative selection criteria applied on the workers. Those criteria cover the declarative profile features of the worker, their rating of their profiles, their time of task completion, their time of profile completion as well as the number of task they achieved. The filtering process has two main goals: First, it helps selecting a subset of the workers based on qualitative criteria to allow studying its characteristics e.g., its average performance of female workers. Second, it allows to clean the data based on behavioral criteria. For instance, a profile filled in less than 20 s is most likely to be inconsistent. That is, it has been most likely filled randomly, which means that the worker associated to this profile is very likely a malicious worker. Consequently, considering only the contributions of workers who spent a reasonable time answering the profile questionnaire would yield a noiseless dataset.

7   CrowdED and CREX Re-usability

7.1   Usability in Quality Control Evaluation

Table 3 shows the usability of CrowdED for evaluating the QC methods reported in Table 1. This usability is judged based on the needs of these methods in terms of information about workers, tasks and contributions and their availability in CrowdED. The majority of the methods that require information about workers and tasks only (regardless the type of the information) are natively supported by CrowdED (blue cells). Others are supported either through *simulation*, i.e., vertically or horizontally splitting the dataset to simulate a real world situation like worker screening or through *augmentation*, i.e., adding more knowledge to the available data without the need for additional crowdsourcing by extracting new features or using external taxonomy to represent tasks and workers (orange cells). Less frequent methods that require more information are not supported natively. Nevertheless, thanks to CREX, they could be supported by extending CrowdED with a minor reconfiguration effort (e.g. changing the reward) or with a more demanding coding effort (red cells).

Table 3. CrowdED's usability for the existing quality control methods: **native**, *simulation/augmentation*, extension.

Methods	Workers	Tasks	Contributions	Ref
Optimize design	ID	**Content**	**Yes**	
Priming	ID	**Content**	**Yes**	Iterative
Train workers	ID	**Content**	**Yes**	Interaction
Reputation	ID	ID	**Yes**	Mobility
OAR	ID	ID	**Yes**	
Skill matching	*Skill profile*	*Skill set*	**Yes**	
Recommender	*Preferences*	**Content**	**Yes**	

Methods	Workers	Tasks	Contributions	Other
Profile selection	**Declar. profile**	**Content**	**Yes**	
Reviewing/Editing	ID	**Content**	**Yes**	Iterative
Test questions	ID	**Content**	**Yes**	
Optimize pay	ID	**Content**	**Yes**	Reward
Fingerprinting	ID	ID	**Yes**	Behavior
Task modeling	ID	ID	**Yes**	
Worker model	ID	ID	**Yes**	*Side info.*

Supporting different QC methods is one important facet of CrowdED's re-usability. Being representative from a task type and worker feature perspective is the other facet. While it already supports a diversity of these types and features, the current iteration of CrowdED does not cover the whole ever growing range of task types and worker features. CREX helps fill this gap. Missing types and features can be gradually added by CREX users by appending their croudsourced data to crowdED. Eventually this collaborative effort will lead to a near full coverage of the tasks, workers and QC methods.

7.2 Compliance with the FAIR Principles

To guarantee the re-usability of those resources by the wide community (which allows a better extension and enrichment of CrowdED), the FAIR principles [41] (*Findable, Accessible, Interoperable, Reusable*) were considered during the design, the creation and the publishing process: CrowdED and CREX are available on Github and Figshare (with an associated DOI) which makes them **F**indable. They are published under CC and GPL licensing respectively to allow their **R**e-usablility and **A**ccessibility. CrowdED data are stored in *csv* files and no proprietary languages were used to develop CREX. This ensures the **I**nteroperability of the resources.

Accessibility. The site http://project-crowd.eu/ provides a demo of CREX-D and CREX-C, a tutorial on installing and using CREX, a full description of the configurable parameters as well as additional materials for this paper such as the full statistical sheet of tasks, profiles, ratings and contributions of CrowdED. Moreover, the site provides links to download both CREX and CrowdED.

8 Conclusion

In this paper we proposed CrowdED and CREX in order to address the lack of evaluation datasets, which is unanimously one of the most challenging aspects facing the research in crowdsourcing QC. The specifications fulfilled by CrowdED allow it to be usable in evaluating and comparing a wide range of existing methods. CrowdED covers a large number of quality control methods as well as different task types and worker features. In order to deal with the methods and task types which are not natively supported and included by CrowdED, and to future-proof it, we proposed CREX. CREX is an open-source framework that allows the collaborative extension of CrowdED to fulfill new qualitative requirements e.g., new worker profile types, and quantitative requirements e.g., more contributions for a given task (**S4**).

References

1. Alsayasneh, M., et al.: Personalized and diverse task composition in crowdsourcing. IEEE Trans. Knowl. Data Eng. **30**(1), 128–141 (2018)

2. Amer-Yahia, S., Gaussier, E., Leroy, V., Pilourdault, J., Borromeo, R.M., Toyama, M.: Task composition in crowdsourcing, pp. 194–203 (2016)
3. Awwad, T., Bennani, N., Ziegler, K., Sonigo, V., Brunie, L., Kosch, H.: Efficient worker selection through history-based learning in crowdsourcing, vol. 1, pp. 923–928 (2017)
4. Baba, Y., Kashima, H.: Statistical quality estimation for general crowdsourcing tasks. In: ACM SIGKDD, NY, USA, pp. 554–562 (2013)
5. Daniel, F., Kucherbaev, P., Cappiello, C., Benatallah, B., Allahbakhsh, M.: Quality control in crowdsourcing: a survey of quality attributes, assessment techniques, and assurance actions. ACM Comput. Surv. (CSUR) **51**(1), 7 (2018)
6. Dawid, A.P., Skene, A.M.: Maximum likelihood estimation of observer error-rates using the EM algorithm. J. R. Stat. Soc. Ser. C (Appl. Stat.) **28**(1), 20–28 (1979)
7. Dempster, A.P., Laird, N.M., Rubin, D.B.: Maximum likelihood from incomplete data via the EM algorithm. J. R. Stat. Soc. Ser. B (Methodol.) **39**, 1–38 (1977)
8. Deng, J., Dong, W., Socher, R., Li, L.J., Li, K., Fei-Fei, L.: Imagenet: a large-scale hierarchical image database, pp. 248–255 (2009)
9. Difallah, D.E., Catasta, M., Demartini, G., Cudré-Mauroux, P.: Scaling-up the crowd: micro-task pricing schemes for worker retention and latency improvement (2014)
10. Ghosh, A., Kale, S., McAfee, P.: Who moderates the moderators?: crowdsourcing abuse detection in user-generated content. In: Proceedings of the 12th ACM Conference on Electronic Commerce, pp. 167–176. ACM (2011)
11. Gil, Y., Garijo, D., Ratnakar, V., Khider, D., Emile-Geay, J., McKay, N.: A controlled crowdsourcing approach for practical ontology extensions and metadata annotations. In: d'Amato, C., et al. (eds.) ISWC 2017. LNCS, vol. 10588, pp. 231–246. Springer, Cham (2017). https://doi.org/10.1007/978-3-319-68204-4_24
12. Quoc Viet Hung, N., Tam, N.T., Tran, L.N., Aberer, K.: An evaluation of aggregation techniques in crowdsourcing. In: Lin, X., Manolopoulos, Y., Srivastava, D., Huang, G. (eds.) WISE 2013. LNCS, vol. 8181, pp. 1–15. Springer, Heidelberg (2013). https://doi.org/10.1007/978-3-642-41154-0_1
13. Ipeirotis, P.G.: Demographics of mechanical turk (2010)
14. Jin, Y., Carman, M., Kim, D., Xie, L.: Leveraging side information to improve label quality control in crowd-sourcing (2017)
15. Jung, H.J., Lease, M.: Improving consensus accuracy via z-score and weighted voting. In: Human Computation (2011)
16. Kamar, E., Horvitz, E.: Incentives for truthful reporting in crowdsourcing. In: Proceedings of the 11th International Conference on Autonomous Agents and Multiagent Systems, vol. 3, pp. 1329–1330. International Foundation for Autonomous Agents and Multiagent Systems (2012)
17. Kanoulas, E., Carterette, B., Hall, M., Clough, P., Sanderson, M.: Overview of the TREC 2011 session track (2011)
18. Kazai, G., Kamps, J., Milic-Frayling, N.: The face of quality in crowdsourcing relevance labels: Demographics, personality and labeling accuracy. In: CIKM, pp. 2583–2586 (2012)
19. Le, J., Edmonds, A., Hester, V., Biewald, L.: Ensuring quality in crowdsourced search relevance evaluation: the effects of training question distribution. In: 2010 Workshop on Crowdsourcing for Search Evaluation, pp. 21–26 (2010)
20. Le, Q., Mikolov, T.: Distributed representations of sentences and documents, pp. 1188–1196 (2014)
21. Li, H., Yu, B., Zhou, D.: Error rate analysis of labeling by crowdsourcing. In: Machine Learning Meets Crowdsourcing Workshop (2013)

22. Li, H., Yu, B., Zhou, D.: Error rate bounds in crowdsourcing models. arXiv preprint arXiv:1307.2674 (2013)
23. Li, H., Zhao, B., Fuxman, A.: The wisdom of minority: Discovering and targeting the right group of workers for crowdsourcing. In: WWW, NY, pp. 165–176 (2014)
24. Mashhadi, A.J., Capra, L.: Quality control for real-time ubiquitous crowdsourcing. In: UbiCrowd, NY, USA, pp. 5–8 (2011)
25. Mavridis, P., Gross-Amblard, D., Miklós, Z.: Using hierarchical skills for optimized task assignment in knowledge-intensive crowdsourcing, pp. 843–853 (2016)
26. Morris, R., Dontcheva, M., Gerber, E.: Priming for better performance in microtask crowdsourcing environments. IEEE Internet Comput. 16(5), 13–19 (2012)
27. Mousa, H., Benmokhtar, S., Hasan, O., Brunie, L., Younes, O., Hadhoud, M.: A reputation system resilient against colluding and malicious adversaries in mobile participatory sensing applications (2017)
28. Oleson, D., Sorokin, A., Laughlin, G.P., Hester, V., Le, J., Biewald, L.: Programmatic gold: targeted and scalable quality assurance in crowdsourcing. Hum. Comput. 11(11), 43–48 (2011)
29. Rahman, H., Roy, S.B., Thirumuruganathan, S., Amer-Yahia, S., Das, G.: Task assignment optimization in collaborative crowdsourcing, pp. 949–954 (2015)
30. Rahmanian, B., Davis, J.G.: User interface design for crowdsourcing systems, pp. 405–408 (2014)
31. Raykar, V.C., et al.: Supervised learning from multiple experts: whom to trust when everyone lies a bit. In: Proceedings of the 26th Annual International Conference on Machine Learning, pp. 889–896. ACM (2009)
32. Rosenberg, A., Hirschberg, J.: V-measure: a conditional entropy-based external cluster evaluation measure (2007)
33. Rousseeuw, P.J., Kaufman, L.: Finding Groups in Data. Wiley, Hoboken (1990)
34. Roy, S.B., Lykourentzou, I., Thirumuruganathan, S., Amer-Yahia, S., Das, G.: Task assignment optimization in knowledge-intensive crowdsourcing. VLDB J. 24(4), 467–491 (2015)
35. Rzeszotarski, J.M., Kittur, A.: Instrumenting the crowd: using implicit behavioral measures to predict task performance. In: UIST, NY, USA, pp. 13–22 (2011)
36. Salton, G., McGill, M.: Modern information retrieval (1983)
37. Sarasua, C., Simperl, E., Noy, N., Bernstein, A., Leimeister, J.M.: Crowdsourcing and the semantic web: a research manifesto. Hum. Comput. (HCOMP) 2(1), 3–17 (2015)
38. Snow, R., O'Connor, B., Jurafsky, D., Ng, A.Y.: Cheap and fast–but is it good?: evaluating non-expert annotations for natural language tasks. In: Conference on Empirical Methods in Natural Language Processing, pp. 254–263. Association for Computational Linguistics (2008)
39. Welinder, P., Branson, S., Perona, P., Belongie, S.J.: The multidimensional wisdom of crowds, pp. 2424–2432 (2010)
40. Whitehill, J., Wu, T.F., Bergsma, J., Movellan, J.R., Ruvolo, P.L.: Whose vote should count more: optimal integration of labels from labelers of unknown expertise. In: NIPS, pp. 2035–2043 (2009)
41. Wilkinson, M.D., et al.: The fair guiding principles for scientific data management and stewardship. Sci. Data 3 (2016)
42. Ye, B., Wang, Y., Liu, L.: Crowd trust: a context-aware trust model for worker selection in crowdsourcing environments, pp. 121–128 (2015)
43. Zhou, D., Basu, S., Mao, Y., Platt, J.C.: Learning from the wisdom of crowds by minimax entropy, pp. 2195–2203 (2012)

Query-Oriented Answer Imputation
for Aggregate Queries

Fatma-Zohra Hannou, Bernd Amann[✉], and Mohamed-Amine Baazizi

Sorbonne Université, CNRS, LIP6, Paris, France
{fatma.hannou,bernd.amann,mohamed-amine.baazizi}@lip6.fr

Abstract. Data imputation is a well-known technique for repairing missing data values but can incur a prohibitive cost when applied to large data sets. Query-driven imputation offers a better alternative as it allows for fixing only the data that is relevant for a query. We adopt a rule-based query rewriting technique for imputing the answers of analytic queries that are missing or suffer from incorrectness due to data incompleteness. We present a novel query rewriting mechanism that is guided by partition patterns which are compact representations of complete and missing data partitions. Our solution strives to infer the largest possible set of missing answers while improving the precision of incorrect ones.

Keywords: Data imputation · Aggregation queries ·
Partition patterns

1 Introduction

Data incompleteness naturally leads to query results of poor quality and repairing missing data is a common data cleansing task. Data imputation consists in estimating missing data values manually or by using statistical or predictive models built from existing available data. Repairing large data sets can lead to important imputation costs which may be disproportional to the cost of certain query workloads. In particular for analytic queries, imputation at the raw data level might be inefficient for repairing aggregated query results. Query-driven imputation allows for rectifying only the values that are relevant for the evaluation of a query and, hence, can significantly reduce the amount of data to be processed. Although this approach is appealing, it has received little attention within the data cleansing community. A notable exception is ImputeDB [2] where data imputation is performed during query execution by injecting statistical estimators into the query plans. While this approach achieves its primary goal of repairing only the data accessed by the queries, it requires the extension of existing query processors to accommodate for the new imputation operators. In this article, we are interested in the imputation of aggregated query results where partition-wise imputation at a higher granularity might be preferred to the tuple-wise imputation at the raw data level. Our approach is based on the

© Springer Nature Switzerland AG 2019
T. Welzer et al. (Eds.): ADBIS 2019, LNCS 11695, pp. 302–318, 2019.
https://doi.org/10.1007/978-3-030-28730-6_19

assumption that incorrect values obtained by incomplete partitions are of lower quality than values obtained by imputation. Instead of instrumenting the query execution plan, we adopt a declarative approach where experts express domain and data specific imputation rules which identify missing or incorrect query answers and estimate their values by aggregating available correct results. This approach allows data analysts to capture the semantics of the underlying data imputation process and facilitates the interpretation of the imputation results.

As an example, consider the following SQL query Q_{kwh} over some table $Energy(B, F, R, W, D, kWh)$ storing daily kWh measures for day D of week W and rooms R in floor F of building B. Query Q_{kWh} computes the average weekly energy consumption for three floors in building 25:

select B, F, W, sum(kWh) **as** kWh
from Energy
where B = 25 **and** F **in** (1,2,3)
group by B, F, W

Table $Energy$ might miss some measures for these floors. This incompleteness can be the result of tuples containing *null* values in attribute kWh or defined with respect to some reference tables (maps, calendars etc.). In both cases, a first step consists in identify missing data tuples, and subsequently trace how they impact the quality of query Q_{kWh}. Table 1 shows respectively a representation of table $Energy$ (ordered by week and floor), and the query answer $Q_{kwh}(Energy)$. Missing measures are indicated by *null* values. For example, for week 1, $Energy$ contains all measures of floors 1 and 3 (we consider that each floor has only one room and a week has three days) and misses one measure for floor 2. It misses all measures for floor 1 and week 2. Each tuple in the query answer is obtained by aggregating a partition of the input data and annotated **correct** if

Table 1. *Energy* table and $Q_{kwh}(Energy)$

Energy

B	F	R	W	D	kWh
25	1	1	1	1	12.3
25	1	1	1	2	10.1
25	1	1	1	3	9.6
25	2	1	1	1	8.3
25	2	1	1	2	6.4
25	2	1	1	3	null
25	3	1	1	1	5.3
25	3	1	1	2	7.2
25	3	1	1	3	6.1
25	1	1	2	1	null
25	1	1	2	2	null
25	1	1	2	3	null
...	...	...	...	...	...

$Q_{kWh}(Energy)$

B	F	W	kWh	label
25	1	1	32.0	correct
25	2	1	14.7	incorrect
25	3	1	18.6	correct
25	1	2	null	missing
...	...	...	...	...

the partition is complete, `missing` if the partition is empty and `incorrect` if the partition is not empty by incomplete. Data imputation is a common approach for fixing incomplete data by applying statistical or logical inference methods. Most approaches apply imputation independently of the query at the raw data level. As shown in [2], repairing the entire data may turn expensive when data is large and inefficient when the query only needs to fix a subset of values to improve the query result quality. A second argument we develop in this article is that, for aggregation queries, it might have more sense to estimate missing or incorrect results at the granularity of the query result itself by exploiting specific knowledge about the input data sets at the aggregation level. For this, we propose rule based approach for repairing dirty data similar to [5] where domain experts can define imputation rules using correct available observation to repair missing and incorrect results. For example, an expert can state by the following imputation rule that some missing or incorrect query result (due to incomplete data) for a given floor could be repaired by taking the average of all available correct *results*

$$r_0 : (B : x, F : \_, W : y) \leftarrow (B : x, F : \_, W : y), avg(kWh)$$

More precisely, any missing or incorrect measure for some floor in building x and for week y that matches with the left-hand side of the rule will be estimated by the average over all correct measures for the same building and week (tuples matching the right-hand side of the rule).

Our imputation process is based on a pattern representation of queries and data for identifying correct, incorrect and missing query answers and for selecting and evaluating imputation rules. In summary, the main steps of our imputation approach are the following:

1. experts define imputation rules for repairing aggregate query results using information about correct observations;
2. the system detects and summarizes correct, incomplete and missing query answers in form of partition patterns;
3. the system selects for each missing or incorrect partition pattern one or several rules for imputing data;
4. the selected imputation rules are translated into standard SQL queries that generate correct results for missing or incorrect query answers.

Contribution and Paper Outline: Our imputation model for summarizing, analyzing and repairing aggregate query answers is presented in Sect. 2. Related work is presented in Sect. 3. We describe the imputation process in Sect. 4 and validate it experimentally in Sect. 5. Conclusion and future work directions are presented in Sect. 6.

2   Imputation Model

Our imputation model is defined for a particular sub-class of SQL aggregate queries:

Definition 1. *Let Q be valid SQL aggregate query of the form*

select *S, agg(m)* **from** *T* **where** *P* **group by** *G*

where condition P only uses equality predicates with constants.

Without loss of generality, we assume that P is in disjunctive normal form. For example, query Q_{kWh} can be rewritten by transforming the **where** clause into (B = 25 and F = 1)**or** (B = 25 and F = 2)**or** (B = 25 and F = 3).

We now introduce the notion of query pattern for representing the partitions generated by the **group by** clause of an SQL aggregate query.

Definition 2. *A* query pattern *is a tuple $q = (a_1 : x_1, ..., a_n : x_n)$ where for each attribute a_i, its values $x_i \in dom(a_i) \cup V \cup \{*\}$ is (1) a constant in the domain of attribute a_i or (2) a distinct variable in a set of variables V or (3) the wildcard symbol $*$.*

For example, $(B : 25, F : x, R : *)$ is a query pattern where $x \in V$ is a variable.

Definition 3. *Let Q be some valid SQL aggregate query as defined in Definition 1 and A be a key of the input table containing all attributes in Q except the aggregated attribute. We can then define a set of query patterns $\mathcal{Q}$ over A which contains a query pattern $q_i \in \mathcal{Q}$ for each disjoint d_i in the **where** clause such that (1) all attributes A_j in d_i are represented by the corresponding constants c_j in q_i, (2) all other attributes in the **group by** clause are distinct variable attributes and (5) all attributes in A and not in Q are wildcard attributes.*

For example, suppose that (B, F, R, W, D) is a key of table *Energy* (see Sect. 1). Then, the SQL query Q_{kWh} generates the query pattern set $\mathcal{Q} = \{(B : 25, F : 1, R : *, W : \_, D : *), (B : 25, F : 2, R : *, W : \_, D : *), (B : 25, F : 3, R : *, W : \_, D : *)\}$. Observe that all query patterns of a query share the same wildcard attributes and if q does not contain any wildcard attributes, then the corresponding SQL query corresponds to a simple conjunctive query which returns the measure values of the matching tuples (without aggregation and group-by clause).

The instance of a query pattern q defines a subset of the partitions generated by the **group by** clause over the tuples filtered by the **where** clause of the corresponding SQL query. This filtered partitioning can formally be defined by a partial equivalence relation over the query input tuples:

Definition 4. *A* tuple t matches *a* query pattern q, *denoted $match(t, q)$, if $t.a_i = q.a_i$ for all constant attributes in q. Two tuples t and t' matching some query pattern q are* equivalent *in q, denoted $t \equiv_q t'$, if $t.a_j = t'.a_j$ for all variable attributes a_j in q (t and t' only can differ for wildcard attributes).*

A pattern p defines for each matching tuple t an equivalence class $\Phi_q(t) = \{t'|t \equiv_p t'\}$.

Definition 5. *The instance of a query pattern q in some table M, denoted $I(M,q)$, contains all equivalence classes (partitions) of tuples in M.*

It is easy to see that (1) when p does not contain any wildcard attribute, then $I(M,q) = \{\{t\}|t \in I(q,M)\}$ contains a singleton for each matching tuple in M, and (2) when q does not contain any variable, $I(q,M) = \{\Phi_q\}$ contains a single partition $\Phi_q \subseteq M$. For example, the equivalence class $\Phi_q(t)$ of tuple $t = (B:25, F:1, R:1, W:2, D:3)$ defined by pattern $q = (B:25, F:1, R: *, W:x_w, D:*)$ contains all tuples of building 25, floor 1 and week 2. The equivalence class $\Phi_q(t')$ with the same pattern q for tuple $t' = (B:26, F:1, R: 1, W:2, D:3)$ is empty. Finally, $q' = (B:25, F:1, R:*, W:2, D:*)$ defines a unique equivalence class for all tuples of floor 1 in building 25 and week 2.

Constrained Tables and Partition Patterns: We follow the approach of *relative information completeness* [7] for modeling the completeness of a data table M.

Definition 6. *Let M be a relational table and A be a key in M. Table R is called a* reference table *for data table M if R contains all keys of M: $\pi_A(M) \subseteq R$. The pair $T = (M, R)$ is called a* constrained table.

Observe that any table $M'(\underline{A}, val)$ with key A and *with null values for attribute val* can be decomposed into an equivalent constrained table $T = (M, R)$ where measure table $M = \sigma_{val\ is\ not\ null}(M')$ contains all tuples in M *without null values* and $R = \pi_A(M')$ contains all key values in M'. At the opposite, we can build from any constrained table $T = (M, R)$ a relational table $S = R \bowtie M$ with *null* values (left outer join of R with M). For example, table *Energy* has key (building, floor, room, week, day) and we can decompose *Energy* into a measure table M defined by: **select** $*$ **from** Energy **where** kWh **is not null** and a reference table R defined by **select** B, F, R, W, D **from** Energy.

In order to reason about the completeness of data *partitions*, we introduce the concept of *partition pattern* [14].

Definition 7. *Let $T = (M, R)$ be a constrained table with reference attributes A. A* partition pattern *p is a query pattern without variables over attributes A. A partition pattern p is*

- complete in table T *if its instance (partition) in R is equal to its instance (partition) in M: $I(p, R) = I(p, M)$.*
- incomplete in table T *if its instance (partition) in T is strictly included in its instance (partition) in R: $I(p, M) \subset I(p, R)$.*
- missing in table T *if it is incomplete and its instance in M is empty: $I(p, R) \neq \emptyset \wedge I(p, M) = \emptyset$.*

For example, table *Energy* in Table 1 contains all measures for (1) all rooms and days of the first week and the first floor and (2) misses all measures for floor 1 and week 2. These observations are respectively summarized by pattern pattern c_1 in table C and e_1 in in table $\mathcal{E}$ (Table 2). More exactly, pattern tables C and $\mathcal{E}$ define the complete and missing partitions in the subset of keys M identifying all non-null tuples in $Energy$ with respect to the reference table R, which contains all keys of $Energy$.

Both pattern tables are covering and minimal in the sense that they contain respectively all *complete* and all *empty* patterns of T and no pattern that is subsumed by another pattern in the same table. We call C and $\mathcal{E}$ the complete and empty pattern (fragment) summaries of T and an algorithm for generating these summaries is described in [11].

Imputation rules and imputation queries: Imputation rules repair the results of aggregation queries by estimating missing tuple values. They are defined using *query patterns* characterizing the tuples that should be repaired and the tuples that should be used for their reparation.

Table 2. Complete and missing pattern tables for $Energy$ data table

C	B	F	R	W	D
c_1	25	1	*	1	*
c_2	25	2	*	1	*
c_3	25	2	*	2	*
c_4	25	3	*	1	*
c_5	25	3	2	2	*
c_6	25	5	*	1	*
c_7	25	5	*	2	*
c_8	26	*	*	*	*

$\mathcal{E}$	B	F	R	W	D
e_1	25	1	*	2	*
e_2	25	*	*	3	*
e_3	25	3	1	2	*
e_4	25	4	*	*	*

Definition 8. *An imputation rule for some set of reference attributes A and some measure attribute val is an expression of the form $r : q_m \leftarrow q_a, imp$ where (1) q_m and q_a are query patterns over A without wildcards, (2) all variables shared by q_m and q_a are bound to the same attribute in q_m and q_a and (3) imputation expression imp is an aggregation function transforming a set of values in the domain of m into a single value in the domain of m.*

In the following we use the anonymous variable _ for denoting non-shared variables. For example, we can define the following imputation rules for missing kWh values:

$$r_1 : (B:x, F:\_, W:y) \leftarrow (B:x, F:\_, W:y), (max(kWh) + min(kWh))/2$$
$$r_2 : (B:x, F:y, W:3) \leftarrow (B:x, F:y, W:2), kWh$$
$$r_3 : (B:\_, F:4, W:x) \leftarrow (B:\_, F:4, W:x), avg(kWh)$$

Imputation rule r_1 produces an estimation of the weekly electricity consumption of some floor in some building by the midrange of all *correct* consumption values over *other* floors of the same building and the same week. Rule r_2 takes the correct consumption of a floor in week 2 for estimating the value of the same floor at week 3 (the aggregation function returns the value of the generated singleton). Finally, rule r_3 takes the average of all correct values for floor 4 in all buildings to repair the value of the same floor in some building for the same week.

The formal semantics of an imputation rule is defined with respect to a query Q, a table $ImputeD$ of result tuples to be repaired by the rule and a table $AvailableD$ of all correct values which can be used for imputation. Observe that table $ImputeD$ contains all results generated by incomplete partitions and all missing results corresponding to empty partitions whereas table $AvailableD$ contains all possible correct tuples that are returned by a new query Q' which is obtained by removing the **where** clause of Q (imputation rules can use all correct results of Q but also other correct values that are generated by the same aggregation function over complete partitions).

Definition 9. *Let ImputeD and AvailableD be two tables that contain the tuples to be repaired and the tuples which can be used for reparation. Then, the semantics of an imputation rule $r : q_m \leftarrow q_a, imp$ is defined by the following imputation query $Q(r)$ over ImputeD and AvailableD where A contains all attributes in q_m and S is the set of variable attributes shared by q_m and q_a:*

select *x.A, imp* **as** *m*
from *ImputeD x, AvailableD y*
where $match(x, q_m)$ **and** $match(y, q_a)$ **and** $x.S = y.S$
group by *x.A*

The previous imputation query joins all tuples $x \in ImputeD$ matching q_m with the set of tuples y in $AvailableD$ matching q_a over the shared attributes S, partitions the obtained table over all rule attributes and finally applies the imputation expression imp to estimate a value for m. For example, the previous imputation rule r_2 can be rewritten into the following SQL query :

select x.B, x.F, x.W, y.kWh
from *ImputeD* x, *AvailableD* y
where x.W = 3 **and** y.W = 2 **and** x.B = y.B **and** x.F = y.F ;

where $AvailableD$ corresponds to the correct results generated by query Q_{kWh} without its filtering condition. We will show in Sect. 4 how to exploit partition patterns for identifying missing and incorrect query results and evaluating imputation rules.

3   Related Work

Missing data is an ancient problem in statistical analysis [1] and data imputation has been studied for many decades by the statisticians [1] and more recently by the database community [3]. Different imputation approaches have been developed including Mean Imputation [9], which is still one of the most effective techniques, K-Nearest Neighbor [9] and Hot Deck Imputation [18] (k=1) aggregating the k most similar available samples, Clustering-Based Imputation [12] estimating missing values by the nearest neighbor in the same cluster built over the non-missing values. More recently, Learning-Based Imputation models adapt machine learning techniques and represent data imputation as a classification and regression problem [15]. Crowdsourcing techniques are also used for achieving data completion or repairing tasks and achieve high-quality results balanced by an increased cost of human effort. Rule based cleaning is a middle-ground solution that allows referring to human for drawing cleaning rules, but automates the reparation process. A notable example is the work of [5], introducing a user interface for editing cleaning rules, and running automatic cleaning tasks. The solution presented in [13] allies an SQL-like languages and machine learning techniques to offer non expert users an interface to achieve data imputation. Several classes of data integrity constraints also have been applied for detecting and repairing missing or incorrect data [6]. For example, [8] proposes a formal model based on master data to detect missing and incorrect data and editing rules for repairing this data. The more recent work of [17] introduces the notion of fixing rules to capture semantic errors for specific domains and specify how to fix it. Our work adopts the same idea of using rules for repairing aggregate query answers using a different approach for identifying missing and incorrect results generated by empty or incomplete data partitions.

For large datasets, data imputation can become very expensive and inefficient. In contrast, query result estimation techniques consider repairing query answers, obtained by applying queries to incomplete (or incorrect) data. Sampling methods have been first used for run-time optimization through approximate queries [10,16] over representative sample instead of the entire dataset. In the same spirit, [16] integrates sampling techniques together with a cleaning strategy to optimize during run-time. The price to pay for the reduced time and cleaning effort is a bias introduced by the cleaning and sampling. [4] tackles the problem in the absence of any knowledge about the extent of missing tuples and estimates the impact of unknown unknowns on query results. The work introduces statistical models for quantifying the error in query results due to missing tuples. The goal of dynamic imputation [2] is to incorporate the missing data imputation into the query optimization engine. Physical query plans are augmented with two additional operators, that delete tuples with missing values and replace them with new values. Only data involved in the query evaluation are imputed, and the replacement is achieved at different query execution steps.

4   Query Imputation Process

Our imputation process is decomposed into four steps. The first step consists in identifying the set of all partition patterns $ImputeP(Q)$ summarizing the partitions to be repaired and the set of partition patterns $AvailableP(Q)$ of partitions that can be used for reparation. Step 2 consists in identifying the set of all rules that can be used for repairing $ImputeP(Q)$ by using $AvailableP(Q)$.

In this step, a rule is chosen if and only if it can repair at least one answer tuple and if there exists at least one correct value that can be used for imputation. The result of this step is a set of *candidate imputations*. The third step consists in creating a *sequence of candidate imputations* which repair the missing and incorrect tuples. Observe that a tuple might be repaired by several queries and we assume that each imputation query overwrites conflicting repaired tuples generated by the previous queries. Finally, step 4 consists in generating the imputation queries following the imputation strategy of step 2.

Step 1: Identify correct, incorrect and missing result tuples: For identifying correct, incorrect and missing answers, we first extend the notion of pattern matching from tuples to query patterns.

Definition 10. *A query pattern q matches a partition pattern p, denoted by $match(q, p)$, if for all constant attributes $q.a_i$ in q, $q.a_i = p.a_i$ or $p.a_i = *$. If $match(q, p)$, we can define a mapping ν from the variable attributes a_i in q to the attributes in p such that $\nu(q.a_i) = p.a_i$. Then, a query pattern q (1) fully matches partition pattern p, denoted by $full(q, p)$, if $\nu(q)$ matches p and (2) partially matches p, denoted by $partial(q, p)$, otherwise. Partition pattern $\nu(p)$ is called the matching pattern of q for p.*

For example, query pattern $q = (25, \_, *, \_, *)$ matches all patterns in $\mathcal{C}$ and $\mathcal{E}$ except pattern $c_8 : (26, *, *, *, *)$[1]. It fully matches patterns $c_1 : (25, 1, *, 1, *)$, $c_2 : (25, 2, *, 1, *)$, $c_3 : (25, 2, *, 2, *)$, $c_5 : (25, 3, 2, 2, *)$, $c_6 : (25, 5, *, 1, *)$, $c_7 : (25, 5, *, 2, *)$, $e_1 : (25, 1, *, 2, *)$ and $e_2 : (25, *, *, 3, *)$ and partially matches pattern $e_3 : (25, 3, 1, 2, *)$ and $c_4 : (25, 3, *, 1, *)$.

Let P be a set of partition patterns and $\mathcal{Q}$ be a set of query patterns of some query Q. Then we denote by $match(P, \mathcal{Q})$, $partial(P, \mathcal{Q})$, and $full(P, \mathcal{Q})$ the sets of partition patterns in P that are matched, partially matched and fully matched by query patterns $q \in \mathcal{Q}$. By definition, $partial(P, \mathcal{Q}) = match(P, \mathcal{Q}) - full(P, \mathcal{Q})$.

Definition 11. *Let $W(Q)$ be the set of wildcard attributes in the query patterns $\mathcal{Q}$ of some query Q and q be a query pattern over all variable and constant attributes in the query patterns of Q. Then we denote by q^* the query pattern where all attributes in W are wildcard attributes. Pattern q^* is called the extension of q in Q.*

The extension of query pattern $q = (B : 25, F : 2, W : \_)$ is $q^* = (B : 25, F : 2, R : *, W : \_, D : *)$ and the extension of tuple $t = (B : 25, F : 1, W : 1)$ in Q is pattern $t^* = (B : 25, F : 1, R : *, W : 1, D : *)$.

Proposition 1. *Given a query Q over some constrained table $T = (M, R)$ with complete pattern summary $\mathcal{C}$ and missing pattern summary $\mathcal{E}$. Let $\mathcal{Q}$ be the query pattern set of Q. Then, for any tuple t in the reference table of Q the following holds:*

---

[1] We omit attribute names when they're not necessary for understanding.

- t *is in the result of* Q *and* correct *iff* t^* *matches a pattern* $p \in full(C, Q)$;
- t *is in the result of* Q *and* incorrect *iff* t^* *matches a pattern* $p \in partial(\mathcal{E}, Q)$ *(or, equivalently* $p \in partial(C, Q)$*);*
- t *is* missing *in the result of* Q *iff a pattern* $p \in full(\mathcal{E}, Q)$ *matches* t^*.

Since query pattern $q = (25, 1, *, \_, *)$ fully matches patterns c_1, all answer tuples t of Q_{kWh} where t^* matches c_1 are correct. For example, for tuple $t = (25, 1, 1)$, its extension $t^* = (25, 1, *, 1, *)$ matches $c_1 : (25, 1, *, 1, *)$. The same argument holds for pattern $q' = (25, 2, *, \_, *)$ and tuples $(25, 2, 1)$ and $(25, 2, 2)$ On the opposite, since q also fully matches patterns $e_1 : (25, 1, *, 2, *)$ and $e_2 : (25, *, *, 3, *)$, all answer extensions matching these patterns are missing in the result. Finally, $q'' = (25, 3, *, \_, *)$ partially matches pattern $e_3 : (25, 3, 1, 2, *)$ (or, equivalently, pattern $c_4 : (25, 3, *, 1, *)$), all extended answer patterns matched by these patterns, like for example $(25, 3, 1)^* = (25, 3, *, 1, *)$, are incorrect.

By Proposition 1 and Definition 10, the set of missing or incorrect tuples exactly corresponds to the set of the corresponding extended tuples matching $\mathcal{E}$.

Definition 12. *Given query* Q *over some table* $T = (M, R)$ *with pattern tables* C *and* $\mathcal{E}$ *and query pattern set* Q. *We can then define the following sets of patterns for* Q:

- $ImputeP(Q) = full(\mathcal{E}, Q) \cup partial(\mathcal{E}, Q) = full(\mathcal{E}, Q)$
- $AvailableP(Q) = \{p | p \in C \land \forall A \in W(Q) : p.A = *\}$

$ImputeP(Q)$ contains all patterns describing incomplete or missing partitions (to be repaired) in the result of Q whereas $AvailableP(Q)$ describes all complete partitions that can be used for repairing Q. In the following step, we explain how we can use these two sets for filtering imputation rules for some aggregation query Q.

Step 2: Generate Candidate Imputations: Missing and incorrect answers of some aggregation query Q (query pattern set Q) are estimated by imputation queries. Each imputation query is generated by an imputation rule and repairs some missing and incorrect tuples. We assume that the complete and missing data partitions are represented by a complete and missing pattern summary as defined before. We first define the notion of candidate imputation.

Definition 13. *Let* $ImputeP(Q)$ *be the imputation pattern set and* $AvailableP(Q)$ *the reparation pattern set of* Q. *A rewriting* ω *for* $p_m \in ImputeP(Q)$ *is an expression* $\omega : p_m \leftarrow^r P_a$ *where there exists an imputation rule* $r : q_m \leftarrow q_a, f_{imp}$ *such that the extended query pattern* q_m^* *matches* p_m *with* ν *and* $P_a \subseteq C$ *is a non-empty set of complete patterns in* C *that are matched by* $\nu(q_a^*)$.

We say that rule r generates rewriting ω and call $\nu(q_m^*)$ the *imputation pattern* of ω and $\nu(q_a^*)$ the *repair pattern* of ω. All rules r where there exists at least one rewriting are called *candidate imputations* for Q. For example, $\omega_1 : e_1 \leftarrow^{r_1}$ $\{c_3, c_7\}$ is a candidate imputation for $e_1 : (25, 1, *, 2, *)$ generated by rule r_1 with imputation pattern $\nu(q_m^*) = e_1 : (25, 1, *, 2, *)$, repair pattern $\nu(q_a^*) = (25, \_, *, 2, *)$ and $P_a = \{c_3 : (25, 2, *, 2, *), c_7 : (25, 5, *, 2, *)\}$, Similarly, $\omega_2 : e_2 \leftarrow^{r_2} \{c_3, c_7\}$ is a candidate imputation for $e_2 : (25, *, *, 3, *)$ using rule r_2 with imputation pattern $\nu(q_m^*) = e_2 : (25, *, *, 3, *)$, repair pattern $\nu(q_m^*) = (25, *, *, 2, *)$ and $P_a = \{c_3 : (25, 2, *, 2, *), c_7 : (25, 5, *, 2, *)\}$ and Finally, $\omega_3 : e_2 \leftarrow^{r_3} \{c_8\}$ is second a candidate imputation for $e_2 : (25, *, *, 3, *)$ using rule r_3 with imputation pattern $\nu(q_m^*) = e_2 : (25, *, *, 3, *)$, repair pattern $\nu(q_a^*) = (\_, 4, *, 2, *)$ and $P_a = \{c_8 : (26, *, *, *, *)\}$.

Step 3: Imputation strategy: The result of step 2 is a set of candidate imputation rules where there exists at least one rewriting. Given a set of candidate imputations $\mathcal{R}$ for some aggregation query Q, the goal is to define an ordered sequence of candidate imputations for repairing the answer of Q. This sequence is called an *imputation strategy*. The goal of a strategy is to solve two kinds of conflicts. First there might exist several candidate imputations for the same partition pattern $p_m \in ImputeP(Q)$ as shown in the example above for pattern e_2. Second, patterns in $ImputeP(Q)$ might not be disjoint and repair a subset of shared tuples. For example, missing patterns $e_2 : (25, *, *, 3, *)$ and $e_4 : (25, 4, *, *, *)$ might share the partition $(25, 4, 3)$. A standard way for solving such conflicts is to apply a multiple-imputation strategy which consists in applying all candidate imputations and combining the estimated results through some statistical methods. In this article, we adopt a different strategy which consists in regrouping all candidate imputations for each rule and evaluating these imputation groups following a static priority order defined over the imputation rules. We can show that this process is deterministic since each imputation rule generates at most one imputation value for any missing tuple.

Imputation rules can be ordered in different ways. For example, one might prefer "specialized" rules to more "generic" rules where specialization can be expressed by the number of constants in the and shared variables. For example, rule r_3 is then considered more specialized than rule r_1 since it contains more constants whereas rule r_2 is more specialized than r_3 since it contains more shared variables (with the same number of constants). Another strategy is to order the rules using some statistical estimations about data distribution, bias and completeness or domain specific expert knowledge about the system generating the data. For example, if the kWh values for floor 4 are quite similar over all buildings for a given week, rule r_d might be preferable to rule r_c. Rule r_1 might be preferred to the other rules if the kWh values do not vary over the floors of the same building.

Step 4: Imputation query generation: As shown in Definition 9, each candidate imputation $r : q_m \leftarrow q_a, f_{imp}$ generates an imputation query joining the table $ImputeD$ of values to be repaired with the table $AvailableD$ containing all correct values. As explained in Sect. 2, table $AvailableD$ is shared by all imputation queries, and can be obtained by removing the filter condition (where clause) of query Q and matching the result with the pattern table $AvailableP(Q)$ (see Definition 12). For performance reasons we precompute this table once and store the result, and reuse it for all imputation queries. Table $ImputeD$ can be obtained by matching the result Q with pattern table $ImputeP(Q)$. Each rule $r : q_m \leftarrow q_a, f_{imp}$ then generates the following imputation query over tables $ImputeP(Q)$, the result table $Result$ of Q and $AvailableD(Q)$ where S is the set of variable attributes shared by q_m and q_a and A is the set of remaining attributes in q_m:

select x.A, x.S, f_{imp} **as** m
from $ImputeP(Q)$ p, Result x, $AvailableD$ y
where match(x,q_m) **and** match(y,q_a) **and** x.S = y.S **and** match(x,p)
group by x.A, x.S

In the experiments we use a variation of imputation queries which returns the pattern cover, for partitions to be repaired. This is more efficient since partitions covered by the same pattern are imputed with the same value. An example of such a rewriting is shown in the expriments.

5   Experiments

In this section we investigate the effectiveness and efficiency of our pattern-based approach for repairing analytic queries answers. We consider a real dataset of temperature measure collected by the sensor network at our university campus. The data table $Temp(building, floor, room, year, month, day, hour, value)$ used in our study contains temperatures collected in 12 buildings during one year. $Temp$ features both spatial and temporal incompleteness since sensors only partially cover the campus buildings and operate erratically. In addition to the temperature measures, we consider a second data set $Occ(building, floor, room, occupation, area)$ that records campus rooms areas and occupations. Complete and empty pattern summaries are computed by a pattern generation algorithm described in [11]. This algorithm produces pattern summaries with respect to the campus map table and the calendar. Data and pattern tables cardinalities are reported in Table 3. For example data table D_{Temp} contains $1,321,686$ tuples for a reference table R_{Temp} of $24,615,600$ tuples ($Temp$ only covers about 5% of reference R_{Temp}), and generates $11,268$ complete partition patterns and $10,777$ missing partition patterns. D_{Occ} is almost complete and generates $1,109$ complete partition patterns and 263 missing partition patterns for 611 tuples.

We designed a set of imputation rules over attributes in Tem and Occ. Rules have variable schemas, allowing to match with different query patterns. The set of rules in Table 4 is listed in priority order, the first rule is more pertinent (accurate) than the next one when both apply.

Some implicit (expert) knowledge about campus locations allowed us to define the priority order for some rules. Take the example of rules r_2 and r_3

Table 3. Data and pattern tables cardinalities

Variant x	Data D_x	Reference R_x	Complete P_x	Missing $\bar{P}_x$
Temp	1,321,686	24,615,600	11,268	10,777
Occ	10,131	10,742	1,109	263

Table 4. Imputation rules for sensor dataset $Temp$

building, floor, room:

rule	b	f	r	b	f	r	agg
r_0	3334	x_f	x_r	-	x_f	x_r	min(temp)
r_1	-	x_f	-	-	x_f	-	avg(temp)

building, floor, room, month, day:

rule	b	f	r	m	d	b	f	r	m	d	agg
r_2	x_b	x_f	10	x_m	x_d	x_b	x_f	12	x_m	x_d	temp
r_3	x_b	x_f	11	x_m	x_d	x_b	x_f	13	x_m	x_d	temp
r_4	x_b	x_f	x_r	8	-	x_b	x_f	x_r	-	-	max(temp)
r_5	x_b	x_f	x_r	x_m	-	x_b	x_f	x_r	x_m	-	avg(temp)
r_6	-	-	-	x_m	x_d	-	-	-	x_m	x_d	avg(temp)

building, floor, room, month, occupation:

rule	b	f	r	m	o	b	f	r	m	o	agg
r_7	x_b	x_f	x_r	-	"TD"	x_b	x_f	-	-	"TP"	avg(temp)
r_8	x_b	x_f	-	-	"TD"	x_b	x_f	-	-	-	avg(temp)

in Table 4: in the same floor, rooms are named sequentially in each side: one side with odd numbers, and the other with even numbers. Room 12 is then next room 10 (and not 11). The room planning allow rules such as r_9 since the usage of "TP" rooms is nearly the same (temporal scale) as for "TD", which explains the correlation between their temperature measures.

For our experiments, we define a set of aggregation queries over data tables $Temp$ and $TempOcc = Temp \bowtie Occ$ (Table 5). All queries aggregate temperature measures along different attributes with variable filtering conditions (spatial, temporal, occupation, area).

Query Result Annotation: The query result annotation step consists in classifying each answer tuple as *correct*, *incorrect* and *missing*. We run an identification algorithm that implements functions *strict_match* and *weak_match* of Proposition 1 in Sect. 4. Table 6 classifies result patterns and partitions of each query in Table 5 into missing and correct categories. The answer data partitions are distributed between two classes *correct* and *incorrect*. Missing data is by definition not part of the query answer, since they do not belong to the data

Table 5. Analytical queries over sensor datasets

Q_1	select b, f, r, avg(temp) from Temp where b = 3334 group by b, f, r	Q_2	select b,f,r,m,d,max(temp) from Temp where m in (6,7,8) group by b, f, r, m, d
Q_3	select b,f,r,m,d avg(temp) from Temp where f in (4,5) and r in (10,11) group by b,m,d	Q_4	select b,f,r m, avg(temp) from TempOcc where b in (5354,5455) and o = "TD" group by b, f, r, m

Table 6. Correct, incorrect and missing patterns and data

	|Answer|	Correct		Incorrect		Missing		Time (sec)
		Patts	Data	Patts	Data	Patts	Data	
Q_1	8	0	0	8	8	24	108	1.6×10^{-2}
Q_2	1,012	119	1 012	0	0	132	256,588	10.0×10^{-2}
Q_3	1,602	4	377	7	1,225	116	5,333	2.9×10^{-2}
Q_4	44	19	22	22	22	66	220	4.3×10^{-2}

table (when using *null* values for representing missing information, missing data would correspond to *null* values in the result).

Observe that the number of patterns does not represent the number of corresponding data partitions. Pattern summarize completeness of data partitions at different sizes ([25,*,*] covers much more data than [25,1,10] which corresponds to one room partition). More generic patterns belong to a category set, wider they cover data partitions, and imputing this single patterns extends to all subsumed data. The running the running time is not impacted by the data table size (Q_3 vs. Q_4).

Query Result Imputation: The imputation strategy algorithm generates an ordered set of imputation queries to apply for each query "to repair" pattern set. Since the pattern summaries are shared by all imputation queries, we precompute the join between both data tables and the corresponding pattern tables and use the result in the imputation queries. Recall that rules are applied in the inverse order of their definition order. Take the example of the query Q_2. The ordered set of rules to repair the answer is $\{r_6, r_5, r_4, r_3, r_2\}$. The imputation process described in Sect. 4 is optimized in our experiments. Two imputation queries are executed. First, table *repairedPatt* stores an aggregation estimation obtained by joining the pattern table *torepair* with data table *available*. The obtained pattern table with freshly computed temperature values is then joined with the result table *Result* to generate the final table *repairedResult*. This pre-aggregation at the pattern level improves query performance since it avoids the redundant aggregation of partitions which are covered by the same patterns in the *Result*:

create table repairedPatt **as**
  select a.b, a.f, a.r, 8 **as** r.m, r.d, **max**(a.temp)
  from torepair r, available a
  where (a.b = r.b **or** r.b = '*') **and** (a.f = r.f **or** r.f = '*') **and**
     (a.r = r.r **or** r.r = '*')
  group by b, f, r, m, d

create table repairedResult **as**
  select r.b, r.f, r.r, p.m, r.d, p.temp
  from repairedPatt p, Result r
  where (r.b = p.b **or** p.b = '*') **and** (r.f = p.f **or** p.f = '*') **and**
     (r.r = p.r **or** p.r = '*') **and** (r.d=p.d **or** p.d = '*')

In Table 7, column *match patt.* records the number of patterns that can be repaired and column *cov. part.* shows the number of repaired partitions. The number of imputed partitions (column *imp. part.*) depends on the number of available correct partitions matching the rule's RHS for the repairing process. The number of remaining patterns (column *rem.*) corresponds to patterns that no rule has repaired.

Table 7. Imputation results

Query	rule	match. patt.	cov. part.	imp. part.	rem.	run time (10^{-3} sec)
Q_1	r_1	32	136	109	27	2.40
	r_0	32	136	40		1.58
Q_2	r_6	132	256, 588	256588	0	27, 910.00
	r_5	132	256, 588	9936		720.00
	r_4	132	86459	10261		3, 260.00
	r_3	25	9292	920		1.74
	r_2	25	10212	920		1.84
Q_3	r_6	127	6558	6558	0	13, 890.00
	r_5	127	6558	1084		2, 240.00
	r_4	25	465	10261		3.70
	r_3	123	5333	331		1, 590.00
	r_2	74	1225	342		170.00
Q_4	r_8	88	242	242	0	4.78
	r_7	88	242	66		0.15

Observe from the set of rules that only r_1 and r_0 are applicable for the first query. We start by applying the rule r_1 with less priority, imputing 109 partitions over 136. The rule r_0 repairs less tuples, since it requires repairing a room with the average observed temperature for the same room during the year. Many rooms are not equipped with sensors at all which explains the poor number of imputation update achieved with this rule. At the end, 27 results still remain without any estimation. We found for example that all missing partitions matching the patterns $(3334, JU, *)$, $(3334, SS, *)$ and $(3334, SB, *)$ could not be imputed, since no temperature measure is available for these floors in all campus buildings. Note that both applied rules require a completion using the same floor, but no recording sensor is available for these floors, preventing imputation. All other queries could be repaired completely by applying all matching imputation rules. These experiments demonstrate that the usefulness of imputation rules depends on the existence of correct answers and the expert's knowledge about the sensor network configuration and behavior.

6 Conclusion

We presented a new query-driven imputation approach for repairing analytic query results using imputation rules. We propose a complete query rewriting process that starts from missing and incorrect data identification using completeness patterns to generate imputation strategies for estimating missing and incorrect query results. The current imputation model is limited to aggregation queries with equality predicates in disjunctive normal form and a first possible extension would be to extend matching to inequality predicates. A second extension concerns the introduction of statistical quality criteria like precision in rule selection process. One obvious criteria for choosing a rule might be the coverage of available correct data for estimating missing values. Finally, another research direction concerns the automatic generation of imputation rules by using data mining and machine learning techniques.

Acknowledgement. This work has partially been supported by the EBITA collaborative research project between the Fraunhofer Institute and Sorbonne Université.

References

1. Buck, S.F.: A method of estimation of missing values in multivariate data suitable for use with an electronic computer. J. R. Stat. Soc. Ser. B (Methodol) **22**, 302–306 (1960)
2. Cambronero, J., Feser, J.K., Smith, M.J., Madden, S.: Query optimization for dynamic imputation. Proc. VLDB Endowment **10**(11), 1310–1321 (2017)
3. Chu, X., Ilyas, I.F., Krishnan, S., Wang, J.: Data cleaning: overview and emerging challenges. In: Proceedings of the 2016 ACM SIGMOD International Conference on Management of Data, pp. 2201–2206. ACM, New York (2016)

4. Chung, Y., Mortensen, M.L., Binnig, C., Kraska, T.: Estimating the impact of unknown unknowns on aggregate query results. ACM Trans. Database Syst. (TODS) **43**(1), 3 (2018)
5. Dallachiesa, M., et al.: NADEEF: a commodity data cleaning system. In: Proceedings of the 2013 ACM SIGMOD International Conference on Management of Data, pp. 541–552. ACM (2013)
6. Fan, W.: Dependencies revisited for improving data quality. In: Proceedings of the 2008 ACM SIGMOD-SIGACT-SIGART Symposium on Principles of Database Systems, pp. 159–170. ACM (2008)
7. Fan, W., Geerts, F.: Relative information completeness. ACM Trans. Database Syst. (TODS) **35**(4), 27 (2010)
8. Fan, W., Li, J., Ma, S., Tang, N., Yu, W.: Towards certain fixes with editing rules and master data. Proc. VLDB Endowment **3**(1–2), 173–184 (2010)
9. Farhangfar, A., Kurgan, L., Dy, J.: Impact of imputation of missing values on classification error for discrete data. Pattern Recognit. **41**(12), 3692–3705 (2008)
10. Garofalakis, M.N., Gibbons, P.B.: Approximate query processing: taming the terabytes. In: Proceedings of 27th International Conference on Very Large Databases (VLDB), pp. 343–352 (2001)
11. Hannou, F.Z., Amann, B., Baazizi, A.M.: Exploring and comparing table fragments with fragment summaries. In: The Eleventh International Conference on Advances in Databases, Knowledge, and Data Applications (DBKDA). IARIA (2019)
12. Liao, Z., Lu, X., Yang, T., Wang, H.: Missing data imputation: a fuzzy k-means clustering algorithm over sliding window. In: 2009 Sixth International Conference on Fuzzy Systems and Knowledge Discovery, vol. 3, pp. 133–137. IEEE (2009)
13. Mansinghka, V., Tibbetts, R., Baxter, J., Shafto, P., Eaves, B.: BayesDB: A probabilistic programming system for querying the probable implications of data. arXiv preprint arXiv:1512.05006 (2015)
14. Razniewski, S., Korn, F., Nutt, W., Srivastava, D.: Identifying the extent of completeness of query answers over partially complete databases. In: Proceedings of the 2015 ACM SIGMOD International Conference on Management of Data, Melbourne, Victoria, Australia, pp. 561–576, 31 May–4 June 2015
15. Silva-Ramírez, E.L., Pino-Mejías, R., López-Coello, M., Cubiles-de-la Vega, M.D.: Missing value imputation on missing completely at random data using multilayer perceptrons. Neural Netw. **24**(1), 121–129 (2011)
16. Wang, J., Krishnan, S., Franklin, M.J., Goldberg, K., Kraska, T., Milo, T.: A sample-and-clean framework for fast and accurate query processing on dirty data. In: Proceedings of the 2014 ACM SIGMOD International Conference on Management of Data, pp. 469–480. ACM (2014)
17. Wang, J., Tang, N.: Towards dependable data repairing with fixing rules. In: Proceedings of the 2014 ACM SIGMOD International Conference on Management of Data, pp. 457–468 (2014)
18. Zhu, B., He, C., Liatsis, P.: A robust missing value imputation method for noisy data. Appl. Intell. **36**(1), 61–74 (2012)

Optimization

You Have the Choice: The Borda Voting Rule for Clustering Recommendations

Johannes Kastner[1](✉) and Markus Endres[2](✉)

[1] University of Augsburg, Universitätsstr. 6a, 86159 Augsburg, Germany
johannes.kastner@informatik.uni-augsburg.de
[2] University of Passau, Innstr. 43, 94032 Passau, Germany
markus.endres@uni-passau.de

Abstract. Automatic recommendations are very popular in E-commerce, online shopping platforms, video on-demand services, or music-streaming. However, recommender systems often suggest too many related items such that users are unable to cope with the huge amount of recommendations. In order to avoid losing the overview in recommendations, clustering algorithms like k-means are a very common approach to manage large and confusing sets of items. In this paper, we present a clustering technique, which exploits the *Borda social choice voting rule* for clustering recommendations in order to produce comprehensible results for a user. Our comprehensive benchmark evaluation and experiments regarding quality indicators show that our approach is competitive to k-means and confirms the high quality of our Borda clustering approach.

Keywords: Borda · Clustering · k-means · Recommendations

1 Introduction

Recommender systems are becoming more and more common, because the quantity of data, e.g., in online shopping platforms like Amazon, movie on-demand streaming services like Netflix and Amazon Prime Video, or music-streaming platforms as Spotify, is increasing continuously [19]. In order to handle these large and confusing sets of objects easily, clustering is a very promising approach to encapsulate similar objects and to present only a few representatives of the sets to the user [6,21].

Example 1. Bob wants to watch a movie. He favors timeless old-school movies of the late 70s, 80s and early 90s, prefers action-, adventure-movies and dramas. Since it is later on the evening, he only wants to watch movies, which have a runtime between 90 and 130 min. Furthermore Bob prefers ambitious movies, so the user rating should be higher than 7 on a score from 0 to 10.

The result of such a preference query *(cp. [12]) on a movie data set, e.g., the Internet Movie Database[1] (IMDb) could produce a large, confusing result. In our example the query would return a total of 30 movies, cp. Table 1.*

---

[1] https://www.imdb.com/.

© Springer Nature Switzerland AG 2019
T. Welzer et al. (Eds.): ADBIS 2019, LNCS 11695, pp. 321–336, 2019.
https://doi.org/10.1007/978-3-030-28730-6_20

Table 1. Sample result of Bob's 4-dim. preference query.

ID	movie	rating	running time	release year	genres
0	Star Wars Episode V	8.8	127	1980	Action, Adventure, Sci-Fi
1	Star Wars	8.8	125	1977	Action, Sci-Fi
2	Raiders of the Lost Ark	8.7	115	1981	Action, Adventure
7	Reservoir Dogs	8.4	99	1992	Crime, Drama, Thriller
8	Blade Runner	8.3	117	1982	Drama, Sci-Fi, Thriller
13	The Terminator	8.1	107	1984	Action, Sci-Fi
22	Back to the Future Part II	7.7	108	1989	Adventure, Comedy, Sci-Fi
23	Indiana Jones II	7.6	118	1984	Action, Adventure, Fantasy
27	Die Hard 2	7.1	124	1990	Action, Thriller, Crime
...	...	...	...	...	...

Now, Bob has to select one movie out of a quite confusing set of items. This is in most cases a difficult decision, especially if the user preferences get more and more complex regarding constraints in several dimensions, e.g., for growing number of domains or number of constraints in multi-dimensional use cases.

Clustering approaches like k-means ensure that these large and confusing sets are encapsulated and presented in a clear manner to the user. However, if we consider the individual domains of each dimension, traditional distance measures, which are used commonly in k-means, e.g., Euclidean, stretch to their limits: While the domain of the dimension *rating* in Bob's query from Example 1 yields to a range of almost 2 in the data set of Table 1, the domain of the movies' *runtime* is 28 min between the movies with the shortest and longest runtime (ID 0 vs. 7). The same challenge is noticeable for the dimensions *release year* and *genres*. Since these dimensions have quite diverse domains, using traditional distance measures in k-means meet problems with this use case, because the domains are not set into an equal relation to each other. In our Example, the domains of *release year* and *running time* would have a major influence on the clustering approach. If we want to get a useful clustering, we need to *adjust the domains before the clustering* process (cp. [17,24]). This might be a very challenging task due to various and versatile user preferences.

In this paper we adapt the *Borda social choice voting rule for cluster allocation*, in order to ensure an equal treatment of each dimension. Each object is considered equally in each dimension and receives a "voting", which yields to a more balanced and smooth result compared to a cluster allocation using traditional distance measures. Unlike k-means with traditional distance measures, the Borda social choice voting rule considers each dimension independent from the size of their domains for each object by assigning *votes*.

The rest of the paper is organized as follows: In Sect. 2 we discuss related work. After that, the essential basics for our framework are laid in Sect. 3. In Sect. 4 we explain our Borda social choice clustering algorithm. Thereafter, we discuss synthetic experiments for our approach in Sect. 5. Section 6 presents our recommender system resting upon the IMDb and our comprehensive quality evaluation. We conclude in Sect. 7.

2   Related Work

Thousands of clustering algorithms have been proposed in the literature. We will briefly review some approaches related to our solution.

The basic algorithms of many clustering approaches are k-means and k-means++, cp. [1, 6]. The authors of [9] extended k-means and published an implementation, which filters the data set with a kd-tree in order to ensure a better separation between clusters. Considering multidimensional data sets is a very common research area, hence [25] published an approach using hyper-boxes for partitioning and forming clusters to reach less errors compared to the common k-means clustering, especially in higher dimensions.

The authors of [27] published an approach, which ensures the stability of k-means clustering by adding a heuristic for finding optimal centroids during the cluster allocation. Using a weighting for identifying subsets in k-means ensures better results, which is approached in [7]. In [18] subspace clustering is considered to mask out dimensions in high dimensional data by a so called feature selection, which reduces the dimensions by removing irrelevant and redundant ones.

Both [8, 23] deal with chains as input for clustering algorithms and therefore present solutions for the cluster allocation of chains using orders instead of trivial distances. Gong et al. published a collaborative filtering recommendation algorithm in [26], which considers clustering approaches for user and item clustering using similar ratings. Clustering personalized music recommendations by setting favored music as centroids for the clustering process is published in [13]. The authors of [10] also used preferences in order to minimize the sets, which should be clustered. Their approach is to compute clusterings in high dimensional environments by using a Pareto-dominance criterion for cluster allocation.

Virmani et al. [24] proposed a k-means clustering approach, where a normalization of features is integrated before the clustering process starts. This is done by assigning weights to each attribute value to achieve standardization. Also [17] discuss the effects of domain standardization. They found out that it is important to select a specific standardization procedure according to the data set, in order to obtain better quality results.

In all these cases the question is how to find the right weights and which standardization procedure should be applied? In our paper we overcome the normalization and standardization problem by applying the *Borda social choice voting rule to allocate objects to the clusters.*

3   Background

Before we introduce our novel Borda clustering approach, we explain the most important basics and background knowledge used in this paper. We briefly recap the k-means clustering algorithm, which is an iterative partitioning algorithm with a convergence criterion. Hereby each object gets allocated to one of k clusters iteration by iteration, until a stable configuration is found. Furthermore we review k-means++, a version of k-means where the initial partition is adjusted with seeding.

324   J. Kastner and M. Endres

3.1   The k-means and k-means++ Clustering Algorithms

We present **k-means** as it is defined in [6]. Given a set X consisting of n d-dimensional objects x and k user-desired clusters c_i, k-means works as follows:

(1) *Find an initial partition for the cluster centroids by choosing a random d-dimensional object of X for each of the k centroids.*
(2) *Calculate for each object the distances to all centroids by using a distance-measure, e.g., Euclidean distance, and subsequently allocate each object to the closest centroid.*
(3) *Recalculate each centroid by averaging the contained objects.*
(4) *Proceed with Step (2) until two succeeding clusterings are stable, which means that all clusters from the last iteration contain the equal set of objects as in the current iteration.*

In **k-means++** the initial partition is not chosen arbitrary by random, but by a randomized *seeding technique*. In particular, the possibly best centroids for the initial partition should be found in order to reach more accurate clusterings. Step (1) of the k-means algorithm is replaced as follows, cp. [1]:

(1a) *Arbitrary choose an object of the set X as first cluster centroid c_0.*
(1b) *For each further cluster centroid $c_i \mid i \in \{1, ..., k-1\}$ choose $x \in X$ with a probabiltity of*

$$p(x) = \frac{dist(x, c_i)^2}{\sum_{x \in X} dist(x, c_i)^2} \tag{1}$$

where $dist(x, c_i)^2$ is the shortest squared Euclidean distance from a point x to the already chosen closest centroid c_i.
(1c) *Proceed with Step (2) of the k-means clustering algorithm.*

3.2   Similarity Measures

k-means and its variants use traditional measures like the Euclidean or Canberra norm to calculate the distances between objects and sets [2].

The **Euclidean distance** is the most favored and used distance measure for clustering with k-means. Given two points $x_i = (x_{i_1}, \ldots, x_{i_d})$ and $x_j = (x_{j_1}, \ldots, x_{j_d})$ with d dimensions, the particular squared distances regarding each dimension are summed up and rooted after that, i.e.,

$$dist(x_i, x_j) = \sqrt{\sum_{l=1}^{d} (x_{i_l} - x_{j_l})^2} \tag{2}$$

In order to set the focus on distances using small domains as well, the **Canberra norm** is a very auxiliary measure. It sums up the absolute fractional distances of two d-dimensional points x_i, x_j in relation to the range of the focused dimension for all dimensions:

$$dist(x_i, x_j) = \sum_{l=1}^{d} \frac{|x_{i_l} - x_{j_l}|}{(x_{i_l} + x_{j_l})} \tag{3}$$

Example 2. *Consider Example 1. Assume we want $k = 3$ clusters and the movies with the IDs (1), (7), and (23) are chosen as initial centroids. The movie with ID (27) should be allocated to one of the clusters using k-means with the Euclidean distance on the attributes* rating, running time, release year, *and* genres.

We used the Jaccard coefficient ^{2} to determine the distance for categorical attributes like genre, e.g, $J\ (genres_{ID=1},\ genres_{ID=27}) = \frac{1}{4}$ for movie (1) and (27). The distance between movie (1) and (27) is then given by

$$dist(1, 27) = \sqrt{(8.8 - 7.1)^2 + (125 - 124)^2 + (1990 - 1977)^2 + (1 - 0.25)^2} = 13.2$$

and shows that the (large) domain of the year has a major influence on the calculation of the distance. Finally, movie (27) would be allocated to the cluster with **centroid (23)** *because of the lowest distance of only $dist(23, 27) = 8.5$.*

4   Borda Social Choice Clustering

In this section we present our novel *Borda social choice clustering* approach. Social choice deals with the aggregation of individual preferences for managing social assessments and ruling. The *Borda* social choice voting rule is omnipresent in political or other elections, e.g., the Eurovision Song Contest. Social choice has its foundation back in the 18th century and was published first by Jean-Charles de Borda and Marquis de Condercet [22].

4.1   The Borda Social Choice Voting Rule

As mentioned in [5], the Borda social choice voting rule is a very appealing approach to consider each dimension in a multi-dimensional scenario in an equal manner. This rule can be used for the allocation of objects to one and only one cluster and therefore allows more influence of smaller domains. For our approach, the Borda social choice voting rule is a promising method, because every candidate receives equal weighted votes from each voter.

Definition. Given k candidates C_i, and d voters V_j, where each voter votes for each candidate. Each voter has to allocate the voting $v_{jm} \in \{0, ..., k - 1\}$, $m = 1, ..., k$, where all v_{jm} are pairwise distinct. After all voters assigned their votes, the votes for each candidate are summed up as it can be seen in Eq. 4, while the Borda winner is determined as depicted in Eq. 5.

$$bordaSum_{C_i} = \sum_{l=1}^{d} v_{li} \tag{4}$$

$$bordaWinner = \max\{bordaSum_{C_i} \mid i = 1, ..., k\} \tag{5}$$

---

^{2} Jaccard: $J(A, B) = |A \cap B|/|A \cup B|$ for two sets A and B. $J_\delta(A, B) = 1 - J(A, B)$.

If we apply this approach to our clustering-framework, the *candidates* correspond to the available *clusters* and the *voters* correspond to the *dimensions of the d-dimensional object which should be allocated to a cluster*. Then, for each dimension votes are assigned for the distances between the object and the centroids of the clusters. While the closest distance receives a maximum vote of $k-1$, the second closest gets a vote of $k-2$, etc., the largest distance obtains a vote of 0, where k is the number of desired clusters. After the voting, Eq. 4 determines the sum of all votes for each cluster, and subsequently Eq. 5 identifies the winner.

Therefore, dimensions, which would not be equally considered because of a smaller or larger domain, e.g., by using a distance measure like Euclidean, get equal weighted votes like the other dimensions and have a higher influence on the clustering process.

Example 3. *Reconsider Example 2. Table 2 shows our Borda social choice cluster allocation for movie (27). The centroids of the initial clusters C_1, C_2, C_3 are the movies with the IDs (1), (7), (23).*

*For each dimension the distances between movie (27) and the centroids are calculated. The Borda votes are depicted in parentheses, e.g., the dimension rating is closest to C_3 and therefore gets a vote of $k - 1 = 2$. The second closest centroid is C_2 with vote 1, and C_1 gets the vote 0. Finally, C_2 with **movie (7)** as initial centroid is determined as the Borda winner with a Borda sum of 5, cp. Eqs. 4 and 5. Compared to the Euclidean distance we obtain a more concise result for the cluster allocation, due to ranking the values in each dimensions according to their closeness. Note that a Jaccard coefficient of 1.0 is the best value for the genre.*

Table 2. Cluster allocation for movie (27).

	Movie (27): Die Hard		
Dimension	C_1	C_2	C_3
rating	1.70 (0)	1.30 (1)	**0.50 (2)**
running time	**1.00 (2)**	25.00 (0)	6.00 (1)
release year	13.00 (0)	**2.00 (2)**	6.00 (1)
genre	0.25 (1)	**0.50 (2)**	0.20 (0)
$\sum$	3	5	4

4.2   The Borda Clustering Algorithm

We modified the classic k-means algorithm from Sect. 3.1 to realize a clustering with the Borda social choice voting rule as decision criterion for the cluster allocation. For this, we changed Step (2) of k-means, where the distances of each object to the available clusters are calculated, and used the Borda rule for cluster allocation. This allocation is described in Function 1, which finally returns the id of the centroid the object should be allocated to.

Function 1. Determine Borda Winner

Input: d-dim. object $x = (x_1, ..., x_d)$, centroids C, cluster-id last iteration id_{last}.
Output: id of the closest cluster for object $x = (x_1, ..., x_d)$.

```
1: function GETBORDAWINNER(x, C, id_last)
2:     votes[] ←calculateBordaSum(x, C)          ▷ determine & sum up votes.
3:     id= analyzeBordaWinners(votes[], id_last)  ▷ analyze all Borda winners.
4:     return id
5: end function
```

For managing the Borda social choice voting rule, an object array *votes[]* is used to save the centroid ids and Borda values. As further information for each object $x = (x_1, ..., x_d) \in X$, an identifier id_{last} of the allocated centroid from the previous iteration is necessary.

In Line 2, the array *votes[]* is set to the *bordaSum* values from Eq. 4. In detail, in each dimension the distances between the considered object x and each cluster centroid of C are calculated, saved together with the centroids id in an object based data structure and appended to an object array. Once all distances in the current dimension are calculated, this object array is sorted ascending according to the distances in order to assign the Borda votes from 0 to $k-1$. After the sort, the votes for each cluster are determined and summed up in the array *votes[]* over all dimensions.

Subsequently, we find the Borda winner(s) with the highest score in the array *votes[]* (Line 3) and return the id of the centroid in Line 4. If there is more than one Borda winner, the winner is chosen by random. After the object x got allocated to the centroid with the identifier id, the clustering continues with Step (3) of k-means. Note that we call this approach **Borda**. There are some improvements w.r.t. the convergence, which we will discuss in the next section.

4.3   Convergence

When talking about clustering, *convergence* is a major topic. In [6,16] it was shown that k-means can only converge to a local optimum (with some probability to a global optimum when clusters are well separated). Our algorithm is based on k-means and only uses another "distance measure". Therefore the proof of convergence is similar to that one of k-means.

Proof. (Proof of convergence). There is only a finite number of ways to partition n data points into k clusters [6,16]. For each iteration of our algorithm, we produce a new clustering based *only* on the old clustering. In addition, it holds that

(1) if the old clustering is the same as the new, then the next clustering will again be the same. We have some kind of fixed-point.
(2) if the new clustering is different from the old one, then the newer one has a lower cost (due to a better overall voting).

Since the algorithm iterates a function whose domain is a finite set, the iteration must eventually enter a cycle. The cycle cannot have length greater than 1, because otherwise by (2) one would have some clustering which has a lower cost than itself, which is impossible.   □

Therefore, k-means using the Borda social choice voting rule, converges in a finite number of iterations to a local solution, but does not permit us to eliminate the interesting possibility that a point oscillates indefinitely between two clusters.

Indeed, after some preliminary tests, especially for higher dimensions and higher number of clusters, we noticed that there are some problems regarding the convergence of our approach. In order to solve this problem, we added a decision criterion for the cluster allocation, if there is more than one Borda winner.

In detail: For each iteration we save the ids of cluster objects the object got allocated to. Assume there is more than one Borda winner in the next iteration. We then consult the allocation to the centroid from the last iteration id_{last} (Line 3 of Function 1). If so, the object goes to the same cluster as in the last iteration. As our benchmarks show, this solution ensures that the clusters are becoming stable in a few number of iterations.

Another problem concerns the initial partition, which could result in empty clusters. If the first centroid was randomly chosen, the probability that a quite similar object of the first centroid will be chosen is very small but possible. Especially, if there are, e.g., different movies with almost the same specifications, the possibility is given, that these movies are chosen as cluster centroids. Then, the order of the cluster centroids decides that the first of the regarding clusters will be occupied with objects, while the following cluster with the similar centroid will stay empty. K-means++ minimizes these problems and furthermore cares that the runtime and the number of iterations will decrease. We call this extended approach considering convergence and empty clusters **Borda++**.

4.4   Complexity

The complexity of our algorithm is given by $\mathcal{O}(ndk \cdot k \log(k) + k)$ where each of the n d-dimensional objects should be clustered in k clusters. The algorithm calculates for each object the distances of the dimension d for each cluster k in $\mathcal{O}(ndk)$. Depending on the sorting algorithm, the distances are sorted, e.g., by *Quicksort* in $k \cdot \log(k)$. Finally, we search for the Borda winners in $\mathcal{O}(k)$. Hence we get a complexity of $\mathcal{O}(nd \cdot k^2 \cdot \log(k))$.

5   Synthetic Experiments

The aim of our synthetic experiments was to show that our approach is competitive to k-means w.r.t. runtime and number of iterations.

5.1 Benchmark Settings

We implemented our algorithms in Java 1.8, and ran our experiments on a server (Intel Xeon, 2.53 GHz, 44 GB RAM) running Debian GNU/Linux 7. For our benchmarks we used the data generator described in [3], which creates independent, anti-correlated, and correlated synthetic data. We varied the number of dimensions, the number of objects per set and the number of desired clusters. We investigated the runtime and the number of iterations until a stable clustering is reached. We used the following clustering techniques:

- **Eucl.**: k-means with *Euclidean distance* for cluster allocation.
- **Canb.**: k-means with *Canberra distance* for cluster allocation.
- **Borda**: k-means with *Borda voting rule* for cluster allocation.
- **Borda++**: *k-means++* with *Borda voting rule* for cluster allocation.

5.2 Evaluation

Since Euclidean is the most common distance for k-means, we want to show that our approach terminates at least as fast as k-means and needs the same or less number of iterations until termination. Furthermore, we investigated Canberra to gain useless reference values for our approach, which should be dominated by them of our approach w.r.t. the runtime and the number of iterations as well. In order to receive a faster runtime and less iterations until stable clusterings, we want to show the utility of k-means++ for our Borda approach w.r.t. the runtime and number of iterations.

Experiments on Runtime: In Fig. 1 we varied the number of clusters ($k = 3, 5, 7, 9$) and the data size, i.e., we used 5000, 10000, and 15000 input objects for the clustering. In this 3-dimensional case, increasing the number of clusters and the input data lead to an increasing runtime, too.

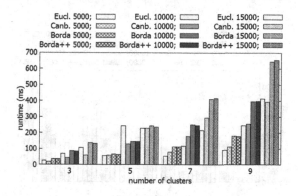

Fig. 1. Runtime w.r.t d=3.

Our approach (**Borda**) works in equal time compared to *k-means with Euclidean* (**Eucl.**) and *Canberra* (**Canb.**) for small numbers of clusters. For 7 and 9 clusters our approach is slower independent of the number of input objects, because of a higher complexity of our approach. Benefits of a faster runtime for *k-means++* (**Borda++**) is in most cases hardly recognizable.

Figure 2 presents our results on a 5-dimensional domain. We see that for increasing numbers of dimensions *Borda* reaches a better runtime compared to *Eucl.* except for a high number of clusters. In some cases *Borda* terminates faster than *Canb.*, e.g., the test series with 7 clusters, but all in all Borda mostly reaches an equal runtime in a 5-dimensional space.

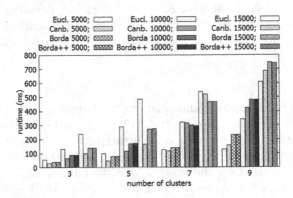

Fig. 2. Runtime w.r.t. d=5.

A similar behavior illustrates the test series for a 9-dimensional set of objects in Fig. 3. While both, *Borda* and *Borda++*, terminate in similar time compared to *Canb.* for 3 and 5 clusters, they are a lot faster than *Eucl.* The trends for growing runtimes w.r.t. the number of clusters and objects can be noticed in 9-dimensional space, too. Further experiments have shown that in higher dimensions the runtime increases with more objects and more clusters.

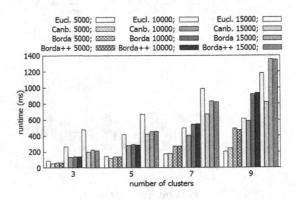

Fig. 3. Runtime w.r.t. d=9.

Experiments on Iterations: In this section we consider the number of iterations necessary to reach a stable clustering, cp. Fig. 4.

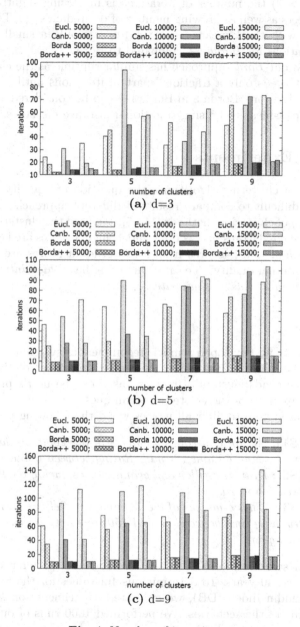

Fig. 4. Number of iterations.

Our experiments indicate that for an increasing number of clusters and an increasing number of objects the number of iterations until termination increases, too. However, *Borda* and *Borda++* reach a stable clustering in clearly less iterations than *Eucl.* and *Canb*. Our experiments have also shown that for higher dimensions ($d > 9$) the number of iterations is increasing slightly for growing number of objects as well as growing number of desired clusters. Thus, the seeding performed with the k-means++ algorithm has only a small effect on the number of iterations for *Borda++* compared to Borda.

In summary, our *Borda* approach has similar runtime as the classic k-means algorithms, but needs only a fractional part of iterations until termination for all test series. Therefore, Borda and Borda++ can be considered as competitive to k-means resp. k-means++ using traditional distance measures.

6   Quality Experiments

In the context of clustering algorithms, the question of "quality" often arises. However, it is difficult to compare results of different approaches and to determine the *quality* of clustering methods [14,15]. We used the clustering evaluation indicators *Silhouette* (scores between -1 and 1, higher scores are better) [20] and the *Davies-Bouldin Index* (scores from 0 to ∞, lower scores are better) [4] to measure the "internal quality", i.e., if a clustering has a *high intra-cluster similarity* and a *low inter-cluster similarity*.

6.1   Settings

To test the quality we developed a movie recommender system [11] based on the JMDb movie database, a Java-based alternative interface[3] of the IMDb. The prototype recommends clusters of movies based on the user's preferences and allows us a comparison of the clustering techniques.

We used this system to filter all movies w.r.t. the following preferences:

- **Scenario (S1):** *Action and comedy movies of the 2000s to the present day. Running time between 60 and 120 min. Rating between 6 and 10.*
- **Scenario (S2):** *Drama, thriller and crime movies during the 90s and 2000s. Rating between 8 and 10.*
- **Scenario (S3):** *Classic-movies of the 70s and 80s. Release year between 1975 and 1989. Running time between 90 and 150 min. Action-, adventure-movies, and dramas as genre.*

The chosen scenarios were build on common user preferences considering movies of the last 40 years. To determine useful values for the Silhouette (Sil.) and Davies-Bouldin Index (DB), we evaluated experiments on $k \in \{3, 5, 7, 9\}$ w.r.t. the results of the scenarios. We performed 1000 runs of our experiments to get a significant mean value.

---

[3] http://www.jmdb.de/.

Note that our Borda clustering approach is not metric in general, but an "assignment function". Therefore, no appropriate numerical distance measure for Silhouette and Davies-Bouldin could be found for our Borda clustering approach. We simply used the Euclidean distance and the Canberra distance in both indicators to evaluate the quality, even though this is not adequate for Borda and leads to a bias. Tables 3, 4 and 5 show our results for the **scenarios (S1), (S2) and (S3)**.

Table 3. Quality measures using Sil. and DB Index for Scenario (S1).

Algorithm	Indicator	with Canberra dist.				with Euclidean dist.			
		k=3	k=5	k=7	k=9	k=3	k=5	k=7	k=9
Canb.	Silhouette	**0.430**	**0.470**	**0.479**	**0.484**	-0.028	-0.119	-0.150	-0.153
Canb.	Davies-Bouldin	**1.080**	**1.029**	**0.896**	**0.839**	10.570	11.230	8.774	7.642
Borda++	Silhouette	0.106	0.091	0.073	0.067	0.062	-0.003	-0.036	-0.050
Borda++	Davies-Bouldin	2.171	2.163	2.279	2.295	2.627	2.784	2.804	2.815
Eucl.	Silhouette	0.135	0.038	-0.024	-0.060	**0.457**	**0.405**	**0.411**	**0.421**
Eucl.	Davies-Bouldin	3.850	4.552	5.115	5.878	**0.775**	**0.881**	**0.836**	**0.799**

Table 4. Quality measures using Sil. and DB Index for Scenario (S2).

Algorithm	Indicator	with Canberra dist.				with Euclidean dist.			
		k=3	k=5	k=7	k=9	k=3	k=5	k=7	k=9
Canb.	Silhouette	**0.591**	**0.605**	**0.650**	**0.679**	-0.050	-0.113	-0.171	-0.221
Canb.	Davies-Bouldin	**0.602**	**0.567**	**0.531**	**0.501**	15.003	18.256	18.195	18.837
Borda++	Silhouette	0.090	0.001	-0.040	-0.065	0.141	-0.010	-0.086	-0.103
Borda++	Davies-Bouldin	2.552	2.892	3.194	3.241	2.394	2.908	3.174	3.155
Eucl.	Silhouette	-0.026	-0.087	-0.192	-0.308	**0.592**	**0.562**	**0.547**	**0.531**
Eucl.	Davies-Bouldin	7.070	13.879	15.768	16.541	**0.513**	**0.551**	**0.606**	**0.618**

Table 5. Quality measures using Sil. and DB Index for Scenario (S3).

Algorithm	Indicator	with Canberra dist.				with Euclidean dist.			
		k=3	k=5	k=7	k=9	k=3	k=5	k=7	k=9
Canb.	Silhouette	**0.507**	**0.541**	**0.591**	**0.612**	0.001	-0.024	-0.034	-0.020
Canb.	Davies-Bouldin	**0.930**	**0.875**	**0.703**	**0.591**	7.493	5.353	4.370	3.928
Borda++	Silhouette	0.132	0.135	0.123	0.133	0.127	0.117	0.116	0.131
Borda++	Davies-Bouldin	1.748	1.730	2.040	2.012	2.022	1.944	2.007	1.881
Eucl.	Silhouette	0.110	0.009	-0.072	-0.138	**0.436**	**0.430**	**0.453**	**0.465**
Eucl.	Davies-Bouldin	4.486	4.811	5.315	5.414	**0.820**	**0.762**	**0.672**	**0.624**

The left most column in all tables represent the algorithm used for clustering, i.e., k-means++ with the Canberra (Canb.), Borda++, or Euclidean (Eucl.)

measure. For each algorithm we computed the *Silhouette* and *Davies-Bouldin* quality indicator, one time with the Canberra distance, the other time with the Euclidean distance.

Unsurprisingly, Canb. performs best with the Canberra distance function, and Eucl. provides the best internal quality using the Euclidean distance for all evaluated scenarios. This is due to the fact that the clusters are computed using the corresponding distance measure, and therefore the quality indicators also compute a high intra-cluster similarity.

For example consider Table 3, where Silhouette leads to an internal quality of 0.484 (Canberra dist.) on the score of -1 to 1 and therefore $k = 9$ would be best in this case. Also Davies-Bouldin having a value of 0.839 is best for $k = 9$. Obviously, this is not the case with Borda due the fact that neither the Euclidean nor the Canberra distance fits to the Borda assignment function. Nevertheless, we observe that the internal quality of Borda always gets values between Canb. and Eucl. In addition, our approach reaches more reasonable values than Canb. using the Euclidean distance or Eucl. with the Canberra distance. Thus our approach is adequate for the intra-cluster and inter-cluster similarity (Davies-Bouldin Index) as well as the coherence of the clusters (Silhouette).

6.2   Lessons Learned

Considering the *internal clustering indicators*, our approach confirms a high quality. That means that Borda++ achieves a high intra-cluster similarity, even if it is between Eucl. and Canb. Note that the quality indicators Silhouette and the Davies-Bouldin-Index are evaluated using the *Euclidean distance* and the *Canberra distance* in their distance calculations. Therefore, it is obvious that Eucl. gets a better quality by using the Euclidean distance, and Canb. is better using the Canberra distance. However, our approach using the Borda++ allocation lies between both quality measures and therefore provides a high internal quality.

Finally, Borda++ is a competitive alternative for the cluster allocation in centroid-based clustering algorithms like k-means due to similar runtimes, less iterations, and the benefit that no normalization is needed before the clustering. Additionally, our comprehensive and thorough experiments considering the internal quality emphasizes the advantages of our alternative clustering approach.

7   Conclusion and Outlook

In our paper, we presented a clustering approach exploiting the Borda social choice voting rule as decision criterion for cluster allocation. Using Borda, users do not need to care about the normalization of domains, because the Borda social choice voting rule for cluster allocation considers each dimension as equally important.

Our experiments show that our approach terminates in comparative runtime to k-means clustering with traditional distance measures, but needs less iterations until a stable clustering is reached. Furthermore, comprehensive quality experiments verify the benefit of our approach in the context of a large and multi-dimensional environment, namely the IMDb movie recommender. Hence, Borda is a novel approach which is competitive to well-known clustering techniques.

In our future work we want to minimize the empty cluster problem by choosing a better initial partition, e.g., populating the centroids initially with the most-preferred movies of the users. Furthermore, since our Borda clustering approach provides very concise results for the Borda winners at the cluster assignment, we want to investigate the possibility to weight dimensions by user-preferences. In addition, we will integrate the Borda voting rule into other clustering techniques like X-Means, EM-Clustering or a density based clustering algorithm like DBSCAN in order to identify the behavior of Borda. Moreover, we want to confirm our quality experiments with an extensive user study using our Demo Recommender [11].

References

1. Arthur, D., Vassilvitskii, S.: K-means++: the advantages of careful seeding. In: ACM-SIAM 2007, SODA 2007, Philadelphia, PA, USA, pp. 1027–1035 (2007)
2. Bandyopadhyay, S., Saha, S.: Unsupervised Classification. Springer, Heidelberg (2013). https://doi.org/10.1007/978-3-642-32451-2
3. Börzsönyi, S., Kossmann, D., Stocker, K.: The skyline operator. In: ICDE 2001, pp. 421–430. IEEE, Washington, DC (2001)
4. Davies, D.L., Bouldin, D.W.: A cluster separation measure. IEEE Trans. Pattern Anal. Mach. Intell. 1(2), 224–227 (1979)
5. Debord, B.: An axiomatic characterization of Borda's k-choice function. Soc. Choice Welfare 9(4), 337–343 (1992)
6. Jain, A.K.: Data clustering: 50 years beyond k-means. Pattern Recogn. Lett. 31(8), 651–666 (2010)
7. Jing, L., Ng, M.K., Huang, J.Z.: An entropy weighting k-means algorithm for subspace clustering of high-dimensional sparse data. IEEE Trans. Knowl. Data Eng. 19(8), 1026–1041 (2007)
8. Kamishima, T., Akaho, S.: Efficient clustering for orders. In: Zighed, D.A., Tsumoto, S., Ras, Z.W., Hacid, H. (eds.) Mining Complex Data. Studies in Computational Intelligence, vol. 165, pp. 261–279. Springer, Heidelberg (2009). https://doi.org/10.1007/978-3-540-88067-7_15
9. Kanungo, T., Mount, D.M., Netanyahu, N.S., Piatko, C.D., Silverman, R., Wu, A.Y.: An efficient k-means clustering algorithm: analysis and implementation. IEEE TPAMI 24(7), 881–892 (2002)
10. Kastner, J., Endres, M., Kießling, W.: A pareto-dominant clustering approach for pareto-frontiers. In: EDBT/ICDT 2017, Venice, Italy, 21–24 March 2017, Workshop Proceedings, vol. 1810 (2017)
11. Kastner, J., Ranitovic, N., Endres, M.: The Borda social choice movie recommender. In: BTW 2019, 4–8 March 2019 in Rostock, Germany, pp. 499–502 (2019)
12. Kießling, W., Endres, M., Wenzel, F.: The preference SQL system - an overview. Bull. Tech. Commitee Data Eng. 34(2), 11–18 (2011)
13. Kim, D., Kim, K.S., Park, K.H., Lee, J.H., Lee, K.M.: A Music Recommendation System with a Dynamic k-means Clustering Algorithm. In: ICMLA (2007)

14. Knijnenburg, B.P., Willemsen, M.C., Gantner, Z., Soncu, H., Newell, C.: Explaining the user experience of recommender systems. User Model. User-Adap. Inter. **22**(4–5), 441–504 (2012)
15. Kunaver, M., Porl, T.: Diversity in recommender systems a survey. Know. Based Syst. **123**(C), 154–162 (2017)
16. Macqueen, J.: Some methods for classification and analysis of multivariate observations. In: In 5-th Berkeley Symposium on Mathematical Statistics and Probability, pp. 281–297 (1967)
17. Mohamad, I., Usman, D.: Standardization and its effects on K-Means Clustering Algorithm. Res. J. Appl. Sci. Eng. Technol. **6**, 3299–3303 (2013)
18. Parsons, L., Haque, E., Liu, H.: Subspace clustering for high dimensional data: a review. SIGKDD Explor. Newsl. **6**(1), 90–105 (2004)
19. Ricci, F., Rokach, L., Shapira, B.: Recommender systems: introduction and challenges. In: Ricci, F., Rokach, L., Shapira, B. (eds.) Recommender Systems Handbook, pp. 1–34. Springer, Boston, MA (2015). https://doi.org/10.1007/978-1-4899-7637-6_1
20. Rousseeuw, P.: Silhouettes: a graphical aid to the interpretation and validation of cluster analysis. J. Comp. Appl. Math. **20**, 53–65 (1987)
21. Sarstedt, M., Mooi, E.: Cluster analysis. In: A Concise Guide to Market Research. STBE, pp. 273–324. Springer, Heidelberg (2014). https://doi.org/10.1007/978-3-642-53965-7_9
22. Sen, A.: The possibility of social choice. Am. Econ. Rev. **89**(3), 349–378 (1999)
23. Ukkonen, A.: Clustering algorithms for chains. J. Mach. Learn. Res. **12**, 1389–1423 (2011)
24. Virmani, D., Shweta, T., Malhotra, G.: Normalization Based K Means Clustering Algorithm. CoRR abs/1503.00900 (2015)
25. Wan, S.J., Wong, S.K.M., Prusinkiewicz, P.: An algorithm for multidimensional data clustering. ACM Trans. Math. Softw. **14**(2), 153–162 (1988)
26. Wei, S., Ye, N., Zhang, S., Huang, X., Zhu, J.: Collaborative filtering recommendation algorithm based on item clustering and global similarity. In: BIFE 2012, pp. 69–72, August 2012
27. Zhang, Z., Zhang, J., Xue, H.: Improved K-means clustering algorithm. In: Proceedings of the Congress on Image and Signal Processing 2008, CISP 2008, vol. 5, pp. 169–172, May 2008

BM-index: Balanced Metric Space Index Based on Weighted Voronoi Partitioning

Matej Antol and Vlastislav Dohnal[✉]

Faculty of Informatics, Masaryk University, Botanicka 68a, Brno, Czech Republic
{xantol,dohnal}@fi.muni.cz

Abstract. Processing large volumes of various data requires index structures that can efficiently organize them on secondary memory. Methods based on pivot permutations have become popular because of their tremendous querying performance. Pivot permutations can be perceived as a recursive Voronoi tessellation with a fixed set of anchors. Its disadvantage is that it cannot adapt to the data distribution well, which leads to cells unbalanced in occupation and unevenly filled disk buckets.

In this paper, we address this issue and propose a novel schema called the BM-index. It exploits a weighted Voronoi partitioning, which is able to respect the data distribution. We present an algorithm to balance the data partitions, and show its correctness. The secondary memory is then accessed efficiently, which is shown in experiments executing k-nearest neighbors queries on a real-life image collection CoPhIR.

Keywords: Indexing structure · k-nearest neighbor query ·
Approximate search · Metric space · Voronoi partitioning

1 Introduction

Aspects of the Big Data phenomenon require new solutions to data processing systems. In this paper, we focus on indexing structures that model data as a metric space. This inherently provides a solution to the property of variety, while volume and velocity are handled by indexing itself. Since metric space model requires only a distance measure to be defined, it forms an extensible solution. It is also much more resilient to the curse of dimensionality than multi-dimensional indexing structures [6]. For details, we refer the reader to the books [17,20]. Current applications need to compare data items non-rigidly, so a similarity operator must be defined, e.g., k-nearest neighbors query [7]. By analogy to common Internet search engines, an approximate evaluation of similarity queries provides very good quality of results at low processing costs.

Indexing structures organizing data by recursive Voronoi partitioning have become popular in recent years [9,11,14,15]. They vary in details, but they share the common idea of defining a data partition via a pivot permutation. Having a preselected set of data objects (called pivots), we can assign each database object a list of m closest pivots – a prefix of pivot permutation. Next, the objects are grouped by their prefixes and each group forms a separate data partition. The performance study [13] on various real-life data-sets proved outstanding performance of such indexing structures. From the data organization point of

© Springer Nature Switzerland AG 2019
T. Welzer et al. (Eds.): ADBIS 2019, LNCS 11695, pp. 337–353, 2019.
https://doi.org/10.1007/978-3-030-28730-6_21

view, the prefix should not be too long, because it might lead to having low fill factor (as low as one data object per partition). On the other hand, it should not be too short, which would create some data partitions impractically big.

1.1 Problem Definition and Contributions

We take the recursive Voronoi partitioning with a fixed set of pivots as a model of pivot permutations. In this variant, the data is separated by all pivots and each partition has a pivot as its center. On the second level, the partitions are divided by the remaining pivots, so new data partitions have their pivots outside them. As a result, it becomes unfeasible to keep the partitions of approximately equal occupation. This issue has also been confirmed by Skala [18]. He stated that the number of data partitions does not grow exponentially, but far more slowly. Consequently, large data-sets then require long prefixes to "carve" spurious partitions. The by-product of this is a huge number of tiny partitions. Such partitions eventually end up as separate buckets or files in the secondary storage.

In this paper, we tackle this property and apply the weighted Voronoi partitioning into the pivot permutation schema to eliminate under- and over-filled buckets. We show correctness of such modification and its applicability to similarity searching. Experiments on a real-life data set reveal much better I/O performance. We compared our results with M-index, which is an up-to-date structure that uses secondary memory. Its recall performance is comparable to newer proposals like PP-index [9] and NAPP [19].

The rest of the paper is organized as follows. The following section summarizes the related work and the principles of pivot permutations in metric spaces. The new Balanced Metric Index is presented in Sect. 3. Next, performance analysis on the CoPhIR data-set is given in Sect. 4. In Conclusions, we foresee future directions to improve searching performance further.

2   Related Work

The paper focuses on the state-of-the-art data organizations based on Voronoi partitioning [13]. The pioneering work was GNAT [5] that applies Voronoi partitioning recursively. It divides the search space using k sites (pivots) to produce k partitions. Each partition is then recursively split using the same principle but with a new set of pivots. There are also simpler structures like the generalized-hyperplane tree that use only two pivots to partition the space. The disadvantage of GNAT of being an in-memory structure was later solved in EGNAT [16]. The advantage of GNAT is that it applies the recursive Voronoi tessellation in its classical way – choosing a brand new set of pivots on next levels, so data locality and density is respected. However, it increases memory requirements and computational costs due to selecting new sets of pivots repeatedly.

The PP-index is probably the first structure that mines the same set of pivots for data partitioning using pivot permutation [8,9]. A similar approach was used in the M-index [14], where the permutation was encoded into one integer and the B^+-tree was applied for fast retrieval. It was shown that PP-index is outperformed by M-index in k-nearest neighbors queries. The MI-file falls into the same category [1]. It employs inverted file and Spearman Footrule metric to identify posting lists relevant to the query's permutation prefix. By analogy, NAPP [19] uses inverted files, but adds compression of the posting lists. Recently, another pivot-permutation-based technique is the PPP-codes [15], where more independent pivot sets are used to make filtering more efficient. The authors also compress the prefixes by applying Trie prefix trees to optimize memory use. Since these techniques are the most relevant for our proposal, we describe their common principle in the following subsection.

An interesting study on comparing and ordering pivot permutations was presented recently [11]. They improve Spearman Footrule to penalize by a power factor α the permutations that vary in more pivots. This is relevant to our approach too, so we presented it in Sect. 2.2 in detail.

Another stream of research focuses on employing other models to further optimize filtering by creating compact representations of original data objects. Binary sketches were proposed recently [12], where a system of independent hyperplanes is converted to a bit-string. It leads to very efficient and effective filtering with Hamming distance. To speed up search on secondary memory, the sketches must be stored apart from the data objects. Even though the representation is very compact it induces further new I/O operations. They might be kept in memory, so reading buckets with data objects can be optimized.

2.1   Backgrounds and Indexing of Pivot Permutations

A *metric space* $\mathcal{M} = (\mathcal{D}, d)$ is defined over a universe $\mathcal{D}$ of data objects and a distance function $d(\cdot, \cdot)$ that satisfies metric postulates. A database $X \subset \mathcal{D}$ of data objects is organized to speed up similarity query evaluation: (i) the *k-nearest neighbors query* $kNN(q) = \{A : A \subseteq X, |A| = k \land \forall a \in A, o \in X - A : d(q, a) \le d(q, o)\}$, (ii) the *range query* $R(q, r) = \{o \in X : d(q, o) \le r\}$. For the reader's convenience, we summarize the complete notation in Table 1.

Pivot-permutations-based index structures are constructed in four steps: (i) select a set of anchor objects, called pivots; (ii) compute their permutation for each database object (order them by distance to the object), (iii) divide the objects into partitions having the same permutation prefix and (iv) store them into buckets. The details are given in the following definitions.

A set of l pivots $P = \{p_1, ..., p_l\}$ is typically taken at random from the database, where the highly correlated objects are dropped. It follows the data distribution of database objects and is cheap and well-performing. These pivots are fixed for the life-time of index structure. The indexing phase organizes the database by computing a *pivot permutation* for each object $o \in X$:

$$pp(o) = \langle p_o^{[1]}, p_o^{[2]}, \ldots, p_o^{[l]} \rangle,$$

where $p^{[1]}$ is the pivot closest to o (having the smallest value of $d(o, p), \forall p \in P$), $p^{[2]}$ is the second closest pivot, and so on. The bucket b to store the object into is identified by taking the prefix π of $pp(o)$, which we denote as *pivot permutation prefix*:

$$ppp(o) = \langle p_o^{[1]}, p_o^{[2]}, \ldots, p_o^{[\pi]} \rangle.$$

Table 1. Notation

$\mathcal{M}(D, d)$	A metric space over a data domain $\mathcal{D}$ and a pair-wise distance function $d(\cdot, \cdot)$
$X = \{o_1, \dots, o_n\}$	The database of objects o_i; $X \subset \mathcal{D}$, $|X| = n$
$P = \{p_1, \dots, p_l\}$	A set of preselected pivots; $P \subset X$, $|P| = l$
$d(o_1, o_2)$	Distance between two objects o_1 and o_2
$wd(o, p)$	Weighted distance between an object o and a pivot p
$p_o^{[i]}$	i^{th} closest pivot to an object o
$rd(o) = p_o^{[1]} / p_o^{[2]}$	Ratio of distance of o to the closest pivot and to the second closest pivot
$pp(o) = \langle p_o^{[1]}, \dots, p_o^{[l]} \rangle$	Permutation of all pivots for an object o
$ppp(o) = \langle p_o^{[1]}, \dots, p_o^{[\pi]} \rangle$	The π-long prefix of $pp(o)$; $|pp(o)| = \pi$
$ppp(o)[i], pp(o)[i]$	Pivot at the index i in the permutation (also without argument o)
$W = \{W_{\langle \rangle}, W_{\langle p_1 \rangle}, \dots\}$	A set of pivots weights, each specific to a permutation prefix
$W_{\langle p^{[1]}, \dots, p^{[m]} \rangle}$	Pivot weights corresponding to ppp in subscript, consists of $|P|$ floats

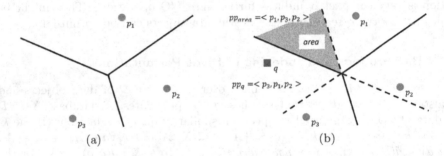

Fig. 1. Voronoi cells for pivots p_1, p_2, p_3: (a) 1-level tessellation, (b) pivot permutations.

If we take the prefix length of 1, we get the Voronoi tessellation [3]. An illustration is provided in Fig. 1.

The major issue of this schema is choosing the right value of π since it defines the coarseness of partitioning. It correlates with the maximum number of data partitions. As stated in [18], the number of partitions is much lower for high-dimensional data than the total number of possible permutations $P(l, \pi)$. High prefix values lead to many almost-empty buckets, thus the partitioning is very unbalanced. The structures surveyed above set π to 7 or 8, typically. It still results in many under-occupied buckets, so M-index [14] uses a dynamic prefix $\leq \pi$ to stop splitting under-occupied buckets prematurely. The down side is that a navigation tree is needed to track the prefixes, which also leads to larger memory requirements.

2.2 Searching of Pivot Permutations

Answering a similarity k-nearest neighbor query is a process of identifying data partitions that contain objects closest to the query object q. The index structure has to construct a priority queue of such promising data partitions, so their permutation prefixes are compared. For illustration of a query object and an area covering a partition, see Fig. 1b. The traditional approaches are Spearman Footrule, Spearman Rho and Kendall Tau metrics [11].

Assume a query object, a data partition and their pivot permutations $pp(q)$ and pp_{area}, Spearman Footrule is defined as the sum of differences in pivot positions in the permutations:

$$S_F(pp(q), pp_{area}) = \sum_{p \in P} \mid \mathrm{idx}(p, pp(q)) - \mathrm{idx}(p, pp_{area}) \mid,$$

$$\mathrm{idx}(p, pp(o)) = i, \text{ where } p_o[i] = p.$$

The optimization presented in [11] amplifies larger index differences by an exponent $\alpha > 1$. It is sufficient to assume that two objects are distant in the metric space if there is just one pivot positioned far away in their permutations. Kendall Tau is based on counting the number of inversions in pivot positions.

$$K_\tau(pp(q), pp_{area}) = \sum_{p_i, p_j \in P} K_{p_i, p_j}(pp(q), pp_{area}),$$

$$K_{p_i, p_j}(pp_1, pp_2) = \begin{cases} 0, \text{if } \mathrm{idx}(p_i, pp_1) < \mathrm{idx}(p_j, pp_1) \Leftrightarrow \mathrm{idx}(p_i, pp_2) < \mathrm{idx}(p_j, pp_2) \\ 1, \text{otherwise.} \end{cases}$$

The effectiveness of these metrics depends highly on the number of pivots we use. For few pivots, they cannot catch the original distance precisely, so at least tens or hundreds of pivots are necessary. These metrics can also be applied to prefixes only [10].

In [15], it was shown that Spearman Footrule and Kendall Tau perform a little worse than the metrics that weigh original distances between the query object and the pivot with decreasing importance as we move to higher positions in the permutation prefixes. The priority queues in M-index and PPP-codes are organized by:

- **Weighted Sum of Pivot Distances (WSPD).** It computes a weighted sum of distances from the query object to the pivots in a partition's pivot permutation prefix (ppp_{area}) as follows:

$$\mathrm{WSPD}(pp(q), ppp_{area}) = \sum_{i=0}^{|ppp_{area}|} d(q, ppp_{area}[i]) * 0.75^i$$

- **Sum of Differences between Pivot Distances (SDPD).** It sums differences between the distance of pivot in the query object's permutation and the distance of pivot in the partition's permutation prefix:

$$\mathrm{SDPD}(pp(q), ppp_{area}) = \sum_{i=0}^{|pp_{area}|} max(0, \ d(q, ppp_{area}[i]) - d(q, ppp(q)[i])).$$

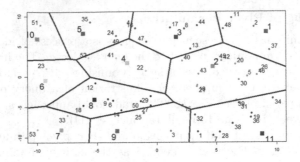

Fig. 2. Voronoi partitioning defined by the pivots denoted by large-font numbers; the other dots are data objects.

To sum up, the pivot-permutation-based indexing techniques suffer from unbalanced partitioning, where many data partitions are under-occupied or even empty, and some are filled with considerably high number of objects. Even though the querying performance is very good, this issue still influences negatively their potential when persisted to secondary memory.

3   Balanced Indexing with Weighted Voronoi Partitioning

In this section, we propose an indexing structure that adopts the weighted Voronoi tessellation [2] for data partitioning. We provide an algorithm to construct it and show its correctness. This principle can be used in any data organization based on pivot permutations, e.g. the M-index [14] and PPP-codes [15].

3.1   Weighted Voronoi Partitioning in Metric Space

Organizing data using Voronoi partitioning depends on the pivots used to define the cells. In metric space, we are not allowed to optimize the position of pivots, since artificial objects cannot be computed anyway [20]. As a result, uneven distribution is very common (see Fig. 2). This problem is escalated at the next levels of recursive Voronoi partitioning for pivot permutations.

We propose the *recursive weighted Voronoi partitioning for metric spaces* that builds on the idea of weighted Voronoi tessellation [2]. A preselected set of pivots P is used in the same way as presented in Sect. 2.1, i.e. omitting already used pivots at the next levels of partitioning. However, we introduce real-value weights that are associated with pivots. This allows us to make a data-driven split, as is exemplified in Fig. 3, where the grey dashed lines denote the original (non-weighted) Voronoi partitioning.

On the first level, we have a vector of weights $W_{<>} = [w_{p_1}, \ldots, w_{p_n}]$ (one weight per pivot), where $w_{p_i} \geq 1$ and the subscript $W_{\langle\rangle}$ stands for the zero-length permutation prefix. Any data object o is then associated with the cell defined by the pivot p_i whose *weighted distance* is shortest, i.e. $\forall p_j \in P : d(o, p_j) \cdot w_{p_j} \geq d(o, p_i) \cdot w_{p_i}$. The uneven distribution is balanced by shrinking overloaded areas,

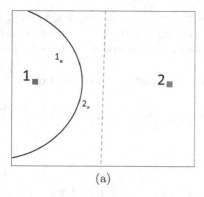
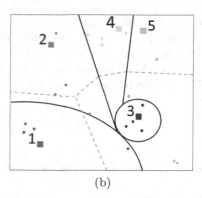

(a)   (b)

Fig. 3. Weighted Voronoi partitioning: (a) for two pivots, and (b) for five pivots; the gray dashed lines delimit cells of Voronoi partitioning whereas the black curves define cells of weighted Voronoi partitioning. Pivots denoted in the large font.

which corresponds to increasing weights of the pivots defining them. Formally, the weighted distance is defined as:

$$wd(o,p) = d(o,p) \cdot w_p \tag{1}$$

On the next level, the same principle is repeated, but we use another vector of weights. Assume we have identified up to now m closest pivots by weighted distances, i.e. the current pivot permutation prefix is $ppp_{aux}(o) = \langle p_o^{[1]}, \ldots, p_o^{[m]} \rangle$. The next pivot is obtained from the remaining ones using the weights associated with this particular prefix, i.e., $W_{ppp_{aux}(o)}$, and the prefix is extended by one pivot. This is repeated until the complete prefix of length π is obtained.

The total number of weight vectors corresponds to the number of nodes in a full l-ary tree: $\frac{l^\pi - 1}{l - 1}$ ($l = |P|$). Since there are few levels, we do not use sparse arrays to store the weights. This results in $l\frac{l^\pi - 1}{l - 1}$ float values being maintained.

3.2   Setting Weights

We start with a motivating example for the case of two pivots (p_1, p_2) and two objects (o_1, o_2) as given in Fig. 3a. The original Voronoi partitioning assigns both objects to p_1. For balanced partitioning, we need to set the weight of pivot p_1 to move the further object o_2 to the other partition, so that the conditions $wd(o_1, p_1) < wd(o_2, p_2)$ and $wd(o_2, p_1) > wd(o_2, p_2)$ hold after the relocation. In metric spaces, it is not possible to measure the distance to the dividing hyperplane exactly, because it is not defined analytically. However, such a measure is necessary to identify and order the objects that are close to the cells' boundary, because these are relocated as first.

We propose the *relative distance* of an object to the hyperplane as the rank of objects for relocation. It is a ratio of distances to the first and the second closest pivots, i.e.

$$rd(o) = \frac{wd(o, p^{[1]})}{wd(o, p^{[2]})} \tag{2}$$

Note that $0 \leq rd(\cdot) \leq 1$ for objects within the cell of $p^{[1]}$, and $1 < rd(\cdot) < \infty$ for objects outside of it. If we order all objects within the cell by their relative distance, we obtain the order in which the objects are moved to other cells. From Fig. 3a, it is clear that $rd(o_1) < rd(o_2)$.

The update to the weight of p_1 that moves o_2 to the cell of p_2 is defined as follows:

$$w_{p_1} = \frac{1}{\mathbf{avg}(rd(o_1), rd(o_2))} \tag{3}$$

Lemma 1. *The object o_2 is assigned to the cell of p_2 now.*

Proof. The weights w_{p_1}, w_{p_2} were initialized to one, so updated weight w_{p_1} is defined using the real distances only:

$$w_{p_1} = \frac{2 \cdot d(o_1, p_2) \cdot d(o_2, p_2)}{d(o_1, p_1) \cdot d(o_2, p_2) + d(o_2, p_1) \cdot d(o_1, p_2)}$$

We will show that $wd(o_2, p_2) < wd(o_2, p_1)$ holds after the update, by contradiction:

$$wd(o_2, p_2) \geq wd(o_2, p_1)$$
$$d(o_2, p_2) \geq \frac{2 \cdot d(o_1, p_2) \cdot d(o_2, p_2) \cdot d(o_2, p_1)}{d(o_1, p_1) \cdot d(o_2, p_2) + d(o_2, p_1) \cdot d(o_1, p_2)}$$
$$1 \geq \frac{2 \cdot d(o_1, p_2) \cdot d(o_2, p_1)}{d(o_1, p_1) \cdot d(o_2, p_2) + d(o_2, p_1) \cdot d(o_1, p_2)}$$
$$d(o_1, p_1) \cdot d(o_2, p_2) + d(o_2, p_1) \cdot d(o_1, p_2) \geq 2 \cdot d(o_1, p_2) \cdot d(o_2, p_1)$$
$$d(o_1, p_1) \cdot d(o_2, p_2) \geq d(o_1, p_2) \cdot d(o_2, p_1)$$
$$\frac{d(o_1, p_1) \cdot d(o_2, p_2)}{d(o_1, p_2) \cdot d(o_2, p_2)} \geq \frac{d(o_1, p_2) \cdot d(o_2, p_1)}{d(o_1, p_2) \cdot d(o_2, p_2)}$$
$$rd(o_1) \geq rd(o_2)$$

This is a contradiction to the assumption that we are relocating the object with higher relative distance. ☐

By analogy, we can show that o_1 stays in the cell of p_1. In the following section, we present an algorithm for settings weights gradually.

3.3   Balancing Cells

Weighting Voronoi partitioning is an iterative process that picks the most populated cell and expels as many objects as prescribed by an occupation constraint. The objects moved from the cell to neighboring cells are identified using the relative distance. We take all objects closest to the cell's borders and update the weight based on Eq. 3, where o_1 stands for the first object staying in the

Algorithm 1. Balancing cells

Require: Weighted Voronoi partitioning $(\mathcal{V})$; weights (W); max. occupation (lim)
  1: **while** $C \leftarrow$ the most populated cell in $\mathcal{V} \wedge |C| > lim$ **do**
  2:     $ppp \leftarrow$ pivot permutation prefix of the cell C
  3:     $p \leftarrow ppp[-1]$ {last pivot in the prefix}
  4:     $W_C \leftarrow W_{\langle ppp[0],\dots,ppp[-2]\rangle}$ {weights for this cell; last pivot is removed from ppp}
  5:     $O \leftarrow$ list of objects in C ordered by increasing $rd(\cdot)$
  6:     {Mark two objects at the split border:}
  7:     $o_1 \leftarrow O[lim]$; $o_2 \leftarrow O[lim+1]$ {$rd(o_1) < rd(o_2)$ holds}
  8:     {Update the weight of p, so o_1 remains and o_2 is the first object moved away:}
  9:     $W_C[p] = \frac{W_C[p]}{\mathbf{avg}(rd(o_1),rd(o_2))}$
  10:    Move objects $O[lim \mathrel{..} |O|]$ to the neighboring cells
  11: **end while**

cell and o_2 for the last being moved out. The correctness of this step, as well as convergence of the algorithm, is discussed in the following two subsections.

The algorithm for redistributing objects which exceed the given capacity is formulated in Algorithm 1. It proceeds until all cells comply with the capacity constraint. The overflowing objects are moved away (line 10), because the weight of the cell's pivot has been increased. This step involves finding the new closest pivot for each moved object. From the local point of view, it is the next pivot in the permutation. However, the weight of such a pivot could have been updated in some other iteration of the algorithm, so the object's pivot permutation must be updated. Since only the weights get updated, the original distances to pivot cannot change. This update is localized to the suffix of the permutation.

3.4   Consistency of One Step in Weight Modification

The update of weight done in line 9 of the algorithm may arouse concerns about unintended shifts of objects between pivots. As the weight of one pivot is increased in this step, no object from any neighboring area can be shifted into the cell. At the same time, the weighted distances of objects within the edited cell must necessarily *increase*.

Lemma 2. *Assume that objects o_1 and o_2 are part of the cell of pivot p and their relative distances hold:*

$$rd(o_1) < rd(o_2) \leq 1. \tag{4}$$

The weight of p is updated to $w'_p > w_p$ and the object o_1 is pushed out from the cell, while o_2 remains there.

Proof. By contradiction, the new relative distances for the objects hold:

$$\frac{d(o_1,p) \cdot w'_p}{d(o_1,p_i) \cdot w_{p_i}} > \frac{d(o_2,p) \cdot w'_p}{d(o_2,p_j) \cdot w_{p_j}}$$

Algorithm 2. Building BM-index.

Require: Set of pivots (P); maximum bucket capacity (cap); data-set (X)
1: $W \leftarrow \{W_{\langle\rangle} = [1 * |P|]\}$ {initialize root weights to ones ($|P|$ in total)}
2: root $\leftarrow$ split$(P, \text{cap}, \langle\rangle, X, W)$ {empty ppp for root node; call Algorithm 3}
3: **return** bmindex(root,P,W)

The update step in line 9 can be simplified to $\exists \delta > 1 : w_p' = w_p \cdot \delta$, so the previous formula is rewritten to:

$$\frac{d(o_1,p) \cdot w_p \cdot \delta}{d(o_1,p_i) \cdot w_{p_i}} > \frac{d(o_2,p) \cdot w \cdot \delta}{d(o_2,p_j) \cdot w_{p_j}}$$

$$\frac{d(o_1,p) \cdot w_p}{d(o_1,p_i) \cdot w_{p_i}} > \frac{d(o_2,p) \cdot w}{d(o_2,p_j) \cdot w_{p_j}}$$

Since the weights w_i, w_j as well as δ and all distances $d(\cdot, \cdot)$ are constants in this update step, we have got the contradiction with Eq. 4. $\qquad\square$

Lastly, in Sect. 3.2, we condition an ability to set a correct weight by the strict non-equality of relative weights of the bordering objects. In practice, the relative weights of two objects can be equal. The weight update should therefore reflect this by yet again modifying the weight computation in line 9 as:

$$w_p = \frac{w_p}{\max(\mathbf{avg}(rd(o_1), rd(o_2)), rd(o_1) + \epsilon)}, \tag{5}$$

where ϵ is equal to the minimum positive value of the given data type.

If the pivot p in the lemma can be replaced by the last pivot of the cell's permutation prefix, it holds for recursive weighted Voronoi partitioning as well.

3.5   Convergence of the Balancing Algorithm

Let us consider two neighboring areas, both exceeding the occupation constraint, where their occupations differ by one object only. The question is whether Algorithm 1 could cycle in exchanging the single object infinitely. Since the weights are increased strictly monotonously, another object will be pushed towards a third cell after several iterations. This happens even if there is a longer loop of cells via object "commutes". Thus the algorithm terminates (assuming the occupation limit is reasonable, i.e. $\#_{buckets} * \text{occupation} > |X|$).

3.6   Indexing with Recursive Weighted Voronoi Partitioning

To complete the description of the proposed BM-index, we add building and search algorithms. The building algorithm uses the permutation prefixes defined earlier and balancing is done on a per-level basis, see Algorithms 2 and 3. The presented search algorithm (Algorithm 4) evaluates k-nearest neighbors queries

---

Algorithm 3. Splitting a node.

Require: Set of pivots (P); maximum bucket capacity (cap); permutation (ppp); data-set (X); weights (W)

1: **if** $|X| \leq$ cap **then**
2:     **return** leaf(ppp, X)   {leaf node with permutation and bucket with objects}
3: **end if**
4: $W \leftarrow W \cup \{W_{ppp} = [1 * |P|]\}$ {initialize weights to ones ($|P|$ in total)}
5: $\mathcal{V} \leftarrow$ voronoi(X,P)   {partition data by the pivots}
6: balance($\mathcal{V}$, W, $\lceil \frac{|X|}{|P|} \rceil$)   {call Algorithm 1}
7: children $\leftarrow \emptyset$
8: **for all** $p \in P$ **do**
9:     $C \leftarrow$ set of objects in the cell defined by p
10:     children $\leftarrow$ children $\cup$ split(P, cap, $\langle$ppp $\cdot p\rangle$, C, W)   {append pivot p to ppp}
11: **end for**
12: **return** node($\langle ppp \rangle$, children)   {ppp, list of pointers to children}

---

in an approximate manner, so it also passes the constraint on the number of distance function calls. The evaluation terminates when this constraint is reached. On line 11, the distance between two pivot permutations is estimated by $WSPD$, but any other metric can be used.

4   Efficiency Evaluation

We examine the performance of our method in three parts. Section 4.2 provides an overview of the efficiency when changing the number of pivots and levels. The costs of balancing are reviewed in Sect. 4.3. Lastly, we compare the performance of the our method with M-index in Sect. 4.4.

4.1   Setup

The CoPhIR data-set [4] used in all experiments is a collection of 282-dimensional vectors, where each was obtained by concatenating five MPEG-7 global visual descriptors extracted from an image. The distance function is a weighted sum of L_1 and L_2 metrics on the corresponding descriptors. We used two sizes of the data-sets: 100,000 and 1,000,000 objects, denoted as 100K and 1M, respectively. Pivots were selected at random from the data-set since they provide sufficient performance [15].

The query evaluation efficiency was measured on 30-nearest-neighbors queries. We used 1,000 query objects that were selected from the 100K data-set at random. The preset stop-condition of approximate evaluation was the number of distance calculations and it included also the distance computations from the query object to the pivots.

4.2   Querying Performance

We summarize the performance of k-nearest-neighbors queries with respect to the data-set size, the number of pivots used to build the BM-index, and two

---

Algorithm 4. Evaluating kNN query.

Require: BM-index (idx); kNN query (q, k); approx. limit (max)
  1: $queue.push(idx.root, 0)$ {priority queue; root has zero (max) priority}
  2: $pp(q) \leftarrow perm(q, idx.P, idx.W)$ {calculate full pivot permutation}
  3: $dc \leftarrow |idx.P|$ {distance computations counter}
  4: $A \leftarrow \emptyset$ {candidates for answer; ordered by distance from q; max. k objects}
  5: **while** $node \leftarrow queue.pop() \wedge dc < max$ **do**
  6:    **if** node is a leaf **then**
  7:       $update(A, node.X)$ {add objects closer than current k^{th} neighbor}
  8:       $dc \leftarrow dc + |node.X|$
  9:    **else**
 10:       **for all** $ch \in node.children$ **do**
 11:          $queue.push(\langle node.ppp \cdot ch.p\rangle, \mathrm{WSPD}(pp(q), \langle node.ppp \cdot ch.p\rangle))$
 12:       **end for**
 13:    **end if**
 14: **end while**
 15: **return** A {final answer; ranked by distance from q}

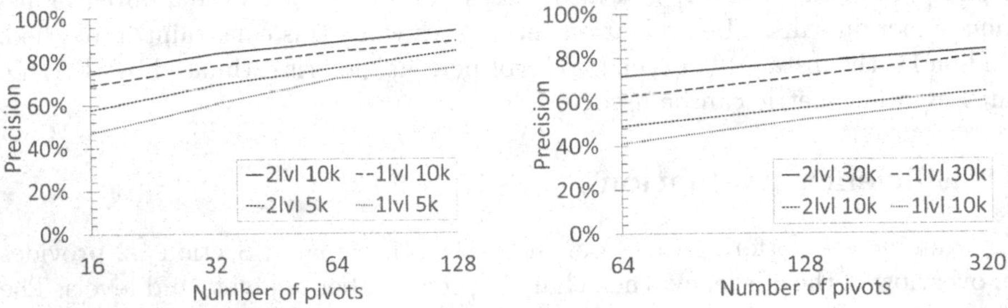

(a) 100K

(b) 1M

Fig. 4. BM-index: average precision of approximate 30NN queries for 100K and 1M data-sets with varying number of pivots (x-axis). The curves are denoted by the number of levels of BM-index and the approximation parameter (stop condition).

different depths of it in Fig. 4. For the priority queue in the query evaluation algorithm, we tested three methods presented in Sect. 2, namely Kendall Tau, WSPD and PSPD. However, since the WSPD method performed slightly better than the others, so we present the result for this method only.

The results show improving precision for increasing number of pivots because the data is split in more smaller parts, which is also supported by better performance of two-level configurations. The number of pivots influences the performance for small pivot sets mainly, where 2-level configuration over 16 pivots (240 buckets in total) exhibits worse performance than 1-level one over 128 pivots – 74% vs. 89% on 100K.

4.3   Construction Costs

The balancing algorithm (Sect. 3.3) is designed as post-processing of cells created by Voronoi partitioning. In Table 2, we compare the overhead of balancing

Table 2. Building costs of 1-level Voronoi part. on 100K data-set and balancing costs.

Number of pivots	Original Voronoi		Balancing process		
	Build time (s)	Distance computations	Time (s)	Optim. steps	Objects moved
16	17	1 600 120	83	3 007	197 950
32	31	3 200 496	71	5 770	226 424
64	65	6 402 016	66	9 047	300 013
128	179	12 808 128	51	15 681	329 904
256	294	25 632 640	65	32 486	416 586
512	467	51 330 816	91	52 667	445 902
1 024	914	102 923 776	140	82 494	577 185

to the original costs in wall-clock time. The most of the time is needed to compute distances to pivots, i.e., to construct the Voronoi tessellation. For higher numbers of pivots, the Voronoi cells become less populated and the balancing time becomes marginal. The costs are presented also in the number of balancing algorithm iterations (steps) and the total number of exchanged objects. The experiments were conducted on the same hardware and the values are averages over five runs. The result clearly shows that costs to compute pivot permutations form the prevalent part, and the balancing process is efficient.

In Fig. 5a, we present occupation of buckets in leaf nodes from unbalanced to balanced structure of one-level BM-index over 1,024 pivots. In particular, we show the buckets ordered by their occupation for the BM-index's state before running balancing, after balancing each leaf node exactly once, after balancing to the max occupation of twice the average occupation, and after balancing to the average occupation (98 objects). This shows the extreme imbalance of the data structure, which negatively influences the querying performance, since all the large cells are scanned completely. After balancing, the distribution is more or less constant. There are a few under-occupied cells, which exist due to ceiling the value of the stop condition to an integer.

The values of pivot weights for the same states are presented in Fig. 5b. We can observe that for the fully balanced structure, 95% of the weights is less than 1.5, and 50% of them is even under 1.25. Next, one-pass balancing was applied to roughly two thirds of the leaf nodes, which is 627 optimization steps (0.8% of complete balancing). The number of objects moved between cells was 84,988, which corresponds to 14.8% of the required exchanges for a fully balanced structure. When the balancing stop condition was set to twice the average occupancy, we observed it to be very efficient as well. It required 1,114 balancing steps (1.4% of complete balancing) and 35,012 object exchanges only.

The main benefit of the looser approaches is that they remove large discrepancies (tens of percents) in bucket occupation for marginal fraction of modifications needed for complete balancing. Figure 6 shows details on trends of exchanged objects during balancing in order to provide an insight into possibilities of

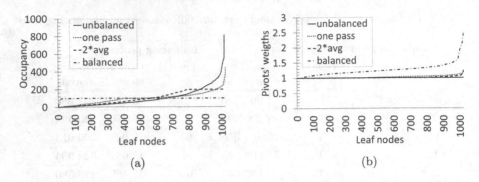

Fig. 5. Influence of balancing from unbalanced to balanced state: (a) on occupation of leaf nodes, and (b) on weights of pivots. (100K, 1,024 pivots, 1-level)

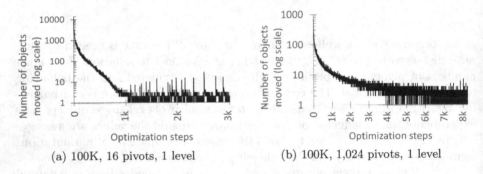

Fig. 6. The number of objects moved out from an area during the balancing process.

setting the stop condition for partial balancing. Setups with 16 and 1,024 pivots are used, where the average occupancy was 7 and 98 objects, respectively. The graphs show that most effort is spent to polish balancing.

4.4 Overall Efficiency of the Proposed Algorithm

In the last set of experiments, we compare BM-index with M-index in precision of 30NN. Figure 7 and Table 3 present results for a one- and a two-level BM-index and M-index with the maximum limit of 8 levels to organize the 100K data set. Leaf node capacity was fixed to 1,564 and 26 objects. We increased the approximation limit gradually and plot the precision and the number of accessed leaf nodes. In Table 3, the average time to answer a query for these limits is shown.

We can see that the M-index has consistently lower demands than BM-index in terms of distance computations. However, our balanced index is significantly more efficient in accessing the secondary memory. It saves up to 85% leaf nodes compared to M-index. For example, 1-level BM-index has precision of around 90–95% at accessing 15 leaves only, where M-index accessed 100 leaf nodes. In Table 3, we can observe that the overall performance is similar in case of

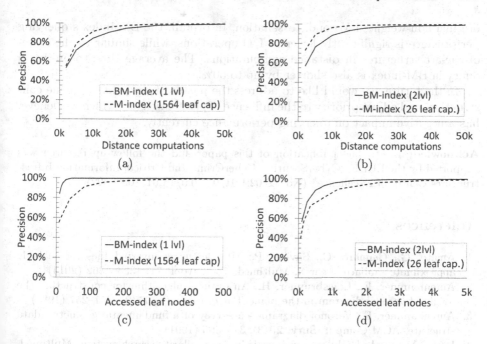

Fig. 7. Comparison of M-index and proposed BM-index in precision of 30NN queries on 100K data-set and 64 pivots: (a,c) 1-level, 1,564 leaf node capacity, and (b,d) 2-level, 26 leaf node capacity.

Table 3. Comparison of M-index and BM-index performance measured as an average time to answer a query in milliseconds.

Distance computations	1 000	5 000	10 000	15 000	20 000
BM-index (1lvl)	20.6	66.8	104.1	195.2	237.5
M-index (1564 leaf cap)	34.6	78.6	125.3	175.6	219.4
BM-index (2lvl)	-	79.5	132.8	194.6	257.1
M-index (26 leaf cap)	133.0	187.7	267.8	400.1	493.9

the setup with greater leaf node capacity. On the other hand, the BM-index outperforms the M-index in the lower leaf node capacity setup, respectively, when the indexing structure has more levels.

5   Conclusions and Future Work

We proposed the BM-index that distributes data equally among buckets by weighting pivot permutations. The balancing algorithm is implemented as the post-processing step that equals the cells' occupation by introducing weighted Voronoi partitioning into metric spaces. We showed its correctness and analyzed its costs. They are reasonably low with respect to the building costs of the original non-weighted Voronoi tessellation. In overall, the BM-index's querying performance is significantly lower in I/O operations, while similar to the state-of-the-art structures in distance computations. The average time to answer a query in BM-index is also shorter by up to 50%.

In the future, we would like to address the means of the priority queue construction, and adjust priority to the influence of pivots' weights, which we believe, has a negative impact on querying performance currently.

Acknowledgment. The publication of this paper and the follow-up research was supported by the ERDF "CyberSecurity, CyberCrime and Critical Information Infrastructures Center of Excellence" (No.CZ.02.1.01/0.0/0.0/16_019/0000822).

References

1. Amato, G., Gennaro, C., Savino, P.: MI-File: using inverted files for scalable approximate similarity search. Multimed. Tools Appl. **71**, 1333–1362 (2014)
2. Aurenhammer, F., Edelsbrunner, H.: An optimal algorithm for constructing the weighted voronoi diagram in the plane. Pattern Recogn. **17**(2), 251–257 (1984)
3. Aurenhammer, F.: Voronoi diagrams - a survey of a fundamental geometric data structure. ACM Comput. Surv. **23**(3), 345–405 (1991)
4. Batko, M., et al.: Building a web-scale image similarity search system. Multimed. Tools Appl. **47**(3), 599–629 (2009)
5. Brin, S.: Near neighbor search in large metric spaces. In: Proceedings of the 21th International Conference on Very Large Data Bases, VLDB 1995, pp. 574–584. Morgan Kaufmann Publishers Inc., San Francisco (1995)
6. Chávez, E., Navarro, G.: A probabilistic spell for the curse of dimensionality. In: Buchsbaum, A.L., Snoeyink, J. (eds.) ALENEX 2001. LNCS, vol. 2153, pp. 147–160. Springer, Heidelberg (2001). https://doi.org/10.1007/3-540-44808-X_12
7. Deepak, P., Prasad, M.D.: Operators for Similarity Search: Semantics, Techniques and Usage Scenarios. Springer, Heidelberg (2015). https://doi.org/10.1007/978-3-319-21257-9
8. Esuli, A.: MiPai: using the pp-index to build an efficient and scalable similarity search system. In: Second International Workshop on Similarity Search and Applications, SISAP 2009, 29–30 2009, Prague, Czech Republic, pp. 146–148 (2009)
9. Esuli, A.: Use of permutation prefixes for efficient and scalable approximate similarity search. Inform. Process. Manage. (IPM) **48**(5), 889–902 (2012)
10. Fagin, R., Kumar, R., Sivakumar, D.: Comparing top k lists. SIAM J. Discrete Math. (2003). https://doi.org/10.1137/s0895480102412856
11. Figueroa, K., Paredes, R., Reyes, N.: New permutation dissimilarity measures for proximity searching. In: Marchand-Maillet, S., Silva, Y.N., Chávez, E. (eds.) SISAP 2018. LNCS, vol. 11223, pp. 122–133. Springer, Cham (2018). https://doi.org/10.1007/978-3-030-02224-2_10
12. Mic, V., Novak, D., Zezula, P.: Binary sketches for secondary filtering. ACM Trans. Inform. Syst. **36**(5), 4:1–4:30 (2018)
13. Naidan, B., Boytsov, L., Nyberg, E.: Permutation search methods are efficient, yet faster search is possible. Proc. VLDB Endow. **8**(12), 1618–1629 (2015)
14. Novak, D., Batko, M., Zezula, P.: Metric index: an efficient and scalable solution for precise and approximate similarity search. Inform. Syst. **36**, 721–733 (2011)

15. Novak, D., Zezula, P.: PPP-codes for large-scale similarity searching. In: Hameurlain, A., Küng, J., Wagner, R., Decker, H., Lhotska, L., Link, S. (eds.) Transactions on Large-Scale Data- and Knowledge-Centered Systems XXIV. LNCS, vol. 9510, pp. 61–87. Springer, Heidelberg (2016). https://doi.org/10.1007/978-3-662-49214-7_2

16. Paredes, R.U., Navarro, G.: EGNAT: a fully dynamic metric access method for secondary memory. In: Second International Workshop on Similarity Search and Applications, SISAP 2009, 29–30 2009, Czech Republic, pp. 57–64 (2009)

17. Samet, H.: Foundations of Multidimensional and Metric Data Structures. The Morgan Kaufmann Series in Data Management Systems. Morgan Kaufmann, Burlington (2006)

18. Skala, M.: Counting distance permutations. J. Discrete Algorithms **7**(1), 49–61 (2009). https://doi.org/10.1016/j.jda.2008.09.011. Selected papers from the 1st International Workshop on Similarity Search and Applications

19. Tellez, E.S., Chavez, E., Navarro, G.: Succinct nearest neighbor search. Inform. Syst. **38**(7), 1019–1030 (2013). https://doi.org/10.1016/J.IS.2012.06.005

20. Zezula, P., Amato, G., Dohnal, V., Batko, M.: Similarity Search: The Metric Space Approach. Advances in Database Systems, vol. 32. Springer, Heidelberg (2005). https://doi.org/10.1007/0-387-29151-2

Theoretical Foundation and New Requirements

ProSA—Using the CHASE
for Provenance Management

Tanja Auge[✉] and Andreas Heuer[✉]

University of Rostock, 18051 Rostock, Germany
{tanja.auge,andreas.heuer}@uni-rostock.de
https://dbis.informatik.uni-rostock.de

Abstract. Collecting, storing, tracking, and archiving scientific data is
the main task of research data management, being the basis for scientific evaluations. In addition to the evaluation (a complex query in the
case of structured databases) and the result itself, the important part
of the original database used has also to be archived. To ensure repro-
ducible and replicable research, the evaluation queries can be processed
again at a later point in time in order to reproduce the result. Being able
to calculate the origin of an evaluation is the main problem in provenance management, particularly in *why* and *how* data provenance. We
are developing a tool called ProSA which combines data provenance and
schema/data evolution using the CHASE for the different database transformations needed. Besides describing the main ideas of ProSA, another
focus of this paper is the concrete use of our CHASE tool ChaTEAU for
invertible query evaluation.

Keywords: Theoretical foundations of databases · Data curation ·
Annotation · Provenance · Temporal databases · CHASE

1 Introduction

Collecting, evaluating, analyzing, archiving, and publishing research data are
the main tasks of research data management. Research institutes all over the
world are producing research data in huge amounts. The processing, analysis,
and storage of a huge amount of data, which is usually generated in research,
can be considerably supported by the use of *data provenance*. With the combination of the CHASE – a universal tool for transforming databases or queries
in database systems – and data provenance, a minimal sub-database of an original research dataset can be computed which is one of the main problems in
minimizing research data [6]. So questions like (1) Where does the data come
from? (2) Why this result? and (3) How is the result being computed? can be
answered in research data management [13,28]. As a use case, we apply our
results in research data management and data provenance within a joint project
of the University of Rostock and the Leibniz Institute for Baltic Sea Research
Warnemünde (IOW) [11,12].

© Springer Nature Switzerland AG 2019
T. Welzer et al. (Eds.): ADBIS 2019, LNCS 11695, pp. 357–372, 2019.
https://doi.org/10.1007/978-3-030-28730-6_22

In this paper, we will introduce our data provenance tool called ProSA, a tool for combining data provenance, schema and data evolution by means of the CHASE in the case of invertible query evaluation (Sect. 5). The CHASE as a basic tool is implemented in ChaTEAU (Sect. 4). Besides the presentation of the general techniques behind ChaTEAU, we will introduce different applications of the CHASE, all being research topics within our Database Research Group at the University of Rostock (DBIS). These applications are *invertible query evaluation* (Sect. 3), *semantic query optimization* and *answering queries using operators* (Sect. 6). First, however, basic Notions and the State of the Art in the research areas research data management, CHASE and data provenance are introduced in the following Sect. 2.

2   Basic Notions and State of the Art

In our data provenance tool ProSA, we are using the CHASE technique to represent all of the database transformations needed for research data management, specifically evaluation queries against the original research data, schema and data evolution operations, as well as data exchange processes (between different research repositories). Therefore, we first introduce the CHASE as a general tool for different database problems. We close this section by introducing the main concepts of data provenance.

2.1   The CHASE Algorithm

Semantic query optimization, answering queries using views (AQuV), data exchange and integration, cleaning as well as invertible query evaluation are basic problems in database theory that can be solved using the CHASE. For this reason, different versions of the algorithm itself as well as various tools that implement or execute the CHASE have been developed over time. Each tool such as LLUNATIC, PDQ or ProvCB is specifically designed for one area of application like data cleaning, semantic optimization and AQuV (see Table 1).

Hence, the idea of this universal tool can be summarized as follows: For an object $\bigcirc$ (e.g. a database, or a query) and a set of dependencies $\star$ (e.g. a set of FDs and JDs) the CHASE incorporates $\star$ into $\bigcirc$, so that $\star$ is implicitly contained in $\bigcirc$. We represent this by:

$$\text{chase}_\star(\bigcirc) = \circledast.$$

Incorporating parameters $\star$ into an object $\bigcirc$ can mean different things. We therefore distinguish between two cases:

1. When a CHASE incorporates dependencies into a query, all interpretations of variables of a query (e.g., in a predicate calculus notation, replacing variables by attribute values) will always satisfy the functional dependencies given;
2. When a CHASE incorporates dependencies into a database with marked null values, the CHASEd database will satisfy all of these dependencies, sometimes replacing null values by attribute values.

Table 1. Overview of CHASE variants

	Parameter ⋆	Object ○	Result ⊛	Goal
0.	dependency	database schema	database schema with integrity constraint	optimized database design
I.	dependency	query	query	semantic optimization
II.	view	query	query using views	AQuV
II'.	operator	query	query using given operators	AQuO
III.	s-t tgds, egds	source database	target database	data exchange, data integration
IV.	tgds, egds	database	modified database	cleaning
V.	tgds, egds	incomplete database	query result	certain answers
VI.	s-t tgds, egds, tgds	database	query result	invertible query evaluation

Database Structures or Queries as a CHASE Object. The original idea of the CHASE was to include dependencies like *functional dependencies* (FD) $X \rightarrow Y$ or *join dependencies* (JD) $\bowtie [R_1, ..., R_n]$ into a given database schema [1, 33]. The result, a database schema with integrity constraints, led to an *optimized database design* (Table 1, case 0) by guaranteeing some desirable database design properties such as the lossless join property. *Semantic query optimization*, on the other hand, requires the incorporation of dependencies into a given query. One of the already existing CHASE tools PDQ [9] has been developed for this purpose. For optimization, the CHASEd query has to be transformed to an equivalent, optimal query by using an additional BACKCHASE phase [15].

Answering Queries using Views (AQuV) and its generalization *answering queries using capabilities of database engines* that we call *Answering Queries using Operators* (AQuO, see Sect. 6) are also possible goals that can be realized with the CHASE. Here a set of views (or operators) will be incorporated as the CHASE parameters ⋆ into a given query, the CHASE object ○ (case II. and II.'). Finding an efficient method of answering a query using a set of previously defined materialized views over the database, rather than accessing the database relations is discussed in [14,27]. Here a given query Q is enriched by its views (CHASE phase) and then reduced to the minimum equivalent view: The CHASE will therefore be extended by a BACKCHASE phase, again. One of the tools for efficiently[1] calculating CHASE&BACKCHASE applied to AQuV is ProvCB [29].

Database Instances with Null Values as a CHASE Object. Instead of queries as CHASE objects ○ we can also use databases with null values as ○. Processing *source-to-target tuple generating dependencies* (s-t tgd's,[16]) using the CHASE is implemented by LLUNATIC [19] and ChaseFUN [10]. S-t tgd's are formulas

---

[1] By the use of Provenance information.

$$\forall \mathbf{x} : (\phi(\mathbf{x}) \rightarrow \exists \mathbf{y} : \psi(\mathbf{x}, \mathbf{y}))$$

where ϕ is a function over source predicates (e.g., relation schemes of the source database), ψ is a function over target predicates (e.g., relation schemes of the target database), $\mathbf{x}$ being domain variables over attributes in the source, and $\mathbf{y}$ being domain variables over (new, additional) attributes in the target. While the CHASE creates a new target database using s-t tgd's as some kind of inter-database dependencies, *tuple generating dependency* (tgd), a s-t tgds on the same database, and *equality generating dependencies* (egd)

$$\forall \mathbf{x} : (\phi(\mathbf{x}) \rightarrow (x_1 = x_2))$$

can be seen as intra-database dependencies representing integrity constraints within a database [7,16]. Egd's replace marked null values by other marked null values or attribute values (also called constants) [16,18].

LLUNATIC and ChaseFUN support data exchange and data integration scenarios as well as database cleaning tasks (case III. and IV.). For the goal of invertible query evaluations, no known tool yet exists (case VI.). In this case, the CHASE result ⊛ is not a (new, integrated or exchanged) database but the result of a database query. The BACKCHASE phase is afterwards used to calculate an inverse of the query or to generate a provenance polynomial [21,22] to be able to determine a (minimal) sub-database that guarantees the reproducibility of the query result.

The CHASE in General. The CHASE applications we are using in our research projects at the University of Rostock are shown as red rows in Table 1. The main application *Invertible Query Evaluation* of the project *ProSA* will be described in Sect. 3. The other CHASE applications of the Data Research Group (DBIS) of the University of Rostock are summarized in Sect. 6. A generalization of CHASE and the development of a general chase tool is therefore a common task for several, different research projects.

A first idea of a universal CHASE tool is given in [8]. Here the authors tested and compared different CHASE tools like PDQ, ChaseFUN and LLUNATIC. They tested for different CHASE strategies like *oblivious CHASE*, *standard/restricted CHASE* and *core CHASE* [20]. The concept of the standard CHASE is shown in Algorithm 1. Eliminating the first If-statement (h is an active trigger) transforms this algorithm to the oblivious CHASE, the core chase is neglected here.

Basically, for a given database instance I, Algorithm 1 provides a modified database I' by incorporating any possible dependency $\sigma \in \Sigma$ into I. While tgd's create new tuples under active trigger, egd's clean the database by replacing null values where an *active trigger* h is a homomorphism from the left-hand side of σ to I which cannot be extended to a homomorphism from the right-hand side to I. In other words, a trigger is active, if a new tuple can (and has to) be generated or null values be replaced.

Algorithm 1 standard CHASE (Σ, I)

Require: Set of dependencies Σ, Database instance I
Ensure: Modified database I'
1: **for all** trigger h for a dependency $\sigma \in \Sigma$ **do**
2:      **if** h is an active trigger **then**
3:          **if** σ is a tgd **then**
4:              Adding new tuples to the database instance I
5:          **else if** σ is an egd **then**
6:              **if** values compared are different constants **then**
7:                  CHASE fails
8:              **else**
9:                  Substituting null values by other null values or constants

---

The problem of this basic algorithm is that the CHASE parameter $\star$ is fixed (a set of dependencies Σ) and the CHASE object $\bigcirc$ is also fixed (a database instance I). In our approach ChaTEAU, we will extend the CHASE to become applicable to other CHASE parameters (such as views, query operators, or privacy constraints, see [24, 25]) and other CHASE objects (such as queries and general database transformations).

In Sect. 3, we will focus on the use of the CHASE for calculating inverses of an evaluation query in research data management, to determine the origin of the evaluation results. This is one of the main tasks in data provenance. We therefore will introduce some basic notions of data provenance in the next subsection.

2.2 Data Provenance

Given a database instance I and a query Q, the central task in data management is to compute the result of the query $K = Q(I)$. However, in many cases the result of the query itself is not sufficient. We are interested, for example, in:

1. Where do the result tuples come from?
2. Why – by means of which parts of the original database I – was a certain result achieved?
3. How – by means of which sequence of operations – was a result actually achieved?
4. Why is an expected value missing in the result?

Thus, in *Data Provenance* we typically distinguish between four Provenance queries (*where*, *why*, *how*, and *why not*) and four Provenance answers (extensional, intensional, query-based, modification-based). In the remainder of this paper, we focus on *how*- and *why*- provenance, and on extensional answers to these queries. An extensional answer to a *why*-query is the sub-database of the original research database I, the result tuples are derived from. The tuples of this sub-database of I are called *witnesses* for the evaluation result K.

The *why*- and *where*-provenance can be derived from the result of the *how*-provenance (see Fig. 1). For this we can define a reduction based on the information content

$$where \preceq why \preceq how.$$

$$where \preceq why \preceq how.$$

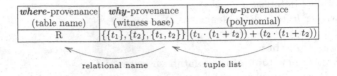

where-provenance (table name)	why-provenance (witness base)	how-provenance (polynomial)
R	$\{\{t_1\}, \{t_2\}, \{t_1, t_2\}\}$	$(t_1 \cdot (t_1 + t_2)) + (t_2 \cdot (t_1 + t_2))$

relational name   tuple list

Fig. 1. Reduction of *why* and *where* from *how*

In our research project ProSA, we mainly focus on **why**- and **how**-provenance with extensional answers to the provenance queries, which are given by *provenance polynomials* [3,21] and *(minimal) witness bases* [13] to guarantee the reproducibility of the query result. The polynomials are defined by a commutative semi-ring $(\mathbb{N}[X], +, \cdot, 0, 1)$ with $+$ for union and projection as well as $\cdot$ for natural join. The selection does not change a tuple, hence it is not represented in the semi-ring. Projection and union both eliminate duplicate tuples, they are therefore represented by an operation in the semi-ring. The second operation $\cdot$ symbolizes the natural join, representing the combination of tuples. The polynomials apply to the positive relational algebra (i.e., without difference or negation), but approaches for the extension by aggregate functions and negation already exist [2,3].

We are interested in specifying a concrete calculation rule (**how**), or at least in specifying all necessary source tuples (**why**). For the given example in Fig. 1 the provenance polynomial is calculated as $(t_1 \cdot (t_1 + t_2)) + (t_2 \cdot (t_1 + t_2))$, where t_1, t_2 and t_3 are tupel identifiers of the source instance I. Reducing each partial polynom t_1^2, $2t_1t_2$, t_2^2 to the set of its identifiers results the witness base $\{\{t_1\}, \{t_2\}, \{t_1, t_2\}\}$. A minimal witness base would be $\{\{t_1\}\}$ or $\{\{t_2\}\}$.

3 Invertible Query Evaluation

In our project ProSA (**Pro**venance management using **S**chema mappings with **A**nnotations) [4], which summarizes the DBIS research interests in the area of research data management, we develop techniques to invert evaluation queries against a research database (case VI. in Table 1) to be able to answer **why**-provenance queries. We use the *witnesses* (results of the **why**-query, see [13]) and the *provenance polynomials* (results of **how**-query, see [21]) to determine the minimal sub-database of the original research database that has to be archived for reproducibility and replicability of research results.

3.1 Research Data Management

In *research data management* research data has to be managed, analyzed, stored, and archived for a long period of time. One of our research interests in this topic is the question of how we can minimize the amount of data to be archived, especially in the case of constantly changing databases or database schemes and permanently performing new evaluations on these data (see [6])? Therefore we have to answer two concrete research questions:

1. Calculation of the minimal part of the original research database (we call it *minimal sub-database*) that has to be stored permanently to achieve replicable research.
2. Unification of the theories behind data provenance and schema (as well as data) evolution.

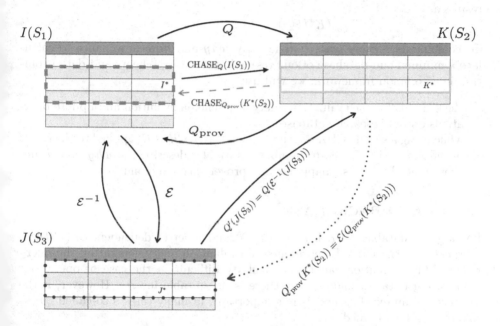

Fig. 2. Unification of provenance and evolution [4]

The idea behind these goals is sketched in Fig. 2: The calculation of an inverse query Q_{prov}, which is used to determine the required minimal sub-database I^* (red dashed box), depends on the original query Q and the result instance (green box) that should be reconstructed:

$$Q_{\text{prov}}(K^*(S_2)) = Q^{-1}(K^*(S_2)).$$

Unfortunately, the specification of a concrete inverse is not always possible, but it is not even necessary in many cases. Sometimes, e.g. due to privacy requirements, we only want to reconstruct some kind of sub-database of the original research database by a quasi-inverted mapping [17]. If there is no direct inverse function, the only possibility to define such an inverse is adding provenance annotations such as provenance polynomials. The aim of our research is to calculate the inverses or quasi-inverses of the evaluation query, if possible, and calculating additional provenance polynomials as annotations in cases, where direct inverses are not possible. The complete inverse mappings for the basic relational algebra operations are introduced in [6].

Nevertheless, the situation in research data management is even more complicated since the research database will evolve over time. There are frequent data evolution operations (updates) and occasional schema changes. Under the schema evolution $\mathcal{E} : S_1 \to S_3$, the query Q' can be directly calculated as a composition of the original query Q and the inverse evolution $\mathcal{E}^{-1}$. So the new provenance query Q'_{prov} of the new minimal sub-database J^* (blue dotted box) results as

$$Q'_{\text{prov}}(K^*(S_2)) = (Q_{\text{prov}} \circ \mathcal{E})(K^*(S_2)).$$

To combine data provenance (particularly **why**-and **how**-provenance to calculate a minimal sub-database of our research database), schema and data evolution in a common framework, we will use

- the CHASE for the evaluation query, the schema evolution and update operations on our research database,
- while using a second CHASE step (called *BACKCHASE*) applied to the result of the first CHASE step to be able to formally describe the **why**- and **how**-provenance by inverse mappings and provenance polynomials.

3.2   CHASE&BACKCHASE

For a given database (*CHASE object* $\bigcirc$) and a set of dependencies (*CHASE parameter* $\star$) the CHASE represents an evaluation query generating the query result. This evaluation can be particularly difficult in the case of operations aggregating data. In many cases, there is no suitable inverse. However, if the existence of an exact inverse is not important, in many cases a so-called *quasi-inverse* can be specified.

Usually three different types of (quasi-)inverse schema mappings created by the CHASE are distinguished: *exact*, *classic* and *relaxed CHASE inverses* [17, 18]. Two more inverse types, the *tuple preserving-relaxed* (*tp-relaxed*) and *result equivalent CHASE inverse* are introduced in [6], which also includes an overview of the inverse types of all relational algebra operations. As one example, the inverse type of the query $\pi_{AD}(\sigma_{C=c}(r_1) \bowtie \pi_{AB}(r_2))$ can be identified as relaxed. With additional provenance polynomials the projection itself can be identified as tp-relaxed. Generally, most basic operations can be optimized to become invertible by adding additional annotations like provenance polynomials [3,21] or provenance games [31]. The inverse type of a composition of operations always corresponds to the weakest type of all partial mappings.

In this context, the CHASE&BACKCHASE procedure [14] can be used to calculate a suitable inverse function. A query Q formulated as an s-t tgd can now be processed using the CHASE and then inverted using the BACKCHASE to calculate a minimal part of the original database. This calculated *minimal sub-database* should be able to reconstruct the query results under various constraints [4,5]. So the questions are: How to automatically compute this minimal sub-database of the primary data? Given an evaluation query and the result of the evaluation query, is it possible to derive the minimal sub-database simply by an inverse mapping without any additional annotations of the data and without any need to freeze the original database? These questions address some exciting problems of research data management, such as traceability, reproducibility and replicability of research results [11,12,28].

3.3   Provenance Using CHASE

Over the years, the CHASE algorithm has been used in many different areas of application. In our ProSA project, the CHASE is used for query evaluation (i.e., data transformation), schema and data evolution, and data exchange. The BACKCHASE is used to invert the query evaluation Q, i.e. to calculate the extensional answer of the ***why***-Provenance as the witness base of the query Q.

Given a schema mapping $\mathcal{M}$ defined by an s-t tgd and a source instance I, Fagin [18] computes a target instance via $\text{CHASE}_{\mathcal{M}}(I)$. He also discussed a target and a source schema evolution and specified some types of CHASE inverse functions (exact, classic, relaxed), which are essential for schema evolution. In ProSA, we use Fagin's schema mappings as queries Q, schema evolution, and data exchange steps.

The result of the evaluation query Q described by extended s-t tgd's and egd's can be calculated using the CHASE algorithm. The calculation of the minimal sub-database I' is computed by inverting Q. This inverse Q_{prov} does not necessarily have to correspond to an inverse in the classical sense $Q \circ Q_{\text{prov}} = \text{Id}$, since a CHASE inverse cannot always be specified [4]. The provenance answer I' can then be calculated using the BACKCHASE.

4   Many Application Areas – One Tool: ChaTEAU

Since different research projects in our group use the CHASE as a basic technique (see Sect. 6), we decided to either use or develop one general CHASE tool that is applicable to all of these applications. A nice overview of already existing CHASE tools is given in [8]. To our knowledge, and after evaluating the CHASE tools being publicly available, none of the existing tools can be applied to all of the scenarios mentioned above (Table 1), or at least to some scenarios of each of both groups (case I. and II.: queries as CHASE objects; case III. to VI.:: databases as CHASE objects).

Here, the basic idea of our universal tool ChaTEAU (**Cha**se for **T**ransforming, **E**volving, and **A**dapting databases and queries, **U**niversal approach) as well as some preliminary implementation work will be presented. Finally, we can construct a tool for each of the three application scenarios based on ChaTEAU (see Fig. 4(b) in Sect. 5). In the case of invertible query evaluation, for example, extended data preparation and a BACKCHASE phase are necessary.

4.1 Theoretical Foundation of ChaTEAU

The main idea of making the various use cases of the CHASE applicable in a single, universal tool is based on the fact that the CHASE objects $\bigcirc$ and parameters $\star$ are essentially interchangeable without significantly changing the way the CHASE is executed. With this in mind, we can abstract Algorithm 1 by replacing the input (a set of dependencies Σ and a database instance I) by a general parameter $\star$ and a general object $\bigcirc$.

CHASE Parameter $\star$. This parameter is always represented as a intra-database or inter-database dependency: In semantic optimization, it corresponds to a set of egd's and tgd's as intra-database constraints, in the case of AQuV, to a set of views (interpreted as dependencies between base and view relations), or to a set of s-t tgd's, tgd's and egd's as in data exchange and data integration.

CHASE Object $\bigcirc$. The CHASE object $\bigcirc$ is either a query Q or a database instance I. While in queries (distinguished or non-distinguished) variables v_i [32] can be replaced by other variables v_j or constants c_i, in database instances null values η_i are replaced by other null values η_j or constants c_i. The variable substitution depends on certain conditions which are shown in Fig. 3. These conditions guarantee a homomorphic mapping of the CHASE object.

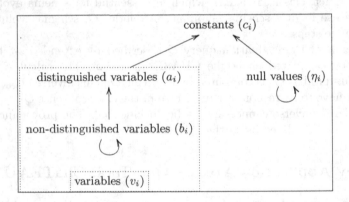

Fig. 3. Possible homomorphisms for the object $\bigcirc$

CHASE Rules. The replacement rules incorporating the CHASE parameter $\star$ into the CHASE object $\bigcirc$ are independent of the selection of $\star$ or $\bigcirc$ itself. Currently these are egd-, tgd- and s-t tgd-rules [18,32]. Currently, we are working on an extension of these rules to handle aggregates in evaluation queries, too: these aggregates have to be integrated similar to Skolem functions in the target part of an s-t tgd. There are also existing rules for views like in case II [14].

4.2   ChaTEAU

The core of ChaTEAU is based on Algorithm 1. A first implementation was given in [30]. The input of the algorithm can be any possible CHASE parameter ⋆ and object ◯ as described above. The output will then be a transformed database schema, a query (satisfying certain constraints) or a modified database instance.

ChaTEAU is based on five classes of objects: terms, homomorphisms, integrity constraints, atoms and instances. Terms form an elementary data structure in ChaTEAU which are represented by the class TERM. As in theory, they can be a variable, a null value or a constant. Constant values are represented by simple data types. However, since they can be strings, numbers, or other values, and the data type is not known at runtime, the approach used here is dynamic typing. The currently supported data types are `string`, `double` and `integer`. For each data type, there is a field variable of the class that stores the corresponding value. A term can only contain one value of one type, which is why only one of these field variables is ever used. Variables and null values are also part of the TERM class. The distinction between these three types is made by an enumeration, which determines what kind of term it is when a term is created. So TERM is elementary for the other classes used in ChaTEAU.

The tool as well as the theory differentiates between relational and equality atoms. Both are the basic building block for instances, queries and integrity constraints. Relational atoms themselves consist of a term and an identifier. Equality atoms contain two terms, expressing an equality relation between them. Both are combined in the class ATOM.

Instances are implemented in INSTANCE and constraints like tgd's as well as egd's in INTEGRITY CONSTRAINTS. An integrity constraint always consists of a head and a body, which again consist of atoms. Apart from the head of an egd, which is formed from an equality atom, relational atoms are used for the heads and bodies. Also instances are based on relational atoms.

If there is a homomorphism between a source and a target term, a term mapping is generated which maps one term to another term like $v_i \rightarrow v_j$, $\eta_1 \rightarrow$ Max and $\eta_2 \rightarrow \eta_1$. Homomorphisms are also fundamental for the use in CHASE rules and defining active triggers (see Algorithm 1).

5   ProSA Using ChaTEAU

The reconstruction of a minimal sub-database I^* like in Fig. 2 is the main task of our project ProSA [4]. To achieve this, a given database instance I as well as the associated schema S_1 are extended by a unique identifier. This ID is a coding usually consisting of the table name and a consecutive number. This modified database, the schema and the evaluation query Q are now input for the CHASE algorithm implemented in ChaTEAU and extended by provenance aspects which are realized in a separate BACKCHASE phase (see Fig. 4(a)). The result of the evaluation query and the result of the provenance query are finally the output of ProSA.

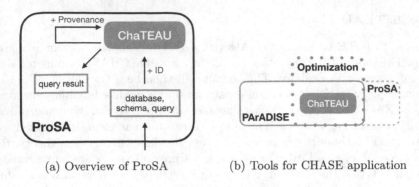

(a) Overview of ProSA   (b) Tools for CHASE application

Fig. 4. Applications of the CHASE at DBIS

ChaTEAU is not only the core of ProSA (red dashed box, Fig. 4(b)), but also the basis of two other tools for Answering Queries using Operators (solid box) as well as for semantic query optimization (dotted box). Both applications will be briefly motivated below in Sect. 6.

CHASE & BACKCHASE. The combination of CHASE and provenance in one application is explained in detail in [6]. There we describe the idea that the *recovered instance*

$$I^* = \mathrm{CHASE}_{Q_{\mathrm{prov}}}(K) = \mathrm{CHASE}_{Q_{\mathrm{prov}}}(\mathrm{CHASE}_Q(I))$$

is thus the result of a query Q_{prov} on the *result instance* K. I^* contains whole tuples from I or tuples restricted to certain attributes of the source schema (and filled with (marked) null values).

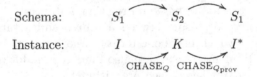

The CHASE&BACKCHASE method for determining a CHASE inverse schema mapping/provenance mapping $Q_{\mathrm{prov}} = (S_2, S_1, \Sigma_2)$ to $Q = (S_1, S_2, \Sigma_1)$ can therefore be described in two phases:

- CHASE phase: Calculate the CHASE of I as the CHASE object ○ with Q as a CHASE parameter ⋆ represented as a sequence of s-t tgd and egd rules which generate new tuples (s-t tgd) and replace (marked) null values with constants or null values with smaller index (egd).
- BACKCHASE phase: Calculate the CHASE of K (or an interesting subset of the result, called K^*, for which the provenance should be checked) as the CHASE object ○ with Q_{prov} as a CHASE parameter ⋆, again represented as a sequence of s-t tgd and egd rules.

An evaluation query Q is processed by the CHASE. So a given database instance $I(S_1)$ delivers a result instance $K(S_2)$ which is shown in Fig. 5 (green). Such a mapping is easy to find for simple queries with SPJU (select, projection, join, union), but difficult for queries with aggregation. For this purpose, the CHASE must be extended by functions which code aggregation functions like MAX, MIN, COUNT, SUM and AVG.

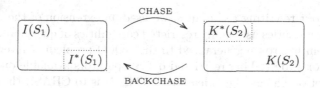

Fig. 5. CHASE&BACKCHASE in ProSA (Color figure online)

Additionally, the provenance query Q_{prov} is processed by the BACKCHASE resulting in an (exact, classic, relaxed, result-equivalent, ...) CHASE inverse adding some annotations like provenance polynomials to guarantee the reconstruction of the recovered instance I^* as a minimal sub-database of the research database I that has been the basis for the evaluation query Q.

6 Other Applications of the CHASE

Within the Database Research Group of the University of Rostock, we work in different areas such as *Semantic Query Optimization, Answering Queries using Operators* (AQuO) and *Invertible Query Evaluation*. However, the different areas of application have one thing in common. All three can be processed by means of the CHASE implemented in ChaTEAU (see Fig. 4(b)).

Semantic Query Optimization. One of the first and classical applications of the CHASE is injecting data dependencies into queries to be able to optimize queries under given dependencies. In one of our research projects, we looked for different kinds of tgd's (such as inclusion dependencies) as inter-object relationships in multimedia databases. Instead of extracting multimedia features for content-based retrieval, we use the inter-object dependencies to optimize queries consisting of longer join paths. The dependencies are CHASEd into the query, and a BACKCHASE phase will generate an optimized query plan minimizing the number of joins used.

Anwering Queries using Operators (AQuO). In our project PArADISE (**P**rivacy-**A**ware assistive **D**istributed **I**nformation **S**ystem **E**nvironment, see [23,25]) we use query rewriting and query containment techniques to achieve an efficient and privacy-aware processing of queries. To achieve this, the whole network structure, from data-producing sensors up to cloud computers, is utilized to create a database machine consisting of billions of devices, the Internet of Things. Based on previous research in the field of query rewriting, we developed a concept to split a query into fragment and remainder queries. Fragment queries can operate on resource limited devices to filter and preaggregate data. Remainder queries take these preprocessed (filtered and aggregated) data and (only) execute the last, complex part of the original queries on more powerful devices like cloud servers. As a result, less data is processed and forwarded to a cloud server and the privacy principle of data minimization is accomplished [24–26].

In one of our rewriting techniques, we used an extension of the AQuV technique to answer queries using only restricted capabilities of query processors. The AQuV problem has to be generalized to the AQuO problem (Anwering Queries using Operators), describing restricted query operators available on the devices in the Internet of Things. The idea for ChaTEAU is to CHASE these operators into the query to be able to BACKCHASE the enhanced query to fragment and remainder queries automatically.

7   Conclusion and Future Work

The aim of our project ProSA is the development of techniques to invert evaluation queries against a research database to be able to answer *how-* and *why*-provenance. We use the CHASE tool ChaTEAU to represent the queries, data and schema evolution, and calculate he provenance answers by the BACKCHASE process. Some open questions in the areas ChaTEAU and ProSA are:

ChaTEAU. We have to complete our implementation to get a tool that can handle a generalized CHASE object ◯ as well as a generalized CHASE parameter ⋆. We are working on extensions of s-t tgd, tgd and egd rules to handle more complex dependencies and query operators such as aggregations. On the other hand, we need to control the CHASE process by evaluating complexity and runtime of the CHASE procedure, and we need to ensure the termination of the CHASE procedure by adapting criteria such as the weak acyclicity introduced for tgd's as a CHASE parameter [16].

ProSA. For a concrete implementation of the BACKCHASE phase, we first need a definition of special BACKCHASE rules. These rules correspond to the CHASE rules with additional annotations like provenance polynomials. Again, the processing of aggregates is a difficulty here, but similar to the processing using the CHASE.

Additionally, we have to decide about the annotations to be stored if an exact or classic inverse does not exist for our evaluation query. While the use of provenance polynomials as an annotation is quite elegant, it is not efficient to process all the polynomials on a large database. This is particularly the case if the research database is very large. Besides this, due to privacy reasons, it is not always appropriate to calculate elements of the original research database exactly. In this case, an intensional answer of the provenance query (only describing the content of the original database in a descriptive, anonymized manner) seems to be more appropriate than the extensional answer we calculated up to now.

Lastly, the ProSA technique is prepared to integrate schema evolution steps and database updates. We have to describe these evolutions by s-t tgd's to incorporate evolving databases into the ProSA tool.

Acknowledgements. We thank our students Fabian Renn and Frank Röger for their comparison of different CHASE tools like LLUNATIC and PDQ as well as Martin Jurklies for the basic implementation of our CHASE tool ChaTEAU.

References

1. Aho, A.V., Beeri, C., Ullman, J.D.: The theory of joins in relational databases. ACM Trans. Database Syst. **4**(3), 297–314 (1979)
2. Amarilli, A., Bourhis, P., Senellart, P.: Provenance circuits for trees and treelike instances (extended version). CoRR abs/1511.08723 (2015)
3. Amsterdamer, Y., Deutch, D., Tannen, V.: Provenance for aggregate queries. In: PODS, pp. 153–164. ACM (2011)
4. Auge, T., Heuer, A.: Combining provenance management and schema evolution. In: Belhajjame, K., Gehani, A., Alper, P. (eds.) IPAW 2018. LNCS, vol. 11017, pp. 222–225. Springer, Cham (2018). https://doi.org/10.1007/978-3-319-98379-0_24
5. Auge, T., Heuer, A.: Inverses in research data management: combining provenance management, schema and data evolution (inverse im forschungsdatenmanagement). In: Grundlagen von Datenbanken. CEUR Workshop Proceedings, vol. 2126, pp. 108–113. CEUR-WS.org (2018)
6. Auge, T., Heuer, A.: The theory behind minimizing research data: result equivalent CHASE-inverse mappings. In: CEUR Workshop Proceedings of the LWDA, vol. 2191, pp. 1–12. CEUR-WS.org (2018)
7. Benczúr, A., Kiss, A., Márkus, T.: On a general class of data dependencies in the relational model and its implication problems. Comput. Math. Appl. **21**(1), 1–11 (1991)
8. Benedikt, M., et al.: Benchmarking the chase. In: PODS, pp. 37–52. ACM (2017)
9. Benedikt, M., Leblay, J., Tsamoura, E.: PDQ: proof-driven query answering over web-based data. PVLDB **7**(13), 1553–1556 (2014)
10. Bonifati, A., Ileana, I., Linardi, M.: ChaseFUN: a data exchange engine for functional dependencies at scale. In: EDBT, pp. 534–537. OpenProceedings.org (2017)
11. Bruder, I., Heuer, A., Schick, S., Spors, S.: Konzepte für das Forschungsdatenmanagement an der Universität Rostock (Concepts for the Management of Research Data at the University of Rostock). In: CEUR Workshop Proceedings of the LWDA, vol. 1917, p. 165. CEUR-WS.org (2017)
12. Bruder, I., et al.: Daten wie Sand am Meer - Datenerhebung, -strukturierung, -management und Data Provenance für die Ostseeforschung. Datenbank-Spektrum **17**(2), 183–196 (2017). https://doi.org/10.1007/s13222-017-0259-4
13. Buneman, P., Khanna, S., Wang-Chiew, T.: Why and where: a characterization of data provenance. In: Van den Bussche, J., Vianu, V. (eds.) ICDT 2001. LNCS, vol. 1973, pp. 316–330. Springer, Heidelberg (2001). https://doi.org/10.1007/3-540-44503-X_20
14. Deutsch, A., Hull, R.: Provenance-directed Chase&Backchase. In: Tannen, V., Wong, L., Libkin, L., Fan, W., Tan, W.C., Fourman, M. (eds.) In Search of Elegance in the Theory and Practice of Computation. LNCS, vol. 8000, pp. 227–236. Springer, Heidelberg (2013). https://doi.org/10.1007/978-3-642-41660-6_11

372   T. Auge and A. Heuer

15. Deutsch, A., Popa, L., Tannen, V.: Query reformulation with constraints. SIGMOD Rec. **35**(1), 65–73 (2006)
16. Fagin, R., Kolaitis, P.G., Miller, R.J., Popa, L.: Data exchange: semantics and query answering. Theor. Comput. Sci. **336**(1), 89–124 (2005)
17. Fagin, R., Kolaitis, P.G., Popa, L., Tan, W.C.: Quasi-inverses of schema mappings. ACM Trans. Database Syst. **33**(2), 11:1–11:52 (2008)
18. Fagin, R., Kolaitis, P.G., Popa, L., Tan, W.C.: Schema mapping evolution through composition and inversion. In: Bellahsene, Z., Bonifati, A., Rahm, E. (eds.) Schema Matching and Mapping. Data-Centric Systems and Applications, pp. 191–222. Springer, Heidelberg (2011). https://doi.org/10.1007/978-3-642-16518-4_7
19. Geerts, F., Mecca, G., Papotti, P., Santoro, D.: That's all folks! LLUNATIC goes open source. PVLDB **7**(13), 1565–1568 (2014)
20. Greco, S., Molinaro, C., Spezzano, F.: Incomplete Data and Data Dependencies in Relational Databases. Synthesis Lectures on Data Management. Morgan & Claypool Publishers, San Rafael (2012)
21. Green, T.J., Karvounarakis, G., Tannen, V.: Provenance semirings. In: PODS, pp. 31–40. ACM (2007)
22. Green, T.J., Tannen, V.: The semiring framework for database provenance. In: PODS, pp. 93–99. ACM (2017)
23. Grunert, H., Heuer, A.: Datenschutz im PArADISE. Datenbank-Spektrum **16**(2), 107–117 (2016)
24. Grunert, H., Heuer, A.: Privacy protection through query rewriting in smart environments. In: EDBT, pp. 708–709. OpenProceedings.org (2016)
25. Grunert, H., Heuer, A.: Rewriting complex queries from cloud to fog under capability constraints to protect the users' privacy. OJIOT **3**(1), 31–45 (2017)
26. Grunert, H., Heuer, A.: Query rewriting by contract under privacy constraints. OJIOT **4**(1), 54–69 (2018)
27. Halevy, A.Y.: Answering queries using views: a survey. VLDB J. **10**(4), 270–294 (2001)
28. Herschel, M., Diestelkämper, R., Ben Lahmar, H.: A survey on provenance: what for? What form? What from? VLDB J. **26**(6), 881–906 (2017)
29. Ileana, I., Cautis, B., Deutsch, A., Katsis, Y.: Complete yet practical search for minimal query reformulations under constraints. In: SIGMOD Conference, pp. 1015–1026. ACM (2014)
30. Jurklies, M.: CHASE und BACKCHASE: Entwicklung eines Universal-Werkzeugs für eine Basistechnik der Datenbankforschung. Master's thesis, Universität Rostock (2018)
31. Köhler, S., Ludäscher, B., Zinn, D.: First-order provenance games. CoRR abs/1309.2655 (2013) http://arxiv.org/abs/1309.2655
32. Maier, D.: The Theory of Relational Databases. Computer Science Press (1983)
33. Maier, D., Mendelzon, A.O., Sagiv, Y.: Testing implications of data dependencies. ACM Trans. Database Syst. **4**(4), 455–469 (1979)

ECHOES: A Fail-Safe, Conflict Handling, and Scalable Data Management Mechanism for the Internet of Things

Christoph Stach[✉][iD] and Bernhard Mitschang

Institute for Parallel and Distributed Systems, University of Stuttgart,
Universitätsstraße 38, 70569 Stuttgart, Germany
{stachch,mitsch}@ipvs.uni-stuttgart.de

Abstract. The Internet of Things (IoT) and Smart Services are becoming increasingly popular. Such services adapt to a user's needs by using sensors to detect the current situation. Yet, an IoT service has to capture its required data by itself, even if another service has already captured it before. There is no data exchange mechanism adapted to the IoT which enables sharing of sensor data among services and across devices.

Therefore, we introduce a data management mechanism for the IoT. Due to its applied state-based synchronization protocol called ECHOES. It is fail-safe in case of connection failures, it detects and handles data conflicts, it is geared towards devices with limited resources, and it is highly scalable. We embed ECHOES into a data provisioning infrastructure, namely the Privacy Management Platform and the Secure Data Container. Evaluation results verify the practicability of our approach.

Keywords: IoT · Data exchange · Synchronization protocol

1 Introduction

The *Internet of Things* (*IoT*) has long left its early stages of development behind in which only technology evangelists and early adopters used *Smart Things*[1]. Due to omnipresent Internet connectivity options and increasing bandwidth speeds, devices with a permanent Internet access grew proliferated in the early 2000s. Yet, it was not until the IoT became more and more invisible, by integrating sensors into everyday objects, that this technology found its way into limelight [7].

The Internet of Things is extremely intriguing for both consumers and service providers as the data collected by Smart Things is extremely valuable. Since these devices are usually permanently close to their users and their sensors are able to record a wide range of data, data scientists can gain profound knowledge

---

[1] We use the term '*Smart Thing*' for any kind of device which is able to connect to other devices in order to exchange data with each other and has the ability either to monitor its environment or to control other devices.

© Springer Nature Switzerland AG 2019
T. Welzer et al. (Eds.): ADBIS 2019, LNCS 11695, pp. 373–389, 2019.
https://doi.org/10.1007/978-3-030-28730-6_23

about the users. Services can thus be tailored to individual customers. Application cases can be found in any domain, such as *Smart Homes*, *Smart Cars*, or *Smart Health*.

So, it is not surprising that analysts predict that by 2020 over 50 billion Smart Things will be in use and the market will be shared among many service providers [13]. Each device vendor and each service provider has its own way of storing and processing the captured data. Therefore, the effort to collect all data required for a certain service is cumbersome—different services have to collect the same data over and over again, as there is no simple data exchange mechanism among services, let alone a direct data exchange across Smart Things. Yet, the IoT can only unfold its true potential if all user data is available to all of his or her Smart Things at any time. Due to the vast amount of data, a smart synchronization mechanism is required, i.e., the volume of data being transmitted has to be as low as possible and the computational costs for Smart Things have to be minimal as these devices often have very limited resources [35].

Since none of the existing data exchange approaches for the IoT provide these characteristics, we introduce a state-based synchronization protocol for the Internet of Things, called *ECHOES*, and apply it to a data exchange mechanism. To this end, we make the following three contributions:

(I) We introduce an approach that enables Smart Things to exchange their data using a central synchronization server. For this purpose, our state-based synchronization protocol is used. Whenever changes to the client's data stock (i.e., adding new data sets, modifying existing data sets, or removing data sets) are done, it calls synchronization server and applies the changes to the remote data stock. Analogously, the changes are then applied to all other Smart Things of this user by the synchronization server. Due to the state-based procedure, our approach is resource-friendly in terms of run time performance and the volume of data being transmitted. *(II)* As for Smart Things it cannot be ensured that they have a permanent connection to the server, our protocol also supports an offline mode. That is, Smart Things are able to gather data locally and synchronize with the server as soon as connectivity is back on. In this mode also, a resource-friendly proceeding is ensured. It has to be mentioned that in this stage conflicts can occur (e.g., if two offline Smart Things modify the same data set) and ECHOES is able to detect and resolve them. *(III)* We embed ECHOES into an existing data provisioning infrastructure for Smart Things, namely the *PMP* (**P**rivacy **M**anagement **P**latform) [27] and the *SDC* (**S**ecure **D**ata **C**ontainer) [29]. The SDC is a shared data storage for IoT services and the PMP provides access to this data stock. By integrating ECHOES, we enable the synchronization of all SDCs and therefore, the PMP is able to provide all collected data of a user to any service on every Smart Thing. As the PMP provides fine-grained access control mechanisms, this data exchange is done in a privacy-aware manner.

The remainder of this paper is as follows: Sect. 2 introduces a real-world application case from the Smart Health domain, in which it is mandatory for Smart Things to exchange their data. From this, we derive a requirement specification in Sect. 3. We discuss related work in Sect. 4. Subsequently, we introduce the protocol of ECHOES in Sect. 5 and detail on its implementation based on the PMP and the SDC in Sect. 6. An evaluation of ECHOES is given in Sect. 7. Finally, Sect. 8 concludes this paper.

2 Application Case

In the following, we depict an IoT use case from the Smart Health domain. Smart Health applications are especially beneficial for patients suffering from chronic diseases such as diabetes. These patients have to visit their physicians regularly in order to do medical screenings and to keep track of their disease progression. Supported by Smart Things however, the patients are able to perform the required medical check-up at home by themselves. This does not only save a lot of money, but it also enables physicians to focus on emergencies [25]. *Smart Phones* are considered as particularly useful in the context of Smart Health application [11]. On the one hand, people always carry their Smart Phones with them and on the other hand, these devices are equipped with a variety of sensors which are relevant for Smart Health applications. For instance, Knöll emphasized the importance of the location where a medical reading was taken for its interpretation [14]. In addition, it is possible to connect medical devices such as glucometers or peak flow meters to Smart Phones via Bluetooth.

An example for such an application is *Candy Castle* [26]. The game is designed for children suffering from diabetes and is intended to help them coping with the continuous blood glucose monitoring. To this end, the children have to defend a virtual castle against attacks by dark forces (as a reflection of their disease). To protect the castle, the children have to build walls which hold the attackers back. They do this by doing a blood glucose reading. As the walls gradually wear down, Candy Castle motivates players to regularly check their blood sugar levels. Additionally, the virtual castle is linked to the location of the children's home and each reading is supplemented with the GPS coordinates where it was taken. Thereby, the walls can be inserted geographically correct in the game and the players are motivated to check their blood sugar levels in many different places. From a medical point of view, the children get used to the continuous blood glucose monitoring and physicians are able to identify healthy and unhealthy places in the children's environment [14]. Further sensors can be integrated to identify unhealthy factors more precisely. For instance, *Smart Bracelets* can be used to detect certain activities (e. g., administering insulin) [16], microphones can be used to determine the child's mood [18], and cameras can be used to calculate the bread units of the children's food [3]. The blood glucose measurement can also be integrated into the game automatically, e. g., via the *Apple Watch* [34].

However, such an application requires a fast and simple data exchange between the patient and his or her physician [6]. And Candy Castle is only one of many Smart Health applications. For example, a COPD application requires similar data, such as the location of a metering [31]. To avoid that every application has to acquire the same data all over again, there is a growing trend towards

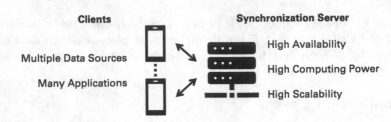

Fig. 1. Client-server model for the IoT.

a *quantified self* [33]. That is, all of a user's data on a specific domain is stored in a central repository [5]. Yet, this is limited to the pre-defined types of data of the repository vendor and it is not possible to use it for a generic use case.

So, it is necessary to have a generic mechanism to easily share data between different Smart Things and applications to face the interconnection and data exchange problems and thereby exploit the full potential of IoT applications [19].

3   Requirement Specification

Although the IoT is designed as a decentralized network of independent devices, the client-server model as shown in Fig. 1 is still predominant [22]. So, a data exchange mechanism for the IoT can capitalize on the advantages of a server, namely its high availability, computing power, and scalability. None of this is provided by the Smart Things. Thus, clients should mainly focus on data gathering and processing and not the management of the data exchange. These management tasks have to meet the following requirements:

R_1 – *Availability of Data:* As the given application case shows, a permanent network connection must not be required to collect and process data. Users have to be able to edit data locally offline (e. g., do a blood glucose reading) and synchronize it with other devices (e. g., their physicians' devices) later when there is a network connection. This guarantees a fail safe usage of IoT applications.

R_2 – *Conflict Handling:* When a user edits the same data set offline on multiple devices, conflicts might occur during synchronization (e. g., a user enters contradictory readings). So, conflicts must be handled without data loss.

R_3 – *Efficiency:* Synchronization has to deal with limited resources of Smart Things, e. g., limited data transfer volume, limited computational power, limited memory, and limited battery. So, the amount of metadata must be minimal and the calculation of the data that has to be transferred must be simple and fast.

R_4 – *Transparency:* The synchronization must be transparent for the user and—as good as possible—also for application developers. That is, user interaction has to be largely avoided and the developers' effort must be minimal.

R_5 – *Genericity:* The mechanism must not be restricted to a certain domain or a specific data schema. The data has to be available to any application.

R_6 – *Scalability:* As users interact with an increasing number of Smart Things and IoT applications, many read and write data accesses must be expected. So, synchronization has to be able to cope with a large number of mobile clients.

R_7 – *Security:* The IoT handles a lot of private data, i. e., security is a key aspect. This applies in particular to data storage, data access, and data sharing.

4 Related Work

With these requirements in mind, we look at related work. We consider work from three different categories: (a) data sharing approaches for Smart Things, (b) Cloud-based solutions for data sharing, and (c) synchronization approaches for the IoT. We focus on Android-based solutions, as this OS is becoming increasingly predominant in the IoT context, e. g., *Android Things*[2] or *emteria.OS*[3].

Smart Thing Approaches. Android's safety concept requires applications to run in sandboxes, strictly isolated from each other. A direct data exchange among applications is therefore not intended. In order to support data sharing, Android introduces so called *Content Providers*, i. e., restricted interfaces to the data sets of an application [12]. Although the interfaces are standardized, it is still cumbersome to use Content Providers. An application has to specify in the program code which Content Providers—i. e., which other applications—it needs to access. As a result, this approach largely lacks genericity. To overcome this flaw, *MetaService* [8] introduces a temporary shared object storage. In this storage, applications can deposit a single data object and another application is able to obtain it. When another object is deposited, any previously stored object is overwritten. In this respect, MetaService works similar to the *Clipboard* which is common on desktop PCs. However, MetaService is not suitable for the distribution of comprehensive data volumes. This is provided by the *SDC* [28,29]. The SDC is a shared database for Android that can be used by any application. Due to its fine-grained access control, users can specify precisely, which application may access which data set. With the *CURATOR* [30] extension, *NoSQL* databases are also supported so that, for instance, objects can be exchanged directly among applications. More information on the SDC is given in Sect. 6. However, none of these approaches supports data exchange across devices. *Mobius* [10] can be used for this purpose. It introduces a system-wide database which is synchronized with a Cloud database to share data across devices. However, Mobius uses locally virtual partitions to realize access control. An application has access to its own partition, only. That is, in order to share data with several applications the respective data has to be added to each respective partition. Although data sharing across devices is very simple, data sharing among applications is still cumbersome.

---

[2] see https://developer.android.com/things/.
[3] see https://emteria.com/.

Cloud-Based Approaches. There is a variety of Cloud services that enable data sharing across Smart Things. The most straightforward approach is a Cloud database in which applications can store and retrieve their data. These databases are usually associated with a specific application or class of applications and have a fixed data schema. For instance, there is the *Glucose Web Portal* [15], which can be used by applications related to diabetes. In addition to the storage of data, the Glucose Web Portal also provides some health services, e. g., the analysis of diabetes data. That is, an application has to collect health data and send it to the service and then the not only the data itself but also the analysis results are available to any other application. Similar services are available for other diseases such as COPD as well [32]. A more generic approach is the *HealthVault* [5]. A user has to create a HealthVault profile. Then s/he can store and link any kind of health data captured by any Smart Thing in his or her profile. This data is available for any application for its analysis and presentation. However, these approaches do not support user-defined types of data, an application automatically has access to all data, and since the data is only stored in the Cloud, it is not available if the Internet connection is interrupted. Many database vendors also provide a mobile version of their databases. These versions often support synchronization with a Cloud-based back-end. *Couchbase Mobile*[4] is such a mobile version. However, its synchronization is designed to ensuring that the data sets of a particular application are kept up to date on all of a user's Smart Things. Data sharing among different applications is not supported. Also, the synchronization only operates with the mobile client from Couchbase—other databases are not supported. *Resilio Sync*[5] improves the availability of such a synchronization by adopting a *P2P* approach instead of a central database [23]. Yet, this enables only the exchange of files and not of single data sets and it has fairness issues, which cause significant slowdowns [20].

Synchronization Approaches. *Syxaw* [17] brings the two aforementioned categories together by introducing a middleware for the synchronization of structured data. It enables multiple users to edit documents and folders collaboratively, and Syxaw takes care of merging the changes. However, since it operates on files, the computation of changes is expensive and locks on at file level are very restrictive. For a use case as presented in Sect. 2, a fine-grained synchronization of data sets is much better suited. This is achieved by *SAMD* [9]. In order to reduce the computational effort for the mobile clients, all expensive operations are carried out server-sided. This includes a multi-layered calculation of hash values for the managed data sets. Thereby it is sufficient to exchange comparatively small hash values for most of the synchronization process instead of the actual data. *SWAMD* [2] follows a quite similar approach, but its focus is on wireless networks, which is common in an IoT environment. Yet, both approaches are designed for a deployment scenario in which synchronization takes place infrequently. A continuous synchronization of data requires a permanent recalculation and exchange of the hash values, which causes high costs.

[4] see https://www.couchbase.com/products/mobile.
[5] see https://www.resilio.com.

Contrary to this, *MRDMS* [24] represents a timestamp-based synchronization approach. The timestamps enable MRDMS to reflect the temporal correlation of changes. In this way, the required data transfer volume can be further reduced compared to SAMD. However, since less data is used for synchronization, conflicts often cannot be resolved. Furthermore, *lost updates* cannot be prevented. By incorporating *snapshots* into such approaches, automatic conflict resolution can be improved [21]. Yet, this increases computational effort and data volumes.

As none of these approaches supports a use case as given in Sect. 2 as well as the requirements from Sect. 3, we introduce a solution in the following.

5   The ECHOES Protocol

In ECHOES, we do not pursue a P2P approach as it is not guaranteed that all Smart Things are permanently available and interconnected. A central, permanently available server component therefore ensures the fastest and most reliable data distribution. Moreover, computation-intensive tasks can be shifted to the server in order to reduce computational effort for the Smart Things. A one-way *push* or *pull* approach is inefficient as data changes can occur on both the server and the clients. Therefore, we apply a two-way state-based approach in our synchronization protocol. The synchronization steps can be simplified, since it is possible to decide based on the respective state which actions are necessary. Furthermore, we introduce version numbers for conflict resolution.

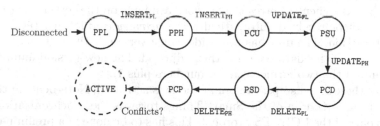

PPL: Primary Pull   PPH: Primary Push   PCU: Primary Client Update   PSU: Primary Server Update
PCD: Primary Client Delete   PSD: Primary Server Delete   PCP: Primary Conflict Pull

Fig. 2. ECHOES offline synchronization process (applies to server and client).

Offline Mode. For the initial synchronization or after a connection failure, the client has to process seven states sequentially (see Fig. 2). This is due to the fact that during the offline period various changes may have been made to the databases of both the client and the server.

First, it is necessary to check whether new data sets have been added. In the *primary pull* (PPL) the client sends all IDs of its data sets to the server and the server calculates the delta to its central database. As a response to the PPL, the server sends all data sets that are not available on the client. The client sets the status of these data sets to RELEASED. The client then performs a *primary push* (PPH) by sending all data sets that were added in the offline stage—i. e., data sets with the status NEW—to the server. When the server acknowledges receipt, their statuses are set to RELEASED.

ECHOES then handles edited data sets. In the *primary client update* (PCU) the client sends all version numbers of data sets with status RELEASED that were not handled by the previous steps to the server. The version number of a data set is incremented when a RELEASED data set is edited. The server sends back updates for all data sets for which a newer version exists. In the *primary server update* (PSU), the client sends then all data sets with the status MODIFIED—i. e., data sets that were edited during offline stage—to the server. The server checks based on the version number whether it can apply the update or whether there is a conflict with a change made by another client. Accordingly, the status on the client is set to RELEASED or the conflict is logged.

From the previous steps, the server already knows all the unmodified data sets on the client (status RELEASED). If these sets do not exist on the server any more, they must be deleted on the client as part of the *primary client delete* (PCD). Accordingly, the *primary server delete* (PSD) synchronizes local deletions. To this end, deletions performed during offline mode are not applied to the data stock immediately, but the status of the affected data sets is set to DELETED and the version numbers are incremented. During PSD, the server checks whether the data can be deleted or whether there is a conflict and gives feedback to the client.

Finally, ECHOES deals with conflict resolution. For all data sets flagged as conflicting, the client receives the versions available on the server. As these conflicts cannot be resolved automatically, the *primary conflict pull* (PCP) requires user interaction. The user has to decide which version is the valid one. The version number of this data record is then adjusted. The new version number is the maximum of the two former version numbers plus 1.

After these seven steps, the online mode is activated. Nota bene, authentication and authorization of the Smart Things towards the synchronization server are not part of the ECHOES protocol. This has to be done in a preliminary step. ECHOES handles synchronization, only. Yet, we tackle both of these issues in our implementation (see Sect. 6).

Online Mode. In online mode (see Fig. 3), client and server mutually send acknowledgement messages periodically as a heartbeat message. In this process, both change their state from ACTIVE to STANDBY and vice versa. That way, no permanent (energy-consuming) connection is necessary.

Each party (i. e., client and server) can continue to work and process data locally, regardless of its current state. To this end, each party adds corresponding tasks to a local queue. This queue is processed as soon as the respective party is active. Each of these tasks refers to a single data set, only. As a result, the processing is significantly less computationally expensive than the synchronization in offline mode and only a single state has to be traversed per task.

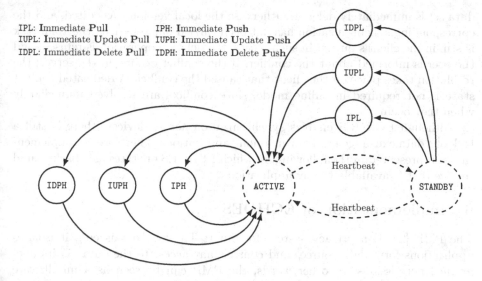

IPL: Immediate Pull   IPH: Immediate Push
IUPL: Immediate Update Pull   IUPH: Immediate Update Push
IDPL: Immediate Delete Pull   IDPH: Immediate Delete Push

Fig. 3. ECHOES online synchronization process (applies to server and client).

A reasonable tradeoff must be achieved in this respect. Long `ACTIVE-STANDBY` cycles cause less communication overhead (for passing the activity token, i. e., the heartbeat message), but more local changes—and therefore potential conflicts—might occur per cycle and it takes more time until the changes are applied to all devices. Short cycles cause increased communication overhead, even if there have not been any changes during the `STANDBY` phase.

Immediate pull (`IPL`) and *immediate push* (`IPH`) are the counterparts in the online mode to the `PPL` and `PPH` state in the offline mode. Conflicts do not have to be considered in these states. The IDs used by ECHOES contain a reference to the Smart Thing that generates the data. This prevents conflicts if new data is simply added.

When data sets are edited, an *immediate update pull* (`IUPL`) or respectively an *immediate update push* (`IUPH`) is triggered. First, the server checks based on the version number whether the changes can be applied immediately to all clients. If this is the case, the status is set to `RELEASED`. If there is a conflict, the server attempts to resolve it by merging the changes. If that is successful, the new version of the data set is distributed to all clients. If the conflict cannot be resolved automatically, the client that submitted the change is notified and the user must resolve the conflict manually. To prevent the synchronization from being blocked by this user interaction, the data set is marked as a conflict and the update task is added to the client's queue again. Once the conflict is resolved, the conflict flag as well as the update task are removed, and the changes are synchronized by the server on all clients.

Finally, *immediate delete pull* (`IDPL`) and *immediate delete push* (`IDPH`) deal with the synchronization of delete operations. The client (or analogously the server) receives the ID and the version number of the data set in question. If the version number is equal to or higher than the one of the local instance, the data set is immediately deleted. Otherwise the local version was edited, and the corresponding synchronization has not been carried out yet—i. e., the IUPH task is still in the client's queue. In such a case, the deletion operation is refused, and the user is informed about the conflict. If the conflict occurs on the server, the resolution takes place on the client that caused the conflict. A dedicated conflict state is not required in online mode, since conflicts are resolved immediately when they occur.

Although ECHOES enables synchronization across devices, there is still a lack of a data exchange mechanism among applications. Therefore, an implementation is presented in the following, in which ECHOES operates as a background service that is available to any application.

6   Implementation of ECHOES

The PMP [27] is a privacy system for Smart Things. To this end, it isolates applications form data sources and controls any access to the data via its fine-grained permissions. In other words, the PMP can be seen as a middleware that operates as an information broker. The PMP's key feature is that it is extendable. New data sources can be added at any time as so-called *Resources*. Subsequently, applications can access these data sources via the PMP.

The SDC [29], a database system based on *SQLite*, is such a Resource. It offers security features, e. g., the stored data is encrypted and it provides a tuple-based access control. In addition, it has a customizable schema to be compatible to any application and stored data can be partitioned to increase performance.

As the SDC is a PMP Resource, it is available to any applications. Thus, data can be exchanged among applications via the fine-grained access control. By integrating ECHOES into the SDC, an exchange across devices can be realized.

Figure 4 shows the data access and synchronization process of an SDC-based ECHOES implementation. Initially, an application requests access to the local SDC instance from the PMP (1). Then, the PMP checks whether the application has the required permissions. If the application is authorized to use the SDC, the PMP enables access. The application can use the SDC like an internal database, i. e., it can query, insert, update, or delete data (2). If data from another application is affected, the SDC checks the required access rights.

Due to the integration of ECHOES, the SDC detects changes to the data stock and synchronizes it (3). Depending on whether the SDC is currently in offline or online mode, this involves different steps. If it is in offline mode, it must establish a connection to the synchronization server and perform a complete synchronization (see Fig. 2). Otherwise, the respective type of alteration can be submitted to the server (see Fig. 3). In this process, the changes are applied to the central database (4). The synchronization server then creates an event to notify all other SDC instances via its event bus (5). These notifications are transformed into tasks and added to the threads of the respective Smart Things (6). Finally, the necessary changes are applied to their local SDC instances (7). The data from Smart Thing$_1$ is then available to applications on Smart Thing$_n$ (8). Again, the PMP handles access control (9).

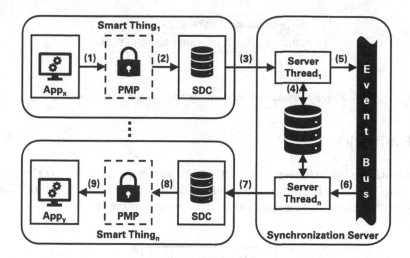

Fig. 4. Data access and synchronization process.

For the integration of ECHOES into the SDC, its data schema has to be adapted, as additional metadata is required for synchronization. The extended relational schema is shown in Fig. 5. First, a new *ID* is added for each data set. This ID contains references to the Smart Thing that created or edited the data set most recently. In addition, the *mode* of each data set has to be logged. In addition to the four modes (NEW, RELEASED, MODIFIED, and DELETED) which are required for synchronization (see Sect. 5), there is a fifth mode OFFLINE. Data sets flagged with this mode are excluded from synchronization. If a conflict cannot be resolved automatically, this is also indicated in the mode entry of the corresponding data sets. The SDC then informs the user about the conflict and s/he can decide which version should be valid. The *VERSION* of the data sets is required by ECHOES to decide which version is valid. The *TIMESTAMP* entry is not necessarily required for the synchronization as the version already represents the chronological order in which the data was edited. Yet, the timestamps help users to resolve conflicts as they are able to track exactly when which data set was edited. *SHARE* is provided by the SDC. It is a foreign key to the maintenance tables of the SDC. These tables have to be synchronized as well in order to enable access control on all devices. Finally, the actual *PAYLOAD* is stored as well. An individual data schema can be specified, just like in the native SDC.

The actual data transfer is realized as a flattened stream of characters. A composer in the SDC converts the database entries into a sequence of key-value pairs and a parser processes such a sequence and inserts the contained values into the SDC. That way, the amount of data that needs to be transferred is minimized as almost solely payload data is transferred. On the synchronization server, there are corresponding counterparts according to the specifications in Sect. 5.

ECHOES_DATA_SETS					
ID	MODE	VERSION	TIMESTAMP	SHARE	PAYLOAD
		INT		FK	DATA

Fig. 5. Relational schema for the mobile database.

7   Evaluation

To evaluate the performance of our ECHOES prototype, we describe the evaluation setup, present the evaluation results, and discuss whether ECHOES fulfills all requirements towards a data exchange mechanism for the IoT.

HealthRecord	
id	: INT
activity	: INT
breadUnits	: REAL
bsl	: INT
condition	: INT
latitude	: REAL
longitude	: REAL
mood	: INT
patient	: FK
timestamp	: INT
freeText	: CHAR(140)

Fig. 6. Candy castle data model used for evaluation.

Evaluation Setup. For the evaluation, we draw on Candy Castle [26]. This application is executed on two Smart Phones and the captured data is synchronized on both devices. For this purpose, we extended the SDC-based data management of Candy Castle by ECHOES—the applied data model of the payload is shown in Fig. 6. In addition, we set up an ECHOES synchronization server. To get a better understanding of how different hardware configurations and different Android versions affect the performance of ECHOES, we perform our measurements on two different types of Smart Phones: on the one hand the *LG Nexus 5X* (S_1) with a current Android version and on the other hand the *Huawei Honor 6 Plus* (S_2) with more memory and more CPU cores but a lower clock speed. Both are intentionally lower middle-class models, since their hardware setup is similar to those of other popular Smart Things, such as the *Raspberry Pi. MariaDB* is used on the server. MariaDB is a highly powerful and scalable database with strong similarities to *MySQL* [1,4]. A detailed evaluation setup is given in Table 1.

Table 1. Evaluation setup.

	Smart Thing S_1	Smart Thing S_2	Server
OS	Android 8.1.0	Android 5.1.1	Debian 9.4
CPU	Snapdragon 808	Kirin 925	$8 * 3.6$ GHz
RAM	2 GB	3 GB	8 GB
Connection	50 Mbit/s	50 Mbit/s	100 Mbit/s
Database	SQLite	SQLite	MariaDB

For the evaluation, we examine four different scenarios for both, the offline mode as well the online mode: (a) Initially, both Smart Phones are disconnected and an ascending number of data sets (from 100 up to 6, 400) is randomly generated on one Smart Phone. Then, both devices are connected. (b) Subsequently, both Smart Phones are disconnected again. Then, on one device 50% of the data sets are edited and synchronization is started. (c) Next, 50% of the data is edited on both devices in offline mode, i. e., ECHOES must resolve conflicts. (d) Finally, 50% of the data sets are deleted on one device and synchronization is started.

In each scenario, we take the time until synchronization is completed. These scenarios are repeated with permanently enabled connectivity to evaluate online mode. In this case, the duration of the synchronization of each individual operation is measured. Each test is performed with a pair of Smart Thing S_1 and S_2. After each run, all databases are reset to avoid side effects caused by warm caches. Each test is carried out for 10 times and the average processing time is considered.

Evaluation Results. All evaluation results for ECHOES's offline mode are shown in Fig. 7. Figure 7a shows the time until newly added data sets are available on all devices (PPL & PPH). The processing time increases nearly linear to the number of data sets. On average, the synchronization of a single newly added data set takes about 84 ms on a pair of S_1 and about 99 ms on a pair of S_2. It is striking that ECHOES is performing very well on the weaker hardware. A considerably more decisive factor is the OS version. These findings are also reflected by the three other scenarios.

The processing time when changes are made to 50% of the data sets (PCU & PSU) is shown in Fig. 7b. This processing time also increases linearly, but it is significantly higher than in the previous scenario. Although only half of the data sets have been changed, the other half must also be cross-checked with the server. Nevertheless, a processing time of 180 ms or 215 ms per edited data set is still reasonable.

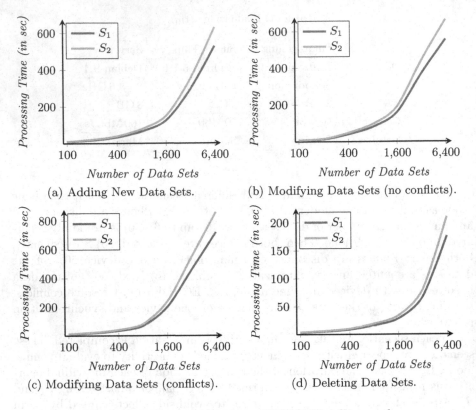

Fig. 7. Overall processing time of ECHOES's offline mode.

The effect of conflict resolving (PCP) on the processing time is shown in Fig. 7c. This conflict resolving increases the costs caused by ECHOES (213 ms or 254 ms per edited data set). Yet, conflicts are unlikely in our application case.

The deletion of data sets (PCU & PSU) is very fast (24 ms or 28 ms per data set, see Fig. 7d).

Table 2. Processing time of ECHOES's online mode.

	Add	Modify	Delete
Smart Thing S_1	365 ms	368 ms	309 ms
Smart Thing S_2	398 ms	402 ms	337 ms

In the online mode, changes (regardless whether it is add, modify, or delete) are available on all devices after about 350 ms. The detailed costs are stated in Table 2. As shown in Sect. 5, conflicts need not to be considered explicitly in online mode since they are handled by the three listed operations already.

The online synchronization takes longer than the synchronization of a single data set in offline mode, as the communication overhead required to initiate the data exchange is generated only once for the total bulk of transferred data. These costs are thus allocated proportionally in offline mode among all data sets contained in the bulk.

Discussion. As ECHOES has an online and an offline mode, it ensures **availability** even in case of connection failures. This enables users to continue working and ECHOES takes care of the synchronization as soon as the connection is reestablished (R_1). *Conflict handling* is ensured, as ECHOES resolves conflicts automatically due to its PCP state (R_2). Evaluation results prove ECHOES **efficiency**. Not only does it cope with limited resources, but also the required metadata is minimal. Based on the memory consumption of an SQL database, the payload in our application case requires 192 bytes per data set, while the metadata occupies only 20 bytes. That is almost a ratio of 10 to 1. Obviously, this is case-specific (R_3). From an application's point of view, **transparency** is achieved. An application interacts with the SDC as if it is a local database (R_4). Also, **genericity** is ensured at the data schema of an SDC instance can be customized (R_5). The server-side **scalability** is ensured by the use of MariaDB and as the SDC can be partitioned, the scalability can be further improved (R_6). Finally, the PMP and the SDC ensures privacy and **security** on the mobile clients (R_7). Therefore, ECHOES meets all requirements towards a data exchange mechanism for the IoT.

8 Conclusion

The IoT is becoming increasingly popular. A growing number of applications are emerging in various domains such as Smart Homes, Smart Cars, or Smart Health. These IoT applications require a mechanism to share data. However, current data sharing approaches do not fulfills all requirements towards such a mechanism.

Therefore, we introduce ECHOES, a state-based synchronization protocol for the IoT. It provides four key features: (1) It supports an online and offline mode to deal with connection failures. (2) It deals with conflicts when several parties edit the same data set. (3) It can be executed on limited resources. (4) It operates with any given data schema. We implement this protocol in a data provisioning infrastructure for Smart Things (PMP & SDC). Thus, our prototype has three further key features: (5) The SDC behaves like a local database. (6) The backend is highly scalable due to MariaDB. (7) The PMP and the SDC provide a wide range of data security features. Evaluation results are very promising as changes are available on all clients in less than 0.4 s.

Acknowledgment. We thank the BW-Stiftung for financing the PATRON research project and the DFG for funding the SitOPT research project.

References

1. Aditya, B., Juhana, T.: A high availability (HA) MariaDB Galera Cluster across data center with optimized WRR scheduling algorithm of LVS - TUN. In: TSSA (2015)
2. Alhaj, T.A., et al.: Synchronization wireless algorithm based on message digest (SWAMD) for mobile device database. In: ICCEEE (2013)
3. Almaghrabi, R., et al.: A novel method for measuring nutrition intake based on food image. In: I2MTC (2012)
4. Bartholomew, D.: MariaDB vs. MySQL. White paper, Monty Program Ab (2012)
5. Bhandari, V.: Enabling Programmable Self with HealthVault: An Accessible Personal Health Record. O'Reilly Media Inc., Newton (2012)
6. Chan, M., et al.: Smart wearable systems: current status and future challenges. Artif. Intell. Med. **56**(3), 137–156 (2012)
7. Chase, J.: The evolution of the Internet of Things. White paper, Texas Instruments (2013)
8. Choe, H., et al.: MetaService: an object transfer platform between Android applications. In: RACS (2011)
9. Choi, M.Y., et al.: A database synchronization algorithm for mobile devices. IEEE Trans. Consum. Electron. **56**(2), 392–398 (2010)
10. Chun, B.G., et al.: Mobius: unified messaging and data serving for mobile apps. In: MobiSys (2012)
11. Dayer, L., et al.: Smartphone medication adherence apps: potential benefits to patients and providers. J. Am. Pharm. Assoc. **53**(2), 172–181 (2013)
12. Enck, W., et al.: Understanding Android security. IEEE Secur. Privacy **7**(1), 50–57 (2009)
13. Hung, M. (ed.): Leading the IoT: Gartner Insights on How to Lead in a Connected World. Gartner (2017)
14. Knöll, M.: Diabetes City: how urban game design strategies can help diabetics. In: eHealth (2009)
15. Koutny, T., et al.: On-line blood glucose level calculation. In: ICTH (2016)
16. Kwapisz, J.R., et al.: Activity recognition using cell phone accelerometers. ACM SIGKDD Explor. Newsl. **12**(2), 74–82 (2010)
17. Lindholm, T., et al.: Syxaw: data synchronization middleware for the mobile web. Mob. Netw. Appl. **14**(5), 661–676 (2009)
18. Mehta, D.D., et al.: Mobile voice health monitoring using a wearable accelerometer sensor and a smartphone platform. IEEE Trans. Biomed. Eng. **59**(11), 3090–3096 (2012)
19. Murnane, E.L., et al.: Mobile health apps: adoption, adherence, and abandonment. In: UbiComp/ISWC 2015, Adjunct (2015)
20. Peng, Z., et al.: On the measurement of P2P file synchronization: Resilio Sync as a case study. In: IWQoS (2017)
21. Phatak, S.H., Nath, B.: Transaction-centric reconciliation in disconnected client-server databases. Mob. Netw. Appl. **9**(5), 459–471 (2004)
22. Ren, J., et al.: Serving at the edge: a scalable IoT architecture based on transparent computing. IEEE Netw. **31**(5), 96–105 (2017)
23. Scanlon, M., et al.: Network investigation methodology for BitTorrent Sync. Comput. Secur. **54**(C), 27–43 (2015)
24. Sethia, D., et al.: MRDMS-mobile replicated database management synchronization. In: SPIN (2014)

25. Silva, B.M.C., et al.: Mobile-health: a review of current state in 2015. J. Biomed. Inform. **56**(August), 265–272 (2015)
26. Stach, C.: Secure Candy Castle – a prototype for privacy-aware mHealth apps. In: MDM (2016)
27. Stach, C., Mitschang, B.: Privacy management for mobile platforms - a review of concepts and approaches. In: MDM (2013)
28. Stach, C., Mitschang, B.: Der Secure Data Container (SDC) - Sicheres Datenmanagement für mobile Anwendungen. Datenbank-Spektrum **15**(2), 109–118 (2015)
29. Stach, C., Mitschang, B.: The Secure Data Container: an approach to harmonize data sharing with information security. In: MDM (2016)
30. Stach, C., Mitschang, B.: CURATOR–a secure shared object store: design, implementation, and evaluation of a manageable, secure, and performant data exchange mechanism for smart devices. In: SAC (2018)
31. Stach, C., et al.: The Privacy Management Platform: an enabler for device interoperability and information security in mHealth applications. In: HEALTHINF (2018)
32. Steimle, F., et al.: Extended provisioning, security and analysis techniques for the ECHO health data management system. Computing **99**(2), 183–201 (2017)
33. Swan, M.: Sensor mania! The Internet of Things, wearable computing, objective metrics, and the Quantified Self 2.0. J. Sens. Actuator Netw. **1**(3), 217–253 (2012)
34. Wakabayashi, D.: Freed from the iPhone, the Apple Watch finds a medical purpose. Report, The New York Times (2017)
35. Walker, M.: Hype cycle for emerging technologies. Market analysis, Gartner (2018)

Transaction Isolation in Mixed-Level and Mixed-Scope Settings

Stephen J. Hegner$^{(\boxtimes)}$

DBMS Research of New Hampshire, PO Box 2153, New London, NH 03257, USA
dbmsnh@gmx.com

Abstract. Modern database-management systems permit the isolation level to be set on a per-transaction basis. In such a *mixed-level* setting, it is important to understand how transactions running at different levels interact. More fundamentally however, these levels are sometimes of different scopes. For example, READ COMMITTED and REPEATABLE READ are of *local scope*, since the defining properties depend upon only the transaction and its relationship to those running concurrently. On the other hand, SERIALIZABLE is of *global scope*; serializability is a property of a schedule of transactions, not of a single transaction. In this work, in addition to formalizing the interaction of transactions at different levels, the meaning of serializability within local scope is also addressed.

Keywords: Database · Transaction · Isolation

1 Introduction

In a modern relational database-management system (RDBMS), there is trade-off between performance via transaction concurrency and adequate isolation of transactions from the operations of other transactions. It has therefore long been held that a single notion of isolation is not adequate. Rather, the level of isolation should be determined by the needs of the application. This philosophy is integral to SQL, the standard of which [10, Part 2, Sect. 4.36] identifies four distinct levels of isolation for transactions, ordered with increasing isolation as READ UNCOMMITTED < READ COMMITTED < REPEATABLE READ < SERIALIZABLE.

Based upon the names of the isolation levels, as well as upon the semantics as defined in the SQL standard, this classification is very confusing, because it mixes isolation levels of distinct scope. The isolation levels READ UNCOMMITTED, READ COMMITTED, and REPEATABLE READ are *local* in scope; the definitions apply to individual transactions, with the relevant isolation properties of a transaction T completely determined in conjunction with the behavior of those transactions which run concurrently with it. On the other hand, the isolation level SERIALIZABLE is *global* in scope; it applies to an entire schedule of transactions. Indeed, it makes no sense to say that an individual transaction is serializable; it only makes sense to say that a set of transactions, organized into a schedule

© Springer Nature Switzerland AG 2019
T. Welzer et al. (Eds.): ADBIS 2019, LNCS 11695, pp. 390–406, 2019.
https://doi.org/10.1007/978-3-030-28730-6_24

S, is serializable; that is, that the results of running the transactions according to S is equivalent to running them in some schedule S' with no concurrency. This raises the obvious question of what it means, in a mixed-level system, to run some but not all transactions with serializable isolation. A main goal of this paper is to address such questions of isolation involving multiple scopes.

It is the apparent intent of the SQL standard that the SERIALIZABLE isolation level serve double duty, with both local and global scope, called a *multiscope* isolation level. On the one hand, it is defined to be a local isolation level, call it DEGREE 3,[1] which is slightly stronger than REPEATABLE READ. On the other hand, it is also defined in the standard to be a *serializable-generating* isolation level, in the sense that if all transactions are run at that level, then the result must be a serializable schedule. Unfortunately, it has been known for some time that DEGREE 3, as defined above, is not serializable generating [2, Sect. A5B], rendering the standard somewhat confusing at best. Nevertheless, the idea of a serializable-generating local isolation level is an important one. In this paper, working within the context of modern MVCC (multiversion concurrency control), a minimal serializable-generating local isolation level called RCX is identified. Interestingly, while it is slightly stronger than READ COMMITTED, it is not order comparable to REPEATABLE READ. Indeed, it is shown that it is not necessary to require repeatable reads (i.e., to require that the transaction read the same value for a data object x regardless of when the read occurs during its lifetime) in order to achieve serializable-generating behavior; rather, the critical requirement is to prohibit so-called backward read-write dependencies. Thus, the SQL standard imposes a condition for the local component of its multiscope isolation level SERIALIZABLE which is not necessary for serialization.

For such a serializable-generating local isolation level, it is natural to ask whether it has any properties related to serializability when run in a mixed-level context, with other transactions running at other levels. The answer is shown to be in the affirmative. RCX, as well as all higher levels of isolation, are *serializable preserving*, in the sense that if transaction T runs at that level, then adding T to the schedule will not result in any new cycles. In particular, if the existing schedule is serializable, then adding T will preserve that property.

The way in which new serializable-generating strategies, including SSI [4, 7] and SSN [15], fit into this picture is also examined. SSI is of particular interest because it is used for implementation of the SERIALIZABLE isolation level in PostgreSQL [12]. Both may be termed *preemptive regional* strategies. They look for certain small structures in the conflict graph which are a necessary part of any cycle, aborting one of the participants when such a structure is found. However, in contradistinction to RCX, neither is serializable preserving, or has any similar property, so they have dubious benefit in a mixed-level context.

---

[1] The SQL standard gives SERIALIZABLE no name to identify its local scope. Since it is sometimes called *Degree 3* isolation in the literature, [9, Sect. 7.6], the moniker DEGREE 3 is introduced here, purely for clarification. Technically, it is REPEATABLE READ which additionally prohibits so-called *phantoms*.

The paper is organized as follows. In Sect. 2, necessary background material on transactions and serializability is summarized. In Sect. 3, local isolation levels are studied, with a particular focus on how they interact with each other in a mixed-mode setting. In Sect. 4, serialization in multi-scope settings is examined. Finally, Sect. 5 contains conclusions and further directions.

2   Transactions, Schedules, and Serialization

In this section, the basic ideas of transactions, schedules, and serialization are summarized. The focus is to provide a precise and ambiguous notation and terminology to use as a foundation for the ideas presented in Sects. 3 and 4.

Context 2.1 (Data objects and the global schema). A database schema **D** is defined by a set $\mathsf{DObj}\langle \mathbf{D} \rangle$ of data objects. Each such object has a single value at any point in time, which may be read or written by a transaction.

In this work, such a schema **D**, called the *global schema*, is fixed. The current instance of the global schema is called the *global database*.

Notation 2.2 (Time). Brackets are used to identify time intervals of the real numbers $\mathbb{R}$, using common conventions. For $a, b \in \mathbb{R}$, $[a,b] = \{c \in \mathbb{R} \mid a \leq c \leq b\}$, $(a,b] = \{c \in \mathbb{R} \mid a < c \leq b\}$, and $[a,b) = \{c \in \mathbb{R} \mid a \leq c < b\}$.

Summary 2.3 (Transactions). A *transaction* T over $\mathsf{DObj}\langle \mathbf{D} \rangle$ is defined by certain *time points* in $\mathbb{R}$, in addition to read and write operations. First, T has a *start time* $t_{\mathsf{Start}}\langle T \rangle$ and an *end time* $t_{\mathsf{End}}\langle T \rangle$, with $t_{\mathsf{Start}}\langle T \rangle < t_{\mathsf{End}}\langle T \rangle$.

The specification of operations on the database follow an *object-level model*, in which it is only known whether a transaction reads and/or writes a given $x \in \mathsf{DObj}\langle \mathbf{D} \rangle$, without knowledge of specific values. The *read set* $\mathsf{ReadSet}\langle T \rangle \subseteq \mathsf{DObj}\langle \mathbf{D} \rangle$ of T consists of all data objects which T reads. Similarly, the *write set* $\mathsf{WriteSet}\langle T \rangle \subseteq \mathsf{DObj}\langle \mathbf{D} \rangle$ of T consists of all data objects which T writes.

The request time assignment of T provides the time at which read and write operations are requested by the transaction. Formally, the *request time assignment* Req for T assigns to each $x \in \mathsf{ReadSet}\langle T \rangle$ a time $t_{\mathsf{Read}\langle x \rangle}^{\mathsf{Req}}\langle T \rangle \in [t_{\mathsf{Start}}\langle T \rangle, t_{\mathsf{End}}\langle T \rangle)$, and to each $x \in \mathsf{WriteSet}\langle T \rangle$ a time $t_{\mathsf{Write}\langle x \rangle}^{\mathsf{Req}}\langle T \rangle \in (t_{\mathsf{Start}}\langle T \rangle, t_{\mathsf{End}}\langle T \rangle]$. Note that the read time $t_{\mathsf{Read}\langle x \rangle}^{\mathsf{Req}}\langle T \rangle$ may be the same as the start time, but that it must occur strictly before the end time. Similarly, the write time $t_{\mathsf{Write}\langle x \rangle}^{\mathsf{Req}}\langle T \rangle$ may be the same as the end time, but that it must occur strictly after the start time. Furthermore, if $x \in \mathsf{ReadSet}\langle T \rangle \cap \mathsf{WriteSet}\langle T \rangle$, then $t_{\mathsf{Read}\langle x \rangle}^{\mathsf{Req}}\langle T \rangle < t_{\mathsf{Write}\langle x \rangle}^{\mathsf{Req}}\langle T \rangle$; that is a write must occur after a read. It is assumed that each transaction T reads and writes a data object x at most once. The set of all *time points* of T is $\mathsf{TimePoints}\langle T \rangle = \{t_{\mathsf{Start}}\langle T \rangle, t_{\mathsf{End}}\langle T \rangle\} \cup \{t_{\mathsf{Read}\langle x \rangle}^{\mathsf{Req}}\langle T \rangle \mid x \in \mathsf{ReadSet}\langle T \rangle\} \cup \{t_{\mathsf{Write}\langle x \rangle}^{\mathsf{Req}}\langle T \rangle \mid x \in \mathsf{WriteSet}\langle T \rangle\}$.

The set of all transactions over **D** is denoted $\mathsf{Trans}\langle \mathbf{D} \rangle$.

Summary 2.4 (Effective time assignments). In early systems using single-version concurrency control (SVCC), the request time of a read or write was often the same as the time at which the global database was actually read or written by the transaction. However, for modern systems, this is almost never the case for writes and often not the case for reads either. Rather, there is an *effective time assignment*, whose values depend upon the isolation protocol. In virtually all cases for a system employing MVCC, the effective time for a write is at the end of the transaction, while the effective time for a read depends upon the isolation protocol. With the *read-request-write-end* time assignment, denoted RRWE, all writes occur at the end of the transaction, while reads occur at their request times. Specifically, for $x \in \mathsf{ReadSet}\langle T \rangle$, $t^{\mathsf{RRWE}}_{\mathsf{Read}\langle x \rangle}\langle T \rangle = t^{\mathsf{Req}}_{\mathsf{Read}\langle x \rangle}\langle T \rangle$, while for $x \in \mathsf{WriteSet}\langle T \rangle$, $t^{\mathsf{RRWE}}_{\mathsf{Write}\langle x \rangle}\langle T \rangle = t_{\mathsf{End}}\langle T \rangle$. With the *read-beginning-write-end* time assignment, denoted RBWE, all reads occur at the start of the transaction, and all writes occur at the end. Specifically, for $x \in \mathsf{ReadSet}\langle T \rangle$, $t^{\mathsf{RBWE}}_{\mathsf{Read}\langle x \rangle}\langle T \rangle = t_{\mathsf{Start}}\langle T \rangle$, while for $x \in \mathsf{WriteSet}\langle T \rangle$, $t^{\mathsf{RRWE}}_{\mathsf{Write}\langle x \rangle}\langle T \rangle = t_{\mathsf{End}}\langle T \rangle$. RRWE is typically associated with variants of read-committed isolation, while RBWE is usually associated with variants of snapshot isolation, as elaborated in Sect. 3. $\mathsf{TASetEff} = \{\mathsf{RRWE}, \mathsf{RBWE}\}$ denotes the set consisting of the two effective time assignments.

Definition 2.5 (Transactions with effective time assignment). It is important to be able to run the same transaction at different levels of isolation, which may have associated with them different effective time assignments (see Definition 3.3). Therefore, rather than building a fixed effective time assignment into a transaction, it is more appropriate to associate such a time assignment as a parameter. Formally, a *transaction with effective time assignment*, or a *Teff-transaction* for short, is a pair $\langle T, \tau \rangle$ with $T \in \mathsf{Trans}\langle \mathbf{D} \rangle$ and $\tau \in \mathsf{TASetEff}$. The set of all Teff-transactions over $\mathbf{D}$ is denoted $\mathsf{TransTeff}\langle \mathbf{D} \rangle$.

The set of time points of $\langle T, \tau \rangle$ is exactly the same as the set of time points of T; $\mathsf{TimePoints}\langle \langle T, \tau \rangle \rangle = \mathsf{TimePoints}\langle T \rangle$. Observe that in an effective time assignment (RRWE or RBWE) each write occurs at $t_{\mathsf{End}}\langle T \rangle$, and each read at either its request time $t^{\mathsf{Req}}_{\mathsf{Read}\langle T \rangle}\langle x \rangle$ or else at $t_{\mathsf{Start}}\langle T \rangle$, so the effective time assignment does not add any new time points, beyond those defined by transaction start and end, plus effective times of reads and writes.

Summary 2.6 (Schedules and temporal relationships between transactions). A pair $\{T_1, T_2\} \subseteq \mathsf{Trans}\langle \mathbf{D} \rangle$ is *time compatible* if $\mathsf{TimePoints}\langle T_1 \rangle \cap \mathsf{TimePoints}\langle T_2 \rangle = \emptyset$. A pair $\{\langle T_1, \tau_1 \rangle, \langle T_2, \tau_2 \rangle\} \subseteq \mathsf{TransTeff}\langle \mathbf{D} \rangle$ is *time compatible* precisely in the case that $\{T_1, T_2\}$ has that property.

A *schedule* over $\mathsf{Trans}\langle \mathbf{D} \rangle$ is a finite (possibly empty) set $S \subseteq \mathsf{TransTeff}\langle \mathbf{D} \rangle$ for which every distinct pair $\{\langle T_1, \tau_1 \rangle, \langle T_2, \tau_2 \rangle\} \subseteq S$ is time compatible. Define $\mathsf{TransOf}\langle S \rangle = \{T \mid \langle T, \tau \rangle \in S\}$.

Two distinct transactions $\{T_1, T_2\} \subseteq \mathsf{TransOf}\langle S \rangle$ are *concurrent*, written $T_1 \parallel T_2$, if both $t_{\mathsf{Start}}\langle T_1 \rangle < t_{\mathsf{End}}\langle T_2 \rangle$ and $t_{\mathsf{Start}}\langle T_2 \rangle < t_{\mathsf{End}}\langle T_1 \rangle$ hold. If $\{T_1, T_2\}$ is not concurrent, then it is *serial* in S. In that case, if $t_{\mathsf{Start}}\langle T_1 \rangle < t_{\mathsf{Start}}\langle T_2 \rangle$, write $T_1 <_s T_2$, and if $t_{\mathsf{Start}}\langle T_2 \rangle < t_{\mathsf{Start}}\langle T_1 \rangle$, write $T_2 <_s T_1$.

The set of all schedules over $\mathsf{Trans}\langle \mathbf{D} \rangle$ is denoted $\mathsf{Sched}\langle \mathbf{D} \rangle$.

Summary 2.7 (Serializable behavior of schedules). Roughly speaking, a schedule S is *serializable* if its transactions may be relocated in time so that no two are concurrent, while preserving the effect of all read and write operations. There are many distinct ways to formalize this idea; in [16, Ch. 3] there are descriptions of no fewer than five major alternatives, many with minor variants. In this work, the notion of *conflict serializability* of a schedule S will be used, owing to its simple characterization in terms of edges in the *conflict graph* of the schedule, also called the *direct serialization graph*, or *DSG*, of S. For comprehensive summaries of conflict serializability, see [11, Sec. 2.6] and [16, Sec. 3.8]. In addition, [1] examines the DSG with an eye towards modern isolation protocols. Here, only the essential notions, will be identified.

The DSG associated with a schedule S is denoted $\mathsf{DSG}\langle S\rangle$. In that graph, the vertices are the members of $\mathsf{TransOf}\langle S\rangle$. There are three types of edges (also called *dependencies*). For $\langle T_1, \tau_1\rangle, \langle T_2, \tau_2\rangle \in S$, there is a *read-write edge*, or rw-*edge*, from T_1 to T_2, denoted $T_1 \xrightarrow{\mathsf{rw}} T_2$, if T_1 reads some data object x for which T_2 is the next writer. More precisely, this means that $t^{\tau_1}_{\mathsf{Read}\langle x\rangle}\langle T_1\rangle <$ $t^{\tau_2}_{\mathsf{Write}\langle x\rangle}\langle T_2\rangle$, and for no other $\langle T_3, \tau_3\rangle \in S$ with $x \in \mathsf{WriteSet}\langle T_3\rangle$ is it the case that $t^{\tau_1}_{\mathsf{Read}\langle x\rangle}\langle T_1\rangle < t^{\tau_3}_{\mathsf{Write}\langle x\rangle}\langle T_3\rangle < t^{\tau_2}_{\mathsf{Write}\langle x\rangle}\langle T_2\rangle$.

Similarly, there is a *write-write edge*, or ww-*edge*, from T_1 to T_2, denoted $T_1 \xrightarrow{\mathsf{ww}} T_2$ if T_1 writes some data object x and T_2 is the next writer of x; i.e., $t^{\tau_1}_{\mathsf{Write}\langle x\rangle}\langle T_1\rangle < t^{\tau_2}_{\mathsf{Write}\langle x\rangle}\langle T_2\rangle$, and for no other $\langle T_3, \tau_3\rangle \in S$ with $x \in \mathsf{WriteSet}\langle T_3\rangle$ is it the case that $t^{\tau_1}_{\mathsf{Write}\langle x\rangle}\langle T_1\rangle < t^{\tau_3}_{\mathsf{Write}\langle x\rangle}\langle T_3\rangle < t^{\tau_2}_{\mathsf{Write}\langle x\rangle}\langle T_2\rangle$.

Finally, there is *write-read edge*, or wr-*edge*, from T_1 to T_2, denoted $T_1 \xrightarrow{\mathsf{wr}} T_2$, if T_1 writes some data object x and T_2 subsequently reads the version of x which T_1 wrote; i.e., $t^{\tau_1}_{\mathsf{Write}\langle x\rangle}\langle T_1\rangle < t^{\tau_2}_{\mathsf{Read}\langle x\rangle}\langle T_2\rangle$, and for no other $\langle T_3, \tau_3\rangle \in S$ with $x \in \mathsf{WriteSet}\langle T_3\rangle$ is it the case that $t^{\tau_1}_{\mathsf{Write}\langle x\rangle}\langle T_1\rangle < t^{\tau_3}_{\mathsf{Write}\langle x\rangle}\langle T_3\rangle < t^{\tau_2}_{\mathsf{Read}\langle x\rangle}\langle T_2\rangle$.

Note that effective time assignments are used throughout these definitions.

For $T_1 \xrightarrow{\mathsf{zz}} T_2$, $\mathsf{zz} \in \{\mathsf{rw}, \mathsf{ww}, \mathsf{wr}\}$ is called the *type* of the edge, which is furthermore *outgoing* from T_1 and *incoming* to T_2.

A schedule S is *conflict serializable* if $\mathsf{DSG}\langle S\rangle$ contains no directed cycles. (Cycles in the DSG are always taken to be directed in this work.) If S is conflict serializable, then an *equivalent serial order* is any (irreflexive) total order $\prec$ of $\mathsf{TransOf}\langle S\rangle$ for which $T_1 \prec T_2$ implies that there is no directed path in $\mathsf{DSG}\langle S\rangle$ from T_2 to T_1. Less formally, if there is an edge of the form $T_1 \xrightarrow{\mathsf{zz}} T_2$, with $\mathsf{zz} \in \{\mathsf{rw}, \mathsf{ww}, \mathsf{wr}\}$, then T_1 must precede T_2 in any equivalent serial order. The order $\prec$ is *commit-order preserving* if for every distinct pair $\{T_1, T_2\} \subseteq \mathsf{TransOf}\langle S\rangle$, $t_{\mathsf{End}}\langle T_1\rangle < t_{\mathsf{End}}\langle T_2\rangle$ implies $T_1 \prec T_2$.

Definition 2.8 (The temporal sense of edges). Given a schedule S, an edge $T_1 \xrightarrow{\mathsf{zz}} T_2$ in $\mathsf{DSG}\langle S\rangle$, with $\mathsf{zz} \in \{\mathsf{rw}, \mathsf{ww}, \mathsf{wr}\}$ is called *(temporally) forward* if T_1 commits before T_2 ($t_{\mathsf{End}}\langle T_1\rangle < t_{\mathsf{End}}\langle T_2\rangle$) and *(temporally) backward* if T_2 commits before T_1 ($t_{\mathsf{End}}\langle T_2\rangle < t_{\mathsf{End}}\langle T_1\rangle$). If $T_1 \xrightarrow{\mathsf{zz}} T_2$ is a forward (resp. backward) edge, this may be noted explicitly via $T_1 \xrightarrow{f:\mathsf{zz}} T_2$ (resp. $T_1 \xrightarrow{b:\mathsf{zz}} T_2$). If the type of an edge (rw, ww, or wr) is unimportant, and only its temporal direction is relevant, this may be denoted via $T_1 \xrightarrow{f:-} T_2$ or $T_1 \xrightarrow{b:-} T_2$. For $T_1 \xrightarrow{d:\mathsf{zz}} T_2$, with $\mathsf{zz} \in \{\mathsf{rw}, \mathsf{ww}, \mathsf{wr}\}$ and $d \in \{f, b\}$, $d:\mathsf{zz}$ is called the *sensed type* of the edge.

Note that a temporally backward edge $T_1 \xrightarrow{b:zz} T_2$ must always connect concurrent transactions, regardless of the type zz. An edge $T_1 \rightarrow T_2$ in which T_2 ends before T_1 begins is never possible in any DSG.

$\mathsf{DSG}\langle S \rangle$ has the *unisense* property if either all edges are temporally forward or else all edges are temporally backward.

Observation 2.9 (Consequences of unisense edges). *Let S be a schedule over* $\mathsf{Trans}\langle \mathbf{D} \rangle$.

(a) *If* $\mathsf{DSG}\langle S \rangle$ *has the unisense property, then it must be acyclic.*
(b) *If* $\mathsf{DSG}\langle S \rangle$ *is conflict serializable, then there is an equivalent serial order $\prec$ which is commit-order preserving iff* $\mathsf{DSG}\langle S \rangle$ *has the unisense property with all forward edges.*

Proof. (a) First assume that all edges are forward. Then, for any cycle $T_1 \xrightarrow{f:-} T_2 \xrightarrow{f:-} \ldots \xrightarrow{f:-} T_n \xrightarrow{f:-} T_1$, it must be the case that $t_{\mathsf{End}}\langle T_1 \rangle < t_{\mathsf{End}}\langle T_2 \rangle < \ldots t_{\mathsf{End}}\langle T_n \rangle < t_{\mathsf{End}}\langle T_1 \rangle$, which is impossible. Thus, no such cycle is possible. The proof for all backward edges is analogous.

(b) This is immediate from the definition of equivalent serial order (see Summary 2.7). $\qquad\square$

Observation 2.10 (Impossible edges). *Let S be a schedule over* $\mathsf{Trans}\langle \mathbf{D} \rangle$.

(a) *In* $\mathsf{DSG}\langle S \rangle$, *edges of sensed type b:ww and b:wr are not possible.*
(b) *In* $\mathsf{DSG}\langle S \rangle$ *with* $\langle T_1, \tau_1 \rangle, \langle T_2, \tau_2 \rangle \in S$ *distinct, if T_2 uses RBWE in S; i.e., if $\tau_2 = $ RBWE, then no edge of the form $T_1 \xrightarrow{f:wr} T_2$ is possible when $T_1 \parallel T_2$.*

Proof. (a) Let $\langle T_1, \tau_1 \rangle, \langle T_2, \tau_2 \rangle \in S$. For an edge of the form $T_1 \xrightarrow{-:ww} T_2$ (resp. $T_1 \xrightarrow{-:wr} T_2$) to exist in $\mathsf{DSG}\langle S \rangle$, it must be the case that $t^{\tau_1}_{\mathsf{Write}\langle T_1 \rangle}\langle x \rangle < t^{\tau_2}_{\mathsf{Write}\langle T_2 \rangle}\langle x \rangle$ (resp. $t^{\tau_1}_{\mathsf{Write}\langle T_1 \rangle}\langle x \rangle < t^{\tau_2}_{\mathsf{Read}\langle T_2 \rangle}\langle x \rangle$) for some $x \in \mathsf{WriteSet}\langle T_1 \rangle \cap \mathsf{WriteSet}\langle T_2 \rangle$ (resp. $x \in \mathsf{WriteSet}\langle T_1 \rangle \cap \mathsf{ReadSet}\langle T_2 \rangle$). Since $t^{\tau_1}_{\mathsf{Write}\langle T_1 \rangle}\langle x \rangle = t_{\mathsf{End}}\langle T_1 \rangle$ for both $\tau_1 = $ RRWE and $\tau_1 = $ RBWE, it follows that $t_{\mathsf{End}}\langle T_1 \rangle = t^{\tau_1}_{\mathsf{Write}\langle T_1 \rangle}\langle x \rangle < t^{\tau_2}_{\mathsf{Write}\langle T_2 \rangle}\langle x \rangle = t_{\mathsf{End}}\langle T_2 \rangle$ (resp. $t_{\mathsf{End}}\langle T_1 \rangle = t^{\tau_1}_{\mathsf{Write}\langle T_1 \rangle}\langle x \rangle < t^{\tau_2}_{\mathsf{Read}\langle T_2 \rangle}\langle x \rangle < t_{\mathsf{End}}\langle T_2 \rangle$); i.e., that T_1 commits before T_2, making any such edge forward.

(b) For an edge of the form $T_1 \xrightarrow{f:wr} T_2$ to exist in $\mathsf{DSG}\langle S \rangle$, there must be an $x \in \mathsf{WriteSet}\langle T_1 \rangle \cap \mathsf{ReadSet}\langle T_2 \rangle$ with $t^{\tau_1}_{\mathsf{Write}\langle T_1 \rangle}\langle x \rangle < t^{\tau_2}_{\mathsf{Read}\langle T_2 \rangle}\langle x \rangle$. Since $\tau_2 = $ RBWE, $t^{\tau_2}_{\mathsf{Read}\langle T_2 \rangle}\langle x \rangle = t_{\mathsf{Start}}\langle T_2 \rangle$, which implies that $t_{\mathsf{End}}\langle T_1 \rangle = t^{\tau_1}_{\mathsf{Write}\langle T_1 \rangle}\langle x \rangle < t^{\tau_2}_{\mathsf{Read}\langle T_2 \rangle}\langle x \rangle = t_{\mathsf{Start}}\langle T_2 \rangle$, and so T_1 and T_2 are not concurrent. $\qquad\square$

3 Concurrency-Based Isolation Levels

In this section, local isolation levels, also called *concurrency-based isolation levels*, are examined in detail, with a focus on how transactions run at different levels of isolation relate to each other. Although the study is formal, the properties of fundamental variants such as read committed (RC) and snapshot isolation (SI) are based upon the way that corresponding levels behave in PostgreSQL [13].

Definition 3.1 (Concurrency-based properties of a transaction). Informally, a *concurrency-based property* (also called *local property*) of a Teff-transaction is one which is based only upon the properties of that transaction, and how it relates to those other transactions in a schedule S with which it is concurrent. Three main ways of characterizing such properties are the following.

Locks: Lock-based characterization of isolation was the first to be studied systematically, [8, 9, Sect. 7.6]. More modern approaches, including S2PL and SS2PL [3], were developed subsequently; however, these approaches have fallen out of favor with the rise of MVCC.

Anomalies: The main ideas (*dirty read, lost update, phantom*) are developed in the early lock-based approach of [8], and are also used in the SQL standard [10, Part 2]. They are somewhat tied to the older SVCC, and due to a lack of rigorous definition, are also subject to multiple interpretations [2].

DSG: In this approach, the properties are based upon edges between concurrent transactions in the DSG. It is well suited to the modern MVCC architecture, providing clean, direct characterizations of isolation levels such as snapshot isolation (SI).

In this work, the focus is upon DSG-based characterization, since the study of serialization in Sect. 4 is based upon it, and a systematic investigation does not appear to have been conducted previously.

Definition 3.2 (Winner and loser transactions). Let $S \in \mathsf{Sched}\langle D \rangle$. As shown in Observation 2.10, edges of (sensed) type b:ww and b:wr are never possible in a $\mathsf{DSG}\langle S \rangle$ when all transactions have effective time assignment RRWE or RBWE. Since those are the only effective time assignments considered, such edge types will not be considered further. Of the four remaining types, f:rw, b:rw, f:ww, and f:wr, any such edge has a *winner* and a *loser*. For all edges of type rw or wr, regardless of sense, the winner is always the first committer. Specifically, in the case of an edge $T_1 \xrightarrow{b:\mathrm{rw}} T_2$, the winner is T_2, while for $T_1 \xrightarrow{f:\mathrm{rw}} T_2$ and $T_1 \xrightarrow{f:\mathrm{wr}} T_2$, the winner is T_1.

For edges of type f:ww, there are two principal variants, *first-committer wins* (FCW) and *first-updater wins* (FUW). With FCW, the winner is the transaction which commits first wins, exactly as for the other three types of edges. With FUW, it is the first transaction which declares a write, according to request, not effective times, which wins. The situation is a bit complex, since there may be a (nonempty) set $X \subseteq \mathsf{DObj}\langle D \rangle$ which each transaction writes. It is the first writer over all such data objects which wins. Formally, for an edge $T_1 \xrightarrow{f:\mathrm{ww}} T_2$, transaction T_1 wins if $\min(\{t^{\mathrm{Req}}_{\mathsf{Write}\langle T_1 \rangle}\langle x \rangle \mid x \in X\}) < \min(\{t^{\mathrm{Req}}_{\mathsf{Write}\langle T_2 \rangle}\langle x \rangle \mid x \in X\})$; otherwise, T_2 wins.

The choice of FCW or FUW is a system-wide policy, since it must be applied to pairs of transactions. Most existing systems use FUW, although Pyrrho [5] (see also Discussion 4.7) is a notable exception.

Definition 3.3 (General local DSG-based isolation levels). A local DSG-based isolation level for a transaction T is defined by three items, the effective time assignment used by T, a set of sensed edge types, and a read-only status. Formally, recall from Summary 2.4 that $\mathsf{TASetEff} = \{\mathsf{RRWE}, \mathsf{RBWE}\}$, and let $\mathsf{CEdges} = \{f\text{:rw}, b\text{:rw}, f\text{:ww}, f\text{:wr}\}$, $\mathsf{RWmode} = \{\mathsf{RW}, \mathsf{RO}\}$. Then, define an *isolation-policy triple* to be an ordered triple $\langle \tau, \Delta, \mu \rangle$ with $\tau \in \mathsf{TASetEff}$, $\Delta \subseteq \mathsf{CEdges}$, and $\mu \in \mathsf{RWmode}$. In $\langle \tau, \Delta, \mu \rangle$, τ identifies the effective time assignment used by the transaction, Δ identifies the types of concurrent edges which are forbidden or impossible to loser transactions, and μ indicates whether the transaction is read-write or read-only. The set of all isolation-policy triples over $\mathbf{D}$ is denoted $\mathsf{PolTr}\langle \mathbf{D} \rangle$. A *local DSG-based isolation level* is defined by such a triple.

In general, a loser transaction with a forbidden edge type must abort in order to satisfy the isolation level. It is very important to understand why only loser transactions may forbid edge types. Consider, for example, an edge of the form $T_1 \xrightarrow{b\text{:rw}} T_2$ in $\mathsf{DSG}\langle S \rangle$. According to the conditions spelled out in Definition 3.2, T_2 is the winner and T_1 is the loser because T_2 commits first (backward edge). Now let $x \in \mathsf{ReadSet}\langle T_1 \rangle \cap \mathsf{WriteSet}\langle T_2 \rangle$. At the time at which T_2 commits, it may not be known that T_1 intends to read x; i.e., it may be the case that $t^{\mathsf{Req}}_{\mathsf{Write}\langle T_2 \rangle}\langle x \rangle \leq t_{\mathsf{End}}\langle T_2 \rangle < t^{\mathsf{Req}}_{\mathsf{Read}\langle T_1 \rangle}\langle x \rangle$. Since committed transactions cannot be rolled back, there is no reasonable way that such a policy could be enforced, other than by delaying the commit of T_2. As such delays are not part of the model, it is impossible for the winner to enforce an edge-prohibition policy.

Discussion 3.4 (Named DSG-based isolation levels). Using the notion of concurrency-based property of Definition 3.1, six named isolation levels are summarized in Table 1. Column 2 indicates the effective time assignment used, while columns 3–6 indicate the status of members of CEdges for that policy, with "P" indicating that the edge type is prohibited for the loser transaction, "X" indicating that it is impossible for the loser transaction to have such an

Table 1. Concurrency properties of transaction classes

Policy	Eff time assign	Status conc edge type				RW mode	Used in practice?
		f:rw	b:rw	f:ww	f:wr		
RC	RRWE					RW	Y
RCX	RRWE		P			RW	?
SI	RBWE			P	X	RW	Y
SIX	RBWE		P	P	X	RW	Y
RCRO	RRWE	X		X		RO	Y
RCXRO	RRWE	X	P	X		RO	?
SIRO	RBWE	X		X	X	RO	Y
SIXRO	RBWE	X	P	X	X	RO	Y

edge under the indicated policy, and blank indicating allowed. These policies are discussed in detail, including the meaning of the abbreviations, in Definitions 3.5, 3.6, and Discussion 3.7.

Definition 3.5 (RRWE-based isolation levels). The fundamental RRWE-based isolation level is *read committed* RC. Its representation as a policy triple is $\langle \text{RRWE}, \emptyset, \text{RW} \rangle$. This may be taken as the definition of the name; thus $\text{RC} = \langle \text{RRWE}, \emptyset, \text{RW} \rangle$. In accordance with RRWE, all reads are performed at request time, while writes are performed at the end of the transaction. There are no further restrictions on allowable edges of the DSG. RC is very common isolation level in real systems, usually offered via the READ COMMITTED SQL isolation level.

Although not widely used in real systems, an important theoretical variant of RC for this work is *read-committed with excluded backward dependencies*, or $\text{RCX} = \langle \text{RRWE}, \{b\text{:rw}\}, \text{RW} \rangle$. It differs from RC only in that backward rw-edges are not allowed, subject, of course, to the general limitation that only a loser transaction may prohibit an edge. As will be seen in Corollary 4.6, it is the weakest local isolation level which guarantees serializability of schedules.

Definition 3.6 (RBWE-based isolation levels). The fundamental RBWE-based isolation level is *snapshot isolation* $\text{SI} = \langle \text{RBWE}, \{f\text{:ww}, f\text{:wr}\}, \text{RW} \rangle$. All effective reads are performed at the beginning of the transaction, while writes are performed at the end. SI is very common isolation level in real systems, often offered using the REPEATABLE READ[2] or SERIALIZABLE SQL isolation level.

For the reader who has learned that concurrent writes are prohibited under SI, it may seem strange that forward ww-edges are allowed for the winner. To understand this better, consider an edge $T_1 \xrightarrow{f:\text{ww}} T_2$ in the DSG, with $T_1 \parallel T_2$, and suppose that $x \in \text{WriteSet}\langle T_1 \rangle \cap \text{WriteSet}\langle T_2 \rangle$. If both T_1 and T_2 run with isolation SI, then since only one of them can be the winner (as defined in Definition 3.2), the edge is not allowed. However, suppose that T_1 runs under SI but T_2 runs under RC (or RCX) isolation. If FCW is used for conflict resolution, then since T_1 commits first, it is the winner. Although T_2 is the loser, its isolation level permits concurrent writes. As it writes x after T_1 commits, that write is completely outside of the lifetime of T_1. If conflicts are resolved via FUW, then either T_1 or T_2 may be the winner. However, even if the winner runs under SI, if the loser runs under RC or RCX, then by a similar argument, both transactions will write x. One must be very careful when asserting that concurrent writes are prohibited under SI when characterizing a mixed-level setting. A transaction running under RC plays by different rules than one running under SI; the SI transaction cannot impose its rules on its RC neighbor.

Note, however, that $T_1 \xrightarrow{f:\text{wr}} T_2$ is impossible when the loser transaction (which must be T_2) runs under SI, since with RBWE reading from a concurrent transaction cannot occur.

---

[2] Strictly speaking, SI does not provide READ COMMITTED isolation. See [2, Remark 9] for details.

The level *snapshot isolation with backward rw exclusion* is SIX $=$ $\langle$RBWE, $\{b$:rw, f:ww, f:wr$\}$, RW$\rangle$. It is the same as SI, save for that backward rw-edges are not allowed. It bears the same relationship to SI as RCX does to RC. It is the sole mode of isolation of Pyrrho, described in Discussion 4.7. As a simple example, suppose that, in schedule S, T_1 running under SIX reads x and writes y, so $t_{\mathsf{Start}}\langle T_1\rangle = t^{\mathsf{RBWE}}_{\mathsf{Read}\langle T_1\rangle}\langle x\rangle < t^{\mathsf{RBWE}}_{\mathsf{Write}\langle T_1\rangle}\langle y\rangle = t_{\mathsf{End}}\langle T_1\rangle$, and T_2, also running under SIX, writes x, so $t_{\mathsf{Start}}\langle T_2\rangle < t^{\mathsf{RBWE}}_{\mathsf{Write}\langle T_2\rangle}\langle x\rangle = t_{\mathsf{End}}\langle T_1\rangle$. Then $T_1 \xrightarrow{d:\mathsf{rw}} T_2$ in DSG$\langle S\rangle$. If $d = f$; i.e., if the edge is forward, then both transactions may commit. However, if $d = b$; i.e., if the edge is backward, then the loser must abort. Under FCW, as is the case in Pyrrho (see Discussion 4.7), this loser is always T_1.

Discussion 3.7 (Read-only transactions). Since it is possible to define a transaction to be read only in SQL, a read-only mode is also supported in the isolation model presented here. RCRO $=$ $\langle$RRWE, $\{f$:rw, f:ww$\}$, RO$\rangle$ is essentially the same as RC with read-only mode enabled. Similarly, SIRO $=$ $\langle$RBWE, $\{f$:rw, f:ww, f:wr$\}$, RO$\rangle$ is essentially the same as SI, with read-only mode enabled. Analogously, RCXRO $=$ $\langle$RRWE, $\{f$:rw, b:rw, f:ww$\}$, RO$\rangle$ and SIXRO $=$ $\langle$RBWE, $\{f$:rw, b:rw, f:ww, f:wr$\}$, RO$\rangle$.

Observe that an edge of the form $T_1 \xrightarrow{f:\mathsf{rw}} T_2$ is not possible if the loser (which must be T_2) is read only.

Definition 3.8 (Ordering of policy triples). Policy triples admit a natural ordering. For TASetEff, use the order RRWE $<$ RBWE, and for RWmode, use the order RW $<$ RO. Then define $\langle \tau_1, \Delta_1, \mu_1 \rangle \leq \langle \tau_2, \Delta_2, \mu_2 \rangle$ iff $\tau_1 \leq \tau_2$, $\Delta_1 \subseteq \Delta_2$, and $\mu_1 \leq \mu_2$. The idea is that lesser policies in this ordering correspond to lower levels of isolation. The intuition behind the ordering on TASetEff is that RBWE imposes more constraints than does RRWE. For example, even under RRWE, a transaction T could perform all of its reads at the very beginning; this would be the case if $t^{\mathsf{Req}}_{\mathsf{Read}\langle T\rangle}\langle x\rangle = t_{\mathsf{Start}}\langle T\rangle$ for every $x \in$ ReadSet$\langle T\rangle$. Similarly, the intuition behind the ordering on RWmode is that prohibiting writes is a stronger condition than allowing them. Finally, it is clear that prohibiting (or rendering impossible) more types of edges results in a more restrictive policy. For the set CBIso $=$ $\{$RC, RCX, SI, SIX, RCRO, SIRO, RCXRO, SIXRO$\}$, RC $<$ RCX $<$ SIX $<$ SIXRO, RC $<$ SI $<$ SIX, RC $<$ RCRO $<$ SIRO, SI $<$ SIRO, and RCX $<$ RCXRO.

4   Multiscope Serializable Isolation

In this section, the main ideas of multiscope serializable isolation are developed.

Definition 4.1 (Transactions with isolation). A *transaction with isolation* is an ordered pair $\langle T, \iota \rangle$ in which $T \in$ Trans$\langle \mathbf{D} \rangle$ and ι is a local DSG-based isolation level. The isolation level ι may be represented either as a member of CBIso, or else as a policy triple. Thus, $\langle T, \mathsf{RCX} \rangle$ and $\langle T, \langle$RRWE, $\{b$:rc$\}$, RWmode$\rangle\rangle$ have exactly the same meaning. The set of all transactions with isolation over $\mathbf{D}$ is denoted TransIso$\langle \mathbf{D} \rangle$. A transaction with isolation $\langle T, \iota \rangle$ carries strictly more information than a transaction with effective time assignment $\langle T, \tau \rangle$. For $\iota = \langle \tau, \Delta, \mu \rangle \in \mathsf{PolTr}$, define $\pi_{\mathsf{TASetEff}}\langle \iota \rangle = \tau$; then $\langle T, \pi_{\mathsf{TASetEff}}\langle \iota \rangle \rangle = \langle T, \tau \rangle$ is the associated transaction with effective time assignment.

Definition 4.2 (Schedule augmentation strategies). When a transaction is ready to commit, a test must be made to determine whether that commit should be allowed. If so, it is added to the set of committed transactions. If not, it must be rejected. To formalize this, begin by defining $\langle S, \langle T, \iota \rangle \rangle$ with $S \in \mathsf{SchedIso}\langle \mathbf{D} \rangle$ and $\langle T, \iota \rangle \in \mathsf{TransIso}\langle \mathbf{D} \rangle$ to be an *augmentation pair* over $\mathbf{D}$ if adding $\langle T, \pi_{\mathsf{TASetEff}}\langle \iota \rangle \rangle$ to S results in a schedule with the property that each transaction in S must either have committed before T, or else run concurrently with T: for every $T' \in \mathsf{TransOf}\langle S \rangle$, one of $t_{\mathsf{End}}\langle T' \rangle < t_{\mathsf{End}}\langle T \rangle$ or $T \parallel T'$ must hold. Think of S as the collection of existing transactions, with $\langle T, \pi_{\mathsf{TASetEff}}\langle \iota \rangle \rangle$ a candidate to be added to S. An *(augmentation) test routine* is a function $\alpha : \mathsf{AugPr}\langle \mathbf{D} \rangle \to \{0, 1\}$, with $\langle S, \langle T, \iota \rangle \rangle \mapsto 1$ indicating that $\langle T, \pi_{\mathsf{TASetEff}}\langle \iota \rangle \rangle$ should commit and be added to S, and $\langle S, \langle T, \iota \rangle \rangle \mapsto 0$ indicating that it should not.

A central example is $\mathsf{AugTest}_{b:rw}$, defined on elements by $\langle S, \langle T, \iota \rangle \rangle \mapsto 1$ iff $\mathsf{DSG}\langle S \cup \{\langle T, \pi_{\mathsf{TASetEff}}\langle \iota \rangle \rangle\} \rangle$ does not contain an edge of the form $T \xrightarrow{b:rw} T'$ or $T' \xrightarrow{b:rw} T$ for a $\langle T', \tau' \rangle \in S$, with T the loser transaction for that edge. Another is $\mathsf{AugTest}_{\mathsf{PolTr}}$, defined on elements by $\langle S, \langle T, \iota \rangle \rangle \mapsto 1$ iff $\mathsf{DSG}\langle S \cup \{\langle T, \pi_{\mathsf{TASetEff}}\langle \iota \rangle \rangle\} \rangle$ does not contain any edges involving T which are forbidden by ι. More precisely, if $\iota = \langle \tau, \Delta, \mu \rangle$, then no new edge of a type in Δ is allowed in the case that T is the loser transaction associated with that edge. Finally, for $\kappa \in \mathsf{PolTr}\langle \mathbf{D} \rangle$, $\mathsf{AugTest}_{\geq \kappa}$ is defined on elements by $\langle S, \langle T, \iota \rangle \rangle \mapsto 1$ iff $\iota \geq \kappa$ and $\mathsf{AugTest}_{\mathsf{PolTr}}(\langle T, \iota \rangle) = 1$. Thus, $\mathsf{AugTest}_{\geq \kappa}$ allows $\langle T, \pi_{\mathsf{TASetEff}}\langle \iota \rangle \rangle$ to be added to S iff ι provides DSG-based isolation at level κ or greater, and adding $\langle T, \pi_{\mathsf{TASetEff}}\langle \iota \rangle \rangle$ to $\mathsf{DSG}\langle S \rangle$ does not result in new edges which are forbidden for T by ι.

These examples are *local* in scope in that the test conditions depend only upon the transaction $\langle T, \iota \rangle$ to be added and certain properties of those transactions in S which run concurrently with it. The reference routine for serialization, which is *global* in scope, is $\mathsf{AugTest}_{\mathsf{DSG}}$, defined on elements by $\langle S, \langle T, \iota \rangle \rangle \mapsto 1$ iff $\mathsf{DSG}\langle S \cup \{\langle T, \iota \rangle\} \rangle$ does not contain any cycles which include T. Other examples which are not local in scope are considered in Summarys 4.10 and 4.12.

Processing a sequence of transactions, in order to build a schedule, is formalized as follows. An *ordered schedule* over $\mathbf{D}$ is a sequence $C = \langle \langle T_1, \iota_1 \rangle, \langle T_2, \iota_2 \rangle, \ldots, \langle T_k, \iota_k \rangle \rangle$ with the properties that $\{\langle T_i, \pi_{\mathsf{TASetEff}}\langle \iota_i \rangle \rangle \mid 1 \leq i \leq k\} \in \mathsf{Sched}\langle \mathbf{D} \rangle$ and for $1 \leq i < j \leq k$, one of $t_{\mathsf{End}}\langle T_i \rangle < t_{\mathsf{End}}\langle T_j \rangle$ or $T_i \parallel T_j$ must hold. The *stepwise commit-based DSG construction* of S using α begins with the empty schedule $\emptyset$, and adds, in the order specified by C, each pair of the form $\langle T_i, \pi_{\mathsf{TASetEff}}\langle \iota_i \rangle \rangle$ which α classifies as acceptable. Formally, $\mathsf{Step}\langle C, \alpha, 0 \rangle = \emptyset$; $\mathsf{Step}\langle C, \alpha, i+1 \rangle = \mathsf{Step}\langle C, \alpha, i \rangle \cup \{\langle T_{i+1}, \pi_{\mathsf{TASetEff}}\langle \iota_{i+1} \rangle \rangle\}$ if $\alpha(\langle \mathsf{Step}\langle C, \alpha, i \rangle, \langle T_{i+1}, \pi_{\mathsf{TASetEff}}\langle \iota_{i+1} \rangle \rangle \rangle) = 1$; $\mathsf{Step}\langle C, \alpha, i+1 \rangle = \mathsf{Step}\langle C, \alpha, i \rangle$ otherwise.

Remark 4.3 (FUW and delayed commit). In the formalism of Definition 4.2, if adding $\langle T, \pi_{\mathsf{TASetEff}}\langle \iota \rangle \rangle$ to S results in a forbidden edge of the form $T' \xrightarrow{\text{---}} T$ or $T \xrightarrow{\text{---}} T'$, then T is not permitted to commit. With FUW, it may be the case that

T' has not yet committed when the test is performed. To maximize concurrency, many systems will suspend T until it is known whether or not T' commits. If T' does not commit, T may continue. Although space limitations preclude formalizing this idea (which involves introducing suspendable transactions with flexible time points), omitting it does not alter the main results developed here. In any case, this issue does not arise with FCW. Indeed, with FCW, all transactions in S will have committed before T.

Definition 4.4 (Serial properties of augmentation strategies). An augmentation test routine α is *serializable generating* (abbreviated SerGen) if for any ordered schedule C of length k, $\mathsf{Step}\langle C, \alpha, k \rangle$ is conflict serializable. Thus, it produces serializable schedules when only transactions which pass its test are allowed. This is the global-scope meaning of serializability, as intended in the SERIALIZABLE isolation level of SQL. The routine α is *commit-order* SerGen if it is SerGen and some equivalent serial order is commit-order preserving.

The routine α is *serializable preserving* (abbreviated SerPres) if for any augmentation pair $\langle S', \langle T, \iota \rangle \rangle$ over $\mathbf{D}$ with $\alpha(\langle S', \langle T, \iota \rangle \rangle) = 1$, T does not participate in any cycle of $S' \cup \{ \langle T, \pi_{\mathsf{TASetEff}} \langle \iota \rangle \rangle \}$. In particular, if S' is conflict serializable, then so too is $S' \cup \{ \langle T, \pi_{\mathsf{TASetEff}} \langle \iota \rangle \rangle \}$. Observe that SerPres always implies SerGen.

In contrast to SerGen, the property of SerPres is local in scope; it does not depend upon properties of the extant schedule S, except those which result from concurrency of its transactions with $\langle T, \tau \rangle$. (Note that S' is universally quantified in the definition of SerPres; it can be the result of running and committing transactions at any level of isolation.) Thus, as elaborated in Discussion 5.1, SerPres is an appropriate semantics for SERIALIZABLE when applied to a single transaction, since, on the one hand, it provides SerGen behavior when all transactions run at that level, and, on the other hand, it provides a meaningful contribution to serializability even when other transactions run at different levels of isolation.

$\mathsf{AugTest}_{\mathsf{DSG}}$ is always SerPres (and hence SerGen). The interesting question is whether there are simpler, local isolation levels which also provide these properties. This is established in the affirmative below.

Theorem 4.5 ($\mathsf{AugTest}_{b:\mathsf{rw}}$ **is both** SerGen **and** SerPres). *The augmentation test* $\mathsf{AugTest}_{b:\mathsf{rw}}$ *is both serializable generating and serializable preserving.*

Proof. Let $\langle S, \langle T, \iota \rangle \rangle \in \mathsf{AugPr}\langle \mathbf{D} \rangle$. For T to be part of a cycle in $S \cup \{ \langle T, \pi_{\mathsf{TASetEff}} \langle \iota \rangle \rangle \}$, it must have at least one outgoing edge. Since it is the last transaction to commit, that outgoing edge must be backward. However, $\mathsf{AugTest}_{b:\mathsf{rw}}$ prohibits edges of type $b:\mathsf{rw}$ for the loser transaction, and since T commits last, it must be the loser for any rw-edge (see Definition 3.2). Thus, T cannot have outgoing edges of type $b:\mathsf{rw}$. Since outgoing edges of types $b:\mathsf{ww}$ and $b:\mathsf{wr}$ are not possible (see Observation 2.10(a)), T cannot have any outgoing edges at all, so it cannot be involved in a cycle of $\langle S, \langle T, \iota \rangle \rangle \in \mathsf{AugPr}\langle \mathbf{D} \rangle$. Hence $\mathsf{AugTest}_{b:\mathsf{rw}}$ is SerPres, and so also SerGen. $\square$

Corollary 4.6 (AugTest$_{\geq RCX}$). *The augmentation test routines* AugTest$_{\geq RCX}$ *and* AugTest$_{\geq SIX}$ *are both* SerPres*(and hence* SerGen*), with* AugTest$_{\geq RCX}$ *the weakest such test defined by a policy triple.*

Proof. The proof follows immediately from Observation 4.5 and the fact that RCX and SIX prohibit edges of type b:rw. □

Discussion 4.7 (Serialization in Pyrrho). The Pyrrho RDBMS [5,6] employs SIX for its only isolation level. In view of Corollary 4.6, it thus provides a working instance of true `SERIALIZABLE` isolation which is based entirely upon a local property of transactions. A unique feature of Pyrrho is that it uses pure FCW for conflict resolution; transactions are never blocked for any reason.

Discussion 4.8 (Wide cursor stability). Some RDBMSs offer a feature called *cursor stability* [9, Sect. 7.6.2] as part the isolation level `READ COMMITTED`. Suppose that transaction T_1, running under RC, reads and then later writes data object x. Suppose further that transaction T_2 also runs under RC and also writes x, and commits between the two operations of T_1. Formally, $t_{\mathsf{Read}\langle T_1 \rangle}^{\mathsf{RRWE}}\langle x \rangle <$ $t_{\mathsf{Write}\langle T_2 \rangle}^{\mathsf{RRWE}}\langle x \rangle < t_{\mathsf{End}}\langle T_2 \rangle < t_{\mathsf{Write}\langle T_1 \rangle}^{\mathsf{RRWE}}\langle x \rangle < t_{\mathsf{End}}\langle T_1 \rangle$. This behavior is not serializable because in the serialization $T_1 \prec T_2$, the final write of x is by T_2, not by T_1 as it should be; and in $T_2 \prec T_1$, T_1 does not read the initial value of x, before T_2 wrote it. *Cursor stability* prevents this "in between" write by T_2, either by locking x between the read and write of T_1, or else by T_1 rereading x after the commit of T_2. However, it does this only when the read and the write of T_1 are part of the same SQL statement. With serialization, this behavior is not permitted, regardless of the "distance" between $t_{\mathsf{Read}\langle T_1 \rangle}^{\mathsf{RRWE}}\langle x \rangle$ and $t_{\mathsf{Write}\langle T_1 \rangle}^{\mathsf{RRWE}}\langle x \rangle$, since there is an edge $T_1 \xrightarrow{b:rw} T_2$ which is not allowed RCX or any stronger isolation level. Thus, RCX effectively provides *wide cursor stability*, which does not require the read and the write to be part of the same statement. If the loser runs under an isolation which is SerPres, cursor stability is automatic.

If preservation of commit order is desired in the serialization, then AugTest$_{b:rw}$ is actually optimal in the following sense.

Theorem 4.9 (Optimality of AugTest$_{b:rw}$). AugTest$_{b:rw}$ *is a globally optimal commit-order-preserving* SerGen *augmentation test routine, in the precise sense that any other such routine* AugTest$'$ *with the property that* AugTest$'(\langle S, \langle T, \tau \rangle \rangle) = 1$ *but* AugTest$_{b:rw}(\langle S, \langle T, \tau \rangle \rangle) = 0$ *for some* $\langle S, \langle T, \tau \rangle \rangle \in$ AugPr$\langle \mathbf{D} \rangle$ *cannot be commit-order preserving.*

Proof. The proof follows immediately from Observation 2.9(b), since the presence of any backward edge in the DSG implies that commit order must be violated in any serialization.

Summary 4.10 (SSI— a preemptive serializable-preserving strategy). The SerGen strategy SSI (serializable SI) [4,7] is used to implement the `SERIALIZABLE` isolation level of PostgreSQL [12]. Define a *dangerous structure*

(DS) to be a path in the DSG of the form $T_2 \xrightarrow{\;-:-\;} T_1 \xrightarrow{\;b:rw\;} T_0$ in which T_0 commits first and $T_1 \parallel T_2$. (Note that $T_0 \parallel T_1$ is automatic since the edge is backward.) T_0 and T_2 may be the same transaction, in which case $\{T_0, T_1\}$ forms a cycle by itself. As shown in [7, Thm. 2.1], if all transactions run under SI, then every cycle of the DSG contains a DS. To represent this in terms of an augmentation routine, define $\mathsf{AugTest_{SSI}}$ on elements by $\langle S, \langle T, \iota \rangle \rangle \mapsto 0$ iff T is the last transaction to commit in a DS of $\mathrm{DSG}\langle S \rangle$.

It is worth noting that it is not necessary to require that all transactions run under SI; RC is sufficient. However, a proof will not be presented here; only the original SI-based SSI will be evaluated. Unfortunately, while serializable generating, $\mathsf{AugTest_{SSI}}$ is not serializable preserving.

Proposition 4.11 (Serialization properties of SSI). $\mathsf{AugTest_{SSI}}$ is SerGen but not SerPres.

Proof. The proof that $\mathsf{AugTest_{SSI}}$ is SerGen is found in [4,7]. To show that it is not SerPres, it suffices to present a counterexample. In Fig. 1, a DSG cycle $T_0 \xrightarrow{\;f:rw\;} T_4 \xrightarrow{\;b:rw\;} T_3 \xrightarrow{\;f:rw\;} T_2 \xrightarrow{\;b:rw\;} T_1 \xrightarrow{\;b:rw\;} T_0$. consisting of five transaction is shown. Time increases horizontally, with the beginning and end of each transaction marked by a vertical bar; the commit order is $\langle T_0, T_3, T_1, T_2, T_4 \rangle$. The reads and writes of each transaction are depicted by $r\langle - \rangle$ and $w\langle - \rangle$ respectively, Each transaction T_i runs under SSI, so as Teff-transactions, $\langle T_i, \tau_i \rangle = \langle T_i, \mathsf{RBWE} \rangle$. The last transaction to commit, T_4, is not part of any DS. So, letting $S' = \{\langle T_i, \tau_i \rangle \mid 0 \le i \le 3\}$, it is immediate that $\mathsf{AugTest_{SSI}}\langle S', \langle T_i, \tau_i \rangle \rangle = 1$. □

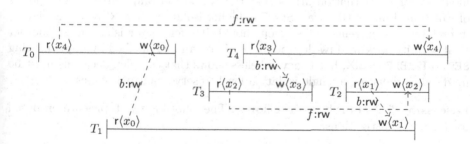

Fig. 1. DSG with no DS involving the last transaction to commit

Summary 4.12 (SSN). Recently, a preemptive SerGen strategy which relies on a more complex "dangerous structure" than does SSI has been developed [15]. Dubbed *serializable safety net*, or SSN for short, it is of particular relevance to this work in that any local level of isolation which is at least as strong as RC may be serialized, thus reinforcing the observation that local isolation and serialization are orthogonal. As is the case with SSI, SSN is not serializable preserving. While space limitations preclude a full proof, the reader familiar with the construction in [15] can verify easily that the schedule of Fig. 1 provides the necessary counterexample.

5   Conclusions and Further Directions

Discussion 5.1 (Conclusions). The semantics of including serializable isolation, global in scope, in a mixed-mode setting with levels of local scope, such as RC and SI, has been investigated. Two alternatives have been identified. In the first, *serializable generating* (SerGen), the serializable level has meaning only when all transactions run at that level. SSI and SSN fall into that category. While highly effective when used exclusively, they revert to a lower level otherwise, with little or no additional benefit. The second is *serializable preserving* (SerPres), which has the property that, regardless of the DSG consisting of all committed transactions, adding a new transaction will never result in a new cycle. Used for SERIALIZABLE, it provides a semantics which is both local and global in scope, in line with the original intent of the SQL standard. The augmentation test AugTest$_{DSG}$ which examines the entire DSG for cycles has this property, although it has large space complexity.[3] Identified in this paper is a far less complex option, in which a standard local complexity level, such as RC or SI, is augmented to disallow all backward rw-dependencies (resulting in RCX or SIX). As elaborated in Discussion 5.2, it is proposed that this alternative be explored more thoroughly, as a suitable implementation of SQL SERIALIZABLE.

An additional issue arises if SI is used to implement REPEATABLE READ in an RDBMS, while RCX is used to implement SERIALIZABLE. The unusual (and likely unwanted) situation arises that the two are incomparable as local levels of isolation. Put another way, SI offers higher isolation than RCX in one way — it prohibits concurrent writes, even though it offers lower isolation in another — it permits backward rw-dependencies. This can be remedied by implementing SERIALIZABLE as SIX, but it nevertheless shows that complex decisions must be made when enlisting a single isolation level to serve multiple scopes.

Discussion 5.2 (Further directions). The following two topics are proposed for further investigations.

PERFORMANCE MEASUREMENT FOR RCX AND SIX: Although SIX is used in the Pyrrho system (see Discussion 4.7), it has not been compared for performance to alternatives such as SSI (used in PostgreSQL). Since SIX appears to perform well in Pyrrho, it may be the case that although it will have a higher number of false positives (aborted transactions due to concurrency conflicts) than SSI, (since every DS must contain a backward dependency), it may nevertheless be completely satisfactory for many transaction mixes. Advantages of RCX

---

[3] It should be noted that one experimental system, called PSSI, has taken exactly the approach of constructing the entire DSG (with all transactions running under SI) to achieve serializable generating behavior, reporting good results [14].

and SIX (over SSI and SSN) include that they are far simpler to implement, and that they provides serializable-preserving isolation. It is thus proposed to study their performance experimentally. In addition, a parallel comparison of FUW and FCW is warranted, given the success of FCW in Pyrrho.

EXTENSION TO LOCK-BASED APPROACHES: Due to space limitations, the local levels of isolation studied in this paper have been limited to those which are DSG based. However, locked-based levels, such as S2PL and SS2PL, are also of importance, as they are the classical local isolation levels which deliver serializable-preserving behavior. An investigation of how they fit into the framework of this paper is therefore warranted.

References

1. Adya, A., Liskov, B., O'Neil, P.E.: Generalized isolation level definitions. In: Lomet, D.B., Weikum, G. (eds.) Proceedings of the 16th International Conference on Data Engineering, San Diego, California, USA, 28 February - 3 March 2000, pp. 67–78 (2000)
2. Berenson, H., Bernstein, P.A., Gray, J. Melton, J., O'Neil, E.J., O'Neil, P.E.: A critique of ANSI SQL isolation levels. In: Proceedings of the 1995 ACM SIGMOD International Conference on Management of Data, San Jose, California, 22–25 May 1995, pp. 1–10 (1995)
3. Breitbart, Y., Georgakopoulos, D., Rusinkiewicz, M., Silberschatz, A.: On rigorous transaction scheduling. IEEE Trans. Softw. Eng. **17**(9), 954–960 (1991)
4. Cahill, M.J., Röhm, U., Fekete, A.D.: Serializable isolation for snapshot databases. ACM Trans. Database Syst. **34**(4), 20 (2009)
5. Crowe, M.: The Pyrrho database management system. https://pyrrhodb.uws.ac.uk/index.htm. Accessed 2019-03-30
6. Crowe, M.: Transactions in the Pyrrho database engine. In: Hamza, M.II. (ed.) IASTED International Conference on Databases and Applications, part of the 23rd Multi-Conference on Applied Informatics, Innsbruck, Austria, 14–16 February 2005, pp. 71–76. IASTED/ACTA Press (2005)
7. Fekete, A., Liarokapis, D., O'Neil, E.J., O'Neil, P.E., Shasha, D.: Making snapshot isolation serializable. ACM Trans. Database Syst. **30**(2), 492–528 (2005)
8. Gray, J., Lorie, R.A., Putzolu, G.R., Traiger, I.L.: Granularity of locks and degrees of consistency in a shared data base. In: Nijssen, G.M. (ed.) Modelling in Data Base Management Systems, Proceeding of the IFIP Working Conference on Modelling in Data Base Management Systems, Freudenstadt, Germany, 5–8 January 1976, North-Holland, pp. 365–394 (1976)
9. Gray, J., Reuter, A.: Transaction Processing: Concepts and Techniques. Morgan Kaufmann, Burlington (1993)
10. Melton, J. (ed.): ISO/IEC 9075:2011, Information Technology – Database Languages – SQL. ANSI, (The 2011 SQL Standard) (2011)
11. Papadimitriou, C.: The Theory of Database Concurrency Control. Computer Science Press (1986)
12. Ports, D.R.K., Grittner, K.: Serializable snapshot isolation in PostgreSQL. Proc. VLDB Endow. **5**(12), 1850–1861 (2012)
13. PostgreSQL: The World's Most Advanced Open Source Relational Database. https://www.postgresql.org. Accessed 30 Mar 2019

14. Revilak, S., O'Neil, P.E., O'Neil, E.J.: Precisely serializable snapshot isolation (PSSI). In: Proceedings of the 27th International Conference on Data Engineering, ICDE 2011, 11–16 April 2011, Hannover, Germany, pp. 482–493 (2011)
15. Wang, T., Johnson, R., Fekete, A., Pandis, I.: Efficiently making (almost) any concurrency control mechanism serializable. VLDB J. **26**(4), 537–562 (2017)
16. Weikum, G., Vossen, G.: Transactional Information Systems. Morgan Kaufmann, Burlington (2002)

Data Warehouses

Data Reduction in Multifunction OLAP

Ali Hassan$^{(\boxtimes)}$ and Patrice Darmon

R&D Umanis, 7-9 rue Paul Vaillant Couturier, 92300 Levallois-Perret, France
{ahassan,pdarmon}@umanis.com

Abstract. Multifunction OLAP allows to associate several types of aggregation functions to the same measure: general, dimensional for each analysis axis, hierarchical for each hierarchy and differentiated for each granularity level. These functions are generally non-commutative, so, an execution order between the functions is predefined. Pivot tables and several diagram types (bars, pies, etc.) are used to visualize interactively the result of an OLAP query. Unfortunately, no works investigate readability issues in multifunction OLAP. Therefore, we propose a post-processing method to reduce data size of the multifunction OLAP query result in order to improve the readability. This method aggregates data at higher granularity levels, i.e., doing a Rollup operation. It starts by studying the current query to find the functions that have already been executed. Then, it finds all possible Rollup operations, which respect the execution order and the aggregation constraints, and it calculates its data size. We propose several strategies to select a Rollup that gives a readable diagram and keeps as many details as possible: looking at the data size only, the number of implicated granularity levels and the number or the type of implicated dimensions. Once a Rollup is selected, we find the functions that realize it and we execute them in the right execution order.

Keywords: OLAP · Multifunction aggregation · Data reduction

1 Introduction

In business, decision support systems are used by decision makers to manage their company. These systems are powered by data from the internal production systems and external environment of the enterprise. However, the exploitation of this distributed and heterogeneous information needs extracting, transforming and loading (ETL) it into a form suitable to the analysis [13]. Therefore, data warehouses are used where the data are modeled according to a multidimensional model. This type of modeling represents the analyzed data (measures) as points in a multidimensional space. These measures are observed according to several dimensions. Each dimension has several levels of granularity/detail organized in hierarchies. Aggregation functions are used to obtain a more global view of measures values at less detailed levels.

Classically, data warehouse provides the aggregation of a measure over all the multidimensional space with an identical aggregation function. In order to overcome this limitation, [7,8] proposed a multidimensional model expressive enough

© Springer Nature Switzerland AG 2019
T. Welzer et al. (Eds.): ADBIS 2019, LNCS 11695, pp. 409–424, 2019.
https://doi.org/10.1007/978-3-030-28730-6_25

to associate several aggregation functions with the same measure. For example, such a model allows to analyze the average annual precipitation, which is calculated by the sum of daily precipitation. However, the average precipitation of a department (a subdivision of regions in the French administrative geographical system) is calculated by the average precipitation of all cities. Since the functions are generally not commutative, the proposed model control calculation validity by planning an execution order between the aggregation functions. For example, it is necessary to aggregate the precipitation by the sum over the dimension "Date" before aggregating them by the average on the dimension "Geography".

Illustration of the Problem

Most of decision support systems rely on an OLAP approach (On Line Analytical Processing) facilitating interactive analysis and data synthesis. Decision makers can use OLAP operators (e.g. Rollup and Drill-down) to modify and refine measures analysis and navigate multidimensional space. The data are therefore grouped according to selected levels and aggregated using aggregation functions. The result of an OLAP query is typically visualized using a pivot table and diagrams (bars, pie, etc.). For example, the pivot table at the top of Fig. 1 shows the average daily precipitation (the first two days of January and February) by department. We can see that such a table has an acceptable readability, although it displays 120 information. The display of the same amount of data in bar chart can significantly reduce the readability (see the bottom left part of Fig. 1) or becomes completely unreadable using a pie chart (see the bottom right part of Fig. 1).

The aim of this paper is to propose a system that automatically adjusts the size of data to display in order to keep the readability of used charts.

Average Precipitation		Year	2010				2011				2012				2013				2014			
		Month	Jan.		Feb.		Jan.		Feb.		Jan.		Feb.		Jan.		Feb.		Jan.		Feb.	
		Day	1	2	1	2	1	2	1	2	1	2	1	2	1	2	1	2	1	2	1	2
Country	Region	Department																						
France	Midi-Pyrénées	Haute-Garonne	2,38	3,75	3,75	2,88	1,63	1,38	1,63	4,38		3,5	2,38	1,38	1,88	3,38	1,75	1,38	3,63	4,13	1,63	1,25
		Hautes-Pyrénées	1,88	0,63	0,88	2,38	4,38	2,25	5,38	2			3,25		2,13	0,5	0,75	6,38	1,75	2,13	1,63	4
	Ile-de-france	Paris			3	3,13	3,25	2,38	4,25	0,38	3,13	1,5	3,5	1,63	0,25	1,75	0,63	2,88	4,25	2,25	3,75	3,75
		Seine-et-Marne	2,75	1,25		1,38	3,75	1,88	3,5	3,5	1,5	4,38	0,75	2	2,38	2	3,5	3,88	5,13	3,63	3,25	2,88
	Rhône-Alpes	Rhône	1,63	1,38		1,88	1,25	2,75	6,13	4	1,38	1,63	1,63	3,38	0,75	2,38	1,63	3,25	1,38	3,13	1,25	2,5
		Isère	2,88	0,75		0,13	3,13	2,88		2,5	1,75	3,13	3,13	2,25	2	1,38	0,63	2,63	2,13		4,5	0,13

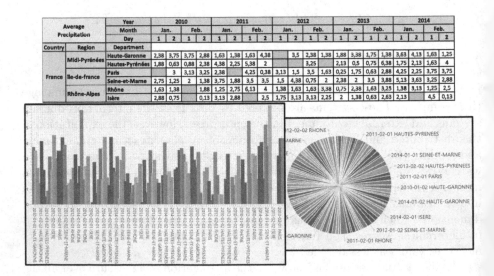

Fig. 1. Visualization of average precipitation by day and by department.

The article is organized as follows. Section 2 reviews related work. Section 3 describes the background of the multifunction multidimensional model and the case study. Section 4 is devoted to the proposed data reduction method. Section 5 presents the implementation of our proposal. We conclude in Sect. 6.

2   Related Work

Several works in the literature [1,17] propose methods to select the "best" visualization technique according to the user's goal and the features of data (e.g. its size, its type and the number of its dimensions). But these works do not discuss how to visualize the data when the data size exceeds the limits of the "best" visualization technique. Thus, several methods have been used to allow the visualization of large size of data. These methods can be classified into two categories [14]: (1) visualization methods. (2) data reduction methods.

2.1   Visualization Methods

These methods aim to visualize all data. Three methods are proposed to improve the visualization of large size of data.

Visual Reduction. Due to the limited size of the display screen, the representation of large size of data will easily clutter visually. To resolve this problem, several methods are proposed, such as using a miniature representation of information [9], pixel-oriented visualization methods that use each pixel to visualize a data value [12], spatial displacement methods to solve overlay problem [24] and methods of dimensions re-ordering that reduce visual clutter without reducing the amount of information [21]. However, all these techniques propose to visualize each data element, which does not necessarily facilitate the reading and observation of the data.

Progressive Methods. Progressive visual analytics [23] uses partial results to display the data. It produces more complete results over time. This allows to respond to queries immediately. The main problem with these methods is that the result remains unconvincing until all progress is complete.

Predictive Methods. Predictive visual analytics [18] uses the techniques of pre-fetching where the performance of a system depends on the predictability of future requests. However, these methods work well only in case of limited user interactions [14].

2.2   Data Reduction Methods

These methods are intended to reduce the size of data before sending them to the visualization procedure. Three strategies are proposed in the state of art.

Sampling and Filtering Methods. The sampling [4] and filtering [2] are traditionally used to select a subset of data before the visualization. However, sampling and filtering may eliminate valuable information, and generally cannot provide an overview of the data distribution [15].

Approximation Methods. These methods propose to aggregate the data so that it has the same visualization (image) as the basic data. *M4* [10] is a specific approximate method for temporal data. It proposes to aggregate the temporal data according to the number of pixels used to display the result. A multidimensional generalization of this method is proposed in *VDDA* [11]. However, the number of dimensions used for displaying the data are predefined (fixed). In other words, this method does not reduce the number of dimensions according to the size of data to display.

Hierarchical Exploration (Data Cube) Methods. Most of the works using these kind of methods propose to use a tree index based on binned aggregation to reduce the size of the data cube. *imMens* [16] and *Nanocubes* [15] propose to pre-aggregate and store in memory all possible aggregations on all dimensions. *Gaussian Cubes* [25] and *TopKube* [19] extended *Nanocubes*. The first one by adding columns containing varied statistics to perform more advanced aggregation operations. The second one by including ranking information to reply to Top-K queries.

However, the previous four proposals have a high memory cost and use a tree index where the highest level concerns the spatial dimension, the lowest level concerns the temporal dimension and the intermediate levels concern the thematic dimensions. Which requires to aggregate the data in the same order for all case studies (first over the temporal dimension, then over the thematic dimensions, and finally over the spatial dimension).

Hashedcubes [20] overcomes the previous problems. It proposes a new structure that allows to quickly query a large quantity of multidimensional data without storing in memory all possible aggregations. The main idea is to integrate all dimensions into a single hierarchy in a carefully selected order. However, this proposal does not allow to treat multiple hierarchies.

The issue of readability of visualized data has been dealt with some works. [22] for spatial data and [3,5] for the readability of pivot tables propose post-processing methods to reduce the size of data to be visualized. These methods consist in grouping and aggregating (clustering) data. After identifying the clusters, a new representation for each cluster is defined.

Unfortunately, none of this work studies how to reduce the size of data in the context of multifunction OLAP. So, in this article, we propose a method to improve the readability of the results of multifunction OLAP queries.

3   Preliminaries

In this section, we present the conceptual model of multifunction multidimensional data base and the case study.

3.1   Multifunction Multidimensional Conceptual Data Model

Let us define a finite set of non-redundant names: $\mathcal{N} = \{n_1, n_2, ...\}$ and a finite set of aggregation functions: $\mathcal{F} = \{f_1, f_2, ...\}$.

Definition 1. A *fact* F_i is defined by (n^{Fi}, M^i):

- $n^{Fi} \in \mathcal{N}$ is the fact name,
- $M^i = \{m_1, ..., m_{pi}\}$ is a set of *measures*.

Definition 2. A *dimension* D_i is defined by (n^{Di}, A^i, H^i):

- $n^{Di} \in \mathcal{N}$ is the dimension name,
- $A^i = \{a_1^i, ..., a_{ri}^i\} \cup \{Id^i, All^i\}$ is the *attributes of the dimension*,
- $H^i = \{H_1^i, ..., H_{si}^i\}$ is the *hierarchies of the dimension*.

The dimension's attributes are organized in hierarchies from the most detailed granularity (Id^i) up to the most general one (All^i).

Definition 3. A *hierarchy* H_j is defined by $(n^{Hj}, P^j, \prec^{Hj})$:

- $n^{Hj} \in \mathcal{N}$ is the hierarchy name,
- $P^j = \{p_1^j, ..., p_{qj}^j\}$ is a set of *attributes* $(P^j \subseteq A^i)$ called *parameters*,
- $\prec^{Hj} = \{(p_x^j, p_y^j) \mid p_x^j \in P^j \wedge p_y^j \in P^j\}$ is a binary relation (antisymmetric and transitive) determining a navigation path on the dimension.
- $\text{Weak}^{Hj} : P^j \rightarrow 2^{A^i \setminus P^j}$ is a function that associates to each parameter a set of attributes of the dimension, called *weak attributes*.

Assuming $M = \bigcup_{i=1}^{n} M^i$, $H = \bigcup_{i=1}^{m} H^i$, $P^i = \bigcup_{j=1}^{s_i} P^j$ and $P = \bigcup_{i=1}^{m} P^i$.

Definition 4. A *multidimensional schema* is defined by (F, D, Star, Aggregate):

- $F = \{F_1, ..., F_n\}$ is a set of facts,
- $D = \{D_1, ..., D_m\}$ is a finite set of dimensions,
- Star: $F \rightarrow 2^D$ is a function relating each fact to its analysis axes (dimensions),
- Aggregate: $M \rightarrow 2^{\mathbb{N}^* \times \mathcal{F} \times 2^D \times 2^H \times 2^P \times \mathbb{N}^-}$ associates each measure to its aggregation functions. It defines 4 types of aggregation functions:
  - General: aggregates measure values with any parameter,
  - Dimensional: aggregates the measure on all the considered dimension,
  - Hierarchical: aggregates the measure on the whole concerned hierarchy,
  - Differentiated: aggregates the measure between two parameters.

To consider the non-commutativity between functions, $\mathbb{N}^*$ gives an *execution order* to each function. $\mathbb{N}^-$ indicates the *constraint*, i.e., if the aggregation is calculated from a specific level (using a negative value) or not (using 0).

3.2   Case Study

The schema in the left part of Fig. 2 (obtained thanks to the "Star" function) illustrates our example to analyze the average precipitation. The fact "Precipitation" is analyzed according to three dimensions: "Geography", "Dates" and "Time". Each dimension is composed of one hierarchy. The "Time" dimension organizes the hourly granularities at which precipitations are recorded during

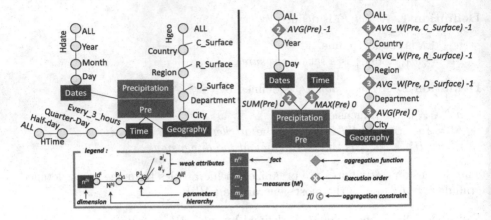

Fig. 2. Multifunction multidimensional conceptual case study schemata.

the day. Here, we should indicate that the precipitation is recorded during the day cumulatively, which means that the precipitation of a day is the last value (MAX) recorded during this day.

Thanks to "Aggregate" function, an aggregation schema can be obtained for each measure. The aggregation schema of our measure of the average precipitation "Pre" is shown in the right part of Fig. 2. Precipitation should be aggregated on the "Time" dimension (having an execution order 1) before the "Dates" dimension (with execution orders of value 2) and finally on the "Geography" dimension (having execution orders equal to 3).

As the precipitation is cumulative during the day, it is aggregated on the "Time" dimension using MAX (dimensional function). In addition, the precipitation is aggregated on the "Dates" dimension to the "Year" level using the SUM (dimensional function) and above using the AVG (differentiated function).

The "Geography" dimension has several differentiated functions AVG and AVG_W which aggregates the data from the level directly below the considered one (constraint -1). The function, weighted average AVG_W(X, Y) returns the average of X weighed by Y. AVG_W(X, Y) $= \frac{\sum(X \times Y)}{\sum Y}$. Thus, the precipitation is weighted by the surface of the current level of the "Geography" dimension for calculating it on the higher level. By example, the regional average precipitation is not calculated from the precipitation of cities directly but it is calculated from the departmental precipitation weighted by the surface of the department.

4   Data Reduction Method

Our proposal is based on a method of post-processing of the results of multifunction OLAP queries. This method reduces the size of data to display according to used diagram type. For example, bar charts can normally display more data than pie charts. So, pie charts need to aggregate data more than bar charts.

This method consists in aggregating the data automatically to a less detailed granularity, i.e., doing a Rollup. This method involves several steps: (1) study the current OLAP query. (2) find possible Rollup operations. (3) calculate the data size for all possible Rollups. (4) choose a Rollup operation based on a selection strategy. (5) realize the chosen Rollup. These steps are detailed in the following:

4.1   Study the Current OLAP Query

Our method examines firstly the current OLAP query, exploiting the aggregation schema of the concerned measure, to find the maximum execution order of aggregation functions that were already executed during the query. Algorithm 1 details this operation. This algorithm takes as input the measure and the current set of parameters, where each parameter belongs to a different dimension and it presents a granularity level selected to observe the measure.

The algorithm searches the function used to aggregate the measure at each parameter. If the parameter is not the finest granularity (line 3), we check firstly whether there is a differentiated aggregation function concerning the parameter (line 4). Otherwise, we look for a hierarchical function (lines 5–6). If there is no result, then we search a dimensional function (lines 7–8). If it is not found, we look for the general function of the measure (lines 9–10). It is important to note that the aggregation functions cover the multidimensional schema completely [8], i.e., for each parameter (granularity level), there is a well-known function for aggregating the measure. Therefore, arriving on line 10, the algorithm necessarily finds an aggregation function. Once the aggregation functions are found, we find the maximum execution order (lines 11–12).

Algorithm 1: Study current query

Input: current measure $m_c \in M$, current set of parameters $Ps_c \subseteq P$
Output: $Max\_execution\_order$

1   $Max\_execution\_order \leftarrow 0$;
2   **foreach** $P_c^{D_i H_j} \in Ps_c$ **do**
3   **if** $(P_c^{D_i H_j} \neq Id^i)$ **then**
4       $Agg \leftarrow$ find_differentiated_aggregation_function$(m_c, P_c^{D_i H_j})$;
5   **if** *(Agg is Null)* **then**
6         $Agg \leftarrow$ find_hierarchical_aggregation_function(m_c, H_j);
7   **if** *(Agg is Null)* **then**
8           $Agg \leftarrow$ find_dimensional_aggregation_function(m_c, D_i);
9   **if** *(Agg is Null)* **then**
10             $Agg \leftarrow$ find_general_aggregation_function(m_c);
11   **if** *(Agg.execution_order > Max_execution_order)* **then**
12         $Max\_execution\_order \leftarrow Agg.execution\_order$;
13 **return** $Max\_execution\_order$;

For example, in the query of Fig. 1, the current parameters of dimensions "Dates" and "Geography" are "Day" and "Department" respectively. The current parameter of a dimension unconsidered in the query is the extremity parameter. Thus, the current parameter of "Time" dimension in Fig. 1 is "All^{Time}".

Using the aggregation schema (the right part of Fig. 2), Algorithm 1 finds that the measure is aggregated at levels "Department" and "All^{Time}" using the aggregation functions AVG and MAX which have the execution orders 3 and 1 respectively. Thus, the max execution order already used is 3. Concerning the parameter "Day", there is no aggregation because it is the finest granularity.

4.2   Find Possible Rollup Operations

To reduce the data size, we propose to aggregate the result of the current OLAP query without re-querying the database. But since the data are already aggregated and the aggregation functions are non-commutative, not all aggregations are possible. By controlling the calculation validity of aggregation functions, Algorithm 2 finds the possible Rollup operations and eliminates those which are forbidden. We can determine the possible Rollups by the Cartesian product of lists of possible parameters on each dimension. A parameter is considered as "possible" if the two following conditions are met:

- **execution order condition:** the function aggregating the measure at this parameter has an execution order equal to or greater than the maximum execution order already executed;
- **constraint condition:** the aggregation at this parameter is feasible from current parameters.

Algorithm 2: Find possible Rollups

Input: current measure $m_c \in M$, current parameters $Ps_c \subseteq P$,
            $Max\_execution\_order$ (from Algorithm 1)
Output: $possible\_Rollups$

1 **foreach** $dimension\ D_i \in D$ **do**
2     $p_c^{D_i H_j} \leftarrow P^i \cap Ps_c;$ /* the current parameter of D_i */
3     $possible\_parameters^{D_i}.\text{add}(p_c^{D_i H_j});$
4   **for** $p^{D_i} \leftarrow (p_c^{D_i H_j} +^{\prec} 1)^1$ **to** All^{D_i} **do**
5        $Agg \leftarrow \text{find\_aggregator}(m_c, p^{D_i});$
6   **if** $(Agg.execution\_order \geq Max\_execution\_order)$ **and**
         $((Agg.constraint{=}0)$ **or**
         $(\exists p \in possible\_parameters^{D_i} \mid p = p^{D_i} +^{\prec} Agg.constraint)^1)$ **then**
7           $possible\_parameters^{D_i}.\text{add}(p^{D_i});$
8 $possible\_Rollups \leftarrow \prod_{i=1}^{m} possible\_parameters^{D_i};$
9 $possible\_Rollups \leftarrow possible\_Rollups \setminus Ps_c;$
10 **return** $possible\_Rollups;$

---

[1] $(p +^{\prec} i)$ returns the parameter at the i-th position relative to p: $(i = 1)$ returns the directly upper parameter; $(i < 0)$ returns a lower parameter.

Algorithm 2 takes three inputs: the measure, current parameters and maximum execution order (resultant of Algorithm 1). To find the possible parameters on a dimension (lines 1 to 7), the algorithm starts by identifying the current parameter of this dimension. To do this it only has to intersect the current parameters (Ps_c) with the parameters of the dimension (P^i) because each current parameter belongs to a different dimension (line 2). This current parameter is considered as "possible" (line 3). All upper parameters (p^{D_i}) are checked if they meet the execution order and constraint conditions (lines 4 to 7). So the function "find_aggregator()", which is equivalent to lines 4 to 10 of Algorithm 1, searches the aggregation function for each upper parameter (line 5). Then, we check the execution order condition (first condition of line 6). We also check:

- if the function has a constraint 0 (second condition of line 6), that means aggregation can be calculated from any lower parameter including the current parameter. In this case the constraint condition is accomplished, or
- if the specific lower parameter ($p^{D_i} +^\prec Agg.constraint$), from which the considered aggregation must be calculated, is "possible" (third condition of line 6). In this case also the constraint condition is accomplished.

The result of the Cartesian product (line 8) includes the current analysis (Ps_c). We must remove it (line 9).

For example, for the query of Fig. 1, on one side, any aggregation (Rollup) on the dimension "Dates" is forbidden because their aggregation functions have an execution order (2) lower than the maximum execution order already executed (3). So, there is only one possible parameter (the current parameter "Day"). On the other side, with an execution order 3, all aggregations on "Geography" dimension are possible. Concerning "Time" dimension, the current parameter "All^{Time}" is the extremity one, so there is no other possible parameter. Thus:
$possible\_parameters^{Dates} = [Day]$
$possible\_parameters^{Geography} = [Department, Region, Country, All^{Geography}]$
$possible\_parameters^{Time} = [All^{Time}]$
By doing the Cartesian product and removing the current analysis [Day, Department, All^{Time}], three Rollups are possible:
[[Day, Region, All^{Time}], [Day, Country, All^{Time}], [Day, $All^{Geography}$, All^{Time}]]

4.3   Calculate the Data Size for All Possible Rollups

In order to choose a Rollup, it is necessary to know the resulting data size of each Rollup operation. To calculate this data size, we distinguish two cases:

- **perfect data:** we have perfect data if the measure has a value for all possible combinations of members of parameters (levels) involved in the query. For example, considering the schema in Fig. 2, we have perfected data if the precipitation of all cities is recorded every three hours every day. In this case, to calculate the data size for each Rollup, we only need to multiply the number of each parameter members in the result of concerned Rollup. We can find the number of parameters' members by analyzing the basic query result headers (in line and in column). For example, looking at the headers of the pivot table of Fig. 1, we find that there are 20 days (20 members), 3 regions and 1 country. Thus, the data size of the three possible Rollups resulting from the previous step is calculated as follows:

- Size of [Day, Region, All^{Time}] = $20 \times 3 \times 1 = 60$
- Size of [Day, Country, All^{Time}] = $20 \times 1 \times 1 = 20$
- Size of [Day, $All^{Geography}$, All^{Time}] = $20 \times 1 \times 1 = 20$

- **imperfect data:** we have imperfect data if the measure does not have a value for all combinations of parameters' members, i.e., it has null values. For example, in the pivot table of Fig. 1, we do not know the daily precipitation for some cities (gray cells). In this case, we have to go through the whole result of the basic query (120 cells of pivot table of Fig. 1) to calculate the exact data size of a Rollup. Thus, the data size of the three possible Rollups concerning Fig. 1 is as follows:

  - Size of [Day, Region, All^{Time}] = 58
  - Size of [Day, Country, All^{Time}] = 20
  - Size of [Day, $All^{Geography}$, All^{Time}] = 20

We can note here that the data size calculation is simpler and faster for perfect data than for imperfect data. Furthermore, the more data are aggregated at higher parameters, the less the difference between perfect and imperfect data sizes is important. Therefore, for a small percentage of null data value, the data can be considered perfect.

4.4   Choose a Rollup Operation Based on a Selection Strategy

Our method reduces the data size according to used diagram type (e.g. bar, pie). It is ruled by parameters expressing the maximum allowed size for each type of diagram. As, the data readability issue is relative (i.e., what is readable for someone is not necessarily readable for the other), we assume that these limits are set by the user himself. For example, a user could determine the maximum size for bar and pie charts by 60 and 20 respectively. In other words, a bar chart can have maximum 60 bars and a pie chart can have up to 20 sectors.

To automate the Rollup selection, we propose five strategies. To explain these strategies, we use Table 1. In this table, the column "Query" identifies the basic query "Current" and the different possible Rollups (from "Rollup1" to "Rollup5"). The columns "Levels" determine the concerned granularity levels (parameters) of the different dimensions. The columns "Changes" determine the number of parameters implicated in the Rollup operation on each dimension (i.e., how many levels we go up). The column "Size" determines the data size of the basic query and Rollup operations.

Table 1. Rollup choice.

Query	Levels			Changes			Size
	Geography	Dates	Time	Geography	Dates	Time	
Current	City	Day	Every_3_hours	0	0	0	135
Rollup1	City	Day	Half-day	0	0	2	62
Rollup2	Region	Year	Every_3_hours	2	2	0	55
Rollup3	Department	Month	Quarter-Day	1	1	1	50
Rollup4	$All^{Geography}$	Day	Every_3_hours	4	0	0	45
Rollup5	City	All^{Dates}	Quarter-Day	0	3	1	40

In Table 1, we notice that, the data size of the basic query (precipitation by city, day and every three hours) is 135. "Rollup1" aggregates the data at "Half-day" parameter on the "Time" dimension. Its data size is 62. It has no changes on "Geography" and "Dates" dimensions, but the number of parameters implicated on the "Time" dimension is 2, i.e., we go up two levels ("Quarter-Day" and "Half-day").

In the following, we present the different selection strategies considering that the maximum allowed size is 60:

1. **"the closest" strategy:** this strategy selects the Rollup with the data size closest to the maximum allowed size even if it is greater than that size. Thus, this strategy chooses "Rollup1" in Table 1 to be executed because it has the closest data size (62) to the maximum allowed size (60).
2. **"the closest less" strategy:** this strategy selects the Rollup having a data size that is both less than and the closest to the maximum allowed size. Thus, it chooses, in Table 1, "Rollup2" having a data size (55).
3. **"the most detailed" strategy:** this strategy selects the Rollup having a data size smaller than maximum allowed size and which changes the granularity levels the least, in other words, that has the minimum number of implicated parameters (the minimum sum of "Changes"). For example, all "Rollup2", "Rollup3", "Rollup4" and "Rollup5" have a data size less than 60. The number of implicated parameters (Nb_P_{imp}) is calculated by $\sum_{i=1}^{m} Changes_{D_i} = Changes_{Geography} + Changes_{Dates} + Changes_{Time}$. So, Nb_P_{imp} of "Rollup2"= $2 + 2 + 0 = 4$
Nb_P_{imp} of "Rollup3"= $1 + 1 + 1 = 3$
Nb_P_{imp} of "Rollup4"= $4 + 0 + 0 = 4$
Nb_P_{imp} of "Rollup5"= $0 + 3 + 1 = 4$
Thus, this strategy chooses "Rollup3" having the minimum Nb_P_{imp}.
4. **"grouped Rollups" strategy:** this strategy selects the Rollup which has on one side a data size lower than the maximum allowed size and on the other side the least number of implicated dimensions. In other words, this strategy chooses the Rollup that has the most dimensions having 0 "Changes", i.e., the changes are grouped over few dimensions. For example, "Rollup2", "Rollup3", "Rollup4" and "Rollup5" have respectively 2, 3, 1 and 2 implicated dimensions. Therefore, this strategy chooses "Rollup4" to be executed.

5. **"by dimension preferences" strategy:** this strategy allows users to specify whether they prefer to make aggregations ("Changes") over some dimensions more than others. So, this strategy selects Rollup having the least sum of changes over the non-preferred dimensions. For example, let's assume that the user prefers doing the aggregations over "Dates" dimension. The sums of "Changes" of "Rollup2", "Rollup3", "Rollup4" and "Rollup5" over the non-preferred dimensions ("Geography" and "Time") are respectively 2, 2, 4 and 1. Thus, this strategy chooses "Rollup5" to be executed.

In Fig. 1, two types of diagram are used (bar and pie chart). So, our method should select two Rollups to be executed, one for each diagram. Considering that the maximum size of a bar chart is 60 and of a pie chart is 20, our method chooses the Rollup [Day, Region, All^{Time}] (having data size 58, see Sect. 4.3) to aggregate the data to be displayed in the bar chart and [Day, Country, All^{Time}] (having data size 20, see Sect. 4.3) to aggregate the data to be displayed in the pie chart.

4.5 Realize the Chosen Rollup

Once a Rollup is selected, it must be executed. Algorithm 3 describes how our method finds the aggregation functions needed to realize a Rollup. It takes three inputs: the measure, current parameters and the chosen Rollup's parameters. It finds functions that aggregate the measure on all the dimensions between the current parameter and the Rollup's parameter of the same dimension (lines 1 to 10). It starts by determining the current parameter ($p_c^{D_i H_j}$) and the Rollup's parameter ($p_R^{D_i H_j}$) of the concerned dimension by intersecting the current parameters (Ps_c) and the Rollup's parameters (Ps_R) with the dimension's parameters (P^i) respectively (lines 2 and 3).

Algorithm 3: Realize chosen Rollup

Input: current measure $m_c \in M$, current parameters $Ps_c \subseteq P$,
   chosen Rollup's parameters $Ps_R \subseteq P$
Output: $aggregation\_to\_do$
1 **foreach** $dimension\ D_i \in D$ **do**
2    $p_c^{D_i H_j} \leftarrow P^i \cap Ps_c$;   /* the current parameter of D_i */
3    $p_R^{D_i H_j} \leftarrow P^i \cap Ps_R$;   /* the chosen Rollup's parameter of D_i */
4   **repeat**
5   **if** ($p_c^{D_i H_j} \neq p_R^{D_i H_j}$) **then**
6          $Agg \leftarrow$ find_aggregator($m_c, p_R^{D_i H_j}$);
7          $aggregation\_to\_do$.add(Agg);
8   **if** ($Agg.constraint < 0$) **then**
9             $p_R^{D_i H_j} \leftarrow p_R^{D_i H_j} +^{\prec} Agg.constraint$; /* intermediate level */
10   **until** ($Agg.constraint = 0$) **or** ($p_c^{D_i H_j} = p_R^{D_i H_j}$);
11 **return** $aggregation\_to\_do$

We check if the current parameter and the Rollup's parameter are not identical (line 5), i.e., if there is an aggregation to do. Then, we find the function that aggregates the measure at the Rollup's parameter (line 6). We add it to a list of functions to be executed (line 7). If this function does not aggregate the measure directly from the current parameter (i.e., it has a constraint < 0 (line 8)), then, we find the intermediate lower parameter $(p_R^{D_i H_j} + ^\prec Agg.constraint)$ from which the aggregation is calculated (line 9). It is considered as a new Rullop's parameter $(p_R^{D_i H_j})$. Afterwards, we repeat the steps to find the aggregation function for this new Rullop's parameter until we find a function that aggregates the measure directly from the current parameter (i.e., it has a constraint $= 0$) or the Rullop's parameter becomes identical to the current parameter (line 10).

After executing this algorithm, the found functions should be executed according to their execution order. If there are two functions that have the same execution order on the same dimension, then these functions are ordered from the most detailed level to the most general level.

If the function is algebraic, intermediate values have to be stored [6]. For example, to avoid the non-desired average of the averages, the algebraic function AVG requires storing the intermediate SUM and COUNT.

In our example, to improve the readability of the bar chart, we realize the Rollup [Day, Country, All^{Time}] from the current parameters [Day, Department, All^{Time}] (Fig. 1) for the measure of the average precipitation "Pre". The execution of Algorithm 3 finds that there is no aggregation to do on the "Dates" and "Time" dimensions. It also finds that the measure is aggregated at the parameter "Country" by the weighted average function AVG_W(Pre, R_Surface) (with an execution order 3) from an intermediate parameter "Region" (see Fig. 2). During the second repetition, it finds that the aggregation at the parameter "Region" is realized from the current parameter "Department" by the function AVG_W(Pre, D_Surface) (having an execution order 3). Thus, the found functions are executed in the following order:

1. aggregation by AVG_W(Pre, D_Surface) at [Day, Region, All^{Time}]
2. aggregation by AVG_W(Pre, R_Surface) at [Day, Country, All^{Time}]

In addition, to improve the readability of the pie chart, Algorithm 3 finds that the realization of the Rollup [Day, Region, All^{Time}] is directly feasible from the current parameters by the function AVG_W(Pre, D_Surface).

Thus, these aggregated data for both Rollups can be displayed as in Fig. 3.

5   Implementation

In this section, we demonstrate the feasibility of our proposal. Our method is implemented as an extension of the prototype "OLAP-Multi-Functions" [8] (see Fig. 4). This prototype has an architecture in two levels:

- **Interface level:** this level is developed using Java. It allows to create and visualize the multifunction multidimensional schemata and query the data warehouse.

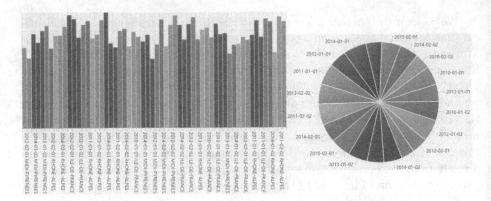

Fig. 3. Improved visualization of average precipitation.

- **Storage level:** an RDBMS (Oracle) is used to store the meta-schema and the data warehouse. An SQL query generator is developed as stored procedures. This generator translates (considering the multifunction context) user interactions (OLAP queries) into SQL queries.

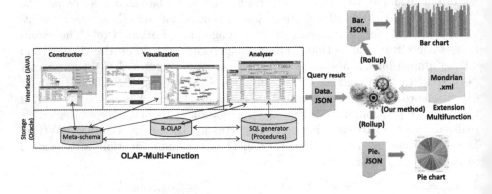

Fig. 4. Method implementation (Prototype).

Our method is developed in Java. It takes the result of an OLAP query in JSON format (Data.JSON file). An extension of the Mondrian schema is used to describe the multifunction schemata (Mondrian.xml file). In this extension, we describe the association of several aggregation functions to the same measure. Our method takes as input these two files (Data.JSON and Mondrian.xml).

The five steps of the method are executed sequentially (one after another). In the case of imperfect data, the execution of step 3 (calculate the data size for all possible Rollups) could be an expensive operation. So, in this case, a parallel execution (multithreading) of this step is preferred. We use one thread for each possible Rollup. If several data visualizations are used (e.g. bar, pie), then a multiple execution of steps 4 (choose a Rollup operation) and 5 (realize the chosen Rollup) is necessary. Thus, a parallel execution (one thread per type of visualization) is used.

The data reduction results are stored by visualization type (e.g. Bar.JSON and Pie.JSON files) before being sent to the corresponding visualization.

6 Conclusion

In this paper, we proposed a method to adjust (reduce) the displayed data size in the context of the multifunction OLAP where a single measure can be associated with several aggregation functions that are not commutative. Our proposal is a data post-processing method to perform Rollup operations to reduce the data size of the OLAP query result, according to the type of diagram in order to improve the readability.

Our method consists of five steps that must be performed sequentially. The **first step** studies the current OLAP query in order to find the maximum execution order already executed. The **second step** eliminates all Rollups that require an aggregation function that has an execution order lower than that found in the first step. The **third step** calculates the data sizes of all possible Rollups. It distinguishes between cases of perfect and imperfect data. We propose for the **fourth step**, which selects a Rollup to be executed, five strategies: (1) "the closest" and (2) "the closest less" that choose according to the data sizes. (3) "the most detailed" that chooses according to the number of implicated parameters. (4) "grouped Rollups" chooses according to the number of implicated dimensions. (5) "by dimension preferences" chooses according to the preferences expressed by the users. Once a Rollup is chosen, the **fifth step** looks for the necessary aggregation functions that will be executed according to their execution order.

We plan to generalize our method by studying the data reduction in the case of a multi-measure analysis. We also envisage to study the data reduction when the measures are analyzed (observed) on several granularity levels of the same dimension.

References

1. Abela, A.: Advanced Presentations by Design: Creating Communication that Drives Action. Wiley, Hoboken (2013)
2. Ahlberg, C., Shneiderman, B.: Visual information seeking: tight coupling of dynamic query filters with starfield displays. In: Readings in Human–Computer Interaction, pp. 450–456. Morgan Kaufmann (1995). ISBN: 978-0-08-051574-8
3. Boschetti, M.A., Golfarelli, M., Graziani, S.: An exact method for shrinking pivot tables. Omega (2019). https://doi.org/10.1016/j.omega.2019.03.002
4. Dix, A., Ellis, G.: By chance enhancing interaction with large data sets through statistical sampling. In: The Working Conference on AVI, pp. 167–176 (2002)
5. Golfarelli, M., Graziani, S., Rizzi, S.: Shrink: an OLAP operation for balancing precision and size of pivot tables. Data Knowl. Eng. **93**, 19–41 (2014)
6. Gray, J., Bosworth, A., Lyaman, A., Pirahesh, H.: Data cube: a relational aggregation operator generalizing group-by, cross-tab, and sub-totals. In: ICDE, pp. 152–159 (1996)

7. Hassan, A., Ravat, F., Teste, O., Tournier, R., Zurfluh, G.: OLAP in multifunction multidimensional databases. In: Catania, B., Guerrini, G., Pokorný, J. (eds.) ADBIS 2013. LNCS, vol. 8133, pp. 190–203. Springer, Heidelberg (2013). https://doi.org/10.1007/978-3-642-40683-6_15

8. Hassan, A., Ravat, F., Teste, O., Tournier, R., Zurfluh, G.: Differentiated multiple aggregations in multidimensional databases. In: Hameurlain, A., Küng, J., Wagner, R., Cuzzocrea, A., Dayal, U. (eds.) Transactions on Large-Scale Data- and Knowledge-Centered Systems XXI. LNCS, vol. 9260, pp. 20–47. Springer, Heidelberg (2015). https://doi.org/10.1007/978-3-662-47804-2_2

9. Jerding, D.F., Stasko, J.T.: The information mural: a technique for displaying and navigating large information spaces. IEEE TVCG 4(3), 257–271 (1998)

10. Jugel, U., Jerzak, Z., Hackenbroich, G.: M4: a visualization-oriented time series data aggregation. Proc. VLDB 7, 797–808 (2014)

11. Jugel, U., Jerzak, Z., Hackenbroich, G., Markl, V.: VDDA: automatic visualization-driven data aggregation in relational databases. VLDB J. 25(1), 53–77 (2016)

12. Keim, D.A.: Pixel-oriented visualization techniques for exploring very large data bases. J. Comput. Graph. Stat. 5(1), 58–77 (1996)

13. Kimball, R.: The Data Warehouse Toolkit: Practical Techniques for Building Dimensional Data Warehouses, vol. 121, 2nd edn. Wiley, Hoboken (2002)

14. Li, M., Choudhury, F., Bao, Z., Samet, H., Sellis, T.: ConcaveCubes: supporting cluster-based geographical visualization in large data scale. Comput. Graph. Forum 37(3), 217–228 (2018)

15. Lins, L., Klosowski, J.T., Scheidegger, C.: Nanocubes for real-time exploration of spatiotemporal datasets. IEEE TVCG 19(12), 2456–2465 (2013)

16. Liu, Z., Jiang, B., Heer, J.: imMens: real-time visual querying of big data. Comput. Graph. Forum 32, 421–430 (2013)

17. Marty, R.: Applied Security Visualization, 1st edn. Addison-Wesley Professional, Boston (2008)

18. Meyer, M., Takahashi, S., Vilanova, A.: The state-of-the-art in predictive visual. Comput. Graph. Forum 36(3), 539–562 (2017)

19. Miranda, F., Lins, L., Klosowski, J.T., Silva, C.T.: TopKube: a rank-aware data cube for real-time exploration of spatiotemporal data. IEEE TVCG 24(3), 1394–1407 (2018)

20. Pahins, C.A., Stephens, S.A., Scheidegger, C., Comba, J.L.: Hashedcubes: simple, low memory, real-time visual exploration of big data. IEEE TVCG 23(1), 671–680 (2017)

21. Peng, W., Ward, M.O., Rundensteiner, E.A.: Clutter reduction in multidimensional data visualization using dimension reordering. In: IEEE Symposium on Information Visualization, pp. 89–96 (2004)

22. Silva, R., Moura-Pires, J., Santos, M.Y.: Spatial clustering in SOLAP systems to enhance map visualization. IJDWM 8(2), 23–43 (2012)

23. Stolper, C.D., Perer, A., Gotz, D.: Progressive visual analytics: user-driven visual exploration of in-progress analytics. IEEE TVCG 20(12), 1653–1662 (2014)

24. Trutschl, M., Grinstein, G., Cvek, U.: Intelligently resolving point occlusion. In: Proceedings of the IEEE Symposium on Information Visualization, pp. 131–136 (2003)

25. Wang, Z., Ferreira, N., Wei, Y., Bhaskar, A.S., Scheidegger, C.: Gaussian cubes: real-time modeling for visual exploration of large multidimensional datasets. IEEE TVCG 23(1), 681–690 (2017)

A Framework for Learning Cell Interestingness from Cube Explorations

Patrick Marcel[1]([✉])(iD), Veronika Peralta[1](iD), and Panos Vassiliadis[2](iD)

[1] University of Tours, Blois, France
{patrick.marcel,veronika.peralta}@univ-tours.fr
[2] University of Ioannina, Ioannina, Greece
pvassil@cs.uoi.gr

Abstract. In this paper, we discuss the problem of organizing the different ways of computing the interestingness of a particular cell derived from a cube in the context of a hierarchical, multidimensional space. We start from an in-depth study of the interestingness aspects in the study of human behavior and include in our survey the approaches taken by computer-science efforts in the area of data mining and user recommendations. We move on to structure interestingness along different fundamental, high level *aspects*, and, due to their high-level nature, we also move towards much more concrete data-oriented definitions of interestingness aspects.

Keywords: Interestingness · Data cube · Cells · OLAP explorations · Novelty · Peculiarity · Surprise · Relevance

1 Introduction

Given a cell of a datacube and a user's exploration over this datacube, how to assign to this cell a score reflecting its interestingness for the exploration?

The **significance** of the answer to the above question, cannot be underestimated. A cell is the most granular piece of information in a BI session, thus, in this paper it is the epicenter of our search, both because it can be of value per se, and because the interestingness of groups of cells can be based on the interestingness of individual cells. We need better systems at recommending questions, data and highlights to the users. Understanding what is important for a user is key to this goal and a cell interestingness score is a pre-requisite for this. If we manage to successfully score (i.e., understand) which cells matter more to each user, this would allow to better understand how users navigate cubes by studying logs of user sessions, categorize these cube explorations, and make on-line recommendations. Apart from the aforementioned practical considerations, from the research point of view, succeeding in structuring the aspects of interestingness will allow to structure our knowledge on existing methods, as well as provide the basis to benchmark and compare such methods and help develop new ones for supporting cube exploration in aspects not successfully covered yet.

ⓒ Springer Nature Switzerland AG 2019
T. Welzer et al. (Eds.): ADBIS 2019, LNCS 11695, pp. 425–440, 2019.
https://doi.org/10.1007/978-3-030-28730-6_26

To the best of our knowledge, there is a **gap in the literature** in answering the motivating question of the paper. Although interestingness measures have attracted a lot of attention in other communities, like for instance Data Mining, for measuring the interestingness of a pattern [9], or Recommender Systems, for measuring the quality of recommendations [16], to our knowledge there exists no principled study or survey for the interestingness of a cell in cube exploration.

We consider the **context of cube exploration** as follows. We assume a user performing exploratory data analysis over a hierarchical, multidimensional nature of data [15]. After a while, the user acquires an overarching informational goal that their exploration tries to address. During the exploration, the user devises queries to acquire new information. Each such query brings in new data and constructs a new cube that is presented to the user. We call this Q&A a transition, and it constitutes a step in the overall exploration of the user. The user, thus, practically covers areas of the hierarchical multidimensional space at each step (possibly at different levels of granularity), and progressively, for each such area, some kind of expectation on its values is constructed or updated (call it a "model" if you will). In each transition the user makes, each new observation, is (0) relevant or not to the user's informational need and either (1) reinforces the expectation or (2) contradicts it, or (3) just creates expectations for newly explored places where none existed before. Each new observation, is therefore, assessed with respect to its novelty, relevance, surprise and peculiarity. Each such criterion covers a different aspect of interestingness.

Our **contributions** in this paper, are structured as follows: In Sect. 2 we discuss earlier proposals of interestingness measures and in Sect. 3, we study the forces that affect interestingness computation and structure them around *high level aspects of interestingness* – specifically, novelty, relevance, surprise and peculiarity. In Sect. 4, we provide exemplary algorithms and methods for assessing the high level aspects of a cell's interestingness, on the basis of low-level measures, and Sect. 5 describes the experiments we ran to showcase the framework. Section 6 concludes the paper and suggests open roads for future work.

2   Related Work

Although there is little work proposing measures for quantifying the interestingness of a cell in a datacube, several measures can be borrowed from close research areas and adapted to cells. In this subsection we discuss interestingness measures proposed for (i) pattern mining, (ii) cube exploration and summaries, and (iii) recommendation.

Interestingness Criteria for Pattern Mining. In [9], the authors point out that interestingness is a broad concept and identify from the literature 9 criteria to determine whether or not a pattern is interesting: conciseness, generality/coverage, reliability, peculiarity, diversity, novelty, surprisingness, utility and actionability/applicability. They categorize these criteria in 3 groups: (i) objective measures, based only on the raw data (generality, reliability, peculiarity, diversity, conciseness), like for instance the classical support, (ii) subjective measures, considering both the data and the user (surprise and novelty), like for instance the informational content [3], and (iii) semantic measures, based on the semantics and explanations of the patterns (utility and actionability), like for instance measures based on user preferences [26]. We note that according to De Bie [3], subjective interestingness is particularly well adapted for *exploratory* data mining, whose goal is to pick patterns that will result in the best updates of the user's belief state, while presenting a minimal strain on the user's resources. One challenge is to define and update the belief of the user. De Bie proposes to model it as a background distribution over patterns representing the belief the user attaches to patterns being present in the data.

Most of the criteria introduced above can be reused in our context, except diversity (that would concern groups of cells) and reliability (since data in cubes are assumed reliable by construction).

Interestingness Criteria for Summaries. In [9], authors also review interestingness measures for what they call summaries, i.e., aggregated cross-tabs corresponding to the result of an OLAP query, where numeric values (i.e., measures) are aggregated by several criteria (i.e., dimensions). Out of the 9 criteria defined for pattern interestingness, 4 are adapted to summaries: diversity (proportional distribution of classes in the summary versus the number of classes), conciseness/generality (level of aggregation), peculiarity (a cell in a summary is peculiar if it is differs from the other cells in the summary) and surprisingness/unexpectedness (a summary is surprising if it deviates from user's expectations). According to the classification of [9], the first three criteria are objective and the last one is subjective. Note that except for peculiarity, and to a lesser extent, conciseness, the criteria concern the interestingness of the whole summary instead of the interestingness of each cell.

To the best of our knowledge, such peculiarity measures are the cornerstone of discovery-driven analysis [22–25] for measuring cell interestingness in the context of cube exploration. Discovery-driven analysis guides the exploration of a datacube by providing users with interestingness values for measuring the peculiarity of the cells in a data cube, according to statistical models, e.g., based on the maximum entropy principle, and leveraging the intrinsic structure of multi-dimensional information. From an initial user query, the system automatically calculates 3 kinds of interestingness values for each cell in the query result: (i) $SelfExp$ measures the difference between the observed and anticipated values (the latter are calculated statistically by computing the mean of subsets of attributes), (ii) $InExp$ is obtained as the maximum of $SelfExp$ over all cells that are under this cell (those that result from a drill down), and (iii) $PathExp$ is calculated as the maximum of $SelfExp$ over all cells reachable by drilling down along a given path. The DIFF, INFORM and RELAX advanced OLAP operators proposed in [22,23,25] use such interestingness values to recommend relevant cells for explaining drops or increases, or for recommending areas of a cube that should surprise the user, based on their history with the cube.

In the context of OLAP, other works propose further measures concerning (or related to) interestingness of a cross-tab, a query result or a set of cells. Without trying to be exhaustive, we mention here some of those works, illustrating the diversity of the proposed measures.

Klemettinen et al. [17] use skewness, as a peculiarity measure of asymmetry in data distribution, for discovering interesting paths and guiding the navigation in a datacube. Given a cuboid, the possible drill-downs are explored, measuring skewness and generating skew-based navigation rules for the more significant paths. Skewness is computed observing the underlying facts (the raw data that is aggregated), looking for outliers or substantial differences with other facts. Based on skewness, Kumar et al. [18] propose interestingness measures based on the unexpectedness of skewness in navigation rules and navigation paths.

Fabris and Freitas [7] defined interestingness measures for attribute-value pairs in a data cube: the I_1 measure reflects the difference between the observed probability of an attribute-value pair and the average probability in the summary and the I_2 measure reflects the degree of correlation among two attributes. Both measures can be seen as value-based conciseness.

Djedaini et al. use supervised classification techniques for learning two interest measures for OLAP queries: *focus*, that indicates to what extent a query is well detailed and related to other queries in an exploration, indicating that the user investigates in details precise facts and learns from this investigation [5], and *contribution*, that highlights to what extent a query is important for an exploration, contributing to its interest and quality [4].

Finally, we mention two recent works [21,27] that are concerned with detecting the validity of insights gained by users when examining query answers. As other works measuring peculiarity by leveraging the nature of OLAP cubes, this is again achieved by statistical tests comparing data at different levels of details.

Interestingness Criteria for Recommendations. There is a long discussion about interestingness in the area of evaluating recommender systems [11,14,16]. We mention [16] as an excellent recent survey on the topic. The survey presents 4 criteria (diversity, serendipity, novelty, and coverage), in addition to the traditional accuracy, for evaluating the quality of a recommendation.

Query recommendation techniques (see e.g., [2,6]) are usually evaluated with interestingness measures coming from the literature on recommender systems exposed above. We mention the more OLAP-specific *foresight* measure [2], that quantifies how distant is the recommendation from the current point of exploration.

3   Interestingness Aspects for Cube Exploration

How can we define interestingness? To the best of our knowledge, there is no formal definition. Online resources[1] propose "Interest is a feeling or emotion that causes attention to focus on an object, event, or process". In contemporary

---

[1] https://en.wikipedia.org/wiki/Interest_(emotion).

psychology of interest, the term is used as a general concept that may encompass other more specific psychological terms, such as *curiosity* [19] and to a much lesser degree *surprise* [20] and *novelty* [8].

In this section, we derive from our study of the literature the criteria of the interestingness of a cell, by listing what influences them. We can conclude from our study of related work that interestingness is a degree attributed to a piece of information, regarding the curiosity and surprise it generates. This piece of information under consideration may spark the will to continue exploring the source of information to close some knowledge gap, or get novel information. But how can we pass from such a high level description of interestingness, to a more concrete one? Our approach is a two level modeling. At the first level, we discuss *high-level aspects* of interestingness, like the ones deduced from the study of human behavior. Second, we provide *data-oriented measures* of interestingness, substantiating the aforementioned high-level aspects, on the grounds of the available information. This section presents the first level, while next section provides examples of concrete measures (the second level of our approach) and describes their computation. A proof of concepts implementing some measures is described in Sect. 5.

3.1   Interestingness Aspects

We now present 4 fundamental, high-level interestingness aspects: relevance, novelty, surprise, and peculiarity.

Relevance as a Measure for the User's Curiosity. Curiosity is the main driver of knowledge acquisition. Data exploration, especially in an environment of Business Intelligence, is primarily related to the answering of an open question. So, it is realistic to assume that the user comes with a question for a particular subset of the multidimensional space, and her exploration has to do with "a walk" within this sub-space in order to answer the question. We will call the aspect of interestingness that pertains to curiosity as the *relevance* of the cell with respect to the exploration and its underlying user goal.

The main force, thus, of the assessment of relevance is the modeling of the user intentions. Basically, we can discriminate between (a) the case where a description of the user intention is given vs. (b) the case where no such knowledge is available. In the former, we deal with an expression of the user's interest as the space of a user goal. In the latter, we need to learn the user goal from the history of past activity, which, in turn, relies on the availability of the coordinates of the cells of the queries in the exploration and the schema of the cube.

Novelty. Novelty is also an aspect of interestingness that mainly pertains to the need of users to learn information previously unknown. The simple reporting of data that have not been previously reported might increase their interestingness.

The main force behind novelty is the existence of a history. A lesser influence is the availability of results (cell coordinates are sufficient to understand if the cell have never been seen). Without the knowledge of the history of the user's queries, novelty is practically a wild guess. When dealing with novelty, we are not primarily interested in the intention of the user, although it can affect the attention that a user pays to a particular cell (in other words, we assume all cells being equally probable to have been observed by the user).

Surprise. Not surprisingly, surprise is a major aspect of interestingness. Surprise occurs when our previous beliefs are disconfirmed or contradicted. This can happen either directly, when the expected value of an event proves to be significantly different than the actual value, or implicitly, when the disconfirmation of a certain fact deduces the disconfirmation of a dependent fact.

Clearly, the main prerequisite for evaluating surprise is the existence of a previous belief of the user. Without the existence of a structured model for the estimation of the previous beliefs, the assessment of surprise is impossible; for this case, it is only possible to measure some objective peculiarity intrinsic to the data (see below). Surprise can be measured using models leveraging the history of the user with the datacube, for instance to estimate belief.

Peculiarity. Consistently with the literature on cubes, we use peculiarity to denote an intrinsic property of the data, i.e., the cell's value, when considered together with other cells related to it.

Peculiarity of a cell cannot be assessed in vacuum. Most typically, it can be assessed against the cells of the same query. Taken to extremes, it can also be evaluated by comparing the cell to all the previous cells of the history of the exploration – or even, to all the cells of the full history of the user with the datacube, i.e., including past explorations. Finally, peculiarity may also be calculated with respect to the unseen cells of the cube. The full instance, i.e., with measure values, of cells considered are prerequisites for this criteria.

3.2   Definition of Interestingness

We define interestingness of a cell as a vector of scores, defined over a set of interestingness measures.

Definition 1 (Cell interestringness). *Given a user's exploration over a datacube, the interestingness of a cell of this exploration is a tuple of scores for a list of interestingness measures.*

We intentionally do not differentiate between high-level and data-oriented criteria. We support an extensible approach towards which criteria would an interestingness assessment tool include, especially as we cannot provide any completeness proof on our list of high-level interestingness aspects.

4 Detecting Interesting Cells in an Exploration

In this Section, armed with the tools of the previous sections, we revisit the originating question of our introduction: How do we compute the different aspects of the interestingness of a cell? To this end, and without trying to be exhaustive, we provide some alternatives per high-level aspect and discuss their computation.

4.1 Relevance

Assessing the relevance of a cell practically answers the question: *how close is this cell to the subset of the multidimensional space that the user intents to explore?* Two fundamental notions hide behind this formulation of the problem, the specification of an area of interest and the understanding of the user's intention.

As already mentioned, we define the *space of a user goal* as the framing of a subspace of the multidimensional space (either intentionally via selection predicates, or explicitly, at the extensional level, as a set of cells) for which the user wants to obtain information. In the former case, we refer to the *intentional specification of a user goal* whereas in the latter to refer to the extensional *area of interest* of the goal, with the explicit set of cells defined by this framing. Then, given a specific exploration, with a user goal as its underlying motive, we define *relevance* as the degree to which the cell overlaps with the area of interest of the exploration's motivating goal.

Concerning the user intentions, as already mentioned, we discriminate between (a) the case we have no such information, and, (b) the case we have an expression of the user's intentions. Let us proceed in exploring both cases.

Relevance Without Knowledge of the User's Intent. Let's start with the case where no model for the user's intent is given a priori. To assess the relevance of a cell, we need to quantify how "close" or "central" the cell is to the subspace induced by the exploration of the user. Practically speaking, we need an algorithm that enumerates the cells that have been visited by the user during her exploration. Due to the hierarchical nature of the space, the easiest way to compare cells is by referring all cells to a common level of granularity (i.e., the node in the lattice of group-by's [13] that is (a) dominated by all the nodes to which history queries correspond, and, (b) the highest among all the candidates of (a)). For simplicity, in this paper, we assume this is the lowest possible node of the group-by lattice, i.e., the level of the facts, that we call C^0.

Now, we need an algorithm that computes the area of interest, starting with its most detailed form, at the level of C^0 (see Algorithm 1). The input to this algorithm is the history of user queries of an exploration. The output is the detailed area of interest. Basically, for every aggregate cell that is part of a query result, the algorithm detects its detailed cells, increases a score for each of the times this cell has contributed to the computation of a query result and adds it to the detailed area of interest, returned by the algorithm.

Algorithm 1: ComputeDetailedAreaOfInterest

Data:

a history of user queries Q

a basic cube C^0

a set of dimension hierarchies defining the multidimensional space set of models $\mathcal{D}$

Result:

a Detailed Area of Interest S^0, with all its cells annotated with a relevance indicator

1 **begin**

2   **for** *every query* $q \in Q$ **do**

3   **for** *every cell* $r \in q.cells$ **do**

4   Let $\mathbf{r}^0$ be the set of descendants of r at the most detailed level, $\mathbf{r}^0 \subseteq C^0$;

5   **for** *every detailed cell* $r_i^0 \in \mathbf{r}^0$ **do**

6   increase $r_i^0.score$ by 1;

7                  $S^0 = S^0 \cup r_i^0$;

8   return the detailed area of interest S^0

Having computed the detailed area of interest of a user goal, we can now proceed to answer the question "What is the relevance of a cell c to an area of interest, say S?" Let S be the area of interest of the session, and $\mathbf{S}^0 = \{c_1^S, \ldots, c_k^S\}$ be the set of cells corresponding to the cells of S at the detailed cube C^0. Let $c = \langle a_1, \ldots, a_n, v_1, \ldots, v_m \rangle$ be the cell we are interested in and $\mathbf{c}^0 = \{c_1^c, \ldots, c_l^S\}$ be the set of descendant cells corresponding to c at the most detailed level. Then, $relevance(c \mid S)$ is a function f_R that calculates the percentage of $\mathbf{c}^0$ that also lies within $\mathbf{S}^0$ (see Algorithm 2).

Algorithm 2: ComputeSimpleRelevance

Data:

a cell c

a history of user queries Q

a basic cube C^0, and a set of dimension hierarchies defining the multidimensional space set of models $\mathcal{D}$

Result:

the relevance of c to Q computed via S^0

1 **begin**

2      $S^0 = $ computeDetailedAreaOfInterest(Q);

3   Let $\mathbf{c}^0$ be the set of descendants of c at the most detailed level, $\mathbf{c}^0 \subseteq C^0$;

4   return $relevance(c|Q) = |S^0 \cap c^0| \, / \, |c^0|$;   /* Other variants of the formula can be envisaged */

Variants. A more liberal definition of relevance can compute a distance function of the two sets. A more strict definition might take the frequency of the visits of the user to each member of S^0 during the exploration. Then, each cell is weighted by how many times it has been visited by the user during the exploration. Then, *relevance* is defined as the fraction of the sum of the weights of the common cells of the two sets over the sum of weights of the cells of S^0.

A side-effect problem, that we leave aside for the moment concerns the most concise description of S^0 by rolling up regions of C^0 completely covered by cuboids at an ancestor level at the lattice of group-by's.

Relevance in the Presence of Knowledge of the User's Intent. Assume now that we have the expression of a user goal. Here, we do not discriminate between an induced goal by a user profile, or a deliberate expression of the goal by the user. We assume that *the goal is expressed as a boolean predicate ϕ* (typically - but not obligatorily - expressed as the conjunction of simple atomic selection formulae). There are several ways to compute the relevance of a cell c to ϕ. Note that ϕ may not be part of the query that retrieves c. The user may (a) compare cells within the area of the original goal with similar/peer cells, or, (b) put the values she observes in context by rolling-up in a way that produces aggregate values broader than the original goal's selection condition.

Variants. The simplest way is to see whether c satisfies the goal ϕ. To do that, both c and ϕ must be converted to the same level of detail – again to their highest common descendant in the lattice of group-by's. Then, *relevance* in its simplest form is Boolean and evaluates to true or false if all descendants of c satisfy ϕ, or numerical, if a percentage is computed. In these variants, the history of queries is not taken into consideration – only the intentional space of the user goal.

If we want to assess relevance given the history too, we can resort to the computation of the previous subsection that did not take the user goal into consideration. Assume now that we convert ϕ to the lowest possible level and obtain ϕ^0 [10]. In this case, we can isolate the subset of the explored space that is relevant to the user goal, via $\sigma_{\phi^0}(S^0)$, with S^0 as previously defined, and then search for its (simple or weighted) intersection with c^0 (also as previously defined).

4.2 Novelty

As already mentioned, novelty refers to the second facet of curiosity, obtaining new knowledge, and for all practical concerns, it deals with whether the user has seen a cell before or not. Due to the hierarchical nature of the multidimensional space, novelty does not only concern the previous appearance of a cell per se, but also, whether the user has been exposed to ancestor or descendant cells too.

Given a datacube C, the history Q of queries of an exploration, and the set H of the cells of the queries of Q, we have several alternatives for the evaluation of novelty, which in all cases is a function f_N assessing *novelty*$(c \mid H)$ or *novelty* $(c \mid H, C)$.

1. We define the *strict novelty* of a cell c as its absence from H or not. Thus, the strict novelty is Boolean, and refers to the cell per se, in the context of the exploration's query history.
2. We define the *coverage novelty* of c based on the fraction of cells of the datacube C covered by c (e.g., all the descendants of c) that the user has seen during the exploration: $1 - \frac{|cov(c,H)|}{|cov(c,C)|}$, where $cov(c, S)$ denotes the cells of S covered by c.
3. We define the *inferred novelty* of a cell c as the extent of overlap of c with the cells of H, even via ancestor or descendant relationships. For each cell of H, say c^H, that is related via an ancestor or descendant relationship with c, we count the complement of the weight of c^H over c. This can be done in many ways, and here we mention the simplest ones. Assume c^H is an ancestor of c, then the respective weight is the fraction of the cell's measures, if the aggregate function is distributive (i.e., not avg). Alternatively, the fraction can be the inverse of the cardinality of c^H's descendants at the level of c. The roles are inverted if the relationship is a descendant rather than an ancestor one. In all these cases, the inferred novelty is a real number that can easily be normalized in the range $[0 .. 1]$
4. We can also define *inferred novelty at the detailed level* by comparing the detailed descendants of c and the descendants of the members of H, say H^0 at the level of C^0. The percentage of descendants of c at the detailed level that also belong to the H^0 define the inferred novelty of c at the detailed level.

4.3 Surprise

Surprise is a fundamental aspect of interestingness. Where relevance describes the general area of data within which the user wants to walk around, and has to do with *why* he is interested in a cell, surprise relates to the divergence of what she sees with her *previous belief* of what she expected to find. Surprise instigates further searches or actions, in order to adapt our challenged beliefs to the new data, and opens new ways of looking at the data. The fundamental premise upon which surprise can be computed is the modeling of the user's previous beliefs.

How then do we structure a model of beliefs for the cells of a multidimensional space? Fundamentally, there are two ways of handling beliefs: (a) the objective way, where there is a function that assigns an expected value to a measure, independently of what the user has seen in her exploration, and, (b) the subjective way, where the expectation of a cell's value is dependent upon the previous cells that the user has seen in her exploration. The objective evaluation is very demanding, in the sense that it requires that the user has full knowledge of the cube - or even, the sub-cube that she explores and some way to express this knowledge as a potential value. The subjective mechanism is more dynamic: it can start with the user being tabula rasa and, progressively, as cells are observed, her beliefs for the next cells that are related to the previously seen ones are updated.

Surprise Assessment. We give two indicative ways to compute surprise, one objective and one subjective.

The *value-based surprise* for a cell c, $surprise(c)$ is the difference between the actual value of a measure M of the cell, say m, compared against its expected value, for instance $\overline{m}$.

The *probability-based surprise* of a cell. Assume a probability distribution P over the set of all potential values for the cells of C. This distribution is used to represent a user's belief, i.e., for a cell $c = \langle a_1, \ldots, a_n, v_1, \ldots, v_m \rangle$ the probability that the user attaches to the statements "the ith measure of c is v_i". The surprise brought by c is a function over this probability, for instance $surprise(c) = -log(P(c))$.

A fundamental aspect of a model for user beliefs is belief refreshment. As the exploration unravels, the beliefs of the user are updated with every new cell he observes. A mechanism for belief update is out of the scope of this paper, but could follow the general principle given in [3]. However, this does not fundamentally alter the mechanism for interestingness assessment that we propose, as, at any time point, when a cell appears, we can assume that the user has an expected value for it.

4.4 Peculiarity

Peculiarity is an intrinsic property of the data: it makes a particular cell to be set apart from its peers, typically due to the divergence of its measure values from a typical value distribution. Peculiarity can be used to estimate surprise in the absence of any other model for the user (e.g., if we know nothing about what the user expects to see, we can possibly assume that very small or high values in the sales, i.e., outliers, could be interesting). Peculiarity is not restricted to naive outlierness, as it can be due to a more complex pattern (e.g., how a cell evolves over time).

Assessing peculiarity can be performed in a plethora of ways (e.g., via isolating extreme values, assessing how close a value is to its "neighboring" values, performing clustering of the values, information theoretic approaches) [1]. It is beyond the scope of this paper to discuss outlier detection methods, either simple or advanced. We refer the interested reader to [1,12] for an extensive coverage.

5   Experiments

This section showcases our framework through preliminary experiments over a small set of real user explorations.

5.1   Experimental Setup

In our experiments, we reuse the dataset described in [5], consisting of navigation traces collected in the context of a French project on energy vulnerability.

Traces consist of logged OLAP sessions[2] of volunteer students of a Master degree in Business Intelligence, answering some high-level information needs defined by their lecturer, using Saiku[3] to ask the queries and see the results. In the present paper, we analyzed 11 sessions, whose sizes range from 12 to 69 queries, 411 queries in total, with an average of 37 queries per session, and an overall of 14,384 cells. Both queries and sessions were manually inspected and labelled by the lecturer. Queries were assigned a binary label regarding their focus on the phenomenon analyzed by the student during the session. The term focus is used as in [5]: "When focused, an analyst would expect more precise queries, related to what she is currently analyzing. On the contrary, when exploring the data, the analyst would prefer more diverse queries, for a better data space coverage".

Sessions were graded from A (lowest) to D (highest grade) with respect to the combination of two characteristics, specifically, (a) the extent to which the queries of the session are semantically linked to their previous query (and not ad-hoc) and (b) the progressive stabilization of an area of interest in the multi-dimensional space (as opposed to everlasting, ad-hoc explorations of the space). Among the 11 sessions analyzed, 4 sessions were labelled B, 3 labelled C and 4 labelled D.

We have developed a prototype *session analyzer* to analyze the logs of the users. Our prototype loads the sessions of each user, and for each of them evaluates the queries one by one, in order. Each time a query is evaluated, the user history is updated, the detailed area of interest (cf. Algorithm 1) is refreshed and the cell interestingness measures are computed. We implemented the extraction of 4 basic measures, one per high level aspect described in the previous section: (i) simple relevance, as of Algorithm 2, (ii) strict binary novelty, i.e., the cell is previously seen or not, (iii) a limited form of surprise, called positional surprise, computed as minus log of the product of the member's probability of appearance in the user history[4], and (iv) simple peculiarity hereafter called outlierness, calculated as z-score w.r.t. the rest of the cells in the query result to which it belongs. Our goal is to confront the measures with the labels assigned to the sessions and queries, looking for correlations between interestingness, user focus, and session quality.

Our prototype is written in Java 8 and ran on a MacBook Pro Core I5 with 16 GB RAM running MacOS Mojave 10.14.3. The average processing time per cell is 1071.55 ms, with a minimum of 376 ms, a maximum of 10663 ms and a standard deviation of 248.11. The computation of relevance constitutes by far the majority of the computation time. The average processing time *per query* is 37.18 s, with a standard deviation of 85.02. Comparatively, the average consideration time (i.e., the time the user took between two consecutive queries) is 29.42 s, with a standard deviation of 65.59.

---

[2] We do not distinguish between the terms session and exploration in what follows.

[3] https://www.meteorite.bi/products/saiku.

[4] In this implementation, the user belief is agnostic of measure values, and the metric therefore characterizes how surprising it is that the user visits this particular cell.

Table 1. Average and standard deviation (in brackets) of measures per query labels

	Relevance	Novelty	Surprise	Peculiarity
Not focused	0.68 (0.43)	0.56 (0.50)	0.77 (0.25)	0.61 (0.90)
Focused	0.78 (0.31)	0.71 (0.46)	0.82 (0.26)	0.66 (0.78)

5.2 Lessons Learned

Our first experiment investigates whether the queries with a higher focus obtain higher values for these interestingness measures compared to the queries with less focus.

The first result comes from Table 1. We average all focused vs non-focused cells and compare the values. The focused category consistently demonstrates higher values for all the measures, with novelty having a 15% difference in the values and relevance a 10%, even though this is nuanced by the standard deviation.

Then, one can refine the above result by assessing whether there is any difference in their behavior of these measures during the progression of the sessions. As session lengths are different, for each query we compute the percentage of progress with respect to the session, as an indicator of how deep the analyst was in her search during that session. To reduce the visual clutter, we organize the demonstration by ranges of 10 steps, where the average value is shown for each category.

Figure 1 shows how the four measures evolve along the progression of the sessions, distinguishing by query labels. Concerning novelty, we see that focused

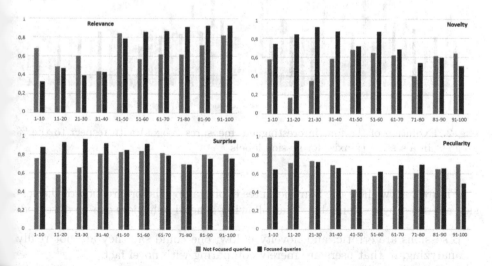

Fig. 1. Evolution of the four interestingness measures (y-axis) with respect to the % progress in a session (x-axis) for focused vs non-focused queries

queries soon demonstrate higher amounts of novelty compared to non-focused ones (which seem to revolve around the same cells). Only very later in the session is this difference equalized or surpassed (and indeed at low levels of novelty anyway). So overall, focused queries demonstrate more novelty than the non-focused ones. The same phenomenon is observed for surprise, but with less variations. Concerning relevance, as already mentioned, we measure relevance as the subset of the detailed multidimensional space that is revisited, as an indicator of what the user is looking at. Practically, this is acting as the counterpart of novelty, albeit here we are found in the detailed multidimensional space rather than the space of the actual aggregated cells. Here, we observe that the non-focused queries, due to the repetition, obtain higher values than the focused ones. Only later in the session, when the focused queries are returning to the well-established area of exploration to finalize conclusions is the situation reversed. For peculiarity, things are pretty much equal throughout the entire session, apart from a few cases where focused queries contain a little bit more outlier cells than non-focused ones. This justifies the small 5% advantage they have in the total scoring of Table 1.

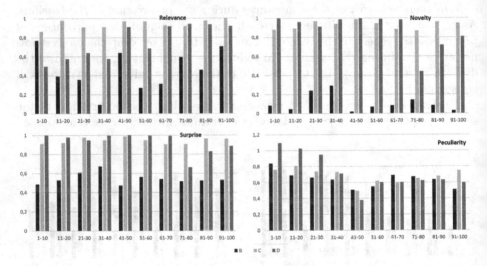

Fig. 2. Evolution of the four interestingness measures (y-axis) with respect to the % progress in a session (x-axis) for session labels

Figure 2 shows how the four measures evolve along the progression of the session arranged by session label. The following general behaviors can be observed:

- B sessions are erratic, and novelty is low, one could say they are not really analyzing, in that users are merely comparing with novel facts.
- In C sessions, all measures are high, there is too much movement, indicating that they are focused, but not enough. The fact that novelty and relevance are

high at the same time is not contradictory: users stay in the same detailed area, but keep rolling-up, drilling-down. In other words, they keep investigating, but seem inconclusive, which is corroborated by the fact that those sessions are often longer than D sessions, that get straight to the point. And also by the fact that outlierness tends to increase in the end.

- In D sessions, relevance keeps increasing, novelty is high then collapses, like surprise, and then start increasing again. This indicates that the sessions are more focused in the end. Outlierness is very high in the beginning, which could have sparked the session.

6 Conclusions

This paper has addressed the problem of measuring the interestingness of the cells of a data cube, analyzed by a user during a session of data exploration. We have assumed a hierarchically-structured multidimensional space and, within this context, we have proposed criteria of interestingness at both a high-level and a data-oriented level.

We have kept our discussion independent from the particular model of OLAP operations that can be applied to the data, or from technological aspects influencing it. We believe that the paper opens the road for a more directed research of interestingness assessment and recommendation algorithms with specific targets among the high-level aspects discussed here. Our experiments provide a proof of concept in this direction, showing how even simple measures can help the analysis of user behavior. Extending the framework beyond the realm of clean, simply structured multidimensional spaces, in the realm of an arbitrarily structured and populated database schema, is a clear path for future work.

References

1. Aggarwal, C.C.: Data Mining - The Textbook. Springer, Cham (2015). https://doi.org/10.1007/978-3-319-14142-8
2. Aligon, J., Gallinucci, E., Golfarelli, M., Marcel, P., Rizzi, S.: A collaborative filtering approach for recommending OLAP sessions. Decis. Support Syst. **69**, 20–30 (2015)
3. Bie, T.D.: Subjective interestingness in exploratory data mining. In: Tucker, A., Höppner, F., Siebes, A., Swift, S. (eds.) IDA 2013. LNCS, vol. 8207, pp. 19–31. Springer, Heidelberg (2013). https://doi.org/10.1007/978-3-642-41398-8_3
4. Djedaini, M., Drushku, K., Labroche, N., Marcel, P., Peralta, V., Verdeau, W.: Automatic assessment of interactive OLAP explorations. Inf. Syst. **82**, 148–163 (2019)
5. Djedaini, M., Labroche, N., Marcel, P., Peralta, V.: Detecting user focus in OLAP analyses. In: Kirikova, M., Nørvåg, K., Papadopoulos, G.A. (eds.) ADBIS 2017. LNCS, vol. 10509, pp. 105–119. Springer, Cham (2017). https://doi.org/10.1007/978-3-319-66917-5_8
6. Eirinaki, M., Abraham, S., Polyzotis, N., Shaikh, N.: QueRIE: collaborative database exploration. IEEE Trans. Knowl. Data Eng. **26**(7), 1778–1790 (2014)

7. Fabris, C.C., Freitas, A.A.: Incorporating deviation-detection functionality into the OLAP paradigm. In: SBBD, pp. 274–285 (2001)
8. Förster, J., Marguc, J., Gillebaart, M.: Novelty categorization theory. Soc. Pers. Psychol. Compass **4**(9), 736–755 (2010)
9. Geng, L., Hamilton, H.J.: Interestingness measures for data mining: a survey. ACM Comput. Surv. **38**(3), 9 (2006)
10. Gkesoulis, D., Vassiliadis, P., Manousis, P.: CineCubes: aiding data workers gain insights from OLAP queries. Inf. Syst. **53**, 60–86 (2015)
11. Gunawardana, A., Shani, G.: A survey of accuracy evaluation metrics of recommendation tasks. J. Mach. Learn. Res. **10**, 2935–2962 (2009)
12. Han, J., Kamber, M., Pei, J.: Data Mining: Concepts and Techniques, 3rd edn. Morgan Kaufmann, Burlington (2011)
13. Harinarayan, V., Rajaraman, A., Ullman, J.D.: Implementing data cubes efficiently. In: SIGMOD, pp. 205–216 (1996)
14. Herlocker, J.L., Konstan, J.A., Terveen, L.G., Riedl, J.: Evaluating collaborative filtering recommender systems. ACM Trans. Inf. Syst. **22**(1), 5–53 (2004)
15. Jensen, C.S., Pedersen, T.B., Thomsen, C.: Multidimensional Databases and Data Warehousing. Synthesis Lectures on Data Management. Morgan & Claypool Publishers, San Rafael (2010)
16. Kaminskas, M., Bridge, D.: Diversity, serendipity, novelty, and coverage: a survey and empirical analysis of beyond-accuracy objectives in recommender systems. TiiS **7**(1), 2:1–2:42 (2017)
17. Klemettinen, M., Mannila, H., Toivonen, H.: Interactive exploration of interesting findings in the telecommunication network alarm sequence analyzer (TASA). Inf. Softw. Technol. **41**(9), 557–567 (1999)
18. Kumar, N., Gangopadhyay, A., Bapna, S., Karabatis, G., Chen, Z.: Measuring interestingness of discovered skewed patterns in data cubes. Decis. Support Syst. **46**(1), 429–439 (2008)
19. Litman, J.: Curiosity and the pleasures of learning: wanting and liking new information. Cogn. Emot. **19**(6), 793–814 (2005)
20. Reisenzein, R., Meyer, W.U., Niepel, M.: Surprise. In: Ramachandran, V.S. (ed.) Encyclopedia of Human Behavior, 2nd edn. Elsevier, London (2012)
21. Salimi, B., Gehrke, J., Suciu, D.: Bias in OLAP queries: detection, explanation, and removal. In: SIGMOD, pp. 1021–1035 (2018)
22. Sarawagi, S.: Explaining differences in multidimensional aggregates. In: Proceedings of VLDB, pp. 42–53 (1999)
23. Sarawagi, S.: User-adaptive exploration of multidimensional data. In: Proceedings of VLDB, pp. 307–316 (2000)
24. Sarawagi, S., Agrawal, R., Megiddo, N.: Discovery-driven exploration of OLAP data cubes. In: EDBT, pp. 168–182 (1998)
25. Sathe, G., Sarawagi, S.: Intelligent rollups in multidimensional OLAP data. In: Proceedings of VLDB, pp. 531–540 (2001)
26. Yao, Y., Chen, Y., Yang, X.D.: A measurement-theoretic foundation of rule interestingness evaluation. In: Young Lin, T., Ohsuga, S., Liau, C.J., Hu, X. (eds.) Foundations and Novel Approaches in Data Mining. SCI, vol. 9. Springer, Heidelberg (2005). https://doi.org/10.1007/11539827_3
27. Zhao, Z., Stefani, L.D., Zgraggen, E., Binnig, C., Upfal, E., Kraska, T.: Controlling false discoveries during interactive data exploration. In: SIGMOD, pp. 527–540 (2017)

Towards a Cost Model to Optimize User-Defined Functions in an ETL Workflow Based on User-Defined Performance Metrics

Syed Muhammad Fawad Ali$^{(\boxtimes)}$ (ID) and Robert Wrembel (ID)

Poznan University of Technology, Poznan, Poland
fawadali.ali@gmail.com, robert.wrembel@cs.put.poznan.pl

Abstract. Today's ETL tools provide capabilities for developing custom code as user-defined functions (UDFs) to extend the expressiveness of standard ETL operators. However, a custom code of an UDF may execute inefficiently due to its poor implementation (e.g., due to the lack of using parallel processing or adequate data structures). In this paper we address the problem of the optimization of UDFs in data-intensive workflows and presented our approach to construct a cost model to determine the degree of parallelism for parallelizable UDFs.

Keywords: ETL workflow · ETL execution optimization ·
User-defined functions · Cost model · Parallelization

1 Introduction

An industry accepted architecture for integrating data sources (DSs) is a data warehouse architecture [32]. The integration is implemented by means of the extract-transform-load (ETL) layer where the so-called ETL processes (workflows) are run. They are responsible for: (1) ingesting data from data sources, (2) transforming heterogeneous data into a common data model and schema, (3) cleaning, normalizing, and eliminating data duplicates, (4) loading data into a central repository - a data warehouse (DW). An ETL process has to finish its work within a given time window. Since, (1) such a process moves large volumes of data between DSs and a DW, (2) executes complex cleaning and de-duplication algorithms, its execution is time consuming and typically takes hours to complete. For this reason, one of the most important and only partially solved problem in ETL management is performance optimization. Despite over two decades of research on this topic, it has been only partially solved [4].

In most of the state-of-the-art commercial ETL engines [11], the performance of an ETL process depends on its designer, i.e., he/she is responsible for using appropriate components and orchestrating them manually into an efficient workflow. In practice, only the simplest optimization techniques have been applied so far, i.e, the so-called *push down* and *balance optimization*, cf. Sect. 5. In research

© Springer Nature Switzerland AG 2019
T. Welzer et al. (Eds.): ADBIS 2019, LNCS 11695, pp. 441–456, 2019.
https://doi.org/10.1007/978-3-030-28730-6_27

(only) approaches two basic techniques have been developed so far, namely: (1) task reordering supported by simple cost functions and reordering heuristics and (2) parallel processing, cf. Sect. 5.

Big data add to the ETL optimization problem more complexity that results from: (1) bigger data volumes, and (2) much more complex and diverse data models and formats that need to be processed by ETL. The most common and industry accepted architecture for big data integration is a data lake (DL) [27, 30]. It is a repository that stores large data collections in their native formats. As a consequence, the integration of data is executed on the fly. This feature calls for yet more efficient execution of ETL processes.

Traditional ETL tools are designed to work well with simple data formats (e.g., table-like) and have limited capabilities to efficiently deal with the volume, variety, and velocity of big data. For example, the messy and noisy nature of big data demands new types of operators for data pre-processing tasks, such as classification, clustering, collaborative filtering, outlier-detection, or de-duplication that specifically fit the ever-changing characteristics of the data.

To overcome the limited expressive power provided by the standard ETL operators, most ETL tools provide the functionality to write custom code as user defined functions (UDFs). A custom UDF code may be written in multiple programming languages, e.g., Java, Scala, Python, PL/SQL, Transact SQL, by the ETL developer. Therefore, an UDF may be more prone to errors and inefficient, which may result in a performance bottleneck due to its poor code and high computational complexity. UDFs are treated as black boxes by an ETL engine. Optimizing the execution of black boxes is challenging as their semantics, internal algorithms, and performance characteristics are unknown. As a consequence, it is difficult to assess the run-time and space complexity for a black-box. However, if an UDF is already optimized, or is configurable to be optimized by an ETL framework without changing its code, it may ease the development of an efficient ETL process.

In the research literature there exist some approaches to optimizing UDFs in ETL processes, cf. Sect. 5. Some of them apply parallel processing to UDFs. Most methods either require manual annotation of an UDF, or use static code analysis of an UDF to understand its semantics. The semantics extracted from the UDF code analysis is then used by a cost-based optimizer for performance optimization. Annotating UDFs manually requires more effort on top of manually writing UDFs and is equally error prone. Moreover, the discussed approaches do not cater whether the parallelism is required at the first place or not, resulting in utilizing computing resources unnecessarily, or do not take into consideration the required degree of parallelism.

This paper **contributes a cost model for UDFs, which are treated as black-box operators**. The cost model enables the optimization of already parallelizable (e.g., MapReduce-based [8] or Spark-based [35]) UDFs in an ETL workflow. Our optimization approach draws upon determining the right degree of parallelism for an UDF (or a set of UDFs) to satisfy user-defined performance metrics. In the current work, we consider execution time and monetary cost as the performance metrics. To determine the right degree of parallelism and to generate an optimal configuration for an UDF to be executed in a distributed environment, the cost model must provide the following functionality. First, it must answer the below questions.

- Is an UDF parallelizable?
- If an UDF is parallelizable, will it profit from parallel processing to satisfy user-defined performance metrics?
- If an UDF profits from parallel processing, what will be the adequate (sub-optimal, optimal) parallelization parameters? In this research we consider the following parameters: (1) the number of data partitions, (2) the number of mapper and reducer tasks (in case of a MapReduce-based UDF), (3) the number of nodes in a cluster, (4) a physical and software configuration of the cluster.

Second, the cost model must support generating an optimized configuration for an already parallelizable UDF to be executed in a distributed framework. Notice that the configuration may also support a sub-optimal execution plan.

This paper is organized as follows. Section 2 outlines a running example. Section 3 discusses the motivation for the cost model. Section 4 presents our cost model. Section 5 discusses the related work. Section 6 concludes the paper and points out steps for the future work.

2   Running Example

In this section we present a use case that is to find similar addresses in a dataset, using an ETL workflow. The use case represents the solution to data de-duplication, duplicate web pages detection in web crawling, plagiarism detection, recommendations based on similar user profiles, and document clustering. Finding similar records often requires re-reading the same dataset multiple times. Detecting similar records in a huge amount of data is a computationally-intensive task and it often requires parallel processing.

2.1   Overview of the Use Case

Set-similarity joins using MapReduce (SSJ-MR) [33] is one of the parallel approaches using the MapReduce framework to detect similar records based on string similarity. SSJ-MR uses three components, namely: join-attribute, set-similarity function, and a similarity threshold.

Figure 1 illustrates the workflow for SSJ-MR approach. The workflow is divided into three stages and each stage processes data using MapReduce. Also each stage can be solved by either of the two approaches, e.g., TO1 or TO2 - for the Token Ordering stage, PG1 or PG2 - for the Pair Generation stage, and RJ1 or RJ2 - for the Record Join stage.

Stage 1 - Token Ordering (TO). It computes data statistics using MapReduce in order to generate partitioning keys called *signatures*, which are used in stage 2. The signatures are generated by tokenizing the incoming record into

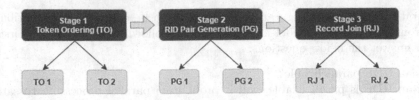

Fig. 1. Set-similarity join workflow

a wordset. For example, the record string "Parallelizing the custom code" is tokenzied into wordset ["Parallelizing", "the", "custom", "code"]. Each element in the wordset is a signature, which is used as a partitioning key instead of using actual join attribute value, because partitioning records using an entire string for partitioning (e.g., hash-based partitioning) is a difficult task.

Stage 2 - RID Pair Generation (PG). IT extracts a record ID (RID) and join-attribute value for each record, computes the similarity of the join-attribute values, and propagates the RID pair of similar records to the next stage. The similarity is computed in the Reduce phase of MapReduce, using signatures from the previous stage as partitioning keys.

Stage 3 - Record Join (RJ), uses the RID pair of similar records from the previous stage and generates actual pairs of joined records.

The explanation of the use case clearly indicates that SSJ-MR is a computationally-intensive process divided into multiple stages (considered as multiple UDFs in this paper).

2.2 Use Case for Running Example

In the example, we use Pentaho Data Integrator (PDI)[1] to implement the SSJ-MR algorithm as an ETL workflow to efficiently detect similar records in a large amount of datasets. The SSJ-MR algorithm is divided into three stages and executed in Amazon EMR Cluster[2] - a Hadoop [6] managed framework that makes it feasible to process vast amounts of data across dynamically scalable Amazon EC2 instances[3]. Each stage of SSJ-MR may be executed in a differently configured Amazon EMR cluster, if proposed by the cost model. Finally, data are stored in a data warehouse.

Our example ETL process is shown in Fig. 2. The first step - *START* indicates the start of the ETL workflow in PDI. The next step - *FetchData* fetches data from a data store. *FetchConfig* fetches the configuration for each of the Amazon EMR jobs *s1:TokenOrdering*, *s2:PairGeneration*, and *s3:RecordJoin* (generated by the cost model) to be executed in parallel in the Amazon EMR cluster. Finally, *StoreData* step stores the resulted data set a DW.

---

[1] https://github.com/pentaho/pentaho-kettle.

[2] https://aws.amazon.com/emr/.

[3] https://aws.amazon.com/ec2/.

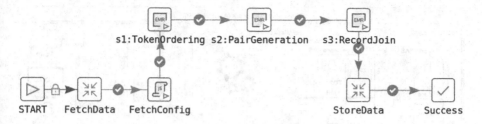

Fig. 2. The running example scenario

3 Motivation

The work presented in this paper is motivated based on our previous work [2], where we presented an extendible theoretical ETL framework. Its architecture is shown in Fig. 3. The extendible ETL framework consists of four modules, namely: (1) an UDFs Component, (2) a Recommender, (3) a Cost Model Library, and (4) a Monitoring Agent.

The *UDF Component* allows ETL developers to easily write parallelizable UDFs by separating parallelization concerns from the code. It contains a library of *Parallel Algorithmic Skeletons* (PASs) or parallelizable code templates. These PASs are designed to be executed in a distributed environment like MapReduce or Spark. The component provides to the ETL developer: (1) the already parallelizable code of some commonly used big data operators (a.k.a *Case-based PASs*) including: sentiment analysis, de-duplication of rows, outlier detection and (2) a list of *Generic PASs* (e.g., worker-farm model, divide and conquer, branch and bound, systolic, MapReduce). The ETL developer chooses either a *Case-based PAS* or a *Generic PAS*, depending on his/her requirements.

The *Recommender* module works together with the *Cost Model Library* and the *Monitoring Agent* to generate optimized ETL workflows. The *Recommender* includes an extendible set of machine learning algorithms (e.g., similarity, recommendation, prediction algorithms) to optimize a given ETL workflow and to generate a more efficient version of the workflow. To this end, it uses performance statistics collected during past ETL executions by a dedicated module, called *Monitoring Agent*. For collecting statistics we rely on standard data profiling methods, e.g., [1] and execution monitoring, e.g., [13]. The statistics are stored in a repository, called *Knoweldge Base*. It also stores optimized configuration plans collected during several ETL workflow executions. It is updated every time when a new configuration (or a new use case) is identified and validated by the cost model.

As shown in Fig. 3, an input to the *UDFs Component* is either the *Generic PAS* or the *Case-based PAS*, along with optional parameters: an *input format* and/or an *output format* of a dataset, a *maximum execution time constraint*, and *distributed machine specifications*.

In our running example, the ETL developer may choose the SSJ-MR algorithm from the *Case-based PAS* for the de-duplication of datasets. The ETL

446   S. M. F. Ali and R. Wrembel

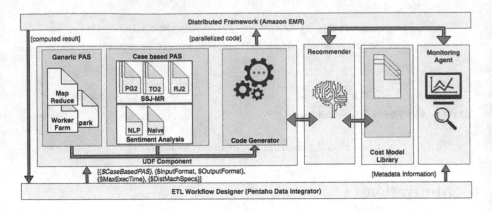

Fig. 3. The architecture of the Extendible ETL Framework

developer may also provide an execution time constraint (e.g., the entire deduplication ETL workflow must execute within a window of, say, 900 s) and a monetary cost constraint (e.g., the execution cost of running an ETL workflow while using the Amazon EMR cluster must not exceed, say, $15).

Based on the input from the ETL developer, the *UDF Component* will choose the optimal code with distributed machine configurations with the help of the *Recommender, Monitoring Agent,* and the cost model provided by the *Extendible ETL Framework*. In the example, the optimal code variant selection is based on 2^3 possible combination of ETL workflows to be executed in the Amazon EMR cluster (i.e., there are three stages, and for each stage there are two code possibilities). The optimal combination of code variants for an ETL workflow having multiple UDFs (stages) is supported by the cost model discussed in Sect. 4. Once the optimal code variant is selected and distributed framework configurations are generated, the code variant chosen by the *Extendible ETL Framework* will be executed in a distributed architecture (e.g., Amazon EMR Cluster).

4   Proposed Cost Model

To optimize the execution of parallelizable UDFs in an ETL workflow according to some user-defined performance metrics, i.e., execution time and monetary costs, the use of the cost model is as follows.

- *Stage 1 - Feasibility*: the cost model will first determine the feasibility to parallelize UDFs, i.e., whether it makes sense to parallelize an UDF, in order to satisfy the user-defined performance metrics.
- *Stage 2 - Degree of Parallelism*: the cost model will reason on the right degree of parallelism, i.e., how much to parallelize (e.g., choosing the appropriate number of partitions to distribute the data to be transformed in parallel).
- *Stage 3 - Optimal Code Generation*: the cost model will guide the creation of an efficient configuration for distributed machines, so that the UDF is

executed optimally in a distributed environment, adhering to the execution performance and monetary cost constraints defined by the developer as an input to the UDF component.

4.1   Stage 1 - Feasibility

The cost model is used by the *Simulator* to simulate an UDF execution in a non-distributed parallel environment. The simulation helps in finding out if it makes sense to execute the UDF in a non-distributed environment by comparing the actual execution time of the UDF with the user-defined performance metrics. If the execution time is lower than or equal to the required execution time, the framework would execute the UDF in the non-distributed environment.

Further extension is planned to the *Simulator* to be able to identify the core aspects about an UDF such as: execution time of the UDF for a given dataset, the number of rows processed per second, the number of bytes processed per second, the size of data, the used distributed machine configuration as well as memory, IO, and CPU usage characteristics, in the spirit of [13]. The performance data extracted from the simulator will then be used to predict the optimal configuration for an UDF execution.

4.2   Stage 2 - Degree of Parallelism

The right degree of parallelism is to assure the user-defined performance metrics and it can be achieved by tuning certain performance parameters depending on the distributed environment and programming paradigm. For example, performance tuning of a MapReduce UDF to be executed in Hadoop is dependent on 190 configurations. Optimal settings for these parameters depend upon a workflow, data characteristics, and distributed machine configurations. However, a fraction of parameters play an important role in achieving the performance optimization and a lack of knowledge of these parameters is mostly the cause of performance problems [14].

The parameters that seem to be *critical* to the optimal execution of UDFs in a distributed framework include:

- The *number of partitions/shards*: represents appropriate number of partitions (neither too few nor too many) for the MapReduce and Spark jobs executing on top of the Hadoop framework.
- *Machine configurations*: represents appropriate processing power of a distributed machine along with optimal configuration of the critical parameters.
- *Parallel processing architecture*: include a degree of parallelism, partitions schuffling scheme (for Spark jobs), the number of Mapper and Reducer tasks (for MapReduce jobs).

4.3   Stage 3 - Optimal Code Generation

The critical parameters (described in Sect. 4.2), are of vital importance in order to satisfy the user-defined performance metrics. The cost model uses obligatory user-defined performance metrics: (1) maximum execution time for an ETL workflow (T) and (2) maximum monetary cost for an ETL workflow to be executed in the distributed environment (B), as an input from the ETL developer. Optional parameters include: (1) the size of a dataset (R) in terms of the number of rows and (2) configuration of a distributed machine (M).

In order to find out the right degree of parallelism, the proposed cost model will be used as follows (also shown in Fig. 4).

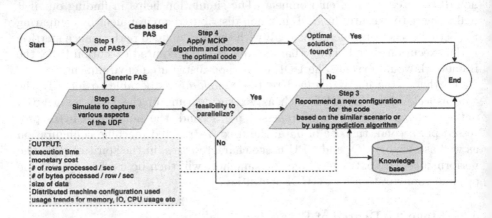

Fig. 4. The processing workflow of the cost model

- *Step 1*: checks for a user code input either as the *Generic PAS* or *Case-based PAS*, (c.f., Sect. 3). If the ETL developer selects the *Generic PAS*, the cost model executes Step 2, otherwise it executes Step 4.
- *Step 2*: if the ETL developer selects the *Generic PAS*, then an UDF provided by the ETL developer is first executed in the *Simulator* to collect run-time execution statistics, e.g., execution time, estimated monetary cost, the number of rows processed per second, the number of bytes processed per second, the size of data, the current configuration of a distributed machine as well as memory, CPU, and IO usage characteristics.
At this point, the cost model is only interested in the execution time and estimated monetary cost. If both are within the user-defined performance metrics constraints, then the processing is stopped, otherwise the processing continues to Step 3.
- *Step 3*: the output from the *Simulator* will be used as an input to this step. First, similar execution statistics are searched in the *Knowledge Base* (c.f., Sect. 3). If there is any optimal configuration found in the *Knowledge Base* similar to the input parameters, then such a code and machine configuration will be used as an optimal code to be executed in the distributed framework. Otherwise, the cost model will be used with new configuration parameters to estimate an optimal code and machine configuration. The new configuration parameters can be suggested by the *Recommender* with the help of prediction algorithms.

- *Step 4*: for the *Case-based PAS*, where a user selects a use case, e.g., SSJ-MR (c.f., Sect. 2) there exists multiple code variants. Therefore, there are n^m possible variants, which correspond to an NP-hard problem and can be mapped to Multiple Choice Knapsack Problem (MCKP) [17] as a special case of our problem. The MCKP is defined as follows:

Given m classes $N_1 ... N_m$ of items to pack in a knapsack of capacity c, where each item $j \in N_i$, $i = 1,2,....,m$, has profit p_{ij} and weight w_{ij}, the problem is to choose exactly one item from each class such that the sum of profits is maximized while the sum of weights does not exceed capacity c.

In order to show MCKP as a special case of the running example, we introduce the following terms. Let *Minimize(Z)* be an optimal solution containing exactly one program from each stage with minimum execution time, while remaining within budget B. Note that we want to calculate the execution time T_{ij} of a program variant N_j from each stage m, such that the total cost C_{ij} of the entire ETL workflow is $\leq B$.

According to the aforementioned definition of MCKP, we can map:

- *m* classes in MCKP definition to *m* stages in the running example,
- *c* weight constraint to *B* budget constraint,
- w_{ij} cost of item for class to c_{ij} cost of variant j at stage i,
- p_{ij} profit of each item to T_{ij} execution time of each variant at each stage.

Then, we can find out the optimal solution as follows:

$$\text{Minimize(Z)} \qquad \sum_{i=1}^{m}\sum_{j\in N_i} T_{i,j} \cdot x_{i,j}$$

subject to

$$\sum_{i=1}^{m}\sum_{j\in N_i} C_{ij} \cdot x_{ij} \leq B,$$

$$\sum_{j\in N_i} x_{ij} = 1, i = 1,\ldots,m,$$

$$x_{ij} \in \{0,1\}, j \in N_i, i = 1,\ldots,m.$$

If the optimal configuration solution is found by the MCKP step, the solution is handed over to the distributed machine to be executed in parallel, otherwise the best possible solution with all the relevant statistics and machine configuration is sent to Step 3 as an input, in order to generate a (sub-)optimal solution.

4.4 Preliminary Results

As a preliminary evaluation of our approach we used the running example (c.f., Sect. 2.2), where each stage has two variants. For stage 1, i.e., Token Ordering, the first variant is called *Basic Token Ordering (BTO)* and the second one is *One Phase Token Ordering (OPTO)*. For stage 2, i.e., RID Pair Generation, the first variant is called *Basic Kernel (BK)* and the second one is *Indexed Kernel*

(PK). For stage 3, i.e., Record Join, the first variant is *Basic Record Join (BRJ)* and the second variant is *One Phase Record Join (OPRJ)*.

In Table 1, column *Stage* stores the number of the stage, cf. Sects. 4.1, 4.2, and 4.3. Column *Algorithm* stores the aforementioned algorithm variants of each stage. Column *#Nodes [exec cost/h]* represents the per hour execution cost associated to a n-node micro-cluster configuration (Amazon Web Service applies per hour billing cycle). For example, column *2 [0.4$/h]* represents execution cost per hour for a 2-nodes micro-cluster, *4 [0.8$/h]* represents execution cost for a 4-nodes micro-cluster, etc.

The costs are estimated based on a machine type mentioned in [33] and near equivalent configuration available for Amazon EC2 instances[4]. We used the Linux machine on t3.2xlarge, 4 vCPUs, and 16 GB RAM, at a main cost of \$0.164/hour plus \$0.036/hour as a buffer cost, which eventually results in \$0.2/hour per node.

Then, each cell under *2 [0.4$/h]*, *4 [0.8$/h]*, *8 [1.6$/h]*, and *10 [2.0$/h]* stores execution time in seconds of a given algorithm in a given micro-cluster in a given stage. Thus, Table 1 includes 24 variants of execution times. The execution times of each variant of each stage are taken from the already carried out evaluation in [33].

Table 1. Execution time in seconds of each stage for self-joining the DBLP dataset on different cluster sizes

Stage	Algorithm	#Nodes [exec cost/h]			
		2 [0.4$/h]	4 [0.8$/h]	8 [1.6$/h]	10 [2.0$/h]
1	BTO	191.98	125.51	91.85	84.02
	OPTO	175.39	115.36	94.82	92.80
2	BK	753.39	371.08	198.70	164.57
	PK	682.51	330.47	178.88	145.01
3	BRJ	255.35	162.53	107.28	101.54
	OPRJ	97.11	74.32	58.35	58.11

In order to evaluate the correctness of our MCKP-based cost model, we mapped our problem on to the *Linear Integer Programming Model*. The cost model is implemented in Java, which utilizes the *lp_solve*[5] library to generate the optimized (i.e., minimum execution time with respect to the allocated budget) combination of available machine configurations to execute each stage of our running example. lp_solve is a mixed integer linear programming (MILP) solver based on the revised Simplex method [9] and the Branch-and-bound method [23] for integers. The implementation of the cost model is accessible via our online git repository[6].

---

[4] https://calculator.s3.amazonaws.com/index.html.
[5] http://lpsolve.sourceforge.net/.
[6] https://github.com/fawadali/MCKPCostModel.

The shaded cells in Table 1 represent execution costs of the selected algorithms at each stage that are minimal (optimal) for a given maximum budget of \$1.6. Thus, for the first stage OPTO is suggested to be executed on 2 nodes, for the second stage PK is selected to be executed on 4 nodes, and for the third stage OPRJ is selected to be executed 2 nodes.

The blacked cells and the shaded one with value 97.11 represent execution costs of the selected algorithms at each stage that are minimal (optimal) for a given maximum budget of \$4.0. That is, for the first stage BTO is suggested to be executed on 8 nodes, for the second stage PK is selected to be executed on 10 nodes, and for the third stage OPRJ is selected to be executed on 2 nodes.

The results show that the cost model provides the best possible configuration for a set of ETL activities to be executed in a cloud based pay-as-you-go environment.

In the future, we will conduct experiments with different use cases in order to fine-tune the MCKP algorithm.

5 Related Work

So far, two promising techniques for an ETL process optimization were proposed, i.e.: task reordering and parallel processing. The first technique applies reordering of tasks in order to produce a more efficient ETL process. In the simplest case, called *push down*, the most selective tasks are moved towards the beginning of an ETL process (towards data sources) to reduce a data volume as soon as possible [19]. IBM extended push down to *balance optimization*, where some tasks are moved towards the end of an ETL process and some are moved towards the beginning [18,24].

In more advanced approaches, an ETL process is assigned an estimated execution cost [13], and next, by using reordering of steps, alternative processes are produced with their estimated costs [20,28]. This technique uses the principles of cost-based query optimization. As the reordering problem is NP-complete, [22,28] propose some reordering heuristics. In [21] the reordering of operators is based on their semantics, e.g., a highly selective operator would be placed (re-ordered) at the beginning of a workflow, similarly as in the push down and balanced optimization.

The second technique applies parallel processing to an ETL process. In the simplest case (available in commercial ETL engines), uploading data into a data warehouse is executed in parallel (e.g., the IBM Netezza *nzload* command, the Oracle *import* command). In a more advanced approach [25], an ETL process is partitioned into linear sub-processes. Next, data parallelization is applied to each of the sub-processes. Finally, all flows are executed with multi-threading.

None of the aforementioned approaches support the optimization of ETL processes with user-defined functions. The reordering of operators is based on the semantics of the operators, which are well known and understood for traditional operators. However, the semantics of UDFs are typically unknown. UFDs have been handled by other approaches, which can be segregated into two classes, i.e.,

(1) that require manual annotation of UDFs, and (2) that perform code-analysis on an UDF to explore the options for optimization.

[15,16] describe a framework for the optimization of data flows, where user defined functions are treated as black boxes. The framework consists of the Nephele execution engine and the PACT compiler to execute UDFs, based on the PACT programming model [5] (PACT is a generalization of the MapReduce framework [8]). The PACT implementation allows the flexibility to parallelize tasks by giving parallel hints in the code. Such hints are later exploited by a cost-based optimizer that generates parallel execution plans. The optimization is based on: (1) re-ordering of UDFs in a workflow and (2) the execution of UDFs in a parallel environment. To optimize UDFs, the optimization program discovers the unknown or hidden algebraic properties of the UDFs, by means of static code analysis. The discovered properties are then used to reorder the UDFs in a workflow. A cost based optimizer (model) is used to compute all the possible alternatives and valid re-orderings to generate an efficient execution plan. A plan with a minimum estimated cost is selected and submitted for a parallel execution.

In [10], the authors propose an inherently parallel UDF framework, called SQL/MR, which enables the parallelization of UDFs in a massively-parallel shared-nothing database. The proposed framework is based on the character- istics of map and reduce functions in the MapReduce framework. That is, the instances of the SQL/MR function will execute in parallel on each node in a parallel database, just like the map and reduce tasks execute in parallel across a cluster in MapReduce. To achieve parallelism, SQL/MR requires a definition of the Row and Partition functions and corresponding execution models for the SQL/MR function instances. The Row function is described as an equivalent to a map function in MapReduce. Row functions perform row-level transformation and processing. The execution model of the Row function allows independent processing of each input row by exactly one instance of the SQL/MR function, thus, allowing the execution engine to control parallelism. The Partition func- tion is similar to the reduce function in MapReduce. Exactly one instance of the SQL/MR functions is used to independently process each group of rows defined by the PARTITION BY clause in a query. Independent processing of each partition allows the execution engine to achieve parallelism at the level of a partition. The dynamic cost-based re-optimizers are used to collect the statistics at run-time and may change the order of UDFs to improve performance.

Another approach discussed in [12] is inspired by [10,16] and it uses anno- tations in an UDF to generate an optimized query plan for relational database operators and UDFs in complex data workflows. The aforementioned approaches achieve optimization by means of rewriting execution plans either by applying reordering of traditional operators in a workflow or by performing code analysis on UDFs.

In [26] an extensible logical optimizer for UDF-heavy workflow (SOFA) is discussed, which rewrites an execution plan based on automatically inferring the semantics of a MapReduce style UDF (instead of manually annotating the UDF) and a small set of rewrite rules. To infer the semantics, SOFA requires: (1) a set of properties, which are either annotated by the developer, e.g., a cost function, resource consumption, the number of input rows or the number of output rows and (2) automatically detectable properties, e.g., parallelization function of an operator (e.g., map and reduce), a schema information, and read/write behavior at the attribute level. The paper introduced the so-called *Operator Property Graph* to infer the properties of UDFs by modeling relationships between a new UDF operator and (pre-defined) properties. Based on the identified properties, SOFA is able to re-write an execution plan.

The approaches discussed above require from the developer to follow certain code-based keywords or parallelism hints (e.g., PARTITION BY, ORDER BY) to enable parallelism. These approaches do not consider an UDF as a black-box and require parsing of an UDF code for parallel execution. Moreover, the discussed approaches tend to utilize all the hardware resources to achieve parallelism and do not cater whether the parallelism is required at the first place or not, or does not take into consideration the required degree of parallelism.

In contrast, the approach proposed in this paper first assesses if the performance of an UDF may increase by applying parallelization. If so, then, the cost model proposes the optimal configuration by using simulation, recommendation, and prediction algorithms for the UDF (or a set of UDFs within an ETL workflow) to be executed in the distributed framework. The main advantage of our approach is that it does not require code annotations, which let the ETL developer not to be restricted to a framework or cost-model specific to a given programming language. Instead, the developer may use any programming language, supported by an ETL engine, to write UDFs, and the proposed cost model will not be dependent on the language to generate the optimal configuration for an UDF.

Finally, for learning a behavioural model of an UDF, apparently, a black-box testing approach is promising. [31] overviews techniques for building models that can be represented as state diagrams and that can be built based on experiments run on a black-box software. The most recent state of the art in this field is presented in [34]. The authors review the approaches that build performance models based on metrics such as execution time, memory usage, and wait times. The authors provide numerous approaches based on: statistics, time series analysis, data mining, and neural networks.

6   Conclusion

This paper is the continuation of our work [2–4], where we presented the Extendible ETL Framework to allow the ETL developer to efficiently write parallelizable UDFs by applying the parallelizable code templates (a.k.a Parallel Algorithmic Skeletons - PAS). The ETL developer may choose either the *Generic PAS* to fill in the PAS with user-defined code or the *Case-based PAS*, as a standard algorithmic code to be executed in a distributed framework.

In this paper, we presented a preliminary contribution towards a cost model to determine the best possible configuration for an UDF generated either via

454   S. M. F. Ali and R. Wrembel

the *Generic PAS* or the *Case-based PAS*. In particular, this paper **contributes**: (1) the cost model for optimizing execution of an UDF (as a black-box), (2) the method for selecting a (sub-)optimal configuration of a parallel execution environment for an UDF. The proposed method uses simulation, recommendation, and prediction algorithms to generate the best possible configuration for an UDF generated by means of the *Generic PAS*. For the *Case-based PAS*, the cost model uses Multiple Choice Knapsack Problem (MCKP) along with the recommendation and prediction algorithms (if required) to generate a (sub-)optimal configuration of a parallel run-time environment for an UDF.

In the next steps, we will conduct experiments with different use cases in order to verify the applicability of the MCKP algorithm to our problem. Furthermore, we will develop a simulator, recommendation, and prediction algorithms for the proposed cost model. Finally, we will incorporate the cost model into our Extendible ETL Framework.

At this stage of our research, a still open issue is to discover the most adequate model of an UDF. To resolve this issue, in the forthcoming future we will also experiment on the techniques for building behavioural models for black boxes, as mentioned in Sect. 5.

Some papers proposed methods for designing ETL processes at a conceptual level, e.g., [7,29]. Other unexplored fields of optimization include conceptual ETL design and transformations from a conceptual to physical design. The latter, while generating implementations could consider some user preferences, envisaged data characteristics, physical parameters of a run-time environment.

Acknowledgements. The work of Fawad Ali is partially supported by the European Commission through the Erasmus Mundus Joint Doctorate project *Information Technologies for Business Intelligence-Doctoral College* (IT4BI-DC).

The work of Robert Wrembel is partially supported by: (1) the grant No. 2015/19/B/ST6/02637 of the National Science Center and (2) the grant of the Polish National Agency for Academic Exchange, within the Bekker programme.

References

1. Abedjan, Z., Golab, L., Naumann, F.: Profiling relational data: a survey. VLDB J. **24**(4), 557–581 (2015)
2. Ali, S.M.F.: Next-generation ETL framework to address the challenges posed by Big Data. In: International Workshop Design, Optimization, Languages and Analytical Processing of Big Data (DOLAP) (2018)
3. Ali, S.M.F., Mey, J., Thiele, M.: Parallelizing user-defined functions in the ETL workflow using orchestration style sheets. Int. J. Appl. Math. Comput. Sci. (AMCS) **29**, 69–79 (2019)
4. Ali, S.M.F., Wrembel, R.: From conceptual design to performance optimization of ETL workflows: current state of research and open problems. VLDB J. **26**, 1–25 (2017)
5. Battré, D., Ewen, S., Hueske, F., Kao, O., Markl, V., Warneke, D.: Nephele/PACTs: a programming model and execution framework for web-scale analytical processing. In: ACM Symposium on Cloud Computing, pp. 119–130 (2010)

6. Borthakur, D.: The Hadoop distributed file system: Architecture and design. Hadoop Project Website, vol. 11, p. 21 (2007)
7. Caruccio, L., Deufemia, V., Polese, G.: Visual data integration based on description logic reasoning. In: International Database Engineering Applications Symposium, pp. 19–28 (2014)
8. Dean, J., Ghemawat, S.: MapReduce: simplified data processing on large clusters. Commun. ACM **51**(1), 107–113 (2008)
9. Evans, J.P., Steuer, R.E.: A revised simplex method for linear multiple objective programs. Math. Program. **5**(1), 54–72 (1973)
10. Friedman, E., Pawlowski, P., Cieslewicz, J.: SQL/MapReduce: a practical approach to self-describing, polymorphic, and parallelizable user-defined functions. VLDB Endowment **2**(2), 1402–1413 (2009)
11. Gartner: Magic Quadrant for Data Integration Tools. https://www.gartner.com/doc/3883264/magic-quadrant-data-integration-tools. Accessed 18 Mar 2019
12. Große, P., May, N., Lehner, W.: A study of partitioning and parallel UDF execution with the SAP HANA database. In: International Conference on Scientific and Statistical Database Management, p. 36. ACM (2014)
13. Halasipuram, R., Deshpande, P.M., Padmanabhan, S.: Determining essential statistics for cost based optimization of an ETL workflow. In: International Conference on Extending Database Technology (EDBT), pp. 307–318 (2014)
14. Herodotou, H., et al.: Starfish: a self-tuning system for big data analytics. In: Conference on Innovative Data Systems Research (CIDR), vol. 11, pp. 261–272 (2011)
15. Hueske, F., et al.: Peeking into the optimization of data flow programs with MapReduce-style UDFs. In: International Conference on Data Engineering (ICDE), pp. 1292–1295 (2013)
16. Hueske, F., et al.: Opening the black boxes in data flow optimization. VLDB Endowment **5**(11), 1256–1267 (2012)
17. Ibaraki, T., Hasegawa, T., Teranaka, K., Iwase, J.: The multiple choice knapsack problem. J. Oper. Res. Soc. Japan **21**(1), 59–93 (1978)
18. IBM: IBM InfoSphere DataStage Balanced Optimization. IBM Whitepaper. Accessed 18 Mar 2019
19. Informatica: How to Achieve Flexible, Cost-effective Scalability and Performance through Pushdown Processing. https://www.informatica.com/downloads/pushdown_wp_6650_web.pdf. Accessed 18 Mar 2019
20. Jovanovic, P., Romero, O., Simitsis, A., Abelló, A.: Incremental consolidation of data-intensive multi-flows. IEEE Trans. Knowl. Data Eng. **28**(5), 1203–1216 (2016)
21. Karagiannis, A., Vassiliadis, P., Simitsis, A.: Scheduling strategies for efficient ETL execution. Inf. Syst. **38**(6), 927–945 (2013)
22. Kumar, N., Kumar, P.S.: An efficient heuristic for logical optimization of ETL workflows. In: VLDB Workshop on Enabling Real-Time Business Intelligence, pp. 68–83 (2010)
23. Lawler, E.L., Wood, D.E.: Branch-and-bound methods: a survey. Oper. Res. **14**(4), 699–719 (1966)
24. Lella, R.: Optimizing BDFS jobs using InfoSphere DataStage Balanced Optimization. https://www.ibm.com/developerworks/data/library/techarticle/dm-1402optimizebdfs/index.html. Accessed 18 Mar 2019
25. Liu, X., Iftikhar, N.: An ETL optimization framework using partitioning and parallelization. In: ACM Symposium on Applied Computing, pp. 1015–1022 (2015)
26. Rheinländer, A., Heise, A., Hueske, F., Leser, U., Naumann, F.: SOFA: an extensible logical optimizer for UDF-heavy data flows. Inf. Syst. **52**, 96–125 (2015)

27. Russom, P.: Data lakes: purposes, practices, patterns, and platforms. TDWI white paper (2017)
28. Simitsis, A., Vassiliadis, P., Sellis, T.K.: State-space optimization of ETL workflows. IEEE Trans. Knowl. Data Eng. **17**(10), 1404–1419 (2005)
29. Skoutas, D., Simitsis, A., Sellis, T.: Ontology-driven conceptual design of ETL processes using graph transformations. J. Data Semant. **13**, 120–146 (2009)
30. Terrizzano, I., Schwarz, P., Roth, M., Colino, J.E.: Data wrangling: the challenging journey from the wild to the lake. In: Conference on Innovative Data Systems Research (CIDR) (2015)
31. Vaandrager, F.: Model learning. Commun. ACM **60**(2), 86–95 (2017)
32. Vaisman, A.A., Zimányi, E.: Data Warehouse Systems - Design and Implementation. Data-Centric Systems and Applications. Springer, Heidelberg (2014). https://doi.org/10.1007/978-3-642-54655-6
33. Vernica, R., Carey, M.J., Li, C.: Efficient parallel set-similarity joins using MapReduce. In: ACM SIGMOD International Conference on Management of Data (2010)
34. Witt, C., Bux, M., Gusew, W., Leser, U.: Predictive performance modeling for distributed batch processing using black box monitoring and machine learning. Inf. Syst. **82**, 34–52 (2019)
35. Zaharia, M., et al.: Apache spark: a unified engine for big data processing. Commun. ACM **59**(11), 56–65 (2016)

Author Index

Aklouche, Billel   105
Alattar, Munqath   33
Ali, Syed Muhammad Fawad   441
Al-Mallah, Mouaz   53
Amann, Bernd   302
Amaral, Glenda   215
Andritsos, Periklis   235, 251
Antol, Matej   337
Auge, Tanja   357
Awwad, Tarek   285

Baazizi, Mohamed-Amine   302
Bennani, Nadia   285
Bernard, Gaël   235, 251
Bielikova, Maria   186
Böhm, Klemens   3
Bounhas, Ibrahim   105
Broneske, David   69
Brunie, Lionel   285

Chen, Xiao   69

Darmon, Patrice   409
Dohnal, Vlastislav   337
dos Santos Mello, Ronaldo   123
Dulai, Tibor   89
Durand, Gabriel Campero   69

ElShawi, Radwa   53
Endres, Markus   321
Evangelidis, Georgios   20

Fouché, Edouard   3

Guerra, Francesco   169
Guizzardi, Giancarlo   215

Hannou, Fatma-Zohra   302
Hardock, Sergey   139
Hassan, Ali   409
Hegner, Stephen J.   390
Heuer, Andreas   151, 357
Hlavac, Patrik   186

Kastner, Johannes   321
Koch, Andreas   139

Kosch, Harald   285
Kozierkiewicz, Adrianna   201

Marcel, Patrick   425
Marten, Dennis   151
Meyer, Holger   151
Mitschang, Bernhard   373

Nagy, Zsuzsanna   89

Ougiaroglou, Stefanos   20

Paganelli, Matteo   169
Peralta, Veronika   425
Petrov, Ilia   139
Pietranik, Marcin   201
Pohl, Constantin   267
Ponos, Pavlos   20
Popovic, Daniel   3

Rehn-Sonigo, Veronika   285
Riegger, Christian   139

Saake, Gunter   69
Sakr, Sherif   53
Sali, Attila   33
Santana, Luiz Henrique Zambom   123
Sattler, Kai-Uwe   267
Sherif, Youssef   53
Simko, Jakub   186
Slimani, Yahya   105
Sottovia, Paolo   169
Stach, Christoph   373

Vassiliadis, Panos   425
Velegrakis, Yannis   169
Vinçon, Tobias   139

Werner-Stark, Agnes   89
Wrembel, Robert   441

Xu, Yinlong   69

Zoun, Roman   69

Printed in the United States
By Bookmasters

Printed in the United States
By Bookmasters